W0234890

Rubber Technology

Volume I

Rubber Technology

Volume I

S. C. Bhatia

BE (Chemical), MBA

Avishek Goel

(B.E. Hons.) in Mechanical Engineering from BITS Pilani, Dubai Campus.
Research Associate at The Energy and Resources Institute (TERI)
in the Renewable Energy Technologies Division

WOODHEAD PUBLISHING INDIA PVT LTD

New Delhi

Published by Woodhead Publishing India Pvt. Ltd.
Woodhead Publishing India Pvt. Ltd.,
303, Vardaan House, 7/28, Ansari Road,
Daryaganj, New Delhi - 110002, India
www.woodheadpublishingindia.com

First published 2019, Woodhead Publishing India Pvt. Ltd.
© Woodhead Publishing India Pvt. Ltd., 2019

Woodhead Publishing India Pvt. Ltd. ISBN: 978-93-88320-00-9
Woodhead Publishing India Pvt. Ltd. e-ISBN: 978-93-88320-01-6

Typeset by Asian Enterprises, New Delhi
Printed and bound by Replika Press Pvt. Ltd.

Contents

Preface

Rubber technology is the subject dealing with the transformation of rubbers or elastomers into useful products, such as automobile tyres and rubber mats. The materials includes latex, natural rubber, synthetic rubber and other polymeric materials, such as thermoplastic elastomers. Rubber processed through such methods are components of a wide range of items. Rubbers - elastomers are polymeric materials characterised by their ability of reversible deformation due to external deforming forces. Their deformation rate depends on the structure and molar mass of the deformed rubber and on external conditions of the deformation. This characteristics, referred to as elastic and/ or hyper elastic deformation, is entropic in nature and results from the ability of rubber macromolecules to form a more organised state under influence of deforming forces without deformation of chemical bonds between atoms of the polymer chain or without deformation of their valence angles.

Rubbers usually have long and regular macromolecule chains without bulk substitutes with spatially oriented structural units. When heated, rubbers change their elastic and/or hyper elastic state to a visco-elastic state and they become plastic and flow above the softening temperature (T_m). Natural rubber comes from a plant. In industrial applications, it is obtained primarily from *Hevea Brasiliensis tree*. Synthetic rubbers are made by constructional polyreactions of chain or grade nature. Rubbers are used most often in the form of vulcanisates - a vulcanised rubber. They can be brought to this form by vulcanisation. This process is based on creation of chemical and physical transverse bonds between rubber macromolecules resulting in a spatial vulcanisate mesh, giving unique properties to the material. Various chemicals and vulcanising agents are used to create the chemical transverse bonds between rubber macromolecules (such as sulphur, peroxides, metal oxides, resins, quinones and others), which can react with appropriate functional rubber groups in the process of vulcanisation to create transverse bonds between them. The cross-linking can be induced also by radiation, however its energy must be sufficient to generate reactive forms of rubber macro-molecules - radicals in most cases. They react with each other giving rise to transverse bonds. Cross-linking can occur also due to microwave energy or ultrasound. Most rubbers require vulcanisation, though it is not inevitable for some type of thermoplastic rubbers. Anyway, the optimum vulcanisate (rubber) properties cannot be achieved only by cross-linking rubber molecules, but other additives must be added. Besides cross-linking agents and antidegradants (used to slow down the process of ageing), they include fillers that have a positive influence on some of the utilisation properties and make them cheaper, as well as additives allowing admixture of all the powdery or liquid additives, often referred to as supplementary processing additives.

This book on Rubber Technology summarises various aspects of natural, synthetic rubbers, vulcanisation mixing and calendering, manufacturing techniques of various rubber such as– tyres, conveyor belt, hose, footwear, sports goods, anti vibration mounts, rollers, rubber to metal bonding, pollution control and energy conservation in rubber industry. Carbon footprints and nanotechnology in rubber industry, applications, safety and testing of rubber products. The book is divided in two volumes. Volume I contains 1 to 22 chapters and Volume II has 23 to 39 chapters.

Section I discusses general consideration and engineering aspects. Chapter 1 deals with basic concepts of polymerisation. Polymerisation is a process of reacting monomer molecules together in a chemical reaction to form polymer chains or three-dimensional networks. There are many forms of polymerisation and different systems exist to categorise them.

Chapter 2 is devoted to natural rubber. Natural rubber as initially produced, consists of polymers of the organic compound isoprene, with minor impurities of other organic compounds plus water.

Section II discusses synthetic rubbers—non-oil resistant, special purpose and other specialty rubbers. Chapter 3 focuses on synthetic rubbers: an overview. Synthetic rubber are white, crumbly, plastic mass which can be processed and vulcanised in the same way as natural rubber. Chapter 4 covers styrene butadiene rubber, its manufacture, compounding, processing and uses. Chapter 5 provides valuable information on preparation, properties, uses and compounding aspects of polybutadiene. Chapter 6 deals with polyisoprene rubber which is a major component of manufacture of rubber and is made synthetically and forms are stereospecific *cis*-1-4 and *trans*-1-4-polyisoprene. Chapter 7 concentrates on butyl and halobutyl rubber. Butyl rubber is a copolymer of isobutylene and isoprene. Chapter 8 focuses on ethylene propylene rubber which is valuable for their excellent resistance to heat, oxidation, ozone and weather ageing due to their stable, saturated polymer backbone structure. Chapter 9 deals with thermoplastic rubbers (elastomers). Chapter 10 concentrates on chloroprene rubber which is polymerised by emulsion polymerisation using potassium persulphate as free radical initiator. Chapter 11 focuses on chlorosulphonated polyethylene rubber noted for its resistance to chemicals, temperature extremes and ultraviolet light. Chapter 12 is devoted to nitrile rubber which is a family of unsaturated copolymers of 2-propenenitrile and various butadiene monomers. Chapter 13 deals with polyacrylic rubber which are copolymers of acrylic esters and a minor proportion (1–5%) of a second monomer with reactive sites for cross-linking. Chapter 14 focuses on fluorocarbon rubber which are elastomeric high polymer which contains fluorine and may be a homopolymer or copolymer. Chapter 15 concentrates on silicone rubber, the remarkable space age performer, various types of silicone rubbers, their synthesis and vulcanisation aspects are also discussed.

Chapter 16 focuses on thermoplastic polyurethane which are flexible polyurethane foams. Chapter 17 is devoted to PEVA, chlorinated polyethylene and ethylene acrylic elastomers. These are speciality rubbers and widely used in various industries such as: hoses, cables and floor materials, etc. Chapter 18 deals with polysulphide, norbornene and polyphosphazene rubbers. These are specialty rubbers and are widely used in various industries such as – oil pipes, drying printing inks and rocket propellants.

Section III discusses vulcanisation mixing and calendering of rubber. Chapter 19 is devoted to chemistry and technology of vulcanisation. Vulcanisation of rubber is a process of improvement of the rubber elasticity and strength by heating it in the presence of sulphur, which results in three-dimensional cross-linking of the chain rubber molecules (polyisoprene) bonded to each other by sulphur atoms. Chapter 20 concentrates on materials for compounding and reinforcement. Various compoundings such as – accelerators, antioxidants, peptisers, plasticiser and fillers, etc., are discussed in details. Chapter 21 focuses on mixing and curing of rubber compounds. Mixing is the first and most critical process. If each of the various components, rubber, fillers, oils and chemicals is not thoroughly distributed and dispersed through the mass of the compounds, then problems cascade down through the subsequent processes of shaping and curing and result in less than optimum physical properties in the end product. Chapter 22 is devoted to calendering, extrusion and molding of rubber compounds. Calendering machines are used to produce continuous sheets from rubber compounds, sometimes incorporating reinforcing materials such as textile or wire cord and for impregnating or coating fabrics with compound.

Section IV discusses manufacturing techniques of rubber products. Chapter 23 is concentrates on tyres. Tyre is a rubber ring placed over the rim of a wheel of a road vehicle to provide traction and reduce road shocks. Manufacturing aspects of tyres and energy conservation are discussed in details. Chapter 24 focuses on rubber conveyor belt. Rubber conveyor belting essentially consists of various fabrics like all cotton, cotton/nylon, rayon/nylon, all synthetic with a cover of various rubber grades like oil resistant, flame resistant, etc. Chapter 25 concentrates on rubber hose. A hose is a flexible hollow tube designed to carry fluids from one location to another. Chapter 26 focuses on rubber footwear. Footwear can be defined as garments that are worn on the feet. The main purpose of footwear is protecting one's feet.

Chapter 27 is devoted to cellular rubber. Cellular rubbers or foam rubbers are porous rubber qualities with all-round closed cells. Chapter 28 focuses on sports goods. Sport goods provide the consumer with satisfaction, entertainment, sociability and achievement. Chapter 29 concentrates on anti vibration mounts. Rubber anti vibration mounts are crucial components in most types of heavy machinery and equipment which help dampen noise levels and vibration frequency while safeguarding fragile components from external vibrations. Chapter 30 is devoted to rubber rollers. The functions of rubber rollers in various applications are diverse, demanding different properties from the rubber that is used. Chapter 31 deals with rubber to metal bonding and has three essential elements—the rubber, the bonding agents and the substrate. Chapter 32 focuses on manufacture of miscellaneous rubber products such as – engine mountings, auto tubes and flaps, rubber cables, rubber gaskets, rubber matting, latex gloves, microcellular rubber sheets and products based on spread fabrics, etc.

Section V discusses pollution control and energy conservation in rubber industry. Chapter 33 concentrates on pollution control in rubber industry. Like all other manufacturing industries, the rubber industry also discharge quite a large number of occupational hazards of different types. This chapter also highlights rubber recovery and treatment of effluents. Chapter 34 deals with energy conservation in rubber industry. Energy saving and reducing plays an important role in rubber industry. Various aspects related to energy saving are discussed in detail.

Section VI discusses carbon footprints and nanotechnology in rubber industry. Chapter 35 is devoted to carbon footprint in rubber industry. The term 'carbon footprint' has become a topic of hot discussion all over the world. Carbon foot print can be described as the extent of damage caused to the environment due to some actions. The carbon footprint of rubber is an estimate of all the emissions caused by the production and delivery to the consumer and the disposal of packaging. Chatper 36 focuses on role of nanotechnology in rubber industry. Rubber nanocomposites based on different nanomaterials have attracted great interest in academia and industry because they exhibit improved mechanical and functional properties with a lower volume fraction of nanomaterials.

Section VII discusses applications, safety and testing of rubber products. Chapter 37 deals with engineering and other applications of rubber products such as – in automobiles, aerospace, railways, cables, bearing and bridges, piers and fenders, machinery and equipments and packaging. The automobile combines many different materials that must be integrated with the ever changing design and manufacturing processes. Without rubber components, the automobile in its current state of technology cannot exist. Chapter 38 concentrates on health and safety aspects in rubber industry. The extensive range of chemicals required and the volume of raw material handled can give rise to substantial quantities of airborne dust which are considered to be general health hazards.

The industry also uses very powerful machinery with the potential to cause fatal and serious injuries. There are established industry safeguarding standards for two-roll mills, internal mixers and calenders. Many serious accidents take place during repairs or to clear blockages, etc., and there must be procedures in place to ensure that safe interventions take place. Chapter 39 is devoted to testing of rubber products. The testing of rubber, rubber products and elastomers is essentially a continuously changing process due to natural instability in testing procedures. To achieve highest standards in rubber products, testing of rubber must contain not only a description of existing methods but also some indications of the directions in which changes are expected or needed.

Appreciations are also extended to Mr Harinder Singh, Senior DTP operator, who drew and labelled the flow diagrams and worked long hours to bring the book in time. I am also thankful to the editorial team of Woodhead Publishing India Pvt. for their wholehearted cooperation in bringing out the book in time.

It may not be wrong to hold that this book on Rubber Technology is essential reading for professionals and students pursuing B. Tech/M. Tech, engineering courses in material sciences, rubber technology, polymer science/technology. Besides students, this book will prove useful to industrialists and consultants in the respective fields.

It has been prepared with meticulous care, aiming at making the book error-free. Constructive suggestions are always welcome from users of this book.

S C Bhatia

Avishek Goel

Section I

General considerations and engineering aspects

Basic concepts of polymerisation

1.1 Introduction

Polymer science or macro-molecular science is a subfield of materials science concerned with polymers, primarily synthetic polymers such as plastics and elastomers. The field of polymer science includes researchers in multiple disciplines including chemistry, physics and engineering.

This science comprises three main sub-disciplines:

1. Polymer chemistry or macro-molecular chemistry is concerned with the chemical synthesis and chemical properties of polymers.
2. Polymer physics is concerned with the bulk properties of polymer materials and engineering applications.
3. Polymer characterisation is concerned with the analysis of chemical structure and morphology and the determination of physical properties in relation to compositional and structural parameters.

1.2 Monomers and polymers

1.2.1 Monomers

Monomers are the building blocks of more complex molecules, called polymers. Polymers consist of repeating molecular units which usually are joined by covalent bonds. Here is a closer look at the chemistry of monomers and polymers. Monomers are small molecules which may be joined together in a repeating fashion to form more complex molecules called polymers.

1.2.2 Polymers

A polymer may be a natural or synthetic macro-molecule comprised of repeating units of a smaller molecule (monomers). While many people use the term 'polymer' and 'plastic' interchangeably, polymers are a much larger class of molecules which includes plastics, plus many other materials, such as cellulose, amber and natural rubber. Polymer and their synthesis are discussed in detail section 1.2.4. Some of the examples of polymers include – plastics such as polyethylene, silicones such as silly putty, biopolymers such as cellulose and DNA, natural polymers such as rubber and shellac and many other important macro-molecules.

Formation of polymers: Polymerisation is the process of covalently bonding the smaller monomers into the polymer. During polymerisation, chemical groups are lost from the monomers so that they may join together. In the case of biopolymers, this is a dehydration reaction in which water is formed.

Monomers and polymers are shown in Table 1.1.

Table 1.1: Monomers and polymers.

Monomers		Polymers	
Propylene	$CH_3CH=CH_2$	Polypropylene	$\left[\begin{array}{c} CH_2 - CH \\ \quad\mid \\ \quad CH_3 \end{array}\right]_n$
Vinyl chloride	$CH_2=CHCl$	Polyvinyl chloride	$\left[\begin{array}{c} CH_2 - CH - \\ \quad\mid \\ \quad Cl \end{array}\right]_n$
Tetrafluoroethylene	$CF_2=CF_2$	Teflon	
Styrene	$H_5CH=CH_2$	Polystyrene	$\left[\begin{array}{c} -CH_2 - CH - \\ \quad\mid \\ \quad C_6H_5 \end{array}\right]_n$

The subscript n in the formula indicates the number of monomeric units contained in the polymer.

1.2.3   Macro-molecule

A macro-molecule is a very large molecule commonly created by polymerisation of smaller subunits. In biochemistry, the term is applied to the three conventional biopolymers (nucleic acids, proteins and carbohydrates), as well as non-polymeric molecules with large molecular mass such as lipids and macrocycles. The individual constituent molecules of polymeric macro-molecules are called monomers.

1.2.4   Polymer and their synthesis

A polymer is a large molecule, because of their broad range of properties, both synthetic and natural polymers play an essential and ubiquitous role in everyday life. Polymers range from familiar synthetic plastics such as polystyrene to natural biopolymers such as DNA and proteins that are fundamental to biological structure and function.

Polymers, both natural and synthetic, are created via polymerisation of many small molecules, known as monomers. Their consequently large molecular mass relative to small molecule compounds produces unique physical properties, including toughness, viscoelasticity and a tendency to form glasses and semicrystalline structures rather than crystals.

Polyisoprene of latex rubber and the polystyrene of styrofoam are examples of polymeric natural/biological and synthetic polymers, respectively. In biological contexts, essentially all biological macro-molecules, i.e., proteins (polyamides), nucleic acids (polynucleotides) and polysaccharides are purely polymeric, or are composed in large part of polymeric components, e.g., isoprenylated/lipid-modified glycoproteins, where small lipidic molecule and oligosaccharide modifications occur on the polyamide backbone of the protein.

1.2.5   Common examples of polymers

Polymers are of two types:

1. Natural polymeric materials such as shellac, amber, wool, silk and natural rubber have been used for centuries. A variety of other natural polymers exist, such as cellulose, which is the main constituent of wood and paper.

2. The list of synthetic polymers includes synthetic rubber, phenol formaldehyde resin (or Bakelite), neoprene, nylon, polyvinyl chloride (PVC or vinyl), polystyrene, polyethylene, polypropylene, polyacrylonitrile, PVB, silicone and many more.

Most commonly, the continuously linked backbone of a polymer used for the preparation of plastics consists mainly of carbon atoms. A simple example is polyethylene, whose repeating unit is based on ethylene monomer.

1.2.6   Polymer synthesis

Polymerisation is the process of combining many small molecules known as monomers into a covalently bonded chain or network. During the polymerisation process, some chemical groups may be lost from each monomer. This is the case, for example, in the polymerisation of PET polyester.

The monomers are terephthalic acid ($HOOC-C_6H_4-COOH$) and ethylene glycol ($HO-CH_2-CH_2-OH$) but the repeating unit is $-OC-C_6H_4-COO-CH_2-CH_2-O-$, which corresponds to the combination of the two monomers with the loss of two water molecules. The distinct piece of each monomer that is incorporated into the polymer is known as a repeat unit or monomer residue as shown in Fig. 1.1.

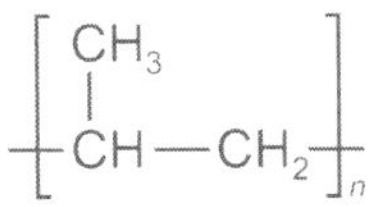

$$\left[\begin{array}{c} CH_3 \\ | \\ CH-CH_2 \end{array} \right]_n$$

Figure 1.1: The repeating unit of the polymer polypropylene.

Biological synthesis

There are three main classes of biopolymers: polysaccharides, polypeptides and polynucleotides. In living cells, they may be synthesised by enzyme-mediated processes, such as the formation of DNA catalysed by DNA polymerase. The synthesis of proteins involves multiple enzyme-mediated processes to transcribe genetic information from the DNA to RNA and subsequently translate that information to synthesise the specified protein from amino acids. The protein may be modified further following translation in order to provide appropriate structure and functioning.

Modification of natural polymers

Naturally occurring polymers such as cotton, starch and rubber were familiar materials for years before synthetic polymers such as polyethene and perspex appeared on the market. Many commercially important polymers are synthesised by chemical modification of naturally occurring polymers. Prominent examples include the reaction of nitric acid and cellulose to form nitrocellulose and the formation of vulcanised rubber by heating natural rubber in the presence of sulphur. Ways in which polymers can be modified include oxidation, cross-linking and end-capping.

Especially in the production of polymers, the gas separation by membranes has acquired increasing importance in the petrochemical industry and is now a relatively well-established unit operation. The process of polymer degassing is necessary to suit polymer for extrusion and pelletising, increasing safety, environmental and product quality aspects. Nitrogen is generally used for this purpose, resulting in a vent gas primarily composed of monomers and nitrogen.

1.2.7   Polymer properties

Polymer properties are broadly divided into several classes based on the scale at which the property is defined as well as upon its physical basis. The most basic property of a polymer is the identity of its constituent monomers. A second set of properties, known as microstructure, essentially describe the arrangement of these monomers within the polymer at the scale of a single chain. These basic structural properties play a major role in determining bulk physical properties of the polymer, which describe how the polymer behaves as a continuous macroscopic material. Chemical properties, at the nano-scale, describe how the chains interact through various physical forces. At the macro-scale, they describe how the bulk polymer interacts with other chemicals and solvents.

Monomers and repeat units

The identity of the repeat units (monomer residues, also known as 'mers') comprising a polymer is its first and most important attribute. Polymer nomenclature is generally based upon the type of monomer residues comprising the polymer. Polymers that contain only a single type of repeat unit are known as homopolymers, while polymers containing a mixture of repeat units are known as copolymers. Poly(styrene), for example, is composed only of styrene monomer residues and is therefore classified as a homo-polymer. Ethylene-vinyl acetate, on the other hand, contains more than one variety of repeat unit and is thus a copolymer.

Some biological polymers are composed of a variety of different but structurally related monomer residues; for example, polynucleotides such as DNA are composed of a variety of nucleotide subunits. A polymer molecule containing ionisable subunits is known as a polyelectrolyte or ionomer.

Microstructure

The microstructure of a polymer (sometimes called configuration) relates to the physical arrangement of monomer residues along the backbone of the chain. These are the elements of polymer structure that require the breaking of a covalent bond in order to change. Structure has a strong influence on the other properties of a polymer. For example, two samples of natural rubber may exhibit different durability, even though their molecules comprise the same monomers.

1.2.8   Degree of polymerisation

The degree of polymerisation, or DP, is usually defined as the number of monomeric units in a macro-molecule or polymer or oligomer molecule. For a homopolymer, there is only one type of monomeric unit and the number-average degree of polymerisation is given by:

$$DP_n \equiv X_n = \frac{M_n}{M_0},$$

where, M_n is the number-average molecular weight and M_0 is the molecular weight of the monomer unit. For most industrial purposes, degrees of poly-merisation in the thousands or tens of thousands are desired.

Some authors, however, define DP as the number of repeat units, where for copolymers the repeat unit may not be identical to the monomeric unit.

For example, in nylon-6,6, the repeat unit contains the two monomeric units —NH(CH$_2$)6NH— and —OC(CH$_2$)$_4$CO—, so that a chain of 1000 monomeric units corresponds to 500 repeat units. The degree of polymerisation or chain length is then 1000 by the first (IUPAC) definition, but 500 by the second.

In step-growth polymerisation, in order to achieve a high degree of poly-merisation (and hence molecular weight), X_n, a high fractional monomer conversion, p, is required, as per Carothers' equation: $X_n = 1/(1-p)$. A monomer conversion of p = 99% would be required to achieve X_n = 100. For chain-growth polymerisation, however, this is not generally true and long chains are formed for much lower monomer conversions.

Backbone chain

In polymer science, the backbone chain or main chain of a polymer is the series of covalently bonded atoms that together create the continuous chain of the molecule.

In polypeptides, the backbone consists of carbon and nitrogen atoms of the constituent amino acids. The backbone does not include the side chains.

The principal chain in the polymer is known as backbone.

For polystyrene:

$$\left[\text{CH}_2\!-\!\text{CH} \right]_n ----$$

the backbone is simply ~C—C—C—C~ ...

For Nylon-6, —(CH$_2$)$_6$—C—NH, the backbone is ~C—C—C—C—C—C—N~.

1.2.9   Linear polymers

Macro-molecules of linear polymers are long chains with a high degree of asymmetry (Fig. 1.2).

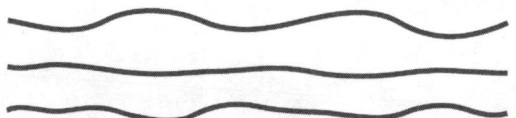

Figure 1.2: Structure of linear polymer.

Some linear polymers are cellulose, natural rubber, certain proteins, e.g., casein, Zein, amylose, etc.

1.2.10   Branched polymers

The macro-molecules of a branched polymer are chains with branches. The number of branches and the ratio of the length of the main chain to that of the side chains may vary (Fig. 1.3).

Amylopectin, glycogen and some mixed natural high molecular mass compounds show such structures. The synthesis of branched polymers has developed extensively in recent years.

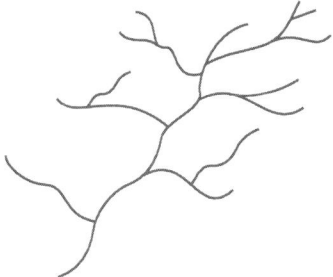

Figure 1.3: Structure of branched polymers.

1.2.11 Cross-linked polymers

Such polymers are composed of macro-molecular chain formed together with covalent bonds or cross-links as shown in Fig. 1.4.

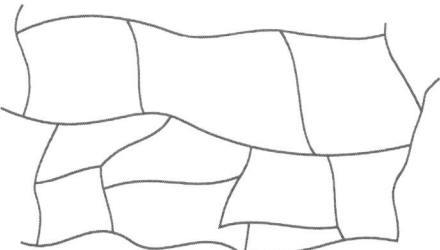

Figure 1.4: Structure of cross-linked polymer.

Cross-linked polymers developed in three directions at right angles are called spatial or three dimensional polymers. Classical examples are diamond and quartz. Cross-linked polymers of two dimensional structure such as graphite are called lamellar polymers.

1.2.12 Copolymers

Polymers having monomer units of different chemical composition are known as copolymers, e.g.,

A—B—B—C—A—C—C—B—A—B—C—

where, A, B and C are monomer units of different chemical composition. Copolymers include many proteins, lignin, nucleic acids, mixed polysaccharides and many synthetic high molecular mass compounds. In some copolymers, monomeric units are arranged either at random or as alternate runs of each of them.

A—A—A—A—B—B—B—B—A—A—A—A

Such copolymers are termed block copolymers.

1.2.13 Graft copolymers

In some cases different monomer units can be added or grafted onto the linear chain of a macro-molecule, such polymers are referred to as graft copolymers, e.g.,

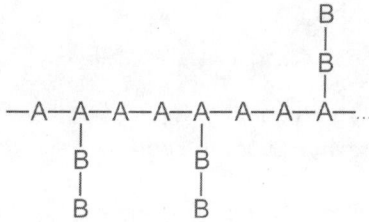

1.2.14 Crystalline polymers

Polymers may be amorphous (completely disordered) or partially ordered. The amorphous phase (Fig. 1.5) consists of linear, branched or cross-linked forming tridimensional networks. Ordered structures have a tendency to crystallisation. Crystalline forms include, fringed, micelles, fibrils, folded lamellae, spherulites.

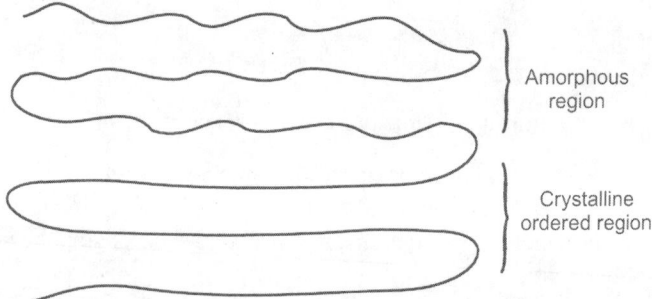

Figure 1.5: Crystalline polymers.

In order to improve mechanical properties, crystallinity is induced in amorphous polymers. Crystalline polymers include polyethylene, polycapro-amide, polyethylene terephthalate, gutta-percha, etc. These are widely used in fabricating various articles.

1.3 Addition and condensation polymers

1.3.1 Condensation polymer

Condensation polymers are any kind of polymers formed through a condensation reaction—where molecules join together—losing small molecules as by-products such as water or methanol, as opposed to addition polymers which involve the reaction of unsaturated monomers. Types of condensation polymers include polyamides, polyacetals and polyesters.

Condensation polymerisation, a form of step-growth polymerisation, is a process by which two molecules join together, resulting loss of small molecules which is often water. The type of end product resulting from a condensation polymerisation is dependent on the number of functional end groups of the monomer which can react.

Monomers with only one reactive group terminate a growing chain and thus give end products with a lower molecular weight. Linear polymers are created using monomers with two reactive end groups and monomers with more than two end groups give three-dimensional polymers which are cross-linked. Dehydration synthesis often involves joining monomers with an –OH (hydroxyl) group and a freely ionised –H on either end (such as a hydrogen from the $-NH_2$ in nylon or proteins). Normally, two or more different monomers are used in the reaction. The bonds between the hydroxyl group, the hydrogen atom and their respective atoms break forming water from the hydroxyl and hydrogen and the polymer. Polyester is created through ester linkages between monomers, which involve the functional groups carboxyl and hydroxyl (an organic acid and an alcohol monomer).

Nylon is another common condensation polymer. It can be manufactured by reacting di-amines with carboxyl derivatives. In this example the derivative is a di-carboxylic acid, but di-acyl chlorides are also used.

The simplest substance, with this type of bonding, is ethylene.

$$\begin{array}{ccc} H & & H \\ \diagdown & & \diagup \\ & C = C & \\ \diagup & & \diagdown \\ H & & H \end{array}$$

Condensation polymerisation is occasionally used to form simple hydrocarbons. This method, however, is expensive and inefficient, so the addition polymer of ethene (polyethylene) is generally used.

Condensation polymers, unlike addition polymers, may be biodegradable. The peptide or ester bonds between monomers can be hydrolysed by acid catalysts or bacterial enzymes breaking the polymer chain into smaller pieces.

The most commonly known condensation polymers are proteins, fabrics such as nylon, silk, or polyester.

1.3.2   Addition polymer

An addition polymer is a polymer which is formed by an addition reaction, where many monomers bond together via rearrangement of bonds without the loss of any atom or molecule. This is in contrast to a condensation polymer which is formed by a condensation reaction where a molecule, usually water, is lost during the formation. An addition polymer is formed by a reaction known as polyaddition or addition polymerisation. This can occur in a variety of ways including free radical polymerisation, cationic polymerisation, anionic polymerisation and coordination polymerisation.

Most of the common addition polymers are formed from unsaturated monomers (usually having a double bond). This includes polythenes, polypropylene, PVC, Teflon, Buna rubbers, polyacrylates, polystyrene and PCTFE. Addition polymers are also formed from monomers that have a closed ring. Through coordination polymerisation, even saturated monomers can form addition polymers.

When two or more types of monomer undergo addition polymerisation, the polymer formed is known as an addition copolymer. Saran wrap, formed from polymerisation of vinyl chloride and vinylidene chloride, is an addition copolymer.

Polymers made in this way are called addition polymers and include:

$-(CH_2-CHCl)_n-$ Polyvinyl chloride (PVC)

$$-(CH_2-CH)_n- \atop \qquad\; | \atop \qquad\; C_6H_5$$ Polystyrene

and $-(CF_2-CF_2)_n-$ Polytetrafluoroethylene (PTFE)

Values of n may run into thousands.

Contrast with condensation polymers

With exception of combustion, the backbone of addition polymers are generally chemically inert. This is due to the very strong C–C and C–H bonds and lack of polarisation within many addition polymers. For this reason they are non-biodegradable and hard to recycle.

This is, again, in contrast to condensation polymers which are biodegradable and can be recycled.

Many exceptions to this rule are products of ring-opening polymerisation, which tends to produce condensation-like polymers even though it is an additive process. For example, poly[ethylene oxide] is chemically identical to polyethylene glycol except that it is formed by opening ethylene oxide rings rather than eliminating water from ethylene glycol.

Nylon 6 was developed to thwart the patent on nylon 6,6 and while it does have a slightly different structure, its mechanical properties are remarkably similar to its condensation counterpart.

1.4   Mechanism of polymerisation

Mechanism of polymerisation of single monomer is given below.

If M represents the monomer and $Init_2$ represents a free radical initiator then we can write:

$$Init_2 \xrightarrow{\ k_1\ } 2\ Init \qquad \text{Chain initiation}$$

$$Init + M \xrightarrow{\ k_p\ } Init\ M$$
$$\cdots \qquad \cdots \qquad \cdots$$
$$\cdots \qquad \cdots \qquad \cdots \qquad \text{Chain propagation}$$
$$Init + M_n + M \xrightarrow{\ k_p\ } Init\ M_{n+1}$$

$$Init + M_n + Init + M_m \xrightarrow{\ k_t\ } Init\ M_{m+n} Init$$
or
$$\xrightarrow{\ k_t\ } Init\ M_n + Init\ M_m$$

Chain termination

Using a steady-state approximation on the concentration of all types of radicals $[R^{\bullet}]$ we obtain,

$$[R^{\bullet}] = \left(\frac{k_i}{k_t}\right)^{1/2} [Init]^{1/2}$$

and rate of propagation

$$= k_p [R^{\bullet}] [M]$$

$$= k_p (k_i/k_t)^{1/2} [Init_2]^{1/2} [M]$$

$$= k [Init_2]^{1/2} [M]$$

The above mechanism is also consistent with the kinetics of copolymerisation.

1.5   Polymer degradation

Polymer degradation is a change in the properties—tensile strength, colour, shape, etc., of a polymer or polymer-based product under the influence of one or more environmental factors such as heat, light or chemicals such as acids, alkalis and some salts. These changes are usually undesirable, such as cracking and chemical disintegration of products or, more rarely, desirable, as in bio-degradation, or deliberately lowering the molecular weight of a polymer for recycling. The changes in properties are often termed 'ageing'.

In a finished product such a change is to be prevented or delayed. Degradation can be useful for recycling/reusing the polymer waste to prevent or reduce environmental pollution. Degradation can also be induced deliberately to assist structure determination.

Polymeric molecules are very large (on the molecular scale) and their unique and useful properties are mainly a result of their size. Any loss in chain length lowers tensile strength and is a primary cause of premature cracking.

1.5.1　Commodity polymers

Today there are primarily seven commodity polymers in use: polyethylene, polypropylene, polyvinyl chloride, polyethylene terephthalate, polystyrene, polycarbonate and poly(methyl methacrylate) (Plexiglas). These make up nearly 98% of all polymers and plastics encountered in daily life. Each of these polymers has its own characteristic modes of degradation and resistances to heat, light and chemicals. Polyethylene, polypropylene and poly(methyl methacrylate) are sensitive to oxidation and UV radiation, while PVC may discolour at high temperatures due to loss of hydrogen chloride gas and become very brittle. PET is sensitive to hydrolysis and attack by strong acids, while polycarbonate depolymerises rapidly when exposed to strong alkalis.

For example, polyethylene usually degrades by random scission—that is by a random breakage of the linkages (bonds) that hold the atoms of the polymer together. When this polymer is heated above 450 Celsius it becomes a complex mixture of molecules of various sizes that resemble gasoline. Other polymers—like polyalphamethylstyrene—undergo 'specific' chain scission with breakage occurring only at the ends; they literally unzip or depolymerise to become the constituent monomers.

1.5.2　Photoinduced degradation

Most polymers can be degraded by photolysis to give lower molecular weight molecules. Electromagnetic waves with the energy of visible light or higher, such as ultraviolet light, X-rays and gamma rays are usually involved in such reactions.

1.5.3　Thermal degradation

Chain-growth polymers like poly(methyl methacrylate) can be degraded by thermolysis at high temperatures to give monomers, oils, gases and water.

1.5.4　Chemically assisted degradation of polymers

Chemically assisted degradation of polymers is a type of polymer degradation that involves a change of the polymer properties due to a chemical reaction with the polymer's surroundings. There are many different types of possible chemical reactions causing degradation however most of these reactions result in the breaking of double bonds within the polymer structure.

Degradation of rubber by ozone

One common example of chemically assisted degradation is the degradation of rubber by ozone particles. Ozone is a naturally occurring atmospheric molecule that is produced by electric discharge or through a reaction of oxygen with solar radiation. Ozone is also produced with atmospheric pollutants reacted with ultraviolet radiation. For a reaction to occur, ozone concentrations only have to be as low as 3–5 parts per hundred million (pphm) and when these concentrations are reached, a reaction occurs with a thin surface layer (5×10^{-7} meters) of the material. The ozone molecules react with the rubber which in most cases is unsaturated (contains double bonds), however a reaction will still occur in saturated polymers (those containing only single bonds). When reaction occurs, scission of the polymer chain (breaking of double covalent bonds) takes place forming decomposition products.

Chain scission increases with the presence of active hydrogen molecules (for example, in water) as well as acids and alcohols. Along with this type of reaction, cross-linking and side branch formations also occur by an activation of the double bond and these make the rubber material more brittle. Due to the increase in brittleness due to the chemical reactions, cracks form in areas of high stress. As propagation of these cracks increases, new surfaces are opened for degradation to occur.

Degradation of poly(vinyl) chloride (PVC)

Degradation can also occur as a result of the formation and then breakage of double bonds, such as solvolysis in PVC(Peacock). Solvolysis occurs when a Carbon-X bond, with X representing a halogen, is broken. This occurs in PVC in the presence of an acid species. Active hydrogen atoms will remove a chlorine atom from the polymer molecule, forming hydrochloric acid (HCl). The HCl produced may then cause dechlorination of adjacent carbon atoms. The dechlorinated carbon atoms then tend to form double bonds, which can be attacked and broken by ozone, just like the degradation of rubbers described above.

Degradation of polyester

Degradation of polyester may occur without the presence of the acidic catalyst that causes degradation of PVC. During hydrolysis water acts as the reactive catalyst instead of the acid. It causes degradation mainly at high temperature and pressure during processing. In this process the water molecule will attack the C–O ester bond, splitting the polymer in half. The water molecule will then dissociate, with one hydrogen atom forming a carboxylic acid group on the carbon atom with the double bonded oxygen, while the remaining atoms form an alcohol on the other chain end. These reactive products may also cause further degradation of the polymer chain. This chain scission lowers average molecular weight of the polymer, decreasing the number and strength of intermolecular bonds as well as the degree of entanglement. This will increase chain mobility, decreasing strength of the polymer and increasing deformation at low stresses.

Protection against chemically assisted degradation

Both physical and chemical barriers can be used to protect a polymer from chemically assisted degradation. A physical barrier must provide continuous protection, must not react with the polymer's environment, must be flexible so that stretching may occur and must also be able to regenerate (after wear processes). A chemical barrier must be highly reactive with the polymer's surroundings so that the barrier reacts with the environmental conditions rather than the polymer itself. This barrier involves addition of a material into the polymer blend during fabrication of the polymer. Due to this, the barrier addition must have a suitable solubility, must be economically feasible and must not hinder the production process. For the barrier to be activated, the addition must diffuse to the surface and so a suitable diffusivity is also required.

There are four theories on how these types of barriers protect the polymer material:

1. Scavenger theory: The protective layer reacts with the ozone rather than the polymer.
2. Protective film theory: The protective layer reacts with the polymer producing a thin film on the polymer surface which is inert and can't be penetrated.
3. Re-linking theory: The protective layer causes broken double bonds to be reformed.
4. Self-healing theory: The protective layer reacts with degraded polymer chains to form low-molecular-weight material which forms an inert film on the surface.

Of these theories, the scavenger theory is the most common and most important. However, more than one theory can act at the same time and the theory that takes place depends on the protective materials, the polymer and surrounding environment.

1.6   Biopolymers

Biopolymers are polymers which are naturally found in nature. Like polymers biopolymers are chain-like molecules made up of repeating chemical blocks and can be very long in length. Biopolymers can be classified in three groups, depending on the nature of the repeating unit they are made of: (i) polysaccharides are made of sugars, (ii) proteins of amino acids and (iii) nucleic acids of nucleotides.

The following substances are example-biopolymers for each group: cellulose (found in plants), myoglobin (muscle tissues) and DNA (genetic material of a given organism).

1.6.1 Biopolymers are complex molecules with biological activity

In contrast to synthetic polymers which have a simpler and more random structure, biopolymers are complex molecular assemblies that adopt precise and defined 3D shapes and structures. This feature is essential because this is what makes biopolymers active molecules *in vivo*. Their defined shape and structure are indeed keys to their function. For example, haemoglobin would not be able to carry oxygen in the blood if it was not folded in a quaternary structure.

1.6.2 Biopolymers have a low environmental footprint

Unlike synthetic polymers which feedstock can be derived from petrochemicals or chemical processes, biopolymers are produced from renewable resources such as plant and/or living organisms. They can be degraded by natural processes, micro-organisms and enzymes down to elemental entities that can be resorbed in the environment. Biopolymers thus offer the possibility to create a sustainable industry and reduce CO_2 emissions.

2

Natural rubber

2.1 Introduction

The economic product of the *Hevea tree*, rubber, is a *cis*-polyisoprenic molecule found in the latex. Latex is found in all the plant parts like bark of the trunk, roots, branches, leaves, flowers, fruits and seeds, though not in the wood. However, it is the latex present in the bark of the trunk of the mature tree that exploited commercially. Latex is harvested from the rubber tree by a process called tapping, during which a thin saving of bark on the tree trunk is removed using a special knife, at regular interval of time.

When tree is tapped for the first time, only a very small quantity of thick latex exudes. Each successive tapping produces more and more latex with lower rubber content until an equilibrium between quantity of latex and rubber content is established.

Tapping means controlled wounding of the bark of rubber trees for extraction of latex. The single cut system of tapping is generally followed now in all well managed rubber plantations. Later, continuous exersion method is proposed by Henery Nicholas Ridley. Under this method, a sloping cut is made with a special knife and the same cut is regularly reopened by removing a thin shaving of bark at each tapping. It avoids injuring of the cambium and thereby prevent bumping of the wounded part.

The trees are brought under tapping only after their a certain girth. As girth of the tree is one of the factors influencing the yield, it is necessary that a satisfactory rate of growth is maintained to obtain sustained yield for a number of years.

Latex is obtained from the bark of rubber tree by tapping. Tapping is a process of controlled wounding during which thin shavings of bark are removed. The aim of tapping is to cut open the latex vessels in the case of trees tapped for the first time or to remove the coagulum which blocks the cut end of latex vessels in the case of trees under regular tapping.

Latex: Hevea latex in the latex vessels of tapped trees contains 30–45% rubber in the form of particles. Latex is a hydrosol in which the dispersed particles are protected by a complex film. It contains more than one disperse phase. Besides rubber, the latex contains certain other particles also, namely lutoids and frey wyssling particles. Lutoids are associated with the process of latex vessel plugging which stops the flow a few hours after tapping.

Time of tapping, task and utensils: It is necessary to commence tapping early in the morning as late tapping reduces the exudation of latex. If used properly, the Michie Golledge knife used in various countries is well adopted for a high standard of tapping with minimum bark consumption. The draw knife or jebong knife common in Malaysia is suitable for both high and low level tapping and bigger task is now popular in various countries. The knives should be sharp. The knives, cups, bucket, etc., should cleaned well to prevent bacterial contamination and spoilage of latex. The tapping task (number of trees tapped on a day by one tapper) in India is around 300–350 trees compared to 400 to 500 trees in other countries. Headlights can be used for early morning tapping during non-rainy season including summer months to extract better crop.

2.2 Tapping techniques

Rubber trees have an economic life of about thirty years and the cost of establishment of these trees are very high. Therefore the selection of the exploitation techniques is very important. The industry emerged into its modern form when systems of tapping based on excision of the outer bark were developed by H. N. Ridley in the last decade of the nineteenth century.

2.2.1 Tapping system

Response to different tapping systems varies from clone to clone. In general budded trees are to be tapped on half spiral alternate daily (s/2d/2) system and seedling trees on half spiral third daily (s/2d/3) system. Alternate daily tapping is the recommended frequency for medium yielding clones (RRIM 600, GT1, PB28/59, etc.). For high yielding clones like RRII 105, PB 217, PB 260, etc., low frequency tapping systems with stimulation may be practised. These clones cannot tolerate daily or alternate daily tapping. However, low frequency tapping can be adopted for medium yielding clones also.

Tapping techniques are low frequency tapping, controlled upward tapping, intensive tapping and high level tapping, among the tapping techniques the barak valley region of non-traditional area is considered and recommended by the rubber board only low frequency tapping and intensive tapping only which is maintained by the growers.

2.2.2 Low frequency tapping (LFT)

Trees under low frequency tapping (d3, d4 and d6) have to be stimulated from opening for maximum sustainable yield. Number of stimulations vary with clone, age of the tree and frequency. In high yielding clones like RRII 105 and PB 217 under third daily (d3) tapping frequency, 15 to 30% sustainable yield increase can be achieved by three annual stimulations (April, September and November). In the case of medium yielding clones like RRIM 600 and GT 1, annual stimulations (April, August, October and December) are recommended under third daily (d3) frequency of tapping. Bark consumption should be restricted to 1.75 mm/tap.

Comparable yield to that under d3 frequency with stimulation can be achieved from d4 frequency. Six annual stimulations (April, June, August, September, November and December) are recommended for clone RRII 105 and five (April, June, August, October and December) for clone PB 217. In the case of clone GT 1, seven annual stimulations (March, April, June, August, October, December and January) have to be given. Under restricted bark consumption of 2 mm/tap, annual consumption will be 14–16 cm. for weekly tapping of clone RRII 105 fortnightly stimulation is to be given in the initial two years after opening and monthly in the subsequent years.

Low frequency tapping (LFT) with stimulation can be practised from the first year of tapping. Trees under higher frequencies of tapping can also be converted to LFT. Conversion to d3, d4 or d6 frequency will result in substantial saving in cost of production. However, when such conversions are done there will be a yield depression for 3–4 months.

2.2.3 Controlled upward tapping (CUT)

Controlled upward tapping (CUT) can be practised for longer duration of crop harvesting from the virgin bark above the basal panel. In CUT, instead of using ladder, a long handled modified gouge knife is used for upward tapping from the ground. Bark consumption is minimised as far as possible. In general 30–50% higher yield can be obtained for many years under CUT.

2.2.4 Intensive tapping

Intensive tapping is generally done on old rubber trees for a few years prior to feeling. The methods depend on condition of trees, previous tapping systems, availability of bark and the period available for harvesting before felling. The methods employed are increased tapping frequency, extension of tapping cut, opening of double cuts and use of stimulants. While opening two cuts should be sufficiently apart atleast by 45 cm to avoid the interference of drainage areas. Slaughter tapping is the last stage in the tapping cycle implemented two to three years before replanting. Since the objective of slaughter tapping is to extract as much latex as possible from the available bark, very little consideration is given to bark consumption, intensity, frequency and quality of tapping as well as intensity and frequency of stimulation.

2.2.5 High level tapping

When tapping of renewed bark on basal panels becomes uneconomic new cuts are opened at high level, 180 cm from bud union or even higher. The tapper uses a small ladder to reach the cut. Since ladder tapping is more strenuous and time consuming usually reduced tapping tasks.

2.3 Latex technology

Latex technology is a highly specialised field that is not too familiar to most polymer chemists and even many rubber compounders. The art and science of handling latex problems is more intricate than regular rubber compounding and requires a good background in colloidal systems. While latex differs in physical form from dry rubber, the properties of the latex polymer differ only slightly from its dry rubber counterpart. Unlike the dry rubber, which must be masticated (mechanically sheared) before use, the latex polymer need not be broken down for application, thus retaining its original high molecular weight which results in higher modulus products.

Other advantages enjoyed by applications involving latex are, lower machinery costs and lower power consumption, since the latex does not have to be further processed into dry form and compounding materials may be simply stirred into the latex using conventional liquid mixing equipment.

2.3.1 Composition of rubber latex

The natural product, which is exuded as a milky liquid by the *Hevea tree*, is a colloidal solution of rubber particles in water, the particle diameters range between 0.05 μ and 5 μ. It is a cytoplasmic system containing rubber and non-rubber particles dispersed in aqueous serum phase.

Freshly tapped *Hevea* latex has a pH of 6.5 to 7.1 and density 0.98 g/cm^3. The total solids of fresh field latex vary typically from 30 to 40 wt % depending on clone, weather, stimulation, age of the tree, method of tapping, tapping frequency and other factors. The dry rubber content is primarily *cis*-1,4,-poly isoprene is shown below:

$$-CH_2-C=CH-CH_2-$$
$$\underset{CH_3}{|}$$

The non-rubber portion is made up of various substances such as sugars, proteins, lipids, amino acids and soluble salts of calcium, magnesium, potassium and copper. The solid phase typically contains 96% rubber hydrocarbon, 1 wt% protein and 3 wt% lipids with traces of metal salts.

2.3.2 Stabilisation of rubber latex

Though fresh rubber latex is nearly neutral and the rubber particles are stabilised by an adsorbed layer of protein and phospholipids, but on exposure to air the latex rapidly develops acidity and within 12 to 24 hr spontaneous coagulation sets in (at an approximate pH of 5). The latex has therefore, to be preserved immediately after collection against rise in acidity by bacterial putrefaction. As already mentioned, ammonia has long been used as preservative of latex owing to certain advantages including the ease of its removal by blowing air or reaction with formaldehyde.

Other preservatives such as sodium penta-chlorophenate, sodium salt of ethylene diamine tetra-acetic acid, boric acid or zinc alkyl dithiocarbamates, may be used with smaller amount of ammonia. This is known as low ammonia latex and has the advantages of lower cost and elimination of the need to deammoniate the latex before processing into products.

2.3.3   Concentration of rubber latex

The ammonia preserved field latex which is known as normal (un-concentrated) latex is not suitable for commercial use as it contains considerable amount of non-rubber constituents which are detrimental to the quality of products and also contains too much water which is costly for transportation.

The latex is, therefore, concentrated to about 60% rubber solids before leaving the plantation. This concentration process is carried out either by centrifuging, creaming, electrodecantation or evaporation.

The first two processes make use of increasing the gravitational force of the rubber particles, by applying centrifugal force on the former or by adding a creaming agent like sodium alginate, gum tragacanth, etc., in the latter process. Both these processes of concentration result in a decrease of non-rubber content, the centrifuging process being superior in this respect.

The concentrated latex obtained by electrodecantation process which utilises the negative charge on the tiny rubber particles, is similar in composition to the centrifuged latex, however cost economics does not favour this process to be exploited on commercial scale.

The evaporated latex contains all the non-rubber constituents present in the original normal latex. It contains a small amount of ammonia. Because of its high stability, evaporated latex is useful in compounding heavily loaded mixes, hydraulic cement, etc.

The centrifuged latex is most widely used in industry. Latex concentrate constitutes slightly more than 8% of the global natural rubber supply and about 90% of this is centrifuge concentrated. Principal outlets for natural rubber latex are foam rubber, dipped goods and adhesives.

2.3.4   Latex compounding

In latex technology, concentrated latex is first blended with the various additives as required for different applications. The blending of different additives is known as latex compounding. Latex compounding involves not only the addition of the proper chemicals to obtain optimum physical properties in the finished product but also the proper control of colloidal properties which enable the latex to be transformed from the liquid state into finished product.

Viscosity control in the latex is very important. The particle size of the latex has a great effect on viscosity. Large particles generally result in low viscosity. Dilution with water is the most common way to reduce viscosity. Certain chemicals such as trisodium phosphate, sodium dinaphthyl methane disulphonate are effective viscosity reducers.

2.3.5 Thickening agents

Thickening may be accomplished with either colloidal or solution thickeners. Small particle size materials such as colloidal silica will thicken latex when added to it. Solutions of such materials as alpha protein, starch, glue, gelatin, casein, sodium polyacrylates and poly(vinyl methyl ether) will also thicken latex.

2.3.6 Wetting agents

Sometimes the addition of a wetting agent to latex mix is necessary for successful impregnation of fabric or fibres with latex. Sulphonated oils have been found to be effective in assisting complete penetration between textile fibres without any danger of destabilising the latex.

2.3.7 Vulcanising agents

Curing or vulcanisation, which involves the chemical reaction of the rubber with sulphur in presence of an activator (such as zinc oxide) and accelerator, manifests itself in an increase in strength and elasticity of the rubber and an enhancement of its resistance to ageing. Vulcanisation of latex may be effected by either of the two ways: (i) the rubber may be vulcanised after it has been shaped and dried or (ii) the latex may be completely vulcanised in the fluid state so that it deposits elastic films of vulcanised rubber on drying. The latter process, however, does not yield products of high quality and is resorted to only in the production of cheaper articles, e.g., toy balloons.

The problem of scorching or premature vulcanisation is rarely encountered in practical latex work and hence ultra accelerators such as zinc diethyl dithiocarbamate (ZDC) alone or in combination with zinc salts of mercapto-benzothiazole (ZMBT), tetramethyl thiuram disulphide (TMTD), polyamines and guanidines are used. The latter two also function as gel sensitisers, or secondary gelling agents, in the preparation of foam rubber. The doses of the vulcanising ingredients are adjusted according to the requirements of the end products. Thus only small amount of sulphur and accelerator (0.5–1.0 phr) with little or no zinc oxide are required in the production of the transparent articles, whereas in case of latex foams the doses are quite high.

2.3.8 Antioxidants

Because of the great surface area exposure of most latex products, protection against oxidation is very important. Many applications involve light coloured products, which must not darken with age or on exposure to light. Non-staining antioxidants such as hindered phenols (styrenated phenols) must be used. Where staining can be tolerated, amine derivatives such as phenylene diamines, phenyl beta-napthylamine, ketone-amine condensates may be used. These have good heat stability and are also effective against copper contamination, which cause rapid degradation of rubber.

2.3.9 Fillers

Fillers may be added to latex to reduce the cost of rubber articles, to prevent spreading mixes leaking through the fabric, to increase the viscosity of the compound or to modify the properties of the rubber. Most of the non black fillers such as china clay, mica powder, whiting (calcium carbonate), Lithopone, Blanc Fixe (barium sulphate) may be used in latex compounds. Carbon black does not reinforce latex in the manner that it does dry rubber and is used only in small amounts in latex for colour, as are various other dyes and pigments.

2.3.10 Softeners

In applications like toy balloons, softeners are added to soften them so that they may be easily inflated. Softening agents in general used are liquid paraffin, paraffin wax and stearic acid.

2.3.11 Dispersing agents

The particle size of solid materials added to latex must usually be made as small as possible to ensure intimate contact with the rubber particles. Solid materials are usually added to latex as dispersion. The material to be added is mixed with dispersing agents in deionised water and ground to a small particle size in a ball mill or attritor. In these devices stones or other hard pebble-sized materials are made to tumble and mix with chemicals reducing them to very small size.

The selection and amount of dispersing agent is determined by the physical properties of the material to be dispersed. The functions of these agents are to wet the powder, to prevent or reduce frothing and to obviate reaggregation of the particles. The concentration of dispersing agents rarely exceeds 2% except in special circumstances. None of the common materials such as gelatin, casein, glue or soap such as ammonium oleate possesses all the requisite properties and hence it is necessary to use mixtures of two or more of them. When putrefiable dispersing agents such as casein, glue and gelatin are used, a small amount of bactericide, such as 0.01% sodium trichlorophenate may be added. Non putrifiable proprietory dispersing agents such as Dipersol F conc. of Indian Explosives Ltd. based on sodium salt of methylenebis [naphthalenesulphonic acid] are also available which are highly efficient dispersing agents with little foaming tendency during milling. Time, equipment and labour can often be saved by dispersing together (in the correct proportion) all the water insoluble ingredients required for a particular compound including sulphur, zinc oxide, accelerator, antioxidant, colour and fillers. Mixed dispersion having excellent storage stability against reaggregation and settling can be prepared by using the following formula and method:

Mixed total solids   –   100 parts

Dispersal F conc.   –   4 parts

Deionised water   –   96 parts

The mixed ingredients are dispersed by ball milling for at least 48 hr.

2.3.12   Emulsifying agents

As in the case of dispersions, deionised water should also be used for the preparation of emulsion of water immiscible liquids to be used in latex compounds. An emulsion is defined as a system in which a liquid is colloidally dispersed in another liquid. The emulsions use in for latex should be the 'oil-in-water' type in which water is the continuous phase.

Simple equipment for the preparation of emulsion consists of a tank and a high-speed stirrer. Very fine and stable emulsions can be prepared by using a homogeniser. In a homogeniser, the liquid mixed with the required amount of water and emulsifying agent is forced through fine orifice under high pressure (1000–5000 psi), the liquid mix is thus subjected to a high shearing force which breaks down the particles to the required size.

Various synthetic emulsifying agents are available in the market, but for use with latex, soaps have been found to be quite satisfactory. For getting a satisfactory emulsion, the soap is produced *in situ* during mixing of the components. In this method, the cationic part of the soap (ammonia, KOH or amine) is dissolved in water and the anionic part (oleic, stearic or rosin acid) is dissolved in the liquid to be emulsified.

Soap is formed when these solutions are mixed. A method of preparation of a typical 50% emulsion of liquid paraffin is given below:

Liquid paraffin   –   50.0 parts

Oleic acid   –   2.5 parts

Concentrated ammonia solution   –   2.5 parts

Deionised water   –   45.0 parts

The oleic acid is mixed with liquid paraffin and the mixture is added to the water containing concentrated ammonia solution. The two phases are mixed by agitation and a stable emulsion is obtained by passing through a homogeniser.

2.3.13 Stabilisers

The stabilising system naturally occurring in ammonia preserved latex is adequate to cope with the conditions normally encountered during concentration, transportation and distribution but fails to withstand the more severe conditions met with during compounding and processing, when additional stability must be ensured by the addition of more powerful agents.

Some degree of stabilisation may be attained by adding simple materials such as soap and proteins (e.g., casein). Casein is liable to putrefy and impart to latex a high initial viscosity, which may yield products having inferior physical properties. Soaps are convenient to use but their behaviour is not always predictable and they have limited applications. Synthetic stabilisers are now available which are free from the limitations associated with soaps and proteins.

An anionic surface-active agent such as sodium salt of cetyl/oleyl sulphate when present in sufficient quantity, stabilises latex against heat, fillers and mechanical working. It has no thickening action on latex compounds, does not alter the rate of cure and has no adverse effect on the vulcanisate. It is most effective in alkaline medium and loses its activity in presence of acids and polyvalent ions. It is, therefore, most suitable for the coagulant dipping process. Its efficiency remains unaffected by the increase in temperature.

A non ionic surface active agent such as an ethylene oxide condensate possesses remarkable stabilising power to protect latex compounds against the effects of mechanical action, acids, polyvalent salts, etc. It differs from anionic stabilisers in its method of functioning. It increases the hydration of the stabiliser film at the rubber/water interface and has little or no effect on the charge. Because of the high chemical stability, its use is not recommended in acid coagulant dipping process. However, it loses its activity at elevated temperature and this property is utilised in heat-sensitive compounds. It affords excellent protection to such compounds during storage at room temperature, but on heating it loses this power and gelling (or setting) of rubber particles takes place.

2.3.14 Compounding criteria

During compounding, it is essential to avoid the addition of any material liable to cause coagulation. As already discussed, the latex compound should be properly stabilised. In general, the addition of water-soluble organic liquids, salts of polyvalent metals and acidic materials are to be avoided. Water-insoluble liquids and solids must be added as emulsions and dispersions respectively, in which the size of the individual particle is of the same order as that of the rubber particles in the latex. Care should be taken to avoid the use of hard water at any stage of latex compounding as it has a destabilising action on latex.

The containers for the latex may be made from stone, enamelled iron, stainless steel and wood lined with rubber or gutta-percha. It is preferably thermostatically controlled against changes in atmospheric temperature and is fitted with water jacket. It is equipped with a mechanical stirrer. During the addition of the compounding ingredients, the mix should be stirred slowly but thoroughly. Slow stirring of the latex mix assists in the removal of bubbles and minimises the formation of a skin, which arises from evaporation of water in the latex. It is important to avoid contact between the stirrer and the container, since latex is readily coagulated by friction.

2.4   Processing of latex compound

After a suitable latex compound has been prepared, the next step is to get the shape of the article to be made, set the shape and then vulcanise. The different latex processes classified according to the method of shaping are: (i) dipping, (ii) casting and molding, (iii) spreading, (iv) spraying and (v) foaming.

Dipping: A variety of thin rubber articles, e.g., toy balloon, teats, gloves, etc., can be prepared from latex by dipping process. The process consists essentially of dipping a former in the shape of the article to be made into the compounded latex. The formers may be made from a variety of materials, including metal, glass, lacquered wood and porcelain. The deposited film is dried, vulcanised in circulating hot air, steam or hot water and then stripped from the former. This is known as 'straight' dipping as against coagulant dipping where the former is first coated by dipping into a chemical coagulating agent. The coagulants may be either salt coagulants or acid coagulants. A typical dipping compound is suitable for balloons, gloves, etc.

Casting and molding: Casting involves the use of a mold on the inside walls of which the rubber article is formed, the pattern on the inside of the mold determining the ultimate shape of the article. The basic principle of latex casting is to 'set' the compound in the mold followed by subsequent drying, removal from the mold and vulcanising. Depending on the technique of 'setting' (gelling) inside the mold, two types of molds are used: (i) Plaster of Paris molds and (ii) metal molds. Gelation in plaster mold is brought about by partial absorption of water by the mold material and in a metal mold by using a heat-sensitising agent.

Both solid and hollow articles can be produced by the process of casting. In the preparation of the solid articles the entire rubber latex content of the mold is gelled and subsequently dried. Non-porous metal molds are used both for hollow and solid articles whereas the porous plaster molds are generally used for hollow articles. Hollow articles are produced by forming the required thickness on the inside wall of the mold. With a well-formulated compound, satisfactory wall thickness can be built up in about 5–10 min. The plaster mold, together with its deposited latex, is then placed in an oven at 40–60°C for several hours. When the deposit is consolidated and partially dry, the mold is removed from the oven, allowed to cool and the article is carefully removed. It is then washed, dried and cured for 30 min at 100°C in air.

Spreading: Spreading of latex is used in the manufacture of proofed fabrics, which consists of applying a suitable latex compound on the fabric with the help of a Doctor's knife. This process has found wide application in the backing of tufted carpets in which the loosely woven piles of wool or jute fibres must be anchored strongly to the base by using a suitable compound. A compound found satisfactory in carpet backing application.

Spraying: The adhesive property of latex has been utilised in the spraying process for bonding paper, cloth, leather, fibre, etc. Spraying of latex is now a days largely used in the manufacture of cushions and mattresses from latex treated coir. Coconut fibres can be bonded by spraying a suitable latex compound to yield latex treated coir, which is a cheap but useful as upholstery material. The process consists of spraying the loose fibres with the latex compound, drying the product, compressing the dried mass in a mold to obtain a desired shape and curing it in an air oven for the permanence of shape.

Foaming: The production of latex foam for mattresses and upholstery is the most important of all the latex processes. Latex foam is a flexible cellular material containing many cells (either open, closed or both) distributed throughout the mass. There are currently two methods of producing latex foam: the Dunlop process and the Talalay process.

In the Dunlop process, sodium silicofluoride is used as the gelling agent. The latex compound is mechanically beaten and/or air blown through it to foam. Then the requisite amount of a dispersion of sodium silicofluoride is added, which in presence of zinc oxide sets the foam into gel in a mold (usually made of aluminium) in which it is poured. The gelled foam is then vulcanised in steam, stripped from the mold, washed and dried. In the compound a secondary gelling agent, diphenyl guanidine (DPG), is added to reduce the gelling time so that no premature foam collapse may occur.

In the Talalay process, partially foamed latex is poured into a mold which is sealed and vacuum is applied so that the foam expands to fill the mold completely. The foam is then frozen by cooling the mold to −35°C. Carbon dioxide is then admitted which penetrates the structure and owing to the pH change, causes gelling. The final stage is heating of the mold to vulcanising temperature to complete the cure. In spite of the high capital cost, this process is currently used because of the excellent quality of the product and the low rejection rate.

2.4.1 Dry rubber technology

A variety of coagulation methods are available to prepare the rubber for dry rubber technology processes. Since the properties of the rubber are affected by trace ingredients and by the coagulating agents used, rubbers of different properties are obtained by using the different methods. The major types of raw rubbers are:

Ribbed smoke sheet (RSS): It is the sheet of coagulum obtained by vertically inserting aluminium partitions into the coagulation tanks containing the latex and the coagulation is effected by adding acetic acid. The sheet is then passed through a series of mill rolls, the last pair of which are ribbed, giving the surface of rubber a diamond pattern, which shortens the drying time of rubber. The sheet is then dried slowly in a 'smoke house' at a temperature gradient of 43–60°C for about four days. The rubber is dark in colour.

Pale crepe: This is a premium grade of rubber, for use where lightness of colour is important as in white side walls of tyres, surgical goods, etc. For pale crepe high quality of latex is used and the lightest colours are obtained by removing a coloured impurity, β-carotene, by a two stage coagulation process, followed by bleaching the latex with xylyl mercaptan and adding sodium bisulphite to inhibit an enzyme catalysed darkening process. The coagulum is machined eight or nine times between grooved differential-speed rollers with liberal washing.

Comminuted and other 'new process' rubbers: In these cases the coagulum is broken up and then dried. The rubber is then packed in flat bales similar in size to those used for major synthetic rubbers unlike the heavier square bales used with smoke sheet and crepe rubbers.

2.4.2 Properties of raw natural rubber

The better types and grades' of natural rubber contain at least 90% of the hydrocarbon *cis*-1,4-polyisoprene, in admixture with naturally occurring resins, proteins, sugars, etc. The raw material of commerce (sheet, crepe. etc.), comprises a molecular weight mainly in the range of 5,00,000 to 10,00,000 which is very high for its processing.

Hence rubber has to be extensively masticated on a mill or in an internal mixer to break down the molecule to a size that enables them to flow without undue difficulty when processing by extrusion or other shaping operations. The break down occurs more rapidly at either high (120–140°C) or moderately low (30–50°C) temperature than it does at temperatures around 100°C. It is now recognised that breakdown at the more elevated temperatures is due to oxidative scission and that at low temperatures due to mechanical ruptures of primary bonds, the free radicals thus produced get stabilised by addition of oxygen.

Because of its highly regular structure, natural rubber is capable of crystallisation, which is substantially increased by stretching of the rubber causing molecular alignment. This crystallisation has a reinforcing effect giving strong gum stock (unfilled) vulcanisates. It also has a marked influence on many other mechanical properties.

The outstanding strength of natural rubber has maintained its position as the preferred material in many engineering applications. It has a long fatigue life, good creep and stress relaxation resistance and is low cost. Other than for thin sections, it can be used to approximately 100°C and sometimes above. It can maintain flexibility down to –60°C if compounded for the purpose. The low 'hysterisis' (heat generation under dynamic condition) and its natural tack make natural rubber ideal for use in tyre building. Its chief disadvantage is its poor oil resistance and its lack of resistance to oxygen and ozone, although these latter disadvantages can be ameliorated by chemical protection. Natural rubber is generally vulcanised using accelerated sulphur system. Peroxides are also occasionally used, particularly where freedom from staining by metals such as copper is important.

Natural rubber is mainly used in passenger tyres, primarily for carcasses and white side walls, the remainder of the tyre usage is in racing cars, aeroplanes, heavy duty trucks and buses, tractors and farm vehicles. Besides, it is used in footwear soles, industrial products such as pump coupling, rail pads, bridge bearings, conveyor belts (cover and friction), hoses, etc.

2.4.3   Applications of latex

The latex is used to make many other products as well, including mattresses, gloves, swim caps, condoms, catheters and balloons. In particular, natural organic latex has become a popular material for mattresses, pillows and other consumer goods, offering a variety of health benefits when compared to synthetic latex and memory foam, most dominant of which being significantly decreased exposure to the effects of treatment chemicals and chemical off-gassing.

Latex from the chicle and jelutong trees is used in chewing gum.

Synthetic latexes are used in coatings (e.g., latex paint) and glues because they solidify by coalescence of the polymer particles as the water evaporates and therefore can form films without releasing potentially toxic organic solvents in the environment. Other uses include cement additives and to conceal information on scratchcards. Latex, usually styrene-based, is also used in immunoassays.

Latex – both synthetic and natural – is used to make mattresses as an alternative to memory foam as it had very similar properties and is preferred as a natural alternative.

Section II

Synthetic rubbers—non-oil resistant, special purpose and other specialty rubbers

Synthetic rubbers: An overview

3.1   Introduction

Rubber is a collective term for macro-molecular substances of natural (natural rubber - NR) or synthetic origin (synthetic rubber - SR). There were a number of reasons responsible for the development of an alternative or substitute for natural rubber. These included volatile or rising prices for natural rubber on the world market in response to the general state of the economy, political events which cut customers off from the suppliers of raw materials, long transport distances, regional constraints with respect to establishing rubber plantations and the increase in global demand for rubber for different applications.

Synthetic rubber is a white, crumbly, plastic mass which can be processed and vulcanised in the same way as natural rubber. Synthetic rubber is produced in different ways. Figure 3.1 illustrates one of the common production processes.

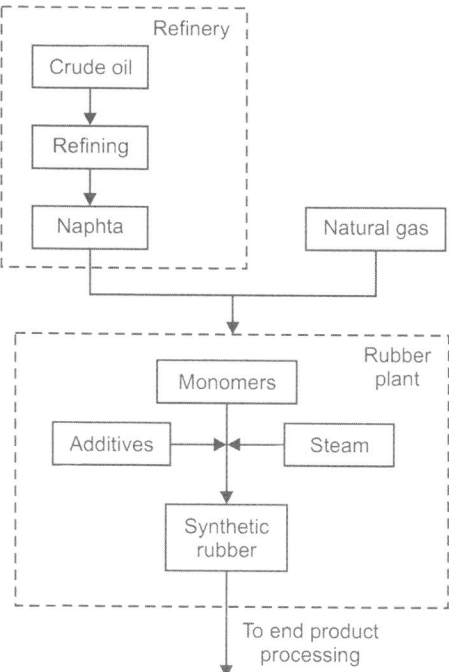

Figure 3.1: Flow diagram showing production process of synthetic rubber.

Synthetic rubbers are artificially produced materials with properties different than natural rubber. Most are obtained by polymerisation or polycondensation of unsaturated monomers. A wide range of different synthetic rubbers have emerged, reflecting the various different applications and the chemical and mechanical properties they require. Co-polymerisation of different monomers allows the material properties to be varied across a wide range. Polymerisation can take place under hot or cold conditions, which result in hot polymers (hot rubber) or cold polymers (cold rubber). Synthetic rubbers are marketed as compressed bales and square blocks. They are also produced in the form of powder rubber, talcum-coated chips, granules and as latex concentrates in liquid form.

Quality/duration of storage: The benefits compared with natural rubber include better oil and temperature resistance and the possibility of a product with an extremely constant quality. Synthetic rubbers made from butadiene (polybutadiene copolymers) rank as the most important synthetic rubbers produced.

3.2   Classification of synthetic rubbers

Synthetic rubber can be classified us under:

1. Non-oil resistance rubber.
2. Special purpose synthetic rubbers.
3. Other speciality rubbers.

3.2.1   Non-oil resistant synthetic rubber

1. Styrene butadiene rubber (SBR).
2. Polybutadiene rubber (BR).
3. Polyisoprene rubber (IR).
4. Butyl (IIR) and halobutyl rubber.
5. Ethylene propylene diene monomer (EPDM) and many more.
6. Thermoplastic rubber.

These are discussed in detail in chapters 4, 5, 6, 7, 8 and 9.

3.2.2   Special purpose synthetic rubbers

1. Chloroprene rubbers.
2. Chlorosulphonated polyetheylene rubbers.
3. Nitrile rubbers.
4. Polyacrylic rubbers.
5. Flouorocarbon rubbers.
6. Silicone rubbers.

These are discussed in detail in chapters 10, 11, 12, 13, 14 and 15.

3.2.3   Other speciality rubbers

1. Thermoplastic polyurethane.
2. Poly(ethylene-co-vinyl acetate).
3. Chlorinated polyethylene.
4. Ethylene acrylic rubber.
5. Polysulphide rubbers.
6. Polynorbornene rubber.
7. Polyphosphazene rubber.

These are discussed in detail in chapters 16, 17 and 18.

3.3   Brief description of important synthetic rubbers

This section briefly discusses some of the important synthetic rubbers along with their properties in comparison with natural rubber in the light of section 3.2 discussed above.

Styrene butadiene rubber (SBR): General purpose rubber made up of different types, better abrasion resistance, lower elasticity, poorer low-temperature behaviour, better heat and ageing resistance, excellent electrical insulation material similar to rubber.

Polybutadiene rubber (BR): Poor processing properties mean that BR is not used on its own, blended with SBR or NR, abrasion-resistant, good elasticity, flexible at low temperatures.

Isoprene rubber (IR): Properties largely comparable with natural rubber, more uniform, cleaner, transparent.

Acrylonitrile butadiene rubber (NBR): Oil and fuel resistant, good heat distortion temperature properties, abrasion resistant.

Chloroprene rubber (CR): Flame retardant, resistant to grease, oil, weathering and ageing, abrasion resistant.

Butyl rubber (IIR): Low permeability to gases, resistant to ageing, ozone and chemicals, good mechanical properties, abrasion resistant, good electrical insulation properties.

3.3.1   Mechanical properties of synthetic rubbers

The mechanical properties of synthetic rubbers can be improved by adding fillers such as carbon black during vulcanisation with sulphur. Temperature resistance, abrasion resistance, ageing resistance, resistance to oxygen and chemicals such as acids and petrol are properties which are improved in this way. Duration of storage varies depending on the type of synthetic rubber. For example, a range of 6–36 months if the ideal conditions recommended by the manufacturer, such as a storage temperature between 10 and 25°C, are observed. Synthetic rubber must be stored dry, some synthetic rubbers must be stored cool and they are to be protected from direct sunlight.

Recommended storage duration for synthetic rubber, in particular SBR.

Initial storage   –   5 years
Extended storage   –   2 years

Intended use

Like natural rubber, synthetic rubber has a wide range of applications, such as in the tyre industry (car, aircraft and bicycle tyres), drive belts, hoses, medical equipment, seals, floor coverings, conveyor belts, molded parts.

Operating temperatures and applications

The following list shows the operating temperatures and applications for some important types of synthetic rubbers.

1. Styrene butadiene rubber (SBR):
   (a) Operating temperature: –40–100°C.
   (b) Applications: Tyre industry (treads and carcasses), conveyor belts, seals, technical rubber products.
2. Polybutadiene rubber (BR):
   (a) Operating temperature: –80–90°C.
   (b) Applications: Tyres, conveyor belts, clutches, engine bearings, technical products of all types, drinking water seals.
3. Isoprene rubber (IR):
   (a) Operating temperatures: –40–130°C.
   (b) Applications: Technical products of all types, especially construction sections, cooling and heating hoses for vehicles, high-performance tyres, foodstuffs utensils
4. Acrylonitrile butadiene rubber (NBR):
   (a) Operating temperature: Up to approximately 110°C.
   (b) Applications: Motor vehicle parts, oil and fuel hoses, technical products of all types, plates and mats, rollers, seals and for foodstuffs such as milk.
5. Chloroprene rubber (CR):
   (a) Operating temperature: –40–110°C.
   (b) Applications: Conveyor belts, clutches, drive belts, technical products of all types, pneumatic suspension systems, cables.
6. Butyl rubber (IIR):
   (a) Operating temperature: –40–150°C.

(b)   Applications: Automotive hoses, tyre inner liners, seals, membranes, rubberised fabrics, steam hoses, cable insulation.

3.3.2   Risk factors and loss prevention

Synthetic rubber requires particular temperature, humidity/moisture and possibly ventilation conditions and storage climate conditions. The travel temperature range is 5–30°C. Synthetic rubber withstands temperatures of up to 30°C. When this temperature is exceeded, cold flow starts. The rubber begins to flow and ruptures the wrapping. The material begins to stick or combine with the packaging. Further consequences of excessive temperatures are softening and artificial ageing, combined with hardening. In this event, the recipient has to pick over the blocks in a separate operation.

If the exposure to sunlight is of longer duration, oxidative cross-linking occurs on the surface of the rubber even at temperatures of 10–20°C (pre-mature ageing). UV radiation also causes this degradation. The temperature threshold for transportation is 5°C, the rubber crystallises at −15°C. This process is irreversible in synthetic rubber, which means that it becomes unusable.

RF humidity/moisture

Synthetic rubber requires particular temperature, humidity/moisture and possibly ventilation conditions (SC VI) (storage climate conditions).

Term	Humidity/water content
Relative humidity	65%
Water content	0.2–0.4%
Critical water content	0.75%
Maximum equilibrium moisture content	65%

The water content of synthetic rubber is 0.2–0.4% and synthetic rubber with a low water content thus belongs to water content class 1 (WCC 1).

Despite its low water content, synthetic rubber is very sensitive to moisture/humidity. The critical water content within the wrapping is 0.75%.

The bales are wrapped in heat-sealed foils and must not be damaged. If moisture penetrates, the bales soak it up like a sponge, with the result that expensive additional operations are required or the material can no longer be approved for processing in certain machines. If water bubbles can be seen inside the wrapper, the goods may have been damaged.

RF ventilation

Synthetic rubber requires particular temperature, humidity/moisture and possibly ventilation conditions (SC VI) (storage climate conditions). If the rubber is dry for shipment or container dry, there is no need for ventilation.

RF biotic activity

Synthetic rubber displays 3rd order biotic activity (BA 3). Due to their processing, synthetic rubbers are goods in which respiration processes are suspended, but in which biochemical, microbial and other decomposition processes still proceed, they thus exhibit 3rd order biotic activity (BA 3).

RF gases

This risk factor has no major influence on transportation of this product.

RF odour

Active behaviour: Despite the packaging, a rubber-like smell is to be expected, so synthetic rubber must not be stored with foodstuffs, semiluxury items and animal feedstuffs.

Passive behaviour: Non-odour sensitive.

RF contamination

Active behaviour: Does not cause contamination.

Passive behaviour: Damaged films allow the raw material to become contaminated with foreign bodies, such as dust, chemicals, metal filings, jute fibres etc. This considerably reduces the quality and can cause high processing costs or render the material completely unusable.

RF mechanical influences

Rubber should be stored in such a way as to avoid the risk of crushing. The bales are wrapped in heat-sealed foils and must not be damaged. Use no hooks.

RF toxicity/hazards to health

This risk factor has no major influence on transportation of this product.

RF shrinkage/shortage

This risk factor has no major influence on transportation of this product.

RF Insect infestation/diseases

When wooden packaging containers or cargo securing materials are used, it may, under certain circumstances, be necessary to comply with the quarantine regulations of the country of destination and a phytosanitary certificate may have to be enclosed with the shipping documents. Information may be obtained from the phytosanitary authorities of the countries concerned.

Styrene butadiene rubber

4.1 Introduction

Styrene butadiene rubber (SBR) is a random copolymer produced, as the name suggests, from styrene and butadiene. These are non-oil resistant synthetic rubbers.

Styrene butadiene rubbers are the most commonly used general purpose synthetic rubbers today. They are produced by the copolymerisation of butadiene and styrene.

Copolymers are the polymer chains obtained by polymerising a mixture of two monomers which are butadiene and styrene in the present case.

The structure of these monomers are given below:

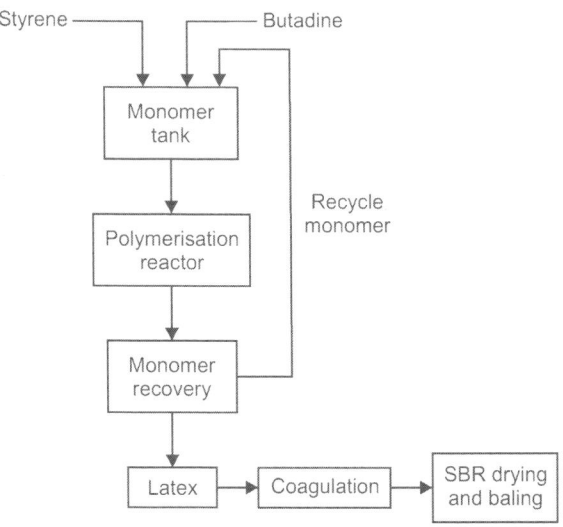

Process flow diagram of SBR manufacture is given in Fig. 4.1.

Figure 4.1: Process flow diagram of SBR manufacture.

SBR is the workhorse of the rubber industry, even though some of its properties do not match those of natural rubber (NR). What it lacks in elongation, hot tear strength, hysteresis, resilience and tensile strength, it makes up for in better processability, slightly better heat ageing and better abrasion resistance than NR. Probably the most important factors in the commercial viability of SBR have been its wide availability, low cost compared with those of all other SRs, ability to accept high filler levels, relatively stable price compared with that of NR and overall properties on a cost/performance basis.

Principal applications are in tyres and tyre products, automotive parts and mechanical rubber goods. SBR technology developments have continued even though the business is mature. Developments are being prompted by tyre producers that are looking for elastomers with improved performance characteristics. The flexibility of solution processes has enabled producers to develop 'tailored' SBR grades, which improves the combination of mechanical properties (e.g., traction—grip, handling—ride, cornering performance, performance at speed and rolling resistance) and processing characteristics. More recently, producers have focused on the capability to use additives to vary the stereochemistry of the diene polymerisation and to control polymer micro- and macro-structure to produce solution SBR grades with optimum properties.

4.2 Chemical structure and properties

The peculiar nature of the insertion of butadiene on the growing chain, i.e., the 1,4- and 1,2-additions, together with the two possible 1,4-addition isomers, *cis* and *trans*, suggests it would be more appropriate to refer to SBR as a four monomer copolymer.

This remark acquires a particular meaning if we consider the physical and rheological characteristics of the finished polymer. The balance between the structural unit content of styrene, 1,4- and 1,2-butadiene along the chain is the most important parameter affecting the glass transition temperature (T_g) of the material. Most interestingly, the concentration of 1,4 *trans* units has a strong influence on the strain-induced crystallisation of the rubber, which means a reinforcing effect on the tensile ultimate properties. Moreover, the relative concentration of 1,4 and 1,2 units may influence the thermal stability of the polymer. The oxidative degradation of the rubber starts from the addition of oxygen on a double bond: if the double bond is part of the main chain, as in the case of 1,4 units, the reaction will lead to a chain scission.

4.2.1 Types of SBR

There are two major types of SBR, based on the manufacturing process adopted:
1. Emulsion SBR (e-SBR).
2. Solution SBR (s-SBR).

At the very beginning of their development, s-SBRs suffered from processing problems, due to their narrow molecular weight distribution (MWD). As a matter of fact, the main problem for s-SBR to overcome has always been the non-interchangeability with the e-SBRs commonly used by tyre producers, the limited number of grades coming from the emulsion process makes it easier to switch from a supplier to the other, without any reconfiguration of the processing machines and procedures.

So one of the main efforts of s-SBR producers has been the search for a better processability, which is obtained by modifying polymer macro-structure (i.e., MWD and long chain branching).

In this sense, the use of coupling agents like $SiCl_4$ and $SnCl_4$ in the batch anionic synthesis provided a way to broaden the MWD, adding a star-shaped structure at the same time.

In order to maximise market share, automotive companies have continuously developed vehicles with superior performance and durability (increased mileage).

This has resulted in increasingly stringent tyre performance specifications. These have been partly met by redesigning the structure components (i.e., tread patterns and reformulating compounds). In spite of their resistance, tyre producers have found no alternative to s-SBR grades for highly specified tyre components for high performance tyres.

The need for reducing fuel consumption led to the definition of specifications also for the rolling resistance of the tread material, in the U.S. the corporate average fuel economy (CAFE) regulations are gradually pushing in this direction. The market is getting more and more demanding in the aforesaid specifications and such performances improvements cannot be achieved with e-SBR, making a gradual trend toward the use of s-SBR inevitable.

s-SBR grades now have comparable styrene content to emulsion types and these grades have superior mechanical properties e-SBR.

Typical e-SBRs contain 18% *cis,* 65% *trans* and 17% vinyl butadiene forms. s-SBRs may be divided into two categories:

1. Random copolymers (80%): The market for the random copolymers is completely dominated by the tyre industry (95%). These types are usually extended with compatible oil. They are blended with other types of rubber, including NR and mixed with reinforcing fillers (carbon black and/or silica), oil and vulcanising chemicals to produce the tread compound where they improve wet grip and decrease rolling resistance, thus improving fuel consumption.

2. Partial block types (20%): The partial block types are used in rubber flooring, carpet underlay, footwear and in many other applications. They also find widespread use in bitumen modification and in adhesives.

4.3 Production technology

There are three conventional routes to producing solid SBR:

1. Hot emulsion polymerisation.
2. Cold emulsion polymerisation.
3. Solution polymerisation.

Each process produces SBR grades with different properties.

4.3.1 Hot emulsion polymerisation

Hot emulsion polymerisation is the original SBR process. The major characteristic of this process is that these grades have exceptional processing characteristics in terms of low mill shrinkage, good dimensional stability and good extrusion characteristics.

However, high levels of micro-gels are also produced, so there is a trend towards using the cold emulsion grades in many applications. However, they are still used in applications such as adhesives and flow modifiers for other elastomers where good flow properties are required.

4.3.2 Cold emulsion polymerisation

Cold emulsion polymerisation produces SBR grades with superior mechanical properties, especially tensile strength and abrasion resistance, compared to the grades produced by the hot emulsion polymerisation process. This process has largely replaced the hot emulsion polymerisation process for production of e-SBR grades.

Basics of emulsion polymerisation

Styrene and butadiene monomers polymerise in the presence of an emulsifier (fatty or rosin acid soaps), an initiator system, a modifier (mercaptan) and water. Initiator agent in case of cold polymerisation is the redox reaction between chelated iron/organic peroxide and sodium formaldehyde sulphoxylate as reducing agent.

The MWD is primarily controlled by addition of mercaptan, which terminates the growing chains, besides initiating the new ones.

Polymerisation takes place at the mild conditions typical of emulsion polymerisation, controlling at the same time the reactants flow rate up to the addition of the shortstop agent – when desired conversion is reached – which rapidly reacts with free radicals, blocking reaction.

Residual butadiene and styrene are then removed from the short-stopped lattice, which is first stabilised by addition of appropriate antioxidants and then coagulated by using an inorganic acid and chemical aids to regulate crumb dimensions.

The coagulated crumb is then washed, dewatered, dried, baled and packaged.

The major cost in e-SBR manufacture is the purchase of monomers. The monomer price is dependent on the crude oil price, but can fluctuate quite widely due to other reasons. About 70 % of e-SBR is used for the manufacture of car tyres, in particular in the tread where it confers a good balance between wear resistance and wet grip. e-SBR is also used to manufacture conveyor belts, flooring and carpet underlay, hoses, seals, sheeting, footwear and a large number of other rubber goods.

4.3.3   Solution polymerisation

Solution polymerisation grades have superior mechanical properties, particularly tensile strength, low rolling resistance and handling (encompasses traction under a variety of conditions and performance at different speeds, when cornering etc.), when used in tyre applications. The ratio of butadiene configurations varies. Generally speaking, s-SBR grades have a lower *trans* and vinyl content and a higher *cis*-butadiene content than e-SBR grades.

s-SBR is of particular importance because of the improved performance and flexibility of the grade range. It is a preferred component with silica filler in the so-called 'green tyres' which exhibit low rolling resistance and therefore improved fuel economy.

In initially making s-SBR grades producers attempted to replicate the stereochemistry of e-SBR grades. However, solution polymerisation differs from emulsion polymerisation due to its flexibility and enables SBR grades with varying styrene/butadiene ratios and *cis*, *trans* and vinyl contents to be produced by varying the catalyst and the monomer ratios and process conditions. This enables s-SBR producers to produce grades specifically tailored for individual applications. Most s-SBR producers have issued patents relating to various aspects of s-SBR production. However, there are basically two commercial processes used to produce conventional s-SBRs.

Phillips process

Commercial Phillips processes are batch (although the patents cover both batch and continuous polymerisation) and produce a branched polymer with comparatively narrow molecular weight distribution. Phillips is no longer in the SBR business, but this process is being utilised by a number of European producers including Repsol, Petrochim and Dow in Italy. The process has been licensed by Petrofina in Belgium in the 1990s.

4.3.4   Firestone process

The firestone process is continuous, producing s-SBR grades with more or less linear chains and a comparatively broad molecular weight distribution.

Basics of solution polymerisation: The basic principles of the two solution processes are the same. s-SBR is made by termination-free, anionic (living) polymerisation initiated usually by alkyl-lithium compounds. The absence of a spontaneous termination step enables the synthesis of polymers possessing a very narrow MWD and less chain branching.

Solution rubber plants are generally integrated into larger production sites, which deliver the required feedstocks (solvent and monomers), electricity, steam, treated water and take back the solvents for purification or combustion. However, some plants generate their own steam and treated water and import the monomers and solvent.

Generally speaking, a solution diene rubber process can be sub-divided into the following process steps:

1. Purification of monomers and solvent.
2. Polymerisation.
3. Hydrogenation (if applicable).
4. Blending section.
5. Solvent removal.
6. Product isolation.
7. Packaging.

The chemicals used are:

1. The monomers (styrene and butadiene, in this case).
2. The catalyst (usually *n*- or s-butyl lithium or Ziegler-Natta catalysts based on transition metals such as neodymium, titanium and cobalt).
3. The solvent (commonly cyclohexane, hexane, heptane, toluene, cyclopentane, isopentane or mixtures thereof).
4. Process additives like coupling agents, structure modifiers, extender oil, killing agents and product stabilisers.

Carbon dioxide, water, oxygen, alcohols, mercaptans and primary/secondary amines interfere with the activity of alkyllithium catalysts, so the polymerisation must be carried out in a clean, near-anhydrous condition. Stirred bed or agitated stainless steel reactors are widely used commercially.

Polymerisation is carried out in a solution of inert aromatic or aliphatic solvent. The polymerisation rate of butadiene in the presence of lithium based catalyst is lower than styrene. However, when butadiene and styrene are mixed, the rate of polymerisation is reversed, resulting in block copolymer production with a high proportion of butadiene blocks. Block formation may be suppressed since the property requirements of traditional SBR markets cannot be met by block copolymers. Random copolymerisation is encouraged by incorporating into the solution 'randomising' agents such as dialkyl and heterocyclid ethers, which act as Lewis base on the catalyst, or by controlled monomer charging (i.e., some of the styrene is added later in the polymerisation cycle).

The resulting copolymer is precipitated, separated, dried and baled.

4.4   Technical developments

Property optimisation of SBR has been achieved, to some extent, by conventional s-SBR technology. By modifying the way in which monomers are added, the polymerisation conditions and the use of co-catalysts and randomising agents, the proportion of *cis* and vinyl isomers and the chain structure of the tailored polymers can be altered.

Nevertheless, the overall properties of the tailored s-SBRs sometimes fall short of expected tyre industry requirements.

The automotive industry is under continuous pressure to improve the environmental performance and the useful life of automotive components. s-SBR producers are responding by modifying conventional s-SBR technology to develop grades with optimum combinations of rolling resistance, wear resistance, blow-out resistance, chipping/chunking resistance, road traction under a variety of weather conditions, handling, noise transmissions and other performance properties for different tyre applications.

There are a number of ways in which these improvements are being achieved:

1. Reformulating compounds using high performance additives in conjunction with tailored s-SBR grades.

2. Developing novel additives/modifiers that can be added to the SBR at the compound stage.

3. Further modification of polymerisation conditions to enable both block and random copolymerisation.

4. Introduction of post-polymerisation steps to enable better control of copolymer end-group structure, thus significantly altering the properties of the resulting SBR

5. Introduction of post-polymerisation steps to facilitate better interaction with the reinforcement system. This is one of the most radical developments affecting the rubber industry, since it enables silica to significantly displace carbon black as the favoured reinforcement for tyre applications.

SBR is generally compounded with a vulcanisation system, reinforcing filler (usually carbon black), processing/extending oil and an antioxidant/ stabiliser package, prior to molding/fabrication. Tyre compounds frequently use a combination of elastomers in order to achieve optimum properties in the final application.

4.5   Compounding and processing of SBR

The compounding of styrene butadiene rubbers is similar to that of natural rubber and other unsaturated hydrocarbon rubbers. The most convenient and effective compounds in large-scale usage, such as tyres, are all based on fillers, such as carbon black, extending oils, zinc oxide, sulphur, accelerators, such as– mercaptobenzothiazole and protective agents, such as antioxidants, antiozonants and waxes. Processing these complex mixtures into smooth compounds that can be quickly pressed, sheeted, calendered, or extruded is a most important step for manufacture. Emulsion SBR is the prototype of a 'general-purpose' rubber because of its ability to be blended with any other such rubber into compounds that process and cure homogeneously. Thus, it is said to have 'cure-compatibility' and excellent 'processing' or 'processability.' Besides T_g and possible chemical attributes that may be important but are poorly understood, the good processing of SBR results from its favourable combination of molecular weight and molecular weight distribution and the considerable proportion of long branches in its molecules. The main advantage of solution SBR is that it can be constructed so as to have just enough branching and molecular weight distribution for adequate processing, while at the same time providing maximised molecular weight for rolling resistance and wear.

Compounding recipes with low sulphur or with only organically bound sulphur, as in thiazoles, lead to vulcanisates with better ageing but slower curing. Zinc stearate, or zinc oxide plus stearic acid, is the most common activator for SBR. There are many accelerators that speed up slow-curing stocks and retarders that slow down 'scorchy' ones. Recipes may also contain plasticisers, softeners, tackifiers and other ingredients that have given evidence of solving some compounding problem or other.

Preparation of SBR compounds is similar to that of the rubbers. The ingredients are mixed in internal mixers or on mills and may then be extruded, calendered, molded and cured in conventional equipment. Mixing procedures vary with the compound. In general, the rubber, zinc oxide, antioxidants and stearic acid are mixed, then the carbon black is added in portions with the oil. This 'non-productive' mix may be considered a black masterbatch. It may be desirable at this point to dump, sheet out and cool the batch. The next phase involves mixing- in all the other ingredients, with the accelerator and sulphur being added last. The mix is now considered 'productive,' since it will 'cure', i.e., cross-link irreversibly—when held for a carefully regulated time at a selected temperature, such as 150°C.

Mixing procedures for rubber stocks vary with different companies. Some plasticise the rubber in a separate operation before blacks and other ingredients are added. Some make a masterbatch first, without a preliminary plasticising step. If additional plasticising is required, the masterbatch is remilled and self-plasticisation of the rubber is avoided. Excessive remilling of SBR compounds may lead to gel formation and poorer extrusions.

Not only does modern styrene butadiene rubber have extrusion properties superior to those of natural rubber, but its stocks have less tendency to scorch in processing. Although cold SBR is often preferable to hot for optimum physical properties, hot SBR can be better for both processing and product properties. Hot SBR breaks down more rapidly to a desirable molecular weight on the mill, develops less heat and accepts more filler in processing. All types of SBR require less sulphur than natural rubber does for curing. The usual range is about 1.5 to 2.0 parts per hundred rubber. On the other hand, SBR requires more accelerator because of the lower unsaturation to achieve the same rate of cure.

Tyre tread-wear and ageing properties are superior to those of natural rubber, resistance to abrasion and resistance to crack initiation are better. Building tack is still poor and dynamic properties are such that heavy-duty tyres become too hot in use. Without reinforcing fillers such as carbon black or silica, the physical properties of SBR are much inferior to those of natural rubber. Similarly, its green strength properties, for example, the tensile strength of the fully mixed compound before cure, are distinctly inferior to those of natural rubber. This is the principal reason it was necessary to go back to a higher proportion of natural rubber in radial tyres, which would otherwise deform in transit from the tyre-building machines to the curing presses.

Nearly, 70% of SBR is consumed by the automobile industry for tyres and tyre products where it is most widely used in the manufacture of tread.

4.5.1 Industry structure

Globally, SBR is a fragmented industry with little leadership. The situation in s-SBR is better than e-SBR, where there is generally more market discipline and technology is harder to obtain. The tyre industry is highly raw material intensive with raw materials accounting for 60% of industry turnover. The sharp volatility in raw material prices, in particular NR prices, has been the bane of the industry.

4.5.2 Raw materials

The tyre industry is highly raw material intensive. Raw material costs accounts for about 63% of tyre industry turnover and 70% of production costs. The cost of rubber is the single largest component of raw material costs. Technical parameters for a typical s-SBR plant are given in Table 4.1.

Table 4.1: Technical parameters for a typical s-SBR plant.

Product type	s-SBR, batch or continuous type
Reactor type and size	Continuously stirred tank reactors in series or batch reactors, 10–100 m^3
Monomer addition	Simultaneous addition of styrene and butadiene
Number of reactors in use	Up to 10 depending on the process
Polymerisation pressure	Up to 5-bar
Polymerisation temperature and control system	30–100°C, control system based in external evaporators, cooling coils, adiabatic
Catalysts/initiators	Various anionic initiators (usually *n*-butyl lithium)
Structure modifiers	Various ethers, e.g., THF, TMEDA
Shortstops	Water and/or fatty acids
Conversion of monomer to polymer	95–99%
Antioxidant	*p*-Phenylenediamine derivatives, phenolic types, phosphite types
Extender oil	Highly aromatic, treated distillate aromatic extract (TDAE), mild extract solvent
Capacity per reactor line	Typically 30,000-tpa

Raw material composition of tyres is given in Table 4.2.

Table 4.2: Raw material composition of tyres.

Raw material	Raw material cost [%]
Natural rubber	42
Nylon tyre cord fabric	16
Carbon black	11
Rubber chemicals	5
Butyl rubber	5
Polybutadiene rubber	6
SBR	5
Others	10
Total	100

Represents cost as a percentage of total raw material cost

Composition of automotive tyres is given in Table 4.3.

Table 4.3: Composition of automotive tyres.

Raw material	Radial tyres[1]	Tyres[2]	Tubes[3]
Natural rubber	34.8	44	5
SBR	0.06	8.60[4]	
Vinyl-pyridine latex	0.020	0.40	

(Cont'd...)

Raw material	Radial tyres[1]	Tyres[2]	Tubes[3]
Polybutadiene rubber	12.36		
Carbon black	22	23	30
Nylon tyre yarn/cord wrap/sheet/fabric	0.393	13	
Bead wire	3.36	4	
Butyl rubber	0		53
Steel tyre cord	15.48		
Rubber chemicals	2	2	2
Zinc oxide	2	2	2
Misc. materials	7.5	8.0	8.0
Total	100	100	100

[1] Automobile steel belted radials per 100-kg tyre

[2] Automobile tyres reinforced with nylon/rayon tyrecord/wrap sheet per 100-kg

[3] Automobile tubes per 100-kg

[4] Includes SBR and PBR

4.5.3   Properties of E-SBR

E-SBR is commercially available in Mooney viscosities ranging from 30 to about 120. Lower Mooney viscosity E-SBR grades band more easily on the mill, incorporate fillers and oil more readily, show less heat generation during mixing, are calendered more easily, shrink less, give higher extrusion rates and have superior extrudate appearance than the higher Mooney viscosity grades. On the other hand, the high Mooney viscosity SBR's have better green strength, less porosity in the vulcanisate and accept higher filler and oil loadings.

As the molecular weight of the SBR increases, the vulcanisate resilience and the mechanical properties, particularly tensile strength and compression set, improve. The processability of SBR improves as its molecular weight distribution broadens. Formation of high molecular weight fractions with the increase in the average molecular weight can however, prevent improvements in the processability. This is due to the fact that the tendency for gel formation also increases at higher molecular weights.

In addition to the polymer viscosity, polymerisation temperature also plays an important role in shaping the processability. E-SBRs produced at low polymerisation temperatures have less chain branching than those produced at higher temperature. At an equivalent viscosity, cold polymerised E-SBR is normally easier to process than hot polymerised E-SBR and this applies particularly to a better banding on mills, less shrinkage after calendering and a superior surface of green tyre compounds. Hot rubbers give better green strength because they have more chain branching.

The styrene content of most emulsion SBR varies from 0% to 50%. The per cent styrene of most commercially available grades of E-SBR is 23.5%. In vulcanisates of SBR, as styrene content increases, dynamic properties and abrasion resistance decrease while traction and hardness increase.

Polymerisation temperature also affects the microstructure of E-SBR. In the cold polymerised E-SBRs, the butadiene component has, on average, about 9% cis-1.4, 54.5% trans-1.4 and 13% of vinyl-1.2 structure. At a 23.5% bound styrene level, the glass transition temperature, T_g, of SBR is about $-50°C$. As the styrene content in the SBR increases, the glass transition temperature also increases. Rubbers with very low T_g values are characterised by a high resilience and good abrasion resistance, but have poor wet traction. By contrast, those rubbers with high T_g, as, for instance, SBR 1721, exhibit a low resilience and poor abrasion resistance with an excellent wet traction. The emulsifier remains in the rubber after coagulation can also have an influence on the processability. Rosin acid emulsifiers impart better knitting, tack and adhesion to the SBR polymer. Generally, polymers emulsified with rosin acid have better extrusion rates, slower cure rates, poorer heat resistance and can cause mold fouling and polymer discolouration. Fatty acid emulsified SBR polymers generally have less tack, faster curing and high tensile properties. A compromise of the above properties is obtained by using a mixed rosin acid/fatty acid emulsifier system.

Mechanical properties

Since SBR lacks the self-reinforcing qualities of natural rubber due to stress induced crystallisation, gum vulcanisates of SBR have lower tensile properties. The tensile property of E-SBR vulcanisates depends in great measure on the type and amount of filler in the compound. Cured gum stocks have only 2.8 to 4.2 MPa tensile strength, while fine particle carbon black loadings can produce tensile strength of 27.6 Mpa. Though the compression set of some of the common E-SBR compounds is high, by proper compounding and blending, it is possible to obtain E-SBR vulcanisates with a low compression set.

Electrical properties

SBR is a non-polar polymer and its vulcanisates are poor conductors of electricity. The electrical properties of E-SBR depend to a large extent on the amount and type of emulsifier and coagulating agent(s) used.

Resistance to fluids

While E-SBR vulcanisates are resistant to many polar solvents such as dilute acids and bases, they will swell considerably when in contact with gasoline, oils, or fats. Due to this limitation, SBR cannot be used in applications that require resistance to swelling in contact with hydrocarbon solvents.

Cure properties

SBR can be cured with a variety of cure systems including sulphur (accelerators and sulphur), peroxides and phenolic resins. Processing of SBR compounds can be performed in a mill, internal mixers or mixing extruders. SBR compounds are cured in a variety of ways by compression, injection molding, hot air or steam autoclaves, hot air ovens, microwave ovens and combinations of these techniques.

4.5.4   Uses of SBR

E-SBR is predominantly used for the production of car and light truck tyres and truck tyre retread compounds. A complete list of the uses of SBR includes houseware mats, drain board trays, shoe sole and heels, chewing gum, food container, sealants, tyres, conveyor belts, sponge articles, adhesives and caulks, automobile mats, brake and clutch pads, hose, V-belts, flooring, military tank pads, hard rubber battery box cases, extruded gaskets, rubber toys, molded rubber goods, shoe soling, cable insulation and jacketing, pharmaceutical, surgical and sanitary products, food packaging, etc.

Polybutadiene rubber

5.1 Introduction

Polybutadiene rubber (PBR) is a homopolymer of 1,3-butadiene ($CH_2 = CH - CH = CH_2$). These are non-oil resistant synthetic rubbers. It is the double bonds in the butadiene molecule that are the key to polymer formation. They are attacked by catalysts to maintain a repetitive chain growth process, which continues until something is added to terminate the reaction at the desired molecular weight. For a typical PBR, the number average molecular weight is usually >100,000 grams per mole. This represents a chain that contains over 2000 butadiene units. Polybutadiene is a synthetic thermoplastic polymer made by polymerising 1,3 butadiene with a stereospecific organometallic catalyst (butyl lithium) through other catalysts such a titanium tetrachloride and aluminium iodide may be used. The *cis*-isomer, which is similar to natural rubber, is used in the tread due to its abrasion and crack resistance and low heat built up. Large quantities are also used as blend in SBR rubber. The *trans*-isomer resembles gutta percha and has limited utility, liquid polybutadiene, which is sodium catalysed has speciality uses as a coating resin. It is cured with organic peroxides. Combustible liquid form is probably toxic by ingestion and inhalation, as well as skin irritant.

5.2 Chemistry and manufacturing process

Most PBRs are made by a solution process, using either a transition metal (Nd, Ni, or Co) complex or an alkyl metal, like *n*-butyl lithium, as catalyst.

Since the reaction is very exothermic and can be explosive, particularly with alkyl lithium catalysts, the reaction is normally carried out in solvents like hexane, cyclohexane, benzene or toluene. The solvents are used to reduce the rate of reaction, control the heat generated by the polymerisation and to lower the viscosity of the polymer solution in the reactor. A typical PBR polymerisation would be run at about 20% monomer and 80% solvent.

5.2.1 Batch vs continuous process

The polymerisation can either be a batch process or a continuous process.

1. In batch mode, monomer, solvent and catalyst are charged to the reactor, heated to initiate the process and then allowed to continue to completion.

The polymer solution is then transferred to another vessel or process unit to remove the solvent.

2. In continuous mode, monomer, solvent and catalyst are continuously fed into the bottom of the first of a series of reactors at a temperature suitable for polymerisation. The polymerisation progresses as the solution flows through the reactors and polymer solution is taken off at the top of the last reactor without stopping the process.

The continuous process is the most economical. In both processes, the finished product is usually in the form of bales, which weigh from 20–35 kg each.

Process research underway includes:

1. High conversion of monomer and low gel formation in the polymerisation.
2. Production of high *cis*-PBR.
3. Improved mechanical properties.
4. Better catalyst systems.
5. Use of safer solvents.

Emulsion vs solution process

During the 1930s and 1940s, PBR was introduced using the emulsion polymerisation system. Although this is not considered to be satisfactory today, it is used in some plants mainly due to historical reasons.

The solution polymerisation process, developed in the 1950s, is the most popular today.

Two types of solution PBR have emerged as important commercial rubbers— the high *cis* and the medium *cis*-PBRs. The production of the former involves co-ordination catalyst systems based on Ti, Co, Ni and Nd, while that of the latter involves alkyl-lithium catalyst systems.

5.3   Types of PBR

The polymerisation of butadiene results in the formation of a number of stereo isomers. Depending on the disposition of the double bonds present in the polymer chain, PBR can be classified into five configurations:

1. *Cis*-1,4.
2. *Trans*-1,4.
3. Vinyl-1,2 isotactic.
4. Vinyl-1,2 syndiotactic.
5. Vinyl-1,2 atactic.

Typical composition of PBR with different catalyst systems is shown in Table 5.1.

Table 5.1: Typical composition of PBR with different catalyst systems [%].

Catalyst type	Cis-1,4	Trans-1,4	Vinyl
Neodymium	99	1	1
Cobalt	96	2	2
Nickel	96	3	1
Titanium	93	3	4
Lithium	36	52	12

Depending on the choice of catalyst system, PBR can be prepared ranging from almost 100% *cis* to 100% *trans* or 100% vinyl. Out of these configurations, medium and high *cis* varieties are of commercial importance and the most important commercial is the *cis*-1,4 whose configuration is similar to that of natural rubber (NR). NR has properties of high tack and high 'green' (unvulcanised) strength, which are hard to replicate in synthetic rubbers.

5.3.1   Catalyst systems used

Five main catalyst types are used to produce PBR and the catalyst type affects the proportion of *cis*-1,4. Commercially, the Ziegler-Natta catalyst system, based on transition metal compositions and on the rare earth neodymium (Nd), is used to manufacture high *cis* types due to their selectivity in promoting polymerisation of this type of stereo-regularity. The properties of high *cis*-PBR vary between the catalyst systems and even for one catalyst system. Products with unbranched chains and narrow molar mass distributions tend to provide high strength, but are difficult to process. Long linear chains tend also to be subjected to greater 'cold flow,' which is when the 'green' rubber flows and distorts in storage.

1. Polymers made with neodymium catalyst are highly linear with broad molar mass distribution, giving good properties, except for extrudability and cold flow. This catalyst system produces the highest *cis* content of about 99% and produces a polymer with the best tensile and hysteresis (low heat build-up) properties of all the high *cis* types.
2. Nickel systems tend to produce rather more highly branched products, with better processability, but lower tensile strength and fatigue resistance.
3. Titanium and cobalt based products are between these two extremes and the degree of branching can in any case be varied in the cobalt catalysed system. The cobalt system produces a highly branched PBR with a low solution viscosity that makes a good polystyrene and ABS modifier.

Branching can be intentionally introduced into high *cis*-PBR by post-polymerisation reactions, such as with sulphur dichloride. High molar mass PBR has improved tensile strength, abrasion resistance and fatigue resistance.

However, mixing and processing is more difficult with high molar mass product. Oil extension with naphthenic or aromatic oil is used to make this material more easily processed.

5.3.2 High cis-PBR

High cis-1,4 polymer has a very low glass transition temperature (T_g), compared to alkyl lithium-based PBR because it has almost no vinyl structure, leading to high resilience, good low temperature properties and low heat build-up on repeated deformation. This makes high cis-PBR ideal for golf ball cores. These cores are cured with peroxides, which tend to 'over cure' the vinyl units, making a very hard and slow golf ball.

High cis-rubber (i.e., 4% or less of 1,2 vinyl) does not cross-link readily and withstands high temperature without increase in melt viscosity, making it preferred for tyres where significant heat build-up can take place. High cis-PBR is used widely in tyres, where its properties are of benefit in sidewalls, carcass stocks and tyre treads.

However, the high cis-PBR has poor wet grip performance to counter the advantages of the low rolling resistance and high abrasion resistance. Blending with other polymers, such as SBR, is typically used to achieve the desired properties of the tyre tread.

High cis-PBR usually have cis content >95%, which gives rise to better 'green' strength and increased cut growth resistance in the cured product. 'Green' strength, which is the strength of the uncured rubber compound, is important for the tyre building process and cut growth resistance is necessary for tyre performance. Cut growth resistance is the resistance to the propagation of a tear or crack during a dynamic operation like the flexing of a tyre in use.

5.3.3 Lithium-based PBR

The alkyl lithium or anionic catalyst system produces a polymer with about 40% cis, 50% trans and 10% vinyl when no special polar modifiers are used in the process. The alkyl lithium process is probably the most versatile, because the growing chain end contains a 'living' anion, which can be further reacted with coupling agents or functional groups to make a variety of modified PBRs. It also produces gelfree PBR making it ideal for plastics modification.

Vinyl increases the T_g of the PBR by creating a stiffer chain structure. Vinyl also tends to cross-link or 'cure' under high heat conditions so the high vinyl polymers are less thermally stable than low vinyl.

5.3.4 High trans-PBR

In the trans configuration, the main polymer chain is on opposite sides of the internal carbon-carbon double bond. High trans-PBR is a crystalline plastic material similar to high *trans* polyisoprene, which was used in golf ball covers. *Trans*-PBR has a melting point of about 80°C. It is made with transition metal catalysts similar to the high *cis* process (La, Nd and Ni). These catalysts can make polymers with >90% *trans*, again using the solution process.

Low *cis*-PBR is primarily used as a modifier for polystyrene.

5.4   Application for styrenics modification

5.4.1   High impact polystyrene (HIPS) resins

High impact polystyrene (HIPS) resins are conventionally prepared by heat-polymerising a styrene-PBR solution at a temperature of about 100–150°C or graft-polymerising using an initiator. Such rubber modified HIPS is a phase in which rubber particles are dispersed in the matrix of PS, the rubber particles having the sizes of 1.5–6.0 μm. When an external impact is applied to the HIPS resin, the rubber particles function to absorb the impact. Accordingly, the dispersion, sizes and size distribution of the rubber particles affect the mechanical properties of the resin such as impact strength, toughness, surface gloss and heat resistance. Rubber based styrenics process scheme (HIPS or ABS) is shown in Fig. 5.1.

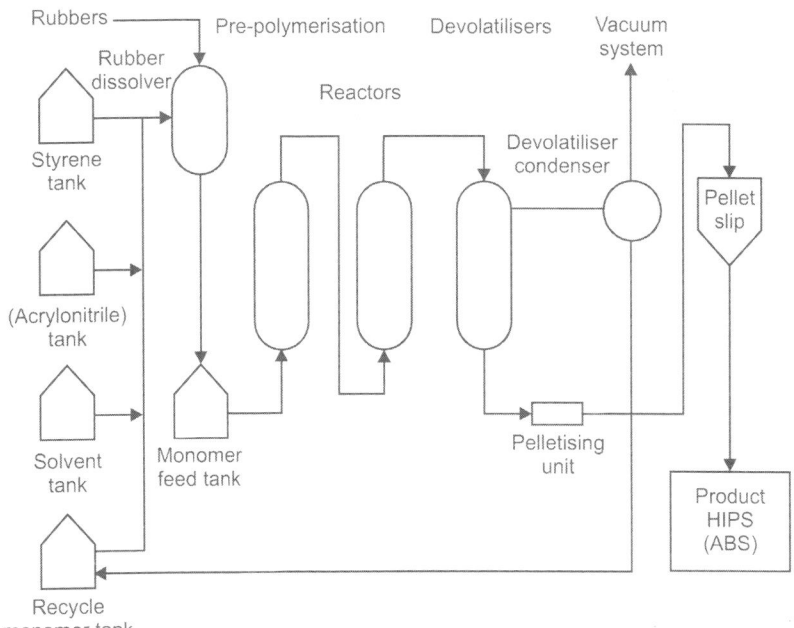

Figure 5.1: Rubber based styrenics process scheme (HIPS or ABS).

Although conventional rubber modified HIPS resins are widely used in electrical household goods and office equipments housing, they are inferior to ABS resin in appearance of the molded articles such as gloss and colour. As molded articles with high gloss are needed in various products, including sheets for packing containers, the weakness of the appearance character of HIPS has been an obstacle for expanding the use of the resin. In general, the smaller the sizes of the rubber particles dispersed in PS, the better the toughness and surface gloss are, but the worse the impact strength. However, the larger the sizes of the rubber particles dispersed in PS, the better the impact strength, but the worse the toughness and surface gloss.

5.4.2 ABS resin

ABS resin is a two-phase polymer system in which a finely dispersed PBR phase is embedded in a continuous matrix of styrene-acrylonitrile (SAN) copolymer. The rubber phase is chemically grafted to SAN during the polymerisation of ABS. Steps shown in modification of properties of styrenics are highlighted in Fig. 5.2. The chemical connection between elastomer molecules and the SAN copolymer is created by polymerising styrene and acrylonitrile in the presence of polymerised elastomer. The resulting grafts serve to compatibilise the rubber and SAN copolymer phases.

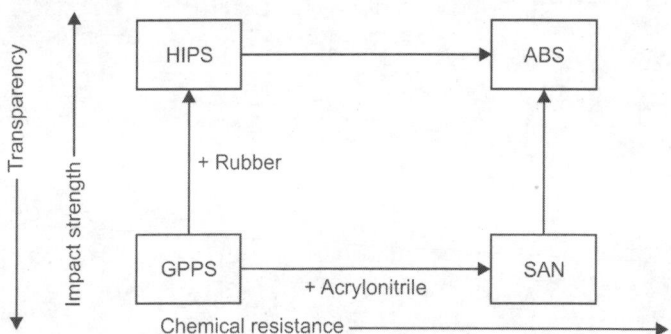

Figure 5.2: Modification of properties of styrenics.

In manufacturing, two important changes have occurred. One is the development and commercial success achieved by the hybrid emulsion/mass process. The second is the refinement of the mass ABS process and the further penetration of grades made by this process into selected market segments.

Emulsion process

In the emulsion process, ABS resins are prepared by emulsifying a mixture of a latex of PBR (or SBR) together with styrene and acrylonitrile in water, in the presence of an initiator, an emulsifier and a chain length regulator/modifier. This emulsion is then heated to effect the graft copolymerisation of styrene-acrylonitrile containing branches onto the elastomer. The copolymerisation of styrene with acrylonitrile can be conducted simultaneously with the grafting, but recently it has become more prevalent for this copolymerisation to be conducted separately at optimal conditions.

Mass or bulk process

The mass or bulk process for ABS production is based on the polymerisation of styrene-acrylonitrile mixtures in the presence of a rubber substrate dissolved in this monomer phase. This is in contrast to the emulsion process, which takes place in an aqueous phase that affords lower viscosity of the reaction medium and better heat transfer.

Rubber is usually introduced to the process in large bales or slabs, which are ground into small particles and dissolved in a mixture of styrene and acrylonitrile monomers. Since no preformed rubber particles are present at the beginning of the grafting reaction as in the emulsion process, a different distribution of elastomer phase in the SAN matrix results from the mass process. To maintain manageable processing viscosities, rubber content of mass polymerised ABS is usually limited to 15%, at most 20%.

Hybrid process

As the name implies, the hybrid emulsion/mass process combines features of the emulsion process, which forms the PBR latex and grafts styrene and acrylonitrile and the mass process, which makes the SAN component.

The recovered graft rubber and SAN are compounded together to give the final ABS product. This approach permits producing a wide range of ABS products using two or three basic SAN polymers and a similar number of high rubber ABS graft resins and compounding them together in various combinations and proportions. In the hybrid emulsion/mass process, the SAN component is prepared by mass polymerisation of styrene and acrylonitrile. The monomers are mixed together with additives and up to about 5% ethylbenzene diluent to lower viscosity, pre-heated and fed to a series of three vertical tower reactors. Reaction heat is rejected to cooling water. The main difference from a pure mass process for ABS is the absence of PBR dissolved in the monomers. Thus, phase inversion is not involved and intense agitation is not needed to tailor the rubber particle size and distribution.

Commercial process

The first commercial grades of ABS polymer, originally introduced in 1948, were simply mechanical blends of rubber with SAN copolymer. These blends reportedly used nitrile rubber since it was found to be more compatible with the SAN resin than was PBR. Although these early ABS polymers were superior to HIPS then on the market, it was recognised that the properties were inferior to graft copolymers where there is an actual chemical linkage between the continuous and elastomeric phases.

5.5 Properties of PBR

All forms of PBR share certain important characteristics. They have very high resilience – in fact they are the only type of synthetic rubber with a higher resilience than NR. The rubber also has outstanding resistance to abrasion and good flexibility at low temperatures.

PBR can tolerate a large proportion of oil extension without a serious loss in properties. Adding oil to the rubber makes it easier to process into down-stream products. PBR can also take heavier loadings of carbon black. This enables manufacturers to obtain increased abrasion resistance in tyre-tread compounds without reducing flexibility or adding to the heat build-up.

The disadvantage of PBR is that it tends to have low tear strength and has relatively poor resistance to cut growth. Therefore it is rarely used alone, but blended with various proportions of SBR, NR and synthetic polyisoprene.

The use of PBR made it possible to increase the proportion of synthetic rubber used for bus and truck tyres. Its greatest usage is for small truck and passenger vehicle tyres—in this application it has improved the wearing properties of these tyres and has eliminated groove cracking. Properties of PBR are shown in Table 5.2.

Table 5.2: Properties of PBR.

Favourable	Unfavourable
Low heat build-up	Very high air permeability
High resilience	Poor tack
Better flex resistance	Poor road grip
Good heat stability	Poor tear and tensile strength

PBR is also the second largest volume synthetic rubber produced, again next to only to SBR. The major use of PBR is in tyres with over 70% of the polymer produced going into treads and sidewalls.

Cured PBR imparts excellent abrasion resistance (good tread wear) and low rolling resistance (good fuel economy) due to its low glass transition temperature. However, this also leads to poor wet traction properties, so PBR is usually blended with other elastomers like NR or SBR for tread compounds.

5.6   Uses of PBR

PBR also has a major application as an impact modifier for polystyrene and acrylonitrile-butadiene-styrene (ABS) resin, with about 17–21% of the total volume going into these applications. Typically about 7% PBR is added to the polymerisation process to make rubber-toughened resins.

Polystyrene modification accounts for the majority of use. In this application, PBR provides toughness and impact resistance for high-impact polystyrene (HIPS). In fact, HIPS is the largest non-tyre end-use for PBR. The next-largest market is ABS, where PBR again adds impact resistance to the copolymer.

Other non-tyre uses – in footwear, other rubber products and golf balls – are a relatively small part of the market, contributing 7–12% of PBR demand.

'High *cis*-PBR' is generally used worldwide for manufacturing of golf balls. This application is growing since the golf ball industry seems to have moved away from the traditional wound ball technology to the two piece, solid core construction.

Specialty applications such as carboxyl-and hydroxyl-terminated PBR grades are used in binders for rocket propellants and in sealants and waterproof membranes.

Tyre and tyre products represent the major use of global PBR, accounting for around 68% of consumption. Polystyrene (GPPS and HIPS) resins and ABS resins modification consume approximately 13% and 4%, respectively, of the total. Specialty applications for PBR include carboxyl- and hydroxyl-terminated PBR grades, golf ball centres, non-tyre rubber products and other specialty adhesives and binders. Certain trends in different regions have also influenced PBR consumption.

5.6.1   Properties and applications of PBR

Polybutadiene (PBR) is the second largest volume synthetic rubber produced, next to styrene butadiene rubber (SBR).

5.6.2   Polybutadiene in tyres

The major use of polybutadiene is in tyres with over 70% of the polymer produced going into treads and sidewalls. Cured BR imparts excellent abrasion resistance (good tread wear) and low rolling resistance (good fuel economy) due to its low glass transition temperature (T_g). The low T_g, typically <−90°C, is a result of the low 'vinyl' content of polybutadiene. However, low T_g also leads to poor wet traction properties, so polybutadiene is usually blended with other elastomers like natural rubber or styrene-butadiene rubber for tread compounds.

5.6.3 Polybutadiene as an impact modifier in other polymers

Polybutadiene also has a major application as an impact modifier for polystyrene and acrylonitrile-butadiene-styrene resin (ABS) with about 25% of the total volume going into these applications. Typically about 7% polybutadiene is added to the polymerisation process to make these rubber-toughened resins.

5.6.4 Polybutadiene in golf balls

Also, about 20,000 MT worldwide of 'high *cis*' polybutadiene is used each year in golf ball cores due to its outstanding resiliency. This application is growing since the golf ball industry seems to be moving away from the traditional wound ball technology to the two-piece, solid core construction.

5.6.5 Molecular weight of polybutadiene

Molecular weight can become quite high. For a typical polybutadiene, molecular weight (M_n = number average) is usually >100,000 grams per mole. This represents a chain that contains over 2,000 butadiene units.

6.1 Introduction

Polyisoprene is a major component of manufacture of rubber. It is made synthetically and forms are stereospecific *cis*-1-4 and *trans*-1-4-polyisoprene. Both can be produced synthetically by the effect of heat and pressure on isoprene in the presence of stereospecific catalyst. These are non-oil resistant synthetic rubbers. Natural rubber is *cis*-1-4 synthetic *cis*-1-4 is some times called synthetic natural rubber. *Trans*-1,4-polyisoprene resembles gutta percha. Polyisoprene is the thermoplastic until mixed with sulphur and vulcanised. It supports combustion and is non-toxic. Isoprene (2-methyl-1,3-butadiene) is an important commodity chemical and is consumed in the production of polyisoprene (via., polymerisation of isoprene). Polyisoprene (isoprene rubber, IR), is similar in structure and properties identical to natural rubber.

Polyisoprene is largely used in the manufacture of vehicle tyres. It is interesting to note that after so many developments in the tyre industry, natural rubber (or polyisoprene) remains a crucial ingredient in the manufacture of quality tyres, mainly because of its ability to resist the heat build-up generated by friction. Styrene-isoprene-styrene copolymer (SIS rubber), is the second largest application of isoprene and it is mainly used as a thermoplastic rubber and as a pressure-sensitive or thermo-setting adhesive. In smaller amounts isoprene is also used in the production of butyl rubber, isobutylene-isoprene copolymer (IIR). Isoprene has some uses as an intermediate for the production of specialty chemicals. Structures of polyisoprene is shown in Fig. 6.1.

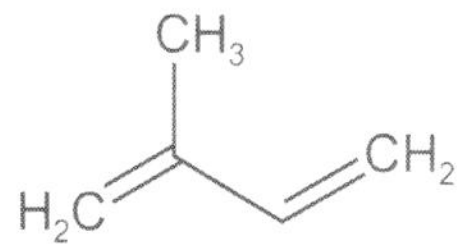

Figure 6.1: Structures of polyisoprene.

Cyclopolyisoprene: Another possible polyisoprene structure can result from the cyclopolymerisation of isoprene. When isoprene is polymerised by certain Ziegler-Natta or cationic catalysts, a cyclic, saturated structure forms that is similar to that obtained by crystallisation of 1,4- and 3,4-polyisoprenes.

Figure 6.2 shows the commonly accepted structure for the cyclic portion of the cyclopolyisoprene chain. These polymers are insoluble, powdery polymers

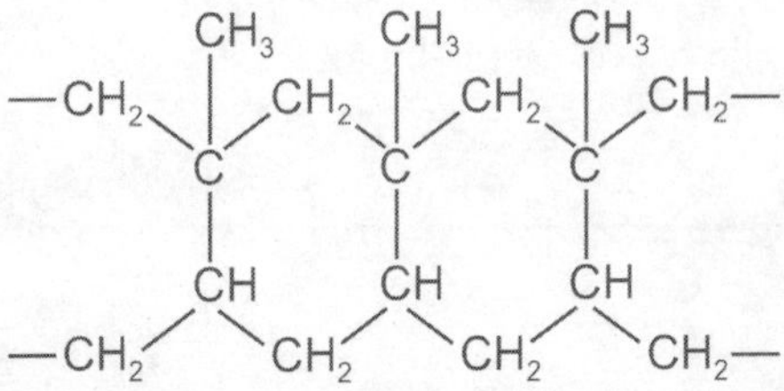

Figure 6.2: Structure of cyclic portion of cyclopolyisoprene.

because of inter and intramolecular cross-linking and also the rigidity resulting from the cyclic segments.

Crystallinity: The more highly stereoregular polyisoprenes demonstrate varying degrees of crystallinity. A high degree of crystallinity requires that the polymer possess long sequences in which the structure is completely stereo-regular, i.e., it must contain linear segments composed exclusively of 1,4-, 1,2-, or 3,4-polyisoprene units. In addition, all the monomer units must be linked in the head tail configuration.

6.2 Production of polyisoprene

Over the years, many process technologies have been investigated for isoprene production. A summary of the process technologies for the production of isoprene is given in the Table 6.1.

Table 6.1: Outline of process technologies for the production of isoprene.

'Routes via C_5 streams'	Isoprene isolation	Extractive distillation
		Fractional distillation
	Dehydrogenation of isoprene precursors	Isopentane dehydrogenation
		Isoamylene dehydrogentation
'On-purpose routes'	$C_3 + C_2 \rightarrow C_5$	Acetone + Acetylene
	$C_3 + C_3 \rightarrow C_6 \rightarrow C_5 + C_1$	Propylene + Propylene (isohexene route)
	$2C_1 + C_4 \rightarrow C_5 + C_1$	Formaldehyde + Isobutene (*m*-dioxane route)
	$2C_1 + C_4 \rightarrow C_5 + C_1$	Isobutane + 2-butene (dismutation route)
'Bioroutes'	Industrial biotechnology	Fermentation

6.2.1 Routes via C_5 streams

The production of isoprene from C_5 streams is currently the preferred route to isoprene. The C_5 feed stream, independent of its source, is a mixture of components with different commercial importance.

The C_5 compounds of commercial interest can be divided into diolefins or dienes and olefins. This division is important because, it corresponds to two distinctive feedstocks in the production of isoprene. Whilst, isoamylenes (e.g., 2-methyl-2-butene, 2-methyl-1-butene) and 1-pentene are the olefins in the C_5 feed stream with higher demand, the C_5 dienes used commercially are isoprene, cyclopentadiene (generally used as its dimer dicyclopentadiene – [i.e., DCPD]), *cis* and *trans*-1,3-pentadiene (*cis* and *trans*-piperylene).

There are three main sources of C_5 feed streams:

1. Unsaturated ('raw') pyrolysis gasoline – by-product of hydrocarbon steam cracking.
2. Product stream from the fluid catalytic cracking unit (FCCU) in refineries.
3. Natural gas liquids (NGLs), (i.e., condensates).

Isoprene cannot be isolated from the C_5 cracking fraction by simply utilising a set of distillation columns because: (i) the C_5 stream comprises many other components which have very similar boiling points to isoprene and (ii) isoprene and *n*-pentane form an azeotropic mixture.

Extractive distillation processes are all based on the same principle, the addition of a solvent that forms no azeotrope with the remaining components in the mixture. The added solvent interacts differently with the different components of the mixture, changing their relative volatilities and consequently, facilitating the distillation of the mixture. There are two main process technologies available to produce isoprene from C_5 streams:

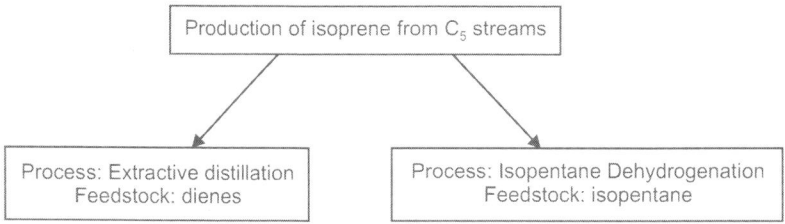

The first-stage hydrotreatment process technology typically employed before sending the C_5 stream to the gasoline pool.

Although, this first-stage hydrogenation process is not relevant in the production of isoprene, it is important to understand the process because of the methodology employed in the development of the process economics (i.e., to determine gasoline blending values).

Recent commercial technologies are given below:

1. Hydrotreatment of the mixed C_5 stream.
2. Extractive distillation process technology.
3. Isopentane dehydrogenation process technology.

Since isoprene is typically produced using a complex series of extractive distillation columns, to separate cyclopentadiene and its dimer dicyclopentadiene (DCPD) and subsequently piperylene.

This is a relatively expensive and energy-intensive process (despite having valuable by-products). Furthermore, with the recent developments in shale gas and fracking technology, most steam crackers in North America, for example, are shifting to lighter feedstocks (i.e., ethane), decreasing the availability of C_5 cracking streams.

6.2.2   On-purpose routes

With the worldwide demand for isoprene expected to increase and with a lower availability of C_5 streams, the cost of isoprene is expected to rise to much higher levels. The current isoprene market growth is constrained by supply. Moreover, the C_5 cut stream has a high octane value making this raw material difficult to secure and costly for petrochemicals.

Consequently, there is a consensus that the petroleum-derived isoprene will be insufficient to meet world demand and that the isoprene price will escalate to unprecedented values.

In summary the on-purpose commercial routes discussed below:

1. Two-stage isobutylene carbonylation process technology (conventional *m*-dioxane route).
2. One-stage isobutylene carbonylation process technology (Eurochim and Kuraray).
3. Acetone-acetylene process technology (Snamprogetti process).
4. Goodyear/scientific design proprietary propylene-based route.

6.2.3   Bioroutes

Industrial biotechnology (or 'white' biotechnology) has never been so eminent to play a key role in chemical processes. The present conjecture of drivers, break throughs in synthetic biology and the steadily increase in oil prices, results in a crucial moment for investments in industrial biotechnologies.

It is increasingly evident that the petrochemical-derived isoprene ('petro-isoprene') faces challenging times with the foreseeable rise in oil price. The recent announcements of partnerships between major downstream rubber players and biochemical companies (e.g., Ajinomoto-Bridgestone, Amyris-Michelin and DuPont-Goodyear) are not surprising. Other companies developing fermentative isoprene processes are GlycosBio, Aemetis and LanzaTech. It is an excellent opportunity for non-petroleum based technologies and even more for industrial biotechnology investments (biomass-based feedstock).

6.2.4   Isoprene polymerisation process

Flow diagram for an isoprene polymerisation process is shown in Fig. 6.3. Before entering the reactors, the solvent, catalyst and isoprene monomer must be free of chemical impurities, moisture and air—all of which are catalyst poisons. The purified streams first enter a chain of reactors in series into which the catalyst is injected and the polymerisation begins.

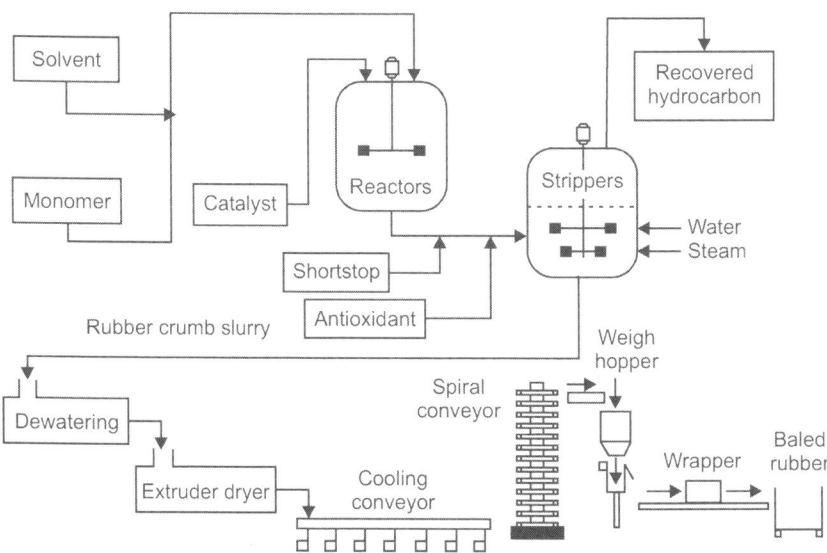

Figure 6.3: Polymerisation process.

After the desired extent of polymerisation has been attained, a short stop or catalyst deactivator is added to the cement so no further linkage of monomer or polymer takes place. A non-staining antioxidant is then added to protect the polymer during finishing and storage.

In the next step, the cement mixture is put through a stripping operation whereby the solvent is recovered and the polymer cement converted to a crumb by hot water and steam. The crumb slurry is processed through extruders to remove water before it is cooled, baled, packaged and placed in storage ready for shipment.

6.3   Applications of polyisoprene

Currently synthetic polyisoprene is being used in a wide variety of industries in applications requiring low water swell, high gum tensile strength, good resilience, high hot tensile and good tack. Gum compounds based on synthetic polyisoprene are being used in rubber bands, cut thread, baby bottle nipples and extruded hose. Black loaded compounds find use in tyres, motor mounts, pipe gaskets, shock absorber bushings and many other molded and mechanical goods. Mineral filled systems find applications in footwear, sponge and sporting goods. In addition, recent concerns about allergic reactions to proteins present in natural rubber have prompted increased usage of the more pure synthetic polyisoprene in some applications.

6.4   Properties of polyisoprene

Typical raw polymer and vulcanised properties of polyisoprene are similar to values obtained for natural rubber. Natural rubber and synthetic polyisoprene both exhibit good inherent tack, high compounded gum tensile, good hysteresis and good hot tensile properties.

The very specific nature of synthetic polyisoprene provides a number of factors that differentiate it from natural rubber. There is minimal variance in physical properties. Polymerisation conditions are narrowly controlled to assure that the polymer is highly specific chemically. There is a low level of non-polymer constituents as compared to natural rubber.

Synthetic polyisoprene's ease of processability is of importance where consistency and quality are major considerations. Since less mechanical work and breakdown are required, shorter mix cycles and the elimination of pre-massing are possible when it is used as a direct replacement for natural rubber. The end results are time and power savings as well as increased throughput. In addition, synthetic polyisoprene exhibits greater compatibility than natural rubber in blends with solution SBR and EPDM. Synthetic polyisoprene's uniformity is a factor where the desire for consistent quality is paramount, as is increasingly the case in many industries with an emphasis on precise dimensional control in processing.

Because of the lower raw polymer viscosity of synthetic polyisoprene, part or the entire break down step normally used for natural rubber should be eliminated. Synthetic polyisoprene compounds at the same plasticity of natural rubber will have less die swell because of having less nerve. Also, at the same plasticity, the synthetic polymer will have significantly faster extrusion rates.

Synthetic polyisoprene compounds can be adapted for curing in any conventional molding operation whether it is compression, transfer, or injection. Synthetic polyisoprene is especially well suited for injection molded compounds. Because of its uniform cure rate, exact time/temperature press cycles can be established with assurance that all pieces will be uniformly cured. In addition, the Mooney of synthetic polyisoprene reduces injection pressures and times with a resultant increase in output.

Some of physical and mechanical properties, chemical resistance, thermal properties and environmental performance of polyisoprene are given below:

Physical and mechanical properties

Physical and mechanical properties	Typical values
Durometer or hardness range	30–95 shore A
Tensile strength range	500–3,500 PSI
Elongation (range%)	300–900%
Abrasion resistance	Good to excellent
Adhesion to metal	Excellent
Adhesion to rigid materials	Excellent
Compression set	Excellent
Flex cracking resistance	Excellent
Impact resistance	Good to excellent
Resilience/rebound	Excellent
Tear resistance	Good to excellent
Vibration dampening	Good to excellent

Chemical resistance

Chemical resistance	Typical values
Acids, dilute	Fair to excellent
Acids, concentrated	Poor to good
Acids, organic (dilute)	Fair to good
Acids, organic (concentrated)	Good
Acids, inorganic	Good
Alcohol's	Good to excellent
Aldehydes	Good
Alkalies, dilute	Fair to excellent
Alkalies, concentrated	Fair to good
Amines	Poor to fair
Animal and vegetable oils	Poor to good
Brake fluids, non-petroleum based	Good
Diester oils	Poor
Esters, alkyl phosphate	Poor
Esters, aryl phosphate	Poor
Ethers	Poor
Fuel, aliphatic hydrocarbon	Poor
Fuel, aromatic hydrocarbon	Poor

Fuel, extended (oxygenated)	Poor
Halogenated solvents	Poor
Hydrocarbon, halogenated	Poor
Ketones	Fair to good
Lacquer solvents	Poor
LP gases and fuel oils	Poor
Mineral oils	Poor
Oil resistance	Poor
Petroleum aromatic	Poor
Petroleum non-aromatic	Poor
Refrigerant ammonia	Good
Refrigerant halofluorocarbons	R-12, R-13
Refrigerant halofluorocarbons w/oil	Poor
Silicone oil	Good
Solvent resistance	Poor

Thermal properties

Thermal properties	Typical values
Low temperature range	$-20°F$ to $-70°F$
Minimum for continuous use (static)	$-60°F$
Brittle point	$-80°F$
High temperature range	$+180°F$ to $+220°F$
Maximum for continuous use (static)	$+180°F$

Environmental performance

Environmental performance	Typical values
Colourability	Poor
Flame resistance	Fair to good
Gas permeability	Fair to good
Odour	Good to excellent
Ozone resistance	Poor
Oxidation resistance	Good
Radiation resistance	Fair to good
Steam resistance	Good
Sunlight resistance	Poor to fair
Taste retention	Fair to good
Weather resistance	Poor to fair
Water resistance	Excellent

Butyl and halobutyl rubber

7.1 Introduction

Butyl rubber is a copolymer of isobutylene (97%) and isoprene (3%), polymerised below –95°C, with aluminium chloride catalyst. It has good abrasion resistance, excellent impermeability of gases, etc. It is used in tyre car cases and linings, electrical wire, steam hose and other mechanical rubber goods. These are non-oil resistant synthetic rubbers.

$$H_2C = \underset{\underset{CH_3}{|}}{\overset{\overset{CH_3}{|}}{C}} \qquad\qquad H_2C = \overset{\overset{CH_3}{|}}{C} - CH = CH_2$$

Isobutylene   2-methyl prop-l-ene

The monomers are polymerised in solvents such as methyl chloride and at low temperature (–80°C) using Friedel Crafts catalysts such as aluminium chloride or barium chloride as mentioned above.

Butyl rubber is commercially produced by cationically copolymerising isobutylene with small amounts of isoprene. The halogen derivatives, chloro- and bromo-, have been commercially available since then. The concept of halogenation to provide more active functionality to the butyl molecule. The halogenated derivatives of butyl rubber provide greater vulcanisation flexibility and enhanced cure compatibility with other, more unsaturated general-purpose elastomers. Butyl polymers are among the most widely used synthetic elastomers in the world, ranking third in total synthetic elastomers consumed.

7.2 Manufacture of butyl rubber

A schematic diagram of a typical butyl plant is shown in Fig. 7.1. The feed, which is a 25% solution of isobutylene (97–98%) and isoprene (2–3%) in methyl chloride, which is the diluent, is cooled to –100°C in a feed tank. At the same time, aluminium chloride is also being dissolved in methyl chloride. Both of these streams are then continuously injected into the reactor. Because the reaction is exothermic and is practically instantaneous, cooling is very important. To remove the heat of reaction, liquid ethylene is boiled continuously through the reactor cooling coils, keeping the reaction at –100°C. As the polymerisation proceeds, a slurry of very small particles is formed in the reactor.

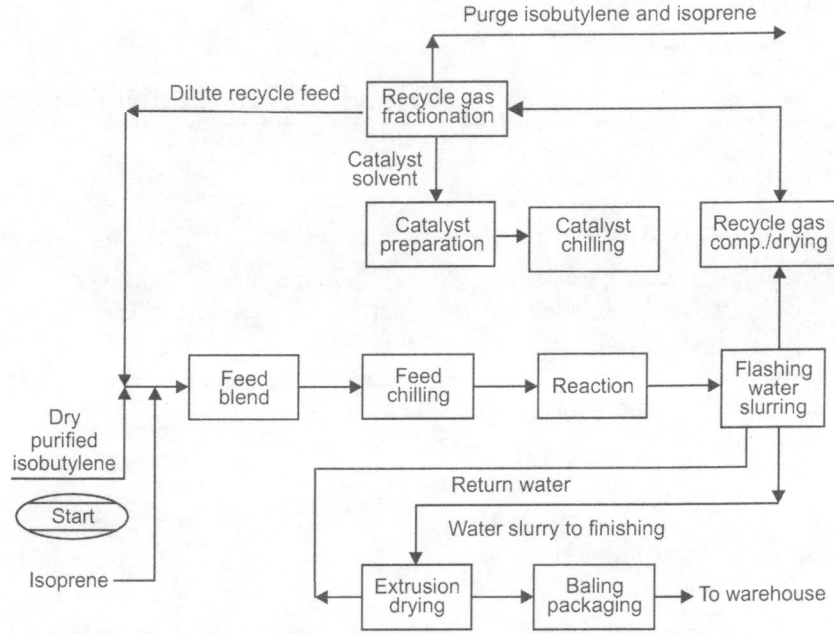

Figure 7.1: Basic components of butyl-rubber process.

This slurry overflows into a flash drum that contains copious quantities of hot water. Here the mixture is vigorously agitated, during which time the diluent and unreacted hydrocarbons are flashed off overhead.

At this point, an antioxidant and zinc stearate are introduced into the polymer. The antioxidant is added to prevent breakdown of the polymer in the subsequent finishing section. Zinc stearate is added to prevent the agglomeration or sticking together, of the wet crumb. The slurry is then vacuum-stripped of residual hydrocarbons.

In the finishing operation, the butyl-rubber slurry is dewatered in a series of extruders to bring the water content to 5–10% in the rubber. Final drying is accomplished in a third extruder by allowing the compressed polymer melt to expand through a die to form an exploded crumb. The crumb is air conveyed to an enclosed fluidised bed conveyor, where water vapour is removed and the crumb is cooled and baled.

7.2.1   Processing and vulcanisation

Like other rubbers, for most applications, butyl rubber must be compounded and vulcanised (chemically cross-linked) to yield useful, durable end use products. Grades of butyl have been developed to meet specific processing and property needs and a range of molecular weights, unsaturation and cure rates are commercially available. Both the end use attributes and the processing equipment are important in determining the right grade of butyl to use in a specific application. The selection and ratios of the proper fillers, processing aids, stabilisers and curatives also play critical roles in both how the compound will process and how the end product will behave.

Care must be taken when processing Halobutyl that premature dehydro-halogenation does not occur due to high temperature. Stabilisers (calcium stearate alone for chlorobutyl, supplemented with an epoxy compound such as epoxidised soyabean oil in the case of bromobutyl) are required to prevent dehydrohalogenation during processing.

Elemental sulphur and organic accelerators are widely used to cross-link butyl rubber for many applications. The low level of unsaturation requires aggressive accelerators such as thiuram or thiocarbamates. The vulcanisation proceeds at the isoprene site with the polysulphidic cross-links attached at the allylic positions, displacing the allylic hydrogen. The number of sulphur atoms per cross-link is between one and four or more. Cure rate and cure state (modulus) both increase if the diolefin content is increased (higher unsaturation). Sulphur cross-links have limited stability at sustained high temperature. Resin cure systems (commonly using alkyl phenol-formaldehyde derivatives) provide for carbon-carbon cross-links and more stable compounds.

In halobutyl, the allylic halogen allows easier cross-linking than does allylic hydrogen alone, because halogen is a better leaving group in nucleophilic substitution reactions. Zinc oxide is commonly used to cross-link halobutyl rubber, forming very stable carbon-carbon bonds by alkylation through dehydro-halogenation, with zinc chloride by-product. Bromobutyl is faster curing than chlorobutyl and has better adhesion to high unsaturation rubbers. As a result, its volume growth rate has exceeded that of chlorobutyl in recent decades as tyre plants have driven to higher productivity operation.

7.3   Properties of butyl rubber

The most widely used butyl rubbers are copolymers of isobutylene and isoprene. Grades are distinguished by molecular weight (Mooney viscosity) and mole % unsaturation. The mole % unsaturation is the number of moles of isoprene per 100 moles of isobutylene. The molecular characteristics of low levels of unsaturation between long segments of polyisobutylene produce unique elastomeric qualities that find application in a wide variety of finished rubber articles. These special properties can be listed as:

1. Low rates of gas permeability.
2. Thermal stability.

3. Ozone and weathering resistance.

4. Vibration damping and higher coefficients of friction.

5. Chemical and moisture resistance.

The more chemically inert nature of butyl rubber is reflected in the lack of significant molecular-weight breakdown during processing. This allows one to perform operations such as heat treatment or high-temperature mixing to alter the vulcanisate characteristics of compound. With carbon black-containing compounds, hot mixing techniques promote pigment-polymer interaction, which alters the stress-strain behaviour of a vulcanisate. The shape of the stress-strain curve of the vulcanisate from a heat-treated mixture is a reflection of a more elastic network and heat treatment has resulted in more flexible vulcanised compounds for a given level of a carbon black type. More flexible butyl rubber compositions have also been prepared with certain types of mineral fillers such as reinforcing clays, talcs and silicas that contain appropriately placed OH groups in the lattice. The heat treatment of filler-butyl rubber masterbatches can be accomplished at 300–350°F for 5 min in a Banbury mixer with the aid of promoters (for example p-quinone dioxime). In general, the processing of butyl rubber follows accepted factory operations. Banbury mixing in conventional formulations requires no longer times than with other natural and synthetic elastomers. Of course, there is no premasticating operation to produce a degree of molecular-weight break down, as in the case of natural rubber, but prewarming of the butyl rubber prior to mixing reduces mixing times.

7.3.1 Gas permeability

The permeability of elastomeric films to the passage of gas is a function of the diffusion of gas molecules through the membrane and the solubility of the gas in the elastomer. The polyisobutylene portion of the butyl molecule provides a low degree of permeability to gases and is a familiar property, leading to an almost exclusive use in inner tubes. For example, the air permeability, at 65°C, of SBR is about 80% that of natural rubber, while butyl shows only 10% permeability on the same scale. The difference in air retention between a natural rubber and a butyl inner tube can be demonstrated by data from controlled road tests on cars driven 60 mph for 100 miles per day. Under these conditions butyl is at least 8 times better than natural rubber in air retention. Other gases such as helium, hydrogen, nitrogen and carbon dioxide are also well retained by a butyl bladder membrane.

7.3.2 Thermal stability

Butyl rubber sulphur vulcanisates tend to soften during prolonged exposure to elevated temperatures of 300–400°F. This deficiency is largely the result of the sulphur cross-link, coupled with low polymeric unsaturation which allows no compensating oxidative (cross-linking) hardening. However, certain cross-linking systems and specifically the resin cure of butyl, provide vulcanised networks of outstanding heat resistance. This has found widespread use in the expandable bladders of automatic tyre-curing presses.

7.3.3   Ozone and weathering resistance

The low level of chemical unsaturation in the polymer chain produces an elastomer with greatly improved resistance to ozone when compared to polydiene rubbers. Butyl with the lowest level of unsaturation produces high levels of ozone resistances, which are also influenced by the type and concentration of vulcanisate cross-links.

For maximum ozone resistance, as in electrical insulation and for weather resistance, as in rubber sheeting for roofs and water management application, the least unsaturated butyl is advantageously used.

7.3.4   Vibration damping

The viscoelastic properties of butyl rubber are a reflection of the molecular structure of the polyisobutylene chain. This molecular chain with two methyl side groups on every other chain carbon atom possesses greater delayed elastic response to deformation. The damping and absorption of shock have found wide application in automotive suspension bumpers. An elastomer with higher damping characteristics also restricts vibrational force transmission in the region of resonant frequencies. Transmissibility is the ratio of output force to input force under impressed oscillatory motion.

7.3.5   Chemical and moisture resistance

The essentially saturated hydrocarbon nature of butyl obviously imparts moisture resistance to compounded articles. This capability is utilised in applications such as electrical insulation and rubber sheeting for external use. This same hydrocarbon nature provides useful solubility characteristics that can be applied to a variety of protective and sealant applications. These useful solubility characteristics are again based upon the hydrocarbon nature of the elastomer back-bone that is expressed as a solubility parameter of 7.8.

7.4   Halogenated butyl rubber

The introduction of chlorine to the butyl molecule, in approximately 1/1 molar ratio of chloride to double bond, achieved a broadening of vulcanisation latitude and rate and enhanced covulcanisation with general-purpose, high-unsaturation elastomers, while preserving the many unique attributes of the basic butyl molecule. Bromination in the same approximate molar ratio further enhanced cure properties and provides greater opportunity for increased covulcanisation or adhesion, or both, to general-purpose elastomers.

These property enhancements are vital to the major applications for halobutyl tubeless tyre innerliners. Thus halogenated butyls contributed the required level of covulcanisability and adhesion to the highly unsaturated tyre elastomers to carry the advantages of butyl inner tubes forward through the development of the radial tubeless tyre.

7.4.1   Halogenation and production

The reactions of elemental bromine and chlorine with the isoprene residue in butyl rubber can be quite complicated. The 'dark' reaction of a solution of butyl in an inert diluent with these halogens can result in the incorporation of butyl in upto about 3 g atoms of halogen per mole of unsaturation originally present in the polymer. It has been found that the overall halogenation occurs substantially as a series of consecutive reactions, each being slower than the preceding one.

Synthesis in the halogenated-butyls production facilities involves the 'dark' reaction of a solution of butyl in hexane with elemental halogens at conventional process temperatures (40–60°C). The target is to produce a product in which no more than 1 halogen atom is introduced into the polymer per unsaturated site initially present, within the constraints of a final product weight per cent halogen specification range.

Under the above conditions, the reaction with chlorine is very fast, probably completed in 15 seconds or less, even at the low molar concentration of reactants employed. The bromine reaction is considerably slower, perhaps about 5 times that of chlorination. In both cases, thorough mixing is a prerequisite to meet the synthesis targets (i.e., to avoid multiple halogenation at a particular site). That these reactions lead primarily to substituted products has been well documented. These fast reactions are presumed to occur via an ionic mechanism. The halogen molecules are polarised at the olefinic sites and undergo heterolytic scission and consequent reaction.

Actual production of butyl rubber is conducted in hexane solvent and the process flow diagram is shown in Fig. 7.2. Butyl rubber in solution is treated with chlorine or bromine in a series of high-intensity mixing steps. Hydrogen chloride or hydrogen bromide is generated during the halogenation step and must be neutralised, usually with a dilute aqueous caustic solution. After neutralisation, the caustic aqueous phase is separated and removed and the halogenated cement is then stabilised and antioxidant is added to protect the halogenated product during the polymer recovery and finishing steps, which are much along similar lines as for butyl recovery.

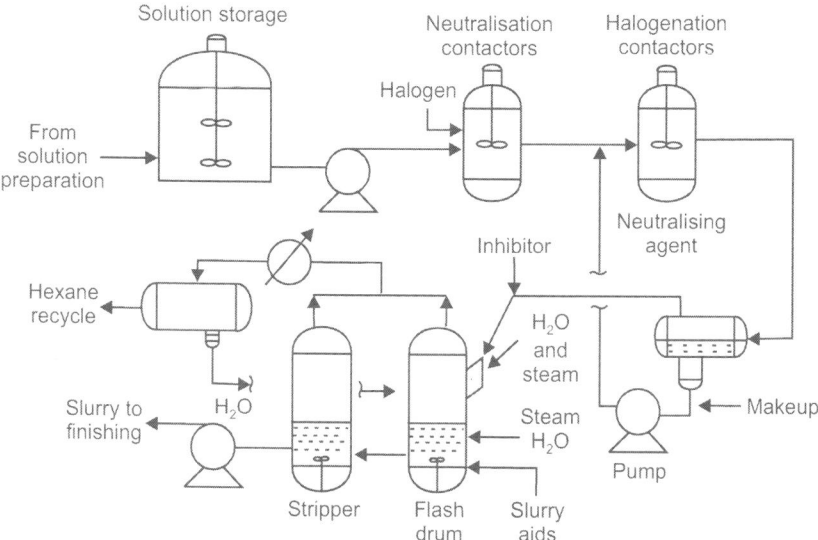

Figure 7.2: Process of manufacture for halogenated butyl rubber.

7.4.2   Stabilisation

As in all hydrocarbon polymers, the presence of an antioxidant is required to protect those elastomers during finishing, storage and compounding. These are chosen on the basis of cost effectiveness, discolouration tendencies and compatibility with the overall process.

These reactive halogens are implanted in the polymer, the principal stabilisation problem is how to preserve them until they can display their main utility in cross-linking reactions. Unstabilised, both elastomers will undergo thermal dehydrohalogenation and simultaneous cross-linking, the brominated product more readily than the chlorinated. The dehydrohalogenation is catalysed by the evolved acid.

Useful stabilisation must meet numerous criteria. The stabiliser or package there of must:

1. Prevent the accumulation of harmless by-products.
2. Not be itself highly reactive with allylic halogens.
3. Have process compatibility.
4. Interact favourably, or at least not unfavourably, with cross-linking systems to be employed.
5. Not introduce a health hazard.
6. Be economical within the foregoing constraints.

Many materials will satisfy one or more of these demands, but very few all of them.

7.4.3   Compounding halobutyl

The following section summarises the influence of compounding ingredients on processing and vulcanisate properties of halobutyls.

Carbon black: Carbon blacks affect the compound properties of halobutyl in a similar way as they affect the compound properties of other rubbers: particle size and structure determine the reinforcing power of the carbon black and hence the final properties of the halobutyl compounds.

1. Increasing reinforcing strength, for example, raises the compound viscosity, hardness and cured modulus.
2. Cured modulus increases with the carbon black level up to 80 phr. Tensile strength goes through a maximum at 50–60 phr carbon black level.

Mineral fillers: Mineral fillers vary not only in particle size but also in chemical composition. As a result, both cure behaviour and physical properties of a bromobutyl compound are affected by the mineral filler used, although to a lesser extent than chlorobutyl compounds.

Generally the common mineral fillers may be used with halobutyl but highly alkaline ingredients and hygroscopic fillers should be avoided.

1. Clays are semireinforcing—acidic clays give very fast cures, therefore extra scorch retarders may be needed. Calcined clay is the preferred filler for pharmaceutical stopper compounds based on halobutyl.
2. Talc is semireinforcing in halobutyl without a major effect on cure.
3. Hydrated silicas even at moderate levels cause compound stiffness and slower cure rate, so their use should be restricted.
4. Silane-treated mineral fillers enhance the interaction between polymer and silicates and hence to improve compound properties, is to add small (1 phr) amounts of silanes. Particularly useful silanes are the mercapto- and amino-derivatives.

Plasticisers: Petroleum-based process oils are the most commonly used plasticisers for halobutyl. They improve mixing and processing, soften stocks, improve flexibility at low temperatures and reduce cost.

Paraffinic/naphthenic oils are preferred for compatibility reasons. Other useful plasticisers are paraffin waxes and low-molecular-weight polyethylene. Adipates and sebacates improve flexibility at very low temperatures.

Process aids:

1. Mineral rubber not only improve the processing characteristics of halobutyl compounds by improving fillers dispersion, but they also enhance compatibility between halobutyl and highly unsaturated rubbers.

2. Tackifying resins should be selected with care. Phenol-formaldehyde resins, even those where the reactive methylol groups have been deactivated, react with halobutyl especially bromobutyl, causing a decrease in scorch time.

3. Stearates, stearic acid should be noted that zinc stearate (which can also be formed via the zinc oxide and stearic acid reaction) is a strong dehydrohalogenation agent and a cure catalyst for halobutyl. Similar effects will be observed with other organic acids such as oleic acid or naphthenic acid. Alkaline stearates, on the other hand such as calcium stearate, have a retarding action on the halobutyl cure.

Anti-degradants—amine-type antioxidants/antiozonants such as Flectol H, mercaptobenzimidazole and especially *p*-phenylene-diamines will react with halobutyl. They should preferably be added with the curatives, not in the masterbatch. Phenol derivative antioxidants are generally preferred.

7.4.4   Processing halobutyl

The following recommended processing conditions are applicable to both chloro and bromobutyl.

Mixing: The mixing is done in two stages. The first stage contains all the ingredients except for zinc oxide and accelerators. The batch weight should be 10–20% higher than that used for a comparable compound based on general-purpose rubbers. A typical mixing cycle and processing for a halobutyl innerliner compound are as follows:

First stage	0 min	Halobutyl, carbon black, retarder.
	1.5 min	Process aids, plasticisers, fillers, stearic aid, etc.
	3.5 min	Dump at 120–140°C.

Higher dump temperatures could result in scorching:

Second stage	0 min	Masterbatch + curatives.
	2 min	Dump at 100°C.

Mill-mixing on a two-roll mill is best accomplished with a roll-speed ratio of 1.25/1 and roll temperatures of 40°C on the slow roll and 55°C on the fast roll.

The following sequence of addition is recommended part of the rubber together with a small amount of a previous mix, 1/4 fillers plus retarder, remainder of polymer, rest of filler in small increments, plasticisers at the end, acceleration below 150°C.

Calendering: Feed preparation can be done either by mill or by extruder. Halobutyl follows the cooler roll, therefore, a temperature differential of 10°C between calender rolls is recommended.

Starting roll temperatures should be:

Cool roll   75–80°C

Warm roll   85–90°C

Normal calendering speeds for halobutyl compounds are between 25 and 30 metres/min.

Rapid cooling of the calendered sheet is beneficial for optimal processability (handling) and maximum tack retention.

Extrusion: Feed temperature should be 75–80°C while the temperature of the extrudate is around 100°C. During calendering and extrusion of halobutyl compounds, the most important problem is blister formation. The reason for this phenomenon is the low permeability of these polymers, which tend to retain entrapped air or moisture. Preventive action should be taken at all stages of the process, for example:

1. Ensure the stock is well mixed in a full mixer to prevent porosity.

2. Avoid moisture at all stages.

3. Keep all rolling banks on mills and calender nips to a minimum.

4. Molding – halobutyl can be formulated to have a fast-cure rate, good mold flow and mold release characteristics and can, therefore, be molded into highly intricate designs with conventional molding equipment. Entrapped air can be removed by bumping of the press during the early part of the molding cycle.

Halobutyl is also very well-suited for injection molding because of its easy flow and fast, reversion-resistant cures. Low-molecular-weight polymer grades may be required for optimum flow and good scorch safety.

7.5   Chlorobutyl vulcanisation and applications

The presence of both olefinic unsaturation and reactive chlorine in chlorinated butyl provides for a great variety of vulcanisation techniques. While conventional sulphur-accelerator curing is possible and useful, here are discussed two general vulcanisation techniques not available to regular butyl.

7.5.1   Zinc oxide cure and modifications

Zinc oxide, preferably with some stearic acid, can function as the sole curing agent for chlorobutyl. After vulcanisation, most of the chlorine originally present in the polymer can be extracted as zinc chloride. A proposed mechanism is based on formation of stable carbon-carbon cross-links through a cationic polymerisation route. When zinc oxide is used as the vulcanising reagent, the necessary initiating amounts of zinc chloride are likely formed as a result of thermal dissociation of some of the allylic chloride to yield hydrogen chloride.

Subsequent reaction of the hydrogen chloride with zinc oxide then provides catalyst.

It is not likely that the propagation step proceeds very far, but for vulcanisation purposes only one step is needed, particularly since both of the termination processes suggested result in the production of more catalyst.

The attainment of the full cross-linking potential of chlorobutyl by the ZnO system is relatively slow. This situation can be remedied by the inclusion of thiurams and thioureas into the curing recipe. It has been observed that chemical compounds with the grouping as shown below:

$$Z-\overset{\displaystyle \overset{S}{\|}}{C}-R$$

(where, Z is some type of activating group) will accelerate the ZnO cure of chlorobutyl. Examples are thiourea and tetramethyl thiuram disulphide.

The vulcanisation of chlorobutyl with this type of accelerator can proceed via a mechanism similar to the mechanism of polychloroprene vulcanisation with substituted thioureas. The increase in cure rate and modulus as compared to the straight ZnO cure is obtained without sacrifice in vulcanisate stability.

Vulcanisation through bis-alkylation

The other unique and valuable curing method for chlorobutyl is that involving *bis*-alkylation reactions. This type of cross-linking reaction is perhaps best illustrated by a cross-linking with primary diamines, a vulcanisation reaction that proceeds rapidly to yield good vulcanisates. This cross-linking reaction is believed to occur by the mechanism shown below:

$$R-NH_2 + \sim CH_2 - C - \overset{\displaystyle \overset{CH_2}{\|}}{\underset{\displaystyle \underset{Cl}{|}}{CH}} - CH_2 \sim \longrightarrow$$

$$\underset{\displaystyle \underset{R-NH_2Cl}{|}}{\sim CH_2 - \overset{\displaystyle \overset{CH_2}{\|}}{C} - CH - CH_2 \sim} \xrightarrow{RNH_2} \underset{\displaystyle \underset{RNH}{|}}{\sim CH_2 - \overset{\displaystyle \overset{CH_2}{\|}}{C} - CH - CH_2 \sim} + RNH_3Cl$$

Obviously, in the presence of a diamine, reaction of both functional amino groups in the molecule with different polymer molecules results in cross-linking. Careful adjustment of the diamine concentration is required for development of the highest cross-link density. Too little could not possibly provide the maximum cross-links, whereas too much would allow for reaction with only one of the amino groups per diamino molecule. In the presence of a hydrogen chloride scavenger, maximum modulus is developed when the ratio of NH_2/Cl is very nearly unity.

Extensions of this *bis*-alkylation vulcanisation technique are numerous and in general any molecule having two active hydrogens can, under the proper catalytic conditions, cross-link the polymer. Typical examples are vulcanisation with dihydroxy aromatics such as resorcinol and with dimercaptans.

7.5.2   Resin cure

Both chlorobutyl and regular butyl are capable of vulcanisation with heat-reactive phenolic resins, which are usually characterised by 6–9 weight per cent of methylol groups. Unlike conventional butyl, no promoter or catalyst other than zinc oxide is needed for efficient vulcanisation and a fast, tight cure is obtained with considerably less reagent.

7.5.3   Scorch control

The modified zinc oxide cures are very fast and tend to be scorchy. As a general rule, acidic materials, such as channel blacks, activate the ZnO cure of chlorobutyl while basic materials hamper or retard it. The retarding effects of some alkaline materials, such as magnesium oxide, can be used in a very practical way to provide processing safety. The addition of 0.25 phr of MgO increases the margin against incipient vulcanisation during processing at 126.5°C (260°F), as judged by Mooney scorch measurements, from 5 to 15 min, without greatly affecting tensile properties at the vulcanisation temperature of 153°C (370°F). However, if the concentration is increased to 0.5 phr, the cure rate is depressed significantly.

As a scorch retarder, MgO is effective with all cure systems except the amine cure. In the latter case, it has the reverse effect since it prevents the hydrogen chloride, generated during cross-linking, from reacting with the curing agent. The choice of magnesium oxide type and concentration depends upon the type of compound used as well as the particular application involved.

7.5.4   Stability of chlorobutyl cross-links

Since chlorobutyl has essentially the same structure as the butyl polymer, all the properties inherent to the butyl backbone are found in chlorobutyl rubber. These properties include low gas and moisture permeability, high hysteresis, good resistance to ozone and oxygen, resistance to flex fatigue and chemical resistance. Additionally, chlorobutyl offers appreciably higher heat resistance than regular butyl cured with conventional sulphur vulcanisation systems.

The chlorobutyl compound accelerated by ZnO/stearic acid displays a constantly increasing modulus or cross-link density with increasing vulcanisation time. The problem of reversion does not exist in properly vulcanised chlorobutyl. The behaviour of regular butyl and TMTDS-ZnO-cured chlorobutyl has been compared with respect to stress relaxation under conditions of fixed strain, which is a common method to follow the degradation of cross-linked networks. Gum vulcanisates are used and the recipes and curing times are adjusted to give approximately equal cross-link densities for both materials. The specimens are held in air at a fixed elongation of 50% and data are taken as a function of time at various temperatures.

7.5.5   Chlorobutyl applications

Chlorobutyl has proven to be highly useful in many commercial rubber products. These products generally take advantage of desirable characteristics generic to butyl polymers, such as the resistance to environmental attack and low permeability to gases. In addition, chlorobutyl offers the superimposed advantages of cure versatility, highly heat-stable cross-links and the ability to vulcanise in blends with highly unsaturated elastomers.

Innerliners for tubeless tyres

The combination of low permeability, high heat resistance, excellent flex resistance and ability to covulcanise with high unsaturation rubbers makes cholorobutyl particularly attractive for use in innerliners for tubeless tyres.

Tyre sidewall components

Tyre sidewall performance is critical from both an appearance and durability standpoint. Longer tyre service life, particularly with the advent of belted bias and radial-ply tyres and increasing amounts of atmospheric degradants are contributing to a higher service severity. Tyre manufacturers find it more difficult and costly to obtain the desired performance requirements with antiozonants. The substitution of chlorobutyl, either alone or in combination with ethylene propylene terpolymer, for a portion of the high unsaturation polymers commonly in use, offers a simple, economical means to upgrade the weathering and flex resistance of tyre-sidewall components.

Heat-resistant truck inner tubes

Chlorobutyl offers improved resistance to heat softening and 'growth' while maintaining the desirable butyl property of excellent air retention in inner tubes. This feature is of particular importance in severe service conditions, for example high-speed, heavy-load trucks and buses for which tyre service temperatures exceed 280°F and sometimes go as high as 300°F. After prolonged exposure, butyl tubes will soften, whereas chlorobutyl tubes will remain serviceable.

Other applications of chlorobutyl rubber

Chlorobutyl is used in many rubber articles in addition to tyres and tubes. In these applications, the cure versatility of chlorobutyl and the stability of its vulcanisates are of particular importance. For example, chlorobutyl can be cured with non-toxic cure systems, such as zinc oxide with stearic acid, for use in products that will contact food. Most cure systems provide fast, reversion-resistant cures with chlorobutyl. This property facilitates the attainment of a uniform cure state in thick articles. Due to its exceptional heat resistance and compression-set properties, properly compounded chlorobutyl will give good service at temperatures upto 150°C (300°F).

Example of typical applications are hose (steam, automotive), gaskets, conveyor belts, adhesives and sealants, tyre-curing bags, tank linings, truck-cab mounts, aircraft engine mounts, rail pads, bridge bearing pads, pharmaceutical stoppers and appliance parts.

8.1 Introduction

Ethylene propylene rubbers and elastomers (also called EPDM and EPM) continue to be one of the most widely used and fastest growing synthetic rubbers having both specialty and general-purpose applications. These are non-oil resistant synthetic rubbers. Polymerisation and catalyst technologies in use today provide the ability to design polymers to meet specific and demanding application and processing needs. Versatility in polymer design and performance has resulted in broad usage in automotive weather-stripping and seals, glass-run channel, radiator, garden and appliance hose, tubing, belts, electrical insulation, roofing membrane, rubber mechanical goods, plastic impact modification, thermoplastic vulcanisates and motor oil additive applications.

Ethylene propylene rubbers are valuable for their excellent resistance to heat, oxidation, ozone and weather ageing due to their stable, saturated polymer backbone structure. Properly pigmented black and non-black compounds are colour stable. As non-polar elastomers, they have good electrical resistivity, as well as resistance to polar solvents, such as water, acids, alkalies, phosphate esters and many ketones and alcohols. Amorphous or low crystalline grades have excellent low temperature flexibility with glass transition points of about minus 60°C. Heat ageing resistance up to 130°C can be obtained with properly selected sulphur acceleration systems and heat resistance at 160°C can be obtained with peroxide cured compounds.

Compression set resistance is good, particularly at high temperatures, if sulphur donor or peroxide cure systems are used. These polymers respond well to high filler and plasticiser loading, providing economical compounds. They can develop high tensile and tear properties, excellent abrasion resistance, as well as improved oil swell resistance and flame retardance. A general summary of properties is shown in Table 8.1.

8.2 Manufacturing processes of EPDM

Ethylene propylene rubbers use the same chemical building blocks or monomers as polyethylene (PE) and polypropylene (PP) thermoplastic polymers. These ethylene (C2) and propylene (C3) monomers are combined in a random manner to produce rubbery and stable polymers. A wide family of ethylene propylene elastomers can be produced ranging from amorphous, non-crystalline to semi-

Table 8.1: Properties of ethylene propylene elastomers.

Polymer properties	
Mooney viscosity, ML 1+4 @ 125°C	5–200+
Ethylene content, wt. %	45 to 80 wt. %
Diene content, wt. %	0 to 15 wt. %
Specific gravity, gm/mL	0.855–0.88 (depending on polymer composition)
*Vulcanisate properties**	
Hardness, shore a durometer	30A to 95A
Tensile strength, MPa	7 to 21
Elongation %	100 to 600
Compression %	20 to 60
Useful temperature range, °C	−50° to +160°
Tear resistance	Fair to good
Abrasion resistance	Good to excellent
Resilience	Fair to good (stable over wide temp. ranges)
Electrical properties	Excellent

* Range can be extended by proper compounding. Not all of these properties can be obtained in one compound.

crystalline structures depending on polymer composition and how the monomers are combined. These polymers are also produced in an exceptionally wide range of Mooney viscosities (or molecular weights). The ethylene and propylene monomers combine to form a chemically saturated, stable polymer backbone providing excellent heat, oxidation, ozone and weather ageing. A third, non-conjugated diene monomer can be terpolymerised in a controlled manner to maintain a saturated backbone and place the reactive unsaturation in a side chain available for vulcanisation or polymer modification chemistry.

The terpolymers are referred to as EPDM (or ethylene propylene-diene with 'M' referring to the saturated backbone structure). An EPDM polymer structure is illustrated in Fig. 8.1. The ethylene propylene copolymers are also called EPM.

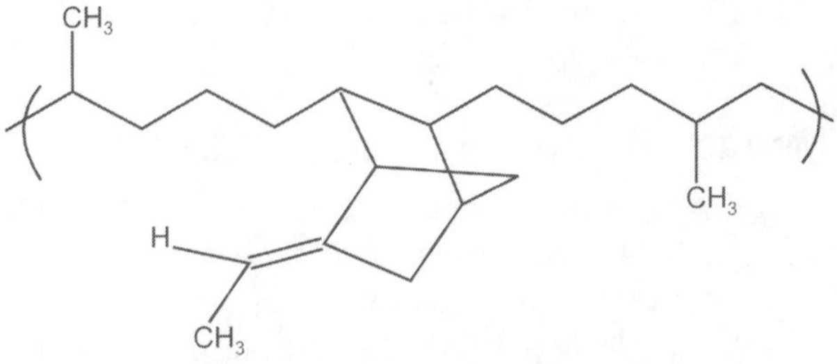

Figure 8.1: Structure of EPDM.

The two most widely used diene termonomers are primarily ethylidene norbornene (ENB) followed by dicyclopentadiene (DCPD). Each diene incorporates with a different tendency for introducing long chain branching (LCB) or polymer side chains that influence processing and rates of vulcanisation by sulphur or peroxide cures.

These characteristics are summarised in Table 8.2.

Table 8.2: Cure-site diene termonomers in ethylene propylene elastomers.

Termonomer	Cure and property features	Long chain branching
ENB	Fastest and highest state of cure	Low to moderate
	Good tensile strength	
	Good compression set resistance	
DCPD	Slow sulphur cure	High
	Good compression set resistance	

Specialised catalysts are used to polymerise the monomers into controlled polymer structures. Since their introduction, ethylene propylene elastomers have used a family of catalysts referred to as Zeigler-Natta named after their initial developers. Improvements in catalysts and processes have provided increased productivity while maintaining control of polymer structure. Most recently a new family of catalysts, referred to as metallocene catalysts, have been developed and are in commercial use.

8.3   Manufacture and processing of EPDM

There are three major commercial processes, solution, slurry (suspension) and gas-phase, for manufacturing ethylene propylene rubbers. The manufacturing systems vary with each of the several producers.

There are differences in the product grade slates made by each producer and process, but all are capable of making a variety of EPDM and EPM polymers. The physical forms range from solid to friable bales, pellets and granular forms and oil blends.

The solution polymerisation process is the most widely used and is highly versatile in making a wide range of polymers. Ethylene, propylene and catalyst systems are polymerised in an excess of hydrocarbon solvent. Stabilisers and oils, if used, are added directly after polymerisation. The solvent and unreacted monomers are then flashed off with hot water or steam, or with mechanical devolatilisation. The polymer, which is in crumb form, is dried with dewatering in screens, mechanical presses or drying ovens. The crumb is formed into wrapped bales or extruded into pellets. The high viscosity, crystalline polymers are sold in loosely compacted, friable bales or as pellets. The amorphous polymers grades are typically in solid bales.

The slurry (or suspension) process is a modification of bulk polymerisation. The monomers and catalyst system are injected into the reactor filled with propylene. The polymerisation takes place immediately, forming crumbs of polymer that are not soluble in the propylene. Slurry polymerisation reduces the need for solvent and solvent handling equipment and the low viscosity of the slurry helps to control temperature and handle the product. The process is not limited by solution viscosity, so high molecular weight polymer can be produced without a production penalty. Flashing off the propylene and termonomer completes the process before forming and packaging.

Gas-phase polymerisation technology was recently developed for the manufacture of ethylene propylene rubbers. The reactor consists of a vertical fluidised bed. Monomers and nitrogen in gas form along with catalyst are fed to the reactor and solid product is removed periodically. Heat of reaction is removed through the use of the circulating gas that also serves to fluidise the polymer bed. Solvents are not used eliminating the need for solvent stripping, washing and drying.

The process is also not limited by solution viscosity, so high molecular weight polymer can be produced without a productivity penalty. Continuous injection of a substantial amount of carbon black used as a partitoning aid is necessary to prevent the polymer granules sticking to each other and to reactor walls. Products are made in a granular form to enable rapid mixing.

8.3.1 Processing of EPDM

Conventional rubber equipment is used for the processing of EPDM compounds. Selection of the polymer or blend of polymers is an essential consideration in order to obtain optimum processability.

8.3.2 Mixing of EPDM

Mixing usually is accomplished with an internal mixer, such as a Banbury mixer. In some instances, it may be desirable or necessary to use the mill mixing process. For practical mill mixing, the low-viscosity and higher-ethylene polymers are considered too difficult. Furthermore, compounds based on these polymers typically contain such high amounts of filler and oil that mill mixing is usually impractical from this standpoint alone.

Although all types of EPDM are suitable for the internal mixer, no single mixing procedure is satisfactory for all compounds. The variety of polymers, the range of compound qualities (which affect choice of filler type) and the condition of the particular mixer are some of the factors that influence the mixing technique. The most popular procedure is to load the ingredients in 'one shot,' and so forth. Depending upon the particular machine (its mechanical condition, speed, etc.,) mixing cycles can range from 3 to 7 min. Using this kind of procedure, the batch is usually dumped when the temperature reaches 115°C to 130°C. This procedure may be limited to compounds that are based on medium and low-ethylene polymers loaded with mineral fillers and/or semi-reinforcing types of carbon black (SRF, GPF, FEF).

For compounds utilising high-ethylene or very-high-molecular-weight EPDM polymers and for compounds containing reinforcing carbon blacks (HAF, ISAF), the 'one-shot' or 'upside-down' loading procedure often results in poor dispersion of black or rubber or both. These polymers and blacks will usually not disperse well if processing oil is added early in the mixing cycle. Better dispersion is attained by an incremental procedure in which rubber and part of the black are loaded into the mixer with little or no oil. The withheld black and oil may then be added in one, two, or more increments.

If allowable from the standpoint of providing required properties, an easy-to-disperse softer black or higher-structure black, should be used in combination with the more difficult-to-disperse, highly reinforcing black. Then the easy-to-diverse black can be added with the plasticiser as the second increment in the mixing procedure. The high ethylene EPDMs should be stored under warm warehouse conditions to minimise dispersion problems.

Other variations in mixing procedures are also successful. For example, a modified 'upside-down' technique in which part of the EPDM polymer is added on top of fillers and plasticisers provides rapid mixing and good dispersion. This procedure may also be further modified by with-holding part of the oil, which is then added incrementally. In addition, two-stage mixing is frequently employed when better dispersion or batch-to-batch consistency is needed.

8.3.3 Extrusion of EPDM

The selection of polymer is a prime consideration if the EPDM compound is to be used for extrusion processing. The high-ethylene types are most useful for this process because they provide good green strength for shape retention. In addition, some green strength seems necessary for good feeding into the extruder; in fact, lack of green strength in the uncured compound is usually associated with poor feeding characteristics. If compounds do not feed easily and if feed is erratic, surging of the extrudate results in its size varying. Hence a well-designed extrusion compound should contain enough high-ethylene polymer to provide the desired processing behaviour. Since excessive green strength (tough compound) is a possibility, one of the lower ethylene types of EPDM should be part of the polymer blend to avoid it.

A wide variety of extrusion equipment is to be found. Most of the older extruders still in use are the short-barrel, hot-feed type and for these the stock must usually be 'warmed' on a mill and then fed to the extruder. This process, therefore, requires a compound that is reasonably easy to band on the warm-up mill. Excessive green strength and/or high compound viscosity will cause difficulty in banding. These results may be avoided by using a lower-ethylene and/or lower-viscosity polymer blended with the high-ethylene type in the compound. In general, the newer cold feed extruder is preferred. Not only does it obviate the need for a warm-up mill, it also provides better control of processing conditions so that more uniform extrudates are obtained. For most compounds, the extruder screw should be operated at a higher temperature than the barrel. Conditions should be adjusted to provide a smooth surface and optimum rate of extrusion. In general, extrusion temperature conditions are somewhat higher for EPDM compounds than for other types of rubber compounds.

8.3.4   Molding of EPDM

Molded rubber parts are made by either a compression, transfer, or injection process, for all of which polymers having medium to high propylene are usually preferred. In applications involving highly extended compounds, however, it may be desirable to use a high-ethylene polymer, either exclusively or in part, to take advantage of its greater extendability. The prime consideration is good flow characteristics and polymer selection should be made accordingly.

EPDM compounds have proved to be particularly suitable for injection molding. When properly compounded, EPDM provides a fast cure rate and good flow and is not prone to reversion at the high curing temperatures common to this process. Compounding considerations for injection molding are not appreciably different from those for compression or transfer molding. The curing system is usually the point of most concern and the plasticiser should be selected for low volatility at the processing temperature.

8.3.5   Calendering of EPDM

The EPDM polymer selected for calendering should be in the medium-to-high propylene range. If the compound is to be lightly loaded, the lowest-viscosity, highest-propylene type is most suitable. Also, a polymer with broad molecular-weight distribution and medium-to-high propylene content is a better choice. For medium to highly loaded compounds, the higher-viscosity types are required. The desirable processing is also often attained by a proper selection of two types of EPDM polymers.

8.4   Elements of compounding EPDM

8.4.1   Curing system for EPDM

Substitution of one EPDM with another in an existing formulation without changing the cure system, the rate of cure and the properties will almost certainly not be the same, particularly if the two polymers utilise different third monomers.

The properties also can be affected by differences in molecular weight and other variations in polymer composition. In general, polymers with DCPD or 1,4 HD will require some what more active accelerators and/or higher levels to provide satisfactory cure rates.

EPDM has pendant unsaturation that allows sulphur vulcanisation. Like most other synthetic rubbers, common accelerators may be used. Since the particular combination selected depends upon many considerations—such as processing methods, the properties desired, cost and compatibility—few generalisations are possible. Usually, however, the cure system will contain a thiazole accelerator (MBT, MBTS, or the like) in combination with a thiuram and/or a dithiocarbamate. Sulphur donor-type accelerators may replace elemental sulphur, if heat resistance and/or compression set requirements are severe. For some molded-goods applications, accelerator bloom may be unacceptable. To assure a non-blooming compound, it is necessary to maintain the levels of the various chemicals below their solubility limits.

Types of accelerators	Parts per hundred (phr)
MBT, MBTS, CBS, ZMBT, ZDBDP	3.0 phr
ZDBDC, DTDM	2.0 phr
ZDEDC, ZDMDC, TDEDC, TMTD	0.8 phr
DPTT, TMTM, TETD, FDMDC	0.8 phr

It has long been known that low sulphur or sulphur donor cure systems give good heat resistance and improved compression set. By using a cure system with 3 to 4 phr of a thiazole (MBT, MBTS, or CBS) in combination with a thiuram and a dithiocarbamate and a level of sulphur below one part, it is possible to obtain outstanding heat resistance. For exposure temperatures above 150°C, a selected antioxidant should be added to enhance the heat resistance of this system. Another sulphur-donor, low-sulphur system that provides good heat resistance but better compression set than that obtained with a 'high thiazole' system entails using 2- to 3-phr levels of a thiuram, of two different dithiocarbamates, of a dithiomorpholine and a low level of sulphur.

In black, steam-cured extrusions, bloom is usually not a problem. Some cure systems that bloom when press cured may not bloom when cured in steam. Since steam curing cycles are typically longer and take place at lower temperatures than molding cycles, the cure system may be quite simple, for example 1 phr MBTS (or MBT), 1.5 phr TMTD and 1.5 phr sulphur. A dithiocarbamate may be added if an increased cure rate is desired.

8.4.2 Reinforcement of EPDM

Like other non-crystallising polymers, EPDM requires reinforcement to be of practical value since the mechanical properties of the unfilled rubber are quite poor. Carbon black is the most useful material for this reinforcement, but silica, clay, talc and some other mineral fillers may also be used. To attain their full effectiveness as reinforcing agents in EPDM rubbers, carbon black and other fillers must be well dispersed. High tensile strength, good tear resistance and improved abrasion resistance are usually associated with good reinforcement. Well-mixed batches also provide better and more uniform processing for extrusion, calendering, molding and so forth.

8.4.3   Plasticisers and processing aids for EPDM

Naphthenic oils have been the most widely used plasticisers for EPDM compounds because they provide the best compatibility at reasonable cost. For applications at higher temperatures or in coloured compounds, paraffinic oils are usually chosen because of their lower volatility and improved UV stability. Some paraffinic oils tend to bleed from cured, high-ethylene EPDM compounds. If such oils are to be used, it is usually advisable to replace part (20–25 phr) of the high-ethylene rubber with one having somewhat lower ethylene content.

Aromatic oils have an adverse effect on some compound properties and, of course, must not be used in conjunction with peroxide curing systems.

Stearic acid, zinc stearate, or other internal lubricants are often included in a compound to aid processing.

EPDM compounds are inherently not tacky. Should there be a need for building tack, it would be necessary to add a tackifier to the compound. Some care must be taken in making the selection because of compatibility problems and the effects on cure rate of tackifiers with excessive unsaturates. Some tackifiers have been developed specifically for EPDM.

8.5   Compounding EPDM for various applications

The inherent properties of EPDM polymers have led to their use in a considerable variety of applications. Some of the more significant uses and formulations to illustrate concepts in compound development for specific applications are discussed below.

8.5.1   Sheeting

Rubber sheeting has been used for some years in roofing, but in rather limited quantities because of its cost compared to that of cheap asphalt. The situation has changed dramatically and the economics for EPDM sheeting are now much more favourable. The formulation suggested for this application is based on a polymer that offers good physical properties at fairly high extension, good green strength and good calendering. The filler loading is designed to provide good processability and reinforcement to meet the required tensile and tear strength properties. Because of long-term ageing conditions, a less volatile, paraffinic oil and a sulphur donor, low-sulphur cure systems are employed. Since the application involves making lap-seam splices in the field, the cured compound, to facilitate adhesion, is non-blooming. Test data show the characteristics of green strength, cured physical properties and heat, water and ozone resistance that are required for good performance in roof sheeting applications.

8.5.2 Automotive hose

Generally EPDM polymers are used in the manufacture of automotive hoses because they offer good physical properties. Typical recipe of hose radiator is given in Table 8.3.

Table 8.3: Radiator hose compound.

Ingredients	
EPDM B	105.0
EPDM A	25.0
Zinc oxide	3.0
N-650 black	130.0
N-762 black	95.0
Ground whiting	40.0
High-viscosity, paraffinic oil type 104B	130.0
Stearic acid	1.0
TMTD	3.0
DTDM	2.0
ZDBDC	2.0
ZDMDC	2.0
Sulphur	0.5
	538.5

8.5.3 Wire and cable

EPDM has found wide acceptance in wire and cable applications because of its inherently excellent electrical properties, combined with resistance to ozone, heat, cold and moisture.

8.5.4 Summary of processing and vulcanisation

The processing, vulcanisation and physical properties of ethylene propylene elastomers are largely controlled by the characteristics of ethylene content, diene content, molecular weight (or Mooney viscosity) and molecular weight distribution. For example, decreasing ethylene content decreases crystallinity and associated properties such as hardness and modulus. General polymer features in rubber compounding are summarised in Table 8.4.

Table 8.4: General features of ethylene propylene elastomers.

Characteristics	*High*	*Low*
Ethylene content	Good green strength	Fast mixing
	Flow at high extrusion temperatures	Low temperature flexibility
	High tensile strength, modulus	Low hardness and modulus
	High loading (reduced cost)	Calendering and milling
Diene content	Cure degree and fast rate	Scorch resistance
	Acceleration versatility	High heat stability
	Good compression set	Low hardness and modulus
	High modulus, low set	
Molecular weight	Good tensile, tear, modulus, set	Fast mixing
	High loading and oil extension	High extrusion rates
	Good green strength	Good calendering
	Collapse resistance	Low viscosity, scorch resistance
MWD	Overall good processing	Low die swell
	Extrusion feed and smoothness	Fast extrusion rate
	Collapse resistance	High cure
	Good calendering and milling	Good physicals

8.5.5   Applications of EPDM

1. Automobiles:
   (a) Tyre sidewalls.
   (b) Inner tubes.
2. Automotive:
   (a) Weather-stripping (sponge and dense) for doors, windows, trunk lids.
   (b) Radiator and heater hose.
   (c) Air emission hose.
   (d) Tubing.
   (e) Brake components.
   (f) Isolators and mounts.
   (g) Grommets, body rubber.

3. Building and construction:
   (a) Glass sealers.
   (b) Curtain wall gaskets and tapes.
   (c) Rubber sheeting for roofing.
   (d) Pond and ditch liners.
   (e) Reservoir liners.
4. Agricultural equipment:
   (a) Hoses.
   (b) Sheet tubes.
   (c) Cushioning.
   (d) Sheeting for grain storage.
5. Mechanical goods:
   (a) Dock fenders.
   (b) Belting.
   (c) Gasketing.
   (d) Seals.
   (e) O-rings.
6. Miscellaneous applications:
   (a) Tank linings.
   (b) Wire and cable.
   (c) Inlet and drain hose.
   (d) Boots.
   (e) Seals.
   (f) Mounts.
   (g) Washers.
   (h) Roll covers.

8.6   Polymer variables and properties

In addition to variations resulting from the introduction of the different termonomers, several other variations are possible in EPDMs. All of them affect the resultant properties of the polymers.

8.6.1   Ethylene propylene (EP) composition

The contents of EP elastomers are generally reported as weight per cent ethylene and vary from 75 to 45%. The monomers are randomly distributed, resulting in amorphous type copolymers. The higher ethylene-containing polymers, which contain some crystallinity, are beneficial in that they possess higher green strength (shape retention), can be more highly loaded with fillers/oil, result in higher compounded tensile strengths, can be more readily pelletised and possess better extrusion properties.

The disadvantages of the higher ethylene contents, however, are their poorer mill processing behaviour at lower temperatures, inferior low-temperature properties and their difficulty in being mixed. Since crystallinity tends to increase with decrease in temperature, it is appropriate to store high ethylene terpolymers under sufficient warmth to minimise dispersion problems.

8.6.2 Molecular weight

The molecular weight of an elastomer is commonly reported as the Mooney viscosity (ML). In the case of EPDMs, these values are obtained at elevated temperatures, usually 125°C. The primary reason for this is to melt out any effect that high ethylene content could produce (crystallisation), thus masking the true molecular weight of the polymer. Mooney viscosities can vary from a low of 20 to a high of 100. Polymers of even higher molecular weight exist but are extended with oil (25 to 100 phr). Such polymers should be stored in the absence of light (UV) since the oils will tend to accelerate oxidative degradation (gelation).

Advantages of higher molecular weights are similar to those of higher ethylene contents in that the tensile and green strength of the polymer improve. In addition, these polymers can be more highly loaded with fillers and oil. Disadvantages are their poorer processing and dispersibility.

8.6.3 Molecular weight distribution (MWD)

Molecular-weight distribution is not normally reported as a polymer variable but in most applications it is a very important property. A measure of the MWD is now commonly obtained through the use of gel permeation chromatography at elevated temperatures (135°C). The reported value is the ratio of the weight-average molecular weight to the number-average molecular weight ($\bar{M}_w/\bar{M}_n$). This value can usually vary from 2 to 5. For broad MWD polymers, there is usually a variation in EP composition as well.

The end with the higher molecular weight possess a higher ethylene content than the end with the low molecular weight.

Polymers with broad molecular weight possess excellent mill processing and calendering capabilities and higher green strength. These polymers are extensively used for applications that do not allow large loadings of filler and oil. Their disadvantages are their slower cure rate and poorer state of cure. Narrow molecular weight distributions are more common for EPDMs since they result in faster cure rates, better state of cure and smoother extrusions.

Ethylidene norbornene (ENB)

ENB is the most widely used termonomer employed even though it is the most expensive, the reasons being that it is the most readily incorporated during copolymerisation and the double bond introduced has the greatest activity for sulphur vulcanisation. This activity is also of such a nature that EPDMs containing ENB have the greatest tendency to be co-cured with diene elastomers. Another unique characteristic of this termonomer is that it makes it possible to prepare linear as well as branched polymers by varying the conditions under which the polymers are synthesised. Branching has an important role in establishing the rheological properties of a polymer. Under proper control, it can introduce properties to the EPDM that are beneficial in certain applications.

1,4 Hexadiene (1,4 HD)

Polymers containing 1,4 HD exhibit a slower cure rate than ENB but possess certain superior properties. One such property is its excellent heat characteristic which is closest to EPM. Such polymers exhibit a good balance of chain scission and cross-linking reactions. Polymers prepared with 1,4 HD are normally linear in structure and possess excellent processing characteristics.

Dicyclopentadiene (DCPD)

The main advantages of DCPD are its low cost and relative ease of incorporation, in which it is similar to ENB. Of the three termonomers, it has the slowest cure rate. All polymers prepared with it are branched as a result of the slight polymerisability of its second double bond. As mentioned previously, this branching can be beneficial, for example, by imparting ozone resistance to diene rubber blends.

Thermoplastic rubbers (Elastomers)

9.1 Introduction

An elastomer is defined by mechanical response not by chemical structure. Elastomers comprise a diverse range of chemical structures although they are characterised as having weak intermolecular forces. An elastomer will undergo an immediate, linear and reversible response to high strain to an applied force. This response has a mechanical analogy with a spring according to Hooke's Law. Non-linear, time dependent mechanical response is distinguished as viscoelasticity. An ideal elastomer will only exhibit an elastic response. Real elastomers exhibit a predominantly elastic response, however they also exhibit viscoelastic and elastic responses especially at higher strains. These are non-oil resistant synthetic rubbers. The chemical structure and molecular architecture of elastomers is tightly related to elastomeric mechanical response. High strain requires a polymer with high molar mass preferred. Many materials can exhibit an elastic response, that is immediate, reversible and linear strain with stress, however only a polymer can exhibit additionally high strain.

High strain is due to uncoiling of random molecular coils into more linear conformations. The limit to elastic response is when molecules are in fully extended conformations. This mechanism is due to uncoiling of chain segments. Molecules do not move relative to each other, there are reversible random coiling not translational motions. Reversibility and immediate response is obtained with macro-molecules that have flexible chains with weak intermolecular forces. Rigid groups such a benzene, bulky side-chains such as isopropyl, polar groups such as ester and hydrogen bonding groups such as hydroxy are not desirable if a polymer is to be an elastomer. This chapter discusses elastomeric properties at ambient temperatures, since at elevated temperatures above the glass transition temperature many polymers become elastomers.

At high extensions and when under strain for longer times viscous flow occurs, known as creep when over longer times. Chemical cross-linking prevents viscous flow, the movement of molecules relative to each other. Elastomers are cross-linked after molding or shaping to fix molecules into their relative positions. Once cross-linked the unstrained shape of an elastomer cannot be altered and the elastomer cannot be reprocessed or recycled. The permanence brought about by cross-linking and the need to perform a cross-linking reaction on elastomers are disadvantages for their applications.

9.2 Thermoplastic elastomer

A thermoplastic elastomer has all the same features as described for an elastomer except that chemical cross-linking is replaced by a network of physical cross-links. The ability to form physical cross-links is the opposite to the chemical and structural requirements of an elastomer just described. The answer to this dilemma is that thermoplastic elastomers must be two-phase materials and each molecule must consist of two opposite types of structure, one the elastomeric part and the second the restraining, physical cross-linking part. Thermoplastic elastomers are typically block copolymers.

The elastic block should have high molar mass and possess all of the others characteristic required of an elastomer. The restraining block should resist viscous flow and creep. One restraining block can be used per macro-molecule, giving a diblock copolymer (AB), or one restraint block at each of the elastomer can be used giving a triblock copolymer (ABA). Specific polymers will be described in the context of these general principles in the following sections. To provide an example of thermoplastic elastomer block copolymer structures the monomers butadiene and styrene are chosen.

Thermoplastics elastomers can be divided into the following groups:

1. Styrene-diene block copolymer.
2. Elastomeric alloys.
3. Thermoplastic urethane elastomers.
4. Thermoplastic ester-ether copolymers (TPE-E).
5. Thermoplastic amide copolymer (TPE-A).

9.2.1 Styrenic thermoplastic elastomers (TPE-S)

Various styrenic thermoplastic elastomers are: SBS (styrene/butadiene copolymer), SIS (styrene/isoprene copolymer), SEBS (styrene/ethylene butylene copolymer), SEPS (styrene/ethylene propylene copolymer). The linear and the radial structure of styrene thermoelasts is shown in Fig. 9.1.

Thermoplastic elastomers based on styrene are block copolymers in which a polydiene unit divides polystyrene blocks. The polydiene may be for example butadiene (SBS), isoprene (SIS), ethylene butylene (SEBS) or ethylene propylene (SEPS). The styrene content varies with different materials, but usually it is 20–40%.

Advantages of styrenic TPEs:

1. High tensile strength and modulus.
2. Good miscibility.
3. Good abrasion resistance.
4. Good electrical properties.

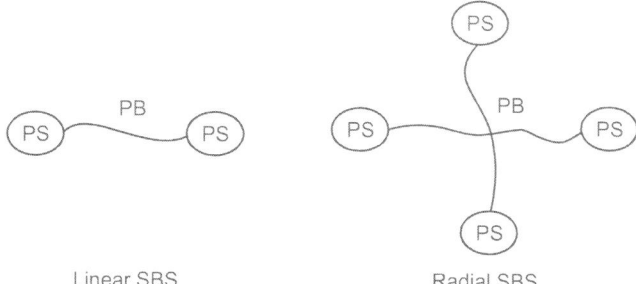

Linear SBS   Radial SBS

Figure 9.1: The linear and the radial structure of styrene thermoelastomers.

5. Large variety in hardness.
6. High friction coefficient (corresponds to that for NR).
7. Colourless, good transparency.

Disadvantages of styrenic TPEs:

1. Poor high temperature resistance (highest operation temperature, SBS 65°C, SEBS 135°C).
2. Weak oxygen, ozone and light resistance of SBS (exception SEBS).
3. Poor oil and solvent resistance.

Applications of styrenic TPEs:

1. Rubber products in car industry
2. Cables and wires
3. Shoe soles
4. Adhesives
5. With thermoplastics in multi-component injection molding and co-extrusion.

9.2.2 Elastomeric alloys

Elastomeric alloys are blends of elastomers and thermoplastics that can be processed using thermoplastic processing methods. Elastomeric alloys are:

1. Thermoplastic olefin elastomers (TPO).
2. Thermoplastic vulcanisates (TPV).
3. Melt processible rubbers (MPR).

Thermoplastic olefin elastomers (TPO, TOE)

Thermoplastic olefin elastomers are most commonly blends of Polypropylene and EPM or Polypropylene and EPDM. Natural rubber and butyl rubber have also been used. A blend can be made in a mechanical mixing unit, e.g., in a twin-screw extruder or in polymerisation reactors.

The properties of thermoplastic olefin elastomers vary according to components, mixture ratio and conditions of alloying.

Properties of thermoplastic olefin elastomers:

1. Good chemical resistance.
2. Excellent weathering resistance.
3. Low density.
4. Good processibility.
5. Low price.

Applications of thermoplastic olefin elastomers:

1. Buffers and outside profiles in car industry.
2. Wire and cable coatings.
3. Hoses.

Thermoplastic vulcanisates (TPE-V, TPV, DVR)

Thermoplastic vulcanisates are blends of thermoplastics and elastomers that have been dynamically vulcanised during their mixing (Fig. 9.2). Those kinds of materials are for example dynamically vulcanised blends of PP and EPDM and PP and NBR. The properties of the material depend greatly on the structure and content of the elastomer. Figure 9.3 shows the effect of rubber particle size in TPE-V (AES).

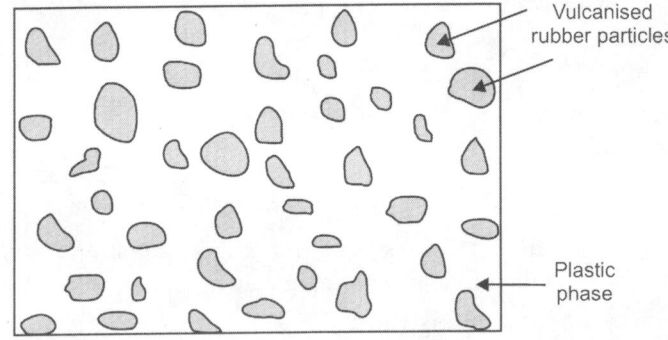

Figure 9.2: The structure of TPE-V, showing finely dispersed vulcanised rubber particles in thermoplastics matrix.

Properties of thermoplastic vulcanisates:

1. Small permanent deformation.
2. Good mechanical properties.
3. Good properties at low temperatures.

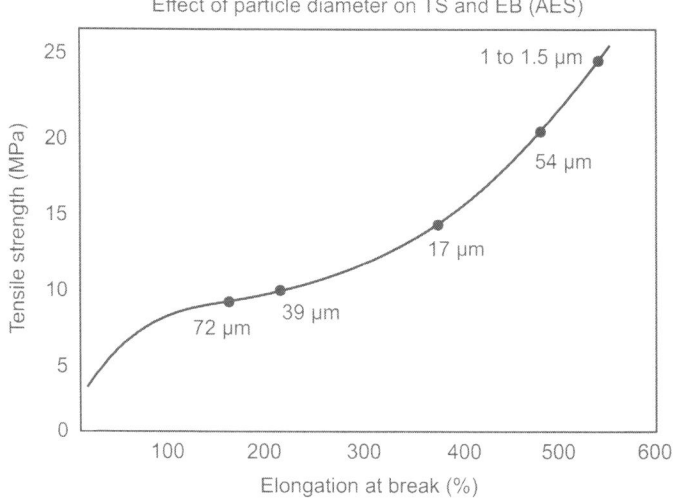

Figure 9.3: The effect of rubber particle size in TPE-V (AES).

4. Fatigue durability.
5. Good liquid and oil resistance

Applications of thermoplastic vulcanisates:
1. Car components.
2. Tubes.
3. Electrical insulators.

9.2.3   Melt-processible rubbers (MPR)

Melt-processible rubbers are very rubbery materials that look and feel like traditional rubbers. However, they can be processed like thermoplastics. Melt-processible rubbers have one phase structure, so they differ from other thermoplastic elastomers that have a two-phase structure.

Properties of melt-processible rubbers:
1. Excellent elasticity.
2. Stress-tensile behaviour corresponds to that of vulcanised rubbers.
3. Softness and flexibility.

9.2.4   Thermoplastic urethane elastomers (TPU, TPE-U)

Polyurethanes are named after the urethane group (Fig. 9.4), which is formed when isocyanate group reacts with the hydroxyl group of the alcohol. Depending on the type and amount of feeding stocks and additives, polyurethanes can be

$$\text{R—NCO} \quad + \quad \text{HO—R}' \quad \longrightarrow \quad \text{R—N—C—O—R}'$$

Forming of urethane group

Figure 9.4: Forming of urethane group.

thermoplastics, rubbers (PUR) or thermoplastic elastomers. Thermoplastic polyurethane elastomers form from long (MW around 600–3000 g/mol) soft segments of linear polyester (TPE-AU) or polyethers (TPE-EU) and short, hard urethane segments that are formed of di-isocyanate and small alcohol molecule chain extender, e.g., butane diol.

The properties of thermoplastic urethane elastomers vary strongly according to feedstocks and the ratio of hard and soft segments in the material. The soft segment component influences especially the low temperature properties of TPE-U, but also many other characteristics.

Depending on whether the soft segment is formed of polyester or polyether, the properties are entirely different.

9.2.5 Advantages and disadvantages of thermoplastic elastomers

Advantages of thermoplastic urethane elastomers

1. Good abrasion resistance.
2. Good tear strength.
3. Good strength and stiffness properties.
4. Low friction coefficient (depends on hardness).
5. Good oxygen, ozone and weather resistance.

Disadvantages of thermoplastic urethane elastomers

1. Poor hydrolysis resistance.
2. Poor resistance to chlorinated and aromatic solvents.
3. Relatively poor UV light resistance.

Applications of thermoplastic urethane elastomers

1. Conveyor belts.
2. Footwear.
3. Cable and wire coatings.
4. Hoses.
5. Components of car industry.

9.2.6 Thermoplastics polyester-ether elastomer (TPE-E)

Polyetherglycols, such as polyethylene, polypropylene or polybutylene ether glycols are soft segments in thermoplastics polyester-ether elastomers. Hard segments are dimethylterephtalate or 1,4-butanediol.

Advantages and disadvantages of TPE-E

Advantages of TPE-E:
1. Good oxygen and ozone resistance.
2. Good oil resistance.
3. Good strength properties.

Disadvantages of TPE-E:
1. Small variety in hardness.
2. Low elongation at break (requires own design principles of products).
3. Poor hydrolysis resistance.
4. Poor UV-light resistance.
5. High price.

Applications of TPE-E

1. Cable and wire coatings.
2. Gaskets.
3. Hoses, tubes.

9.2.7 Thermoplastic polyamide elastomers (TPE-A)

Soft segments of polyesters or polyethers and a rigid block of polyamide form thermoplastic polyamide elastomers. The polyamide can be for example poly-esteramide (PEA), polyetheresteramide (PEEA), polycarbonate-esteramide (PCEA) or polyether-block-amide (PE-b-A). The properties of thermoplastic polyamide elastomers depend strongly on the type of polyamide block, the type of polyol block and the length and amount of blocks. Structure of thermoplastic polyamide elastomers is shown in Fig. 9.5.

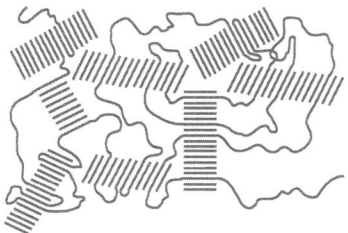

Figure 9.5: The structure of thermoplastic polyamide elastomers.

Properties of thermoplastic polyamide elastomers

1. Good heat resistance (up to 170°C).
2. Good chemical resistance.
3. Good abrasion resistance.

Applications of thermoplastic polyamide elastomers

1. Components in car motors and under the hood.
2. Wire and cable coatings.
3. Hoses.
4. Footballs, skiing boots.
5. Films penetrating water vapour.

Polymerisation of butadiene via, 1,4-addition gives the elastomer poly (1,4-butadiene). This polymer is a hydrocarbon with low intermolecular forces, no rigid or bulky groups and a relatively flexible chain, except for the double bond between carbons 2–3. The *cis* stereoisomer of the double bond is preferred over *trans* since this decreases chain regularity.

The transform is more regular and crystallinity can occur, which will prevent elastomeric response. Poly(butadiene) would need to be cross-linked to be a useful elastomer. Polystyrene is a glassy polymer with glass transition temperature =100°C so it will resist flow and creep at ambient temperatures, but it can flow and be molded at temperatures above T_g. A diblock copolymer of butadiene and styrene will provide the combination of properties required for a thermoplastic elastomer when the butadiene content is higher.

Poly(butadiene-b-styrene) (BS) has two separate phases, a continuous polybutadiene phase with dispersed poly(styrene) phase. The matrix phase gives the overall elastomeric response while the dispersed islands are the restraining physical cross-links.

9.3   Thermodynamics of elasticity

Elastomers extend and contract by conformational change from a compact random coil to extended chain. The random coil can have many possible conformations resulting in a high entropy. A fully extended chain can only have one conformation resulting in low entropy. The extended chain will spontaneously contract into a random coil since the entropy of the process is favourable. Enthalpy is not a contributor for an ideal elastomer since intermolecular forces are minimal. Entropy is over come by a mechanical force deforming the elastomer. Thermodynamic equations are applied to elastomeric deformation and recovery. By analogy with an ideal gas, elastomers that conform to the thermodynamics are called ideal elastomers. As in the case of real gases, real elastomers deviate from ideality. Deviations of elastomers are the result of:

1. At high extensions elastomer chains become fully extended between cross-links, chemical or physical and as the distribution of chains become fully extended the stress-strain response becomes non-linear.

2. At high extensions the extended chains can pack closely forming crystals that cannot be further extended. Crystallisation appears as stress whitening and it is only maintained with the deforming force since the intermolecular interactions are too weak to prevent entropy-controlled recovery.

3. Molecular entanglements prevent free molecular uncoiling and therefore cause deviation from a linear elastic response. Often entanglement caused deviations are more pronounced in a first extension-recovery cycle that differs from subsequent cycles when performed together.

4. In compression deviations from ideality are caused by the finite size of the molecules and the limited free volume available for molecules to occupy.

9.4   Structure of thermoplastic elastomers

Thermoplastic elastomers (TPEs) are defined as a group of polymers that exhibit instantaneous reversible deformation (to be an elastomer). Most of the TPEs consist of continuous phase that exhibit elastic behaviour and dispersed phase that represents the physical cross-links. If the dispersed phase is elastic then the polymer is a toughened thermoplastic, not an elastomer. Elastomer reversibility must have physical cross-links, therefore these cross-links must be reversible. Physical cross-links do not exist permanently and may disappear with the increase of temperature.

Generally, thermoplastic elastomers can be categorised into two groups: multi-block copolymers and blends. The first group is copolymers consist of soft elastomers and hard thermoplastic blocks, such as styrenic block copolymers (SBCs), polyamide/elastomer block copolymers (COPAs), polyether ester/ elastomer block copolymers (COPEs) and polyurethane/elastomer block copolymers (TPUs). TPE blends can be divided into polyolefin blends (TPOs) and dynamically vulcanised blends (TPVs).

Thermoplastic elastomers are known as two-phase system consisting of rubbery elastomeric (soft) component and rigid (hard) component. The soft phase can be polybutadiene, poly(ethylene-*co*-alkene), polyisobutylene, poly(oxyethylene), poly(ester), polysiloxane or any of the typical elastomers while the hard phase are polystyrene, poly(methyl methacrylate), urethane, ionomer–poly(ethylene-*co*-acrylic acid) (sodium, Mg, Zn salt), ethylene propylene diene monomer and fluropolymers.

9.5 Synthesis of thermoplastic elastomers

TPEs are two phase polymers, however they can be synthesised in one reaction step or in two or three steps to create each phase separately.

9.5.1 One-step methods

One-shot method is a commonly used industrial technique to prepare polyurethanes. The urethane reaction involves a diisocyanate (hard segment) and a diol (soft segment) (Fig. 9.6). Generally two diols are required, a chain extender or short chain diol and an elastomeric hydroxyl terminated polymer. Examples of a diisocyanate are methane 4,4′-diphenyl diisocyanate (MDI), 2,4- and 2,6-toluene diisocyanate (TDI) and 1,6-hexane diisocyanate (HDI). A chain extender may be 1,4-butanediol. When MDI and butanediol react they form a polyurethane with alternating monomer units connected by urethane groups, though other functional groups also form as by-products of the reaction.

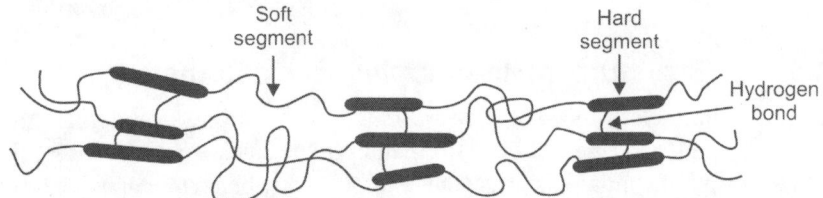

Figure 9.6: Schematic representation of TPUs composed of alternating hard segment and soft segment structures.

This polyurethane is not elastomeric and it constitutes the hard phase of a typical thermoplastic polyurethane (TPU). Hydroxyl terminated elastomers include polyethers: poly(oxyethylene), poly(oxybutylene), polyesters: poly(ethylene succinate), poly(butylene succinate), poly(ethylene adipate), poly(butylene), hydroxyl terminated polybutadiene and hydroxyl terminated poly(butadiene-*co*acrylonitrile).

These polymeric diols react with isocyanate and are linked into the TPU as a complete elastomer block. The hard and soft (elastomeric) chain segments phase separate with the hard segments as a dispersed minor phase, since the soft segments must form a continuous phase if elastomeric properties are to be displayed. The reactivity difference between the –OH groups of the polyol and the chain extender with different isocyanate groups affect the sequence of hard segments in the polymer chain. Thus, polyurethanes obtained by using this method have a more random sequence. However, the polymer is highly crystalline due to the favoured reaction between polyol and diisocyanate before extended polymer growth has occurred.

A chain growth polymerisation can be used to form a TPE in one step. An example is a poly(ethylene-*co*-butene) with a high butane content, polymerised using a single-site metallocene initiator. This polyethylene can undergo phase separation due to crystallisation, crystals are the physical cross-links and the highly branched structure will exhibit elastomeric properties. It may need to be blended with a less branched polyethylene to increase the physical cross-links. Alternatively it can be partially cross-linked by dynamically vulcanising by extrusion with a peroxide initiator. While chemical cross-links are formed this type of polyethylene can still be processed as a thermoplastic. Dynamic vulcanisation can be applied to poly(ethylene-*co*-propylene) rubber (EPR) that may be blended with a thermoplastic polyethylene to provide a binding crystalline phase.

9.5.2   Two-step methods

TPU can be synthesised by a two-step method, which is known as prepolymer method. The reaction may be carried out in two steps where excess diisocyanate is added to the polymeric diol to form an isocyanate terminated pre-polymer with excess diisocyanate monomer that is then reacted or chain extended with the monomeric diol to form the segmented TPU structure. The polymeric diol can be a biopolymer or biodegradable polymer such as castor oil. Comparison has been made between poly(butylene succinate) and poly(butylene adipate) as the soft phase in TPU. The succinate derived TPU exhibited higher soft phase glass transitions and more hard phase to soft phase interactions than the adipate derived TPU, due to higher carbonyl content and hence polar interactions in the succinates. Abrasion resistance was a function of overall hard phase volume fraction.

Sequences are found to be more regular in the polymer obtained via prepolymer method compared to the one shot method. The structural regularity leads to a better packing of hard segments where physical cross-linking points are easier to form. Hence, a two-step method gives a product of better mechanical properties than a one-step method does. Again, the solubility of these two products is different. The polyurethanes obtained from one-method are soluble in some of the common solvents, but the polyurethanes from the prepolymer process could not be dissolved in any common solvents.

Dynamic vulcanisation is a widely used method to prepare thermoplastic elastomers comprising partially or fully cross-linked elastomer particles in melt-processable thermoplastic matrix. Thermoplastic vulcanisates are prepared by melt-mixing the elastomer and thermoplastic in an internal mixer or in a twin-screw extruder. After a well mixed blend has formed, in the second step, vulcanising agents such as cross-linkers or curatives are added. Vulcanisation of the rubber polymer takes place during the continuation of the mixing process under conditions of high temperature and high shear. According to the earlier investigation the best combinations of elastomer and thermoplastic are those in which the surface energies of the two components are matched, the entanglement molecular weight of the elastomer is low and the thermoplastic is at least 15% crystalline. The most common used compositions are based on dynamically vulcanised ethylene propylene diene monomer (EPDM) and polyolefins. Others blends include butyl and halobutyl rubbers and polyolefin resins, polyacrylate rubber and polyolefins and butadiene-acrylonitrile rubber and poly(vinyl chloride).

Anionic polymerisation remains as an important technique for the preparation of welldefined styrene butadiene triblock copolymer. Poly(styrene-b-butadiene-b-styrene) (SBS) is an example of a tri-block copolymer, though di-block copolymers are also formed from the same monomers. Styrene is first initiated with butyl lithium and polymerised until all of the styrene has reacted. The polystyrene has an anionic end group with lithium counter-ion. Butadiene is added and the polymerisation continues forming a butadiene block. After all of the butadiene has reacted, more styrene is added and the polymerisation continues until all styrene has reacted. Then the polymerisation is terminated by addition of a protic substance such as methanol or water.

Termination may be carried out after the second polymerisation step to give a di-block copolymer. The size of blocks is determined by the concentration of initiator and the amounts of monomers added at each step. Molar mass distribution is characteristically low for anionic polymerisation so the macro-molecular architecture is accurately controlled.

Carbocationic polymerisation has a more complex system than the anionic polymerisation described above. It has been used to produce block copolymers with polyisobutylene midsegments, or poly(styrene-b-isobutylene-b-styrene) (SIBS). This polymerisation involves a three-step progression: (i) controlled initiation, (ii) reversible termination (quasi-living systems) and (iii) controlled transfer. The initiators have two or more functionalities.

The polymer segments are produced sequentially from monomers as in anionic polymerisation. The initiator is reacted with isobutylene at the first stage. The product is a difunctional living polymer. It can initiate further polymerisation when more styrene monomers are added. After termination, this gives the block polymer styrene isoprene butadiene (SIBS). Polyisobutylene is the only mid segment that can be produced by this method while there are many aromatic polymers that can form the end segments.

9.6   Processing methods of thermoplastic elastomers

Thermoplastic elastomers are technologically very attractive because they can be processed as thermoplastics, this is their main advantage compared with cross-linkable elastomers. They can be remelted or devitrified and shaped again. Hence, they are generally processed by extrusion and injection molding, which are the most common processing methods used by thermoplastics. A disadvantage is that TPE have an operating temperature below that at which the hard phase becomes dimensionally unstable.

Several factors need to be taken into account during the processing of TPEs, including viscosity or rheology of the two-phase polymer, temperature at which the hard phase can be processed, thermal stability since the complex structures will to potential have several weak chemical links, thermal conductivity since the hard phase is surrounded by soft phase, crystallinity in the hard phase that must be melted with excess enthalpy and moisture that may cause hydrolysis at processing temperatures. Thermoplastic elastomers can be processed by—extrusion, compression, transfer and injection molding. In this chapter a brief outline of these methods is discussed. For detail discussion refer chapter 22.

9.6.1   Extrusion of thermoplastic elastomers

Extrusion is a high volume manufacturing process for fabricating parts from thermoplastic elastomers. This processing technique is essential in the melting of raw materials and shaping them into different continuous profiles. The most common extrusion methods are film and sheet extrusion, blow film extrusion, cast film extrusion, coextrusion, tubing extrusion and extrusion coating. The end products made by extrusion are pipe/tubing, wire insulation, film, sheets, adhesive tapes and window frames. Basically, the extrusion process involves heating a thermoplastic above its melting temperature and forcing it through the die. The extruder is a heating and pressurising device involves one or more screws operating in a heated barrel. The key determinant of an extruder's performance is the screw. It has three main functions to perform: feeding and conveying the raw material feed, melting, compressing and homogenising the material and metering and pumping it through the extrusion die at a constant rate. Raw thermoplastic elastomer material is fed into the barrel of the extruder and comes into contact with the screw. As a melt delivery device, the rotating screw forces the polymer forward into barrel which is heated at a desired temperature. After leaving the screw, the molten travels through a screen pack/ plate breaker, where the contaminants in the melt are removed.

Breaker plate also creates back pressure in the barrel which is needed for uniform melting and proper mixing of polymer. After that, the molten enters the die, where the cross section of the extruded product is determined.

9.6.2   Compression molding of thermoplastic elastomers

Compression molding was among the first method of molding to be used to produce plastic parts. However, it is by far less used than injection molding.

Generally, this method involves four steps. First of all, the raw polymer materials in pellets or powder form are placed in a heated and open mold cavity. The mold is closed with another half of the mold and at the same time, pressure is applied to force the materials into contact with all mold areas. The materials soften under high pressure and temperature, flowing to fill the mold. The part is hardened under pressure by cooling the mold before removal so the part maintains its shape.

There are six important considerations that should be bear in mind, they are the proper amount of material, the minimum energy required to heat the material, the minimum time required to heat the material, the proper heating technique, the force needed to ensure that shots attains the proper shape, the design of the mold for rapid cooling.

Compression molding of TPEs usually requires longer heating and cooling time due to their high melting points. Separate platens can be used to solve this problem where one is hot press that is electrically heated and another one is cold press that is water cooled. The part is hot pressed under pressure and then transferred immediately to the cold press to chill it under pressure. The hot press is usually pre-heated to reduce the total cycle time.

9.6.3 Transfer molding of thermoplastic elastomers

Transfer molding is a process in which the polymer is melted in a separate chamber known as pot then forced into a pre-heated mold through a sprue, taking a shape of the mold cavity. The mold is cooled down before opening. Thermoplastic elastomers usually have high viscosity and longer transfer time is needed. The temperature of the mold should be maintained at above melting temperature of the polymer to avoid premature cooling or freezing before the completion of transfer. The important variables during the process of transfer molding are the type of polymer, melting point of the polymer, pot hold time, transfer pressure, transfer rate and the mold cooling time.

9.6.4 Injection molding of thermoplastic elastomers

Injection molding is by far the most used processing technique of producing parts from thermoplastic elastomers due to its high productivity. Injection molding machines and molds are very expensive because of the high pressures required and complexity of the process control. However, the shortcoming of this technique is balanced by its ability to produce a complex finished part in a single and rapid operation.

The principle of injection molding is very simple. The plastic material is fed into the injection barrel by gravity through hopper. Upon entrance into the barrel, the polymer is heated to the melting temperature. It is then forced into a closed mold that defined the shape of the article to be produced. The mold is cooled constantly to a temperature that allows the molten to solidify and the mold is opened, the finished product is ejected and the process continues.

The injection molding process is capable in producing a variety of parts, from the smallest components to entire body panels of car in a single molding operation. Other part designs obtained from injection molding include threads, springs, storage containers, mechanical parts and automotive dashboards.

9.6.5   Blow molding of thermoplastic elastomers

Blow molding (Fig. 9.7) is a manufacturing process that is used to produce hollow plastic parts. The wide variety of materials can be used in this process, including but not limited to high density polyethylene (HDPE), low density polyethylene (LDPE), polypropylene (PP), Poly(vinyl chloride) (PVC) and Poly(ethylene terephtalate) (PET). The basic process begins with the melting of thermoplastic and extruding it through a die head to form a hollow tube called a parison. The parison is then clamped between two mold halves, which close around it and the parsion is inflated by pressurised air until it conforms to the inner shape of the mold cavity. Lastly, the molds open and the finished part is removed.

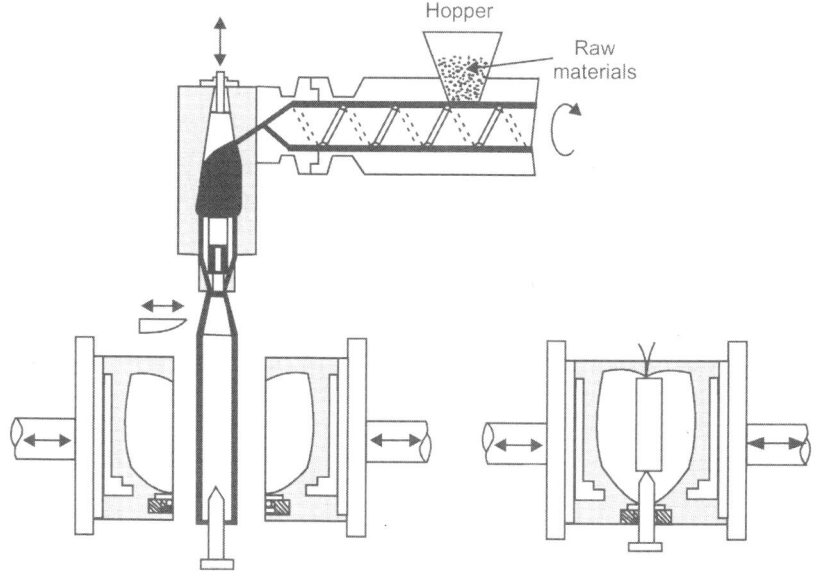

Figure 9.7: Typical phases in blow moulding.

Basically, there are three types of blow molding used to form the parison. In extrusion blow molding, plastic is melted and extruded using a rotating screw to force the molten through a die head that forms the parsion. Injection blow molding is part injection molding and part blow molding where the molten plastic is injection molded around the core pin and then the core pin is transferred to a blow molding station to be inflated. There are two stretch molding techniques. In one-stage process, the preform is injection molded which is then transferred to the blow mold where it is blown and ejected from the machine. In the two stage process, preform is injection-molded, stored for a short period of time and blown into container using a reheat blow machine.

9.6.6   Thermoforming of thermoplastic elastomers

Thermoforming is a process which uses heat and pressure or vacuum to transform thermoplastic flat sheet into a desired three-dimensional parts. The sheet is drawn from large rolls or from an extruder and then transferred to an oven for heating to its softening temperature. The heated sheet is then transferred to a pre-heated, temperature-controlled mold. Vacuum is applied to remove the trapped air and deform the sheet into the mold cavity, where it is cooled to retain the formed shape. After that, a burst of reverse air pressure is applied to break the vacuum and assist the formed part out of the mold. The principal factors in this process include the forming force, mold type, sheet prestretching, the material input form and the process phase condition. These factors have a critical effect on the quality and properties of the final products.

9.6.7   Calendering of thermoplastic elastomers

Calendering is a process where a large amount of molten plastic is fashioned into sheets by passing the polymer between a set of rollers. The rollers are hot and keep the polymer in its semi-molten state. This allows the molten to be rolled many times until the desired thickness is reached. The sheet is then rolled through cold rollers to enable it to go hard and then wound up into rolls. Calender for thermoplastics generally operates in four-roll units made up of three banks, each bank being wider than the preceding one. The advantages of calender over extruder are the possibility of calender to produce embossed films, sheets and laminates and the higher output than extruder. Examples of the final products are cling film, shrink film, clear, translucent rigid sheets for blister packaging and opaque flexible film. Calendering is also discussed in detail in Chapter 22.

9.6.8   Plasticisers and additives

Plasticisers

Plasticisers are the additives added to polymeric materials to improve their flexibility and durability by spacing them apart. A plasticiser-polymer mix contains more free volume than a pure polymer, thus, the plasticised polymer need to be cooled to a lower temperature to reduce its free volume which defines glass transition temperature of the polymer. There are many types of plasticisers which can be used to modify the thermoplastic elastomers and enhancing their utilities. Each of the plasticisers has a compatibility with a specific type of polymer. Among them, ester plasticisers have a well known function in the TPEs due to its exceptional ability to provide improved low-temperature by plasticising the soft phase while allowing the hard phase to stay intact for strength and high temperature properties.

Some examples of ester plasticisers are phthalate esters which are used in situations where good resistance to water and oil is required, adipic esters which are used for low temperature or resistance to UV light and trimellitic esters which are used in automobile interiors and where resistance to high temperature is required. TPU tend to be internally plasticised by choice of monomer combination. For polyolefin elastomers, it is important to identify plasticisers which have low glass transition temperature and high boiling temperature and which are compatible over a broad temperature range with both rubber and polyolefin plastic components.

Additives

Common additives in TPEs include those materials added during or after polymerisation to prevent their degradation, during monomer recovery, drying and compounding and also storage.

Antioxidants: Among the additives, antioxidants are used to prevent oxidation and degradation. Primary or free radical scavenging antioxidants, which have reactive hydroxyl and amine groups, inhibit oxidation via chain terminating reactions. Secondary antioxidants inhibit oxidation of polymers by decomposing hydroperoxides.

Nucleating agents: Nucleating agents are generally used to enhance the formation of nuclei for the growth of crystal in the polymer melt. A higher degree of crystallinity and a more uniform crystalline structure in the hard phase can be obtained by adding a nucleating agent in the polymer. Nucleating agents can be classified as inorganic additives (talc, silica), organic compounds (salts of mono- or polycarboxylic acids) and polymer. Nucleating agents may be used to enhance crystallinity of a hard phase segment.

Colourants: Colourants are often referred to as dyes and pigments. Generally, dyes are soluble in water while pigments are not. The colours from dyes are produced from the light absorption and they are transparent. Pigments produce colours from the dispersion of fine particles throughout the resin. Inorganic pigments are thermally stable than organic pigments. They are less transparent and resistant to migration, chemicals and fading.

Some examples of inorganic pigments are oxides, sulphides, hydroxides and other complexes based on metal.

Flame retardants: Most thermoplastics are flammable, burning easily when heated to high temperature. Flame retardants are added to polymer to delay the ignition and burning of polymer. Char-formers form a foamy porous protective barrier on the polymeric material to shield it for further combustion. Flame retardants acting in the condensed phase deposit a layer on the surface of polymer to prevent it from the heat source while flame retardants acting in gas phase interrupt the combustion chemistry of the fire.

9.6.9   Composites

Various fillers and reinforcements have been introduced into thermoplastic elastomers to enhance their processability and mechanical properties, as well as to reduce material costs.

Most common fillers used in TPEs include cubic and spheroidal fillers (calcium carbonate, silica, carbon black), fibrous fillers (glass fibres, aramid fibres), platy fillers (kaolin, mica, talc) and nanofillers (carbon nanotubes, nanoclays, nanosilica). Reinforcing TPEs with fillers such as silica, clay, carbon black, carbon nanotubes, natural fibre results in better thermal and mechanical properties of the composites.

Carbon black composites with polyether polyurethane exhibited a percolation threshold of 1.25% v/v and significant conductivity at 2% v/v carbon black content. Electric field induced strain was observed due to an increase in dielectric constant.

Polyester thermoplastic elastomers reinforced by mica showed significant increment in the flexural, thermal and electrical properties with an increase in the filler concentration. The improved thermal properties are attributed to the small and uniform crystallite size distribution with the addition of mica. Composites containing silica and poly(styrene-b-ethylene-co-butylene-b-styrene) (SEBS) block copolymer-based thermoplastic elastomer showed an improvement in the mechanical properties such as tear strength due to the strong interaction between the fillers and polymer matrices where the silica particles are wetted by the polymer.

Polypropylene/natural rubber (PP/NR) and poly(propylene-ethylene-propylene-diene-monomer) (PP/EPDM) reinforced by kenaf natural fibre with maleic anhydride polypropylene (MAPP) as a compatibiliser agent has significantly increased the tensile strength, flexural properties and impact strength as compared to unreinforced thermoplastic elastomer.

The improvement achieved in mechanical properties was due to the interaction both matrix system and kenaf fibre. For plasticisers, additive, fillers and pigments also refer Chapter 22.

9.6.10   Morphology of thermoplastic elastomers

The disperse hard-phase in TPE self-agglomerates after processing to form reinforcing domains within the elastomeric matrix. The morphology contribution to elastomeric properties has been investigated using a semi- phenomenological approach. Behaviour under stress for long times resulted from plastic flow, chain pull-out from hard domains and finally disruption of the hard domains. Transmission electron microscopy (TEM) is one of the most powerful equipment to characterise the structure and morphology of thermoplastic elastomers. It is often used to interpenetrate polymer networks (IPNs), morphology and crystallinity of hard blocks, structural evolution of segmented copolymers under strain and blends

9.6.11   Elastomer polymer blends

Elastomers are often used in blends with other polymers. When the elastomer is the minor component it will constitute a disperse phase. A disperse phase elastomer will be a toughening additive for the matrix phase that could be a thermoplastic or a thermoset polymer. When the elastomer is the major component it will be a matrix phase and the overall blend will be an elastomer. The disperse phase blended polymer will contribute to physical cross-links that will prevent creep and assist with reversibility of elastomer deformation.

9.6.12   Mechanical properties of thermoplastic elastomers

Stress-strain

When tested under a tensile stress-strain condition, TPE behave as elastomers until the yield stress after which they can undergo plastic flow, part of which may be viscoelastic (time-dependant recovery) and part will be permanent set. The pre-yield region represents an elastomeric response where the physical cross-links are not deformed. The stress-strain curve of SBS is shown in Fig. 9.8.

Dynamic mechanical analysis

Dynamic mechanical analysis (DMA), uniaxial tensile and microscopic properties were used to observe properties and transitions at lower temperatures to the glass transition of the elastomer phase. Two distinct transitions were detected at the lower temperatures evaluated, an elastomeric to ductile thermo-plastic transition and a thermoplastic to brittle transition.

Creep and recovery

Creep is a problem for TPE because there are only physical cross-links that can dissociate and flow, chemical cross-links are permanent and creep will be low. TPEs soften and melt with increasing temperature, showing creep on

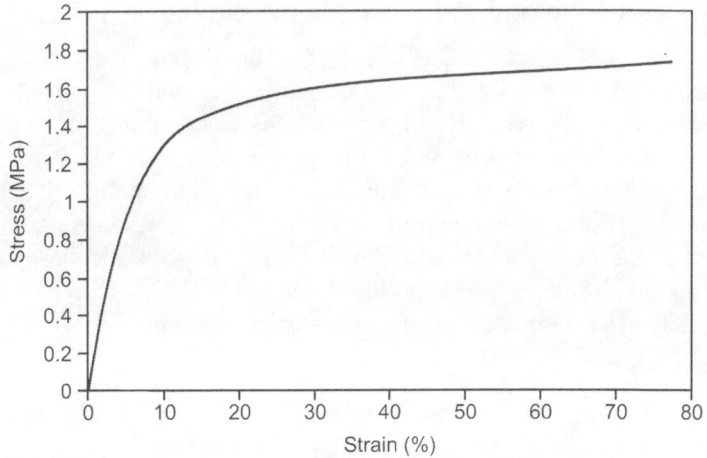

Figure 9.8: The stress-strain curve of SBS.

extended use. Creep resistance and tensile strength are generally directly related. A softer TPE will creep more and have less tensile strength than a harder TPE. Recovery should be elastic with some viscoelasticity, permanent set is not suitable in TPE. The deformation behaviour of TPEs during creep flow can be divided into three strain regimes: linear regime at low strains where the recovery from creep is complete, the transient regime, where the viscosity shows a maximum, flow regime where steady state morphology is formed. Creep and recovery curve of SBS is shown in Fig. 9.9.

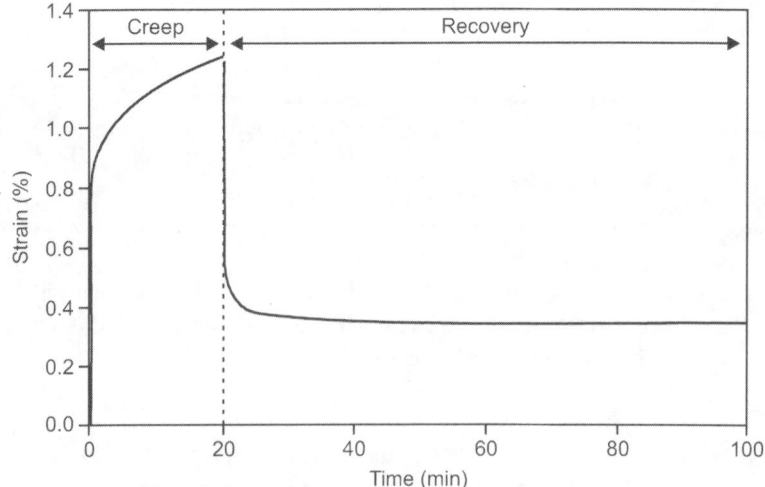

Figure 9.9: The creep and recovery curve of SBS.

Stress relaxation

Stress relaxation is a technique can be used as a physical method to determine the domain structure of the TPEs by studying the time-dependent deformation behaviour of the segmented block copolymer. No stress relaxation occurs in ideal elastomer.

Strain hardening

At high strain, strain hardening takes place and converts TPE from elastic behaviour to leathery or stiffer characteristic so that elasticity is lost while tensile strength and modulus will be increased. In time, the recovery of the deformation occurs suggesting that the part where deformation occurs is viscoelastic.

Tear strength

Tear strength which describes how well the elastomer resists tearing. TPE is stretched and the amount of force required is recorded. Peel strength is a measure of how well a TPE has bonded to a rigid substrate.

9.6.13 Applications of thermoplastic elastomers

Thermoplastic elastomers have been widely used in automotive sector, medical devices, mobile electronics, household appliance sector and construction to replace conventional vulcanised rubber.

1. Automotive: Windshield seal (SEBS), wire/cable (SEBS, TPU), fibre reinforced soft touch surface for interior (TPO), gaskets (TPV) and spoiler (SEBS).
2. Medical devices: Syringe (TPV), medical tubing (TPO), medical wrapping and packaging (TPU).
3. Mobile electronics: Wire/cable (SEBS), earplugs (TPV), cell phone (TPV).
4. Household appliance sector: Sporting goods (TPU), footwear soles (SBS), toys (SEBS), adhesives (SIS).
5. Construction: Gaskets (SEBS).

The only constraint of TPEs is the physical reversible cross-links need to be disrupted by heat to mold, but they maybe disrupted during use.

9.7 Future directions of thermoplastic elastomers

Thermoplastic elastomer structure, performance and specialty applications are interconnected with block copolymer molecular characteristics. Block continuity, molar mass, molar mass distribution and stereochemistry must be controlled. New polymerisation techniques and initiators are expanding the choice of monomers that can be polymerised under controlled conditions. This allows monomer selection and molecular architecture to provide elastomers with chemical resistance, self-healing, abrasion resistance and unique mechanical performance to be prepared for special applications.

Further thermoplastic elastomers can be prepared by creating blends of an elastic polymer with a dimensional stabilising polymer. Inclusion of ionomers as a physically cross-linking phase can be extended with carboxylates, sulphonates and phosphates with various metal ions. In conjunction with polyfluorocarbon elastomers chemical resistance can potentially be improved.

Nanocomposites are being formed with elastomers where the nanoparticles form self assembled clusters or bridges to provide physical cross-links. Nano-particles have traditionally been used to modify elastomeric properties. Such modification will be expanded through increasing knowledge of nano-composite preparation and morphologies.

Carbon black is a much used nano-fillers that is now being assisted or replaced by carbon nanotubes, graphenes and silicas with a diversity of surface modifications. Carbon blacks are known to form reversible clusters and to binder elastomer molecules within the clusters. Nano-silicas remain in multi-particles aggregates while forming reversible agglomerates that enhance absorption of elastic energy. Now carbon nanotubes and graphenes have been found to create reversible networks at low volume fraction.

To sum up, the characteristics of an elastomer require that there is a mechanism to provide reversible deformation. The viscous contribution resulting from molecules sliding past each other, this results in irreversible flow, is eliminated by cross-linking in a thermoset elastomer. Physical cross-links are present in a thermoplastic elastomer as a second vitrified or crystalline phase.

The viscoelastic component can be reduced by minimising chain stiffness and intermolecular interactions in the continuous elastic phase. Thermoplastic elastomers offer ease of processing the same as thermoplastics without the need for a separate curing reaction. Waste material can be reprocessed and production rates will be fast consistent with a thermoplastic. Upper application temperature limitations exist depending upon the glass transition or melting temperature of the hard phase. Stress resistance is limited to the yield stress of the hard phase since permanent deformation will follow distortion or flow of the hard phase. Thermoplastic elastomers are enhanced by fillers, with nano-fillers having particular relevance when small amounts can support the hard phase. In the soft phase fillers will modify the elastic response. Structural diversity is found in thermoplastic elastomers with many chemical structures such as polyurethanes and polyolefins available as both thermoset and thermoplastic elastomers.

New development trends occuring in the field of TPEs

1. Material innovations.
2. New polymerisation techniques, metallocene techniques.
3. Foamed materials, e.g., supercritical gases.
4. Electrical properties, conductivities.
5. Paintability.
6. Blends including nanofillers.
7. Processing.
8. Coextrusion, coinjection, over-molding.
9. Adhesion and joining.
10. Milling, thermoforming, extrusion, injection and blow molding (all processing alternatives).
11. Recycling.
12. Product innovations/development, hybrid products.
13. Product design to maximise the benefits of TPEs.
14. Smart products, functionality.
15. Design.
16. Food and health applications, bioapplications.

Chloroprene rubber

10.1 Introduction

Chloroprene (polychloroprene) or neoprene is one of the special purpose synthetic rubber. The first chloroprene monomers were prepared from acetylene. Now-a-days they are synthesised from butadiene, because it is an easier and safer route. Chloroprene is polymerised by emulsion polymerisation using potassium persulphate as free radical initiator.

The main component of the polymer usually is *trans*-1,4-units. In the vulcanising of chloroprene rubber (CR), zinc oxide and magnesium oxide blend is usually used. Isomeric structure of CR is given below:

$$\sim\sim CH_2 - \underset{\underset{Cl}{|}}{C} = CH - CH_2 \sim\sim \qquad\qquad \sim\sim CH_2 - \underset{\underset{\underset{\underset{CH_2}{\|}}{CH}}{|}}{\underset{|}{\overset{Cl}{|}}{C}} \sim\sim$$

1,4 addition   1,2 addition

Chloroprene rubbers can be divided into G and W-types according to their mechanism for controlling the molecular weight of the polymer during polymerisation. In G-types, sulphur is copolymerised with the chloroprene, when it does not require acceleration during curing. The G-type rubbers have slightly inferior ageing resistance, but resilience and tack are better than in the W-types. The W-types of chloroprene rubbers require an accelerator. The vulcanisation cannot be carried out with sulphur. Suitable accelerators are metal oxides. The W-type rubbers have better ageing properties and thermal resistance than G-type rubbers.

Chloroprene rubbers are manufactured by polymerising 2-chloro-1,3-butadiene in the presence of catalysts, emulsifying agents, modifiers and protective agents. The polymer chain is built up through addition of monomer units, of which approximately 98% add in the 1,4 positions. About 1.5% and in the 1,2 positions and these are utilised in the vulcanisation process since in this arrangement the chlorine atom is both tertiary and allylic. Accordingly, it is strongly activated and thus becomes a curing site on the polymer chain. Neoprene or polychloroprene is a synthetic elastomer available in solid form, as a latex, or as a flexible form. It is vulcanised with metallic oxides rather than with sulphur. With a specific gravity of 1.23, it is resistant to oils, oxygen, ozone, corona discharge and electric current. It is combustible but less so than natural rubber and an isocyanate modified form has high flame resistance.

Emulsion polymerisation is used to makes a number of variations of the product. Varieties are marketed both as solid polymers and as latices, but not all of them are readily available in all trading areas. The solid types vary over a wide viscosity range. In some instances, either non-staining or staining antioxidants are included at the time of manufacture.

10.2   Compounding dry neoprenes

The key to the all-around performance of neoprene products in a variety of services lies in the words 'properly compounded.' Once the service requirements have been documented, the compounder is free to take over, but with his freedom of choice subject to the limitations of process requirements and economics, of course.

Compound development logically begins with the selection of the type of neoprene that best provides the required physical properties within the applicable cost and processing limits. The response of all general-purpose types to compounding ingredients is similar. Compounding of the G-types, however, differs from that of the W and T-types in three ways:

1. They can be softened by the addition of chemical peptisers.
2. They soften under mechanical shear and may, therefore, need less plasticiser to prepare compounds in a workable viscosity range.
3. They can be used with metal oxides alone, organic accelerators are not required.

By peptising, the compounder can achieve workable compound viscosity at high filler loading without the reduction in vulcanisate hardness and strength that usually occurs when additional plasticisers are used to lower compound viscosity.

Minimum requirements for a practical compound include antioxidant, metallic oxide, filler or reinforcing agent, processing aid and vulcanising system. Optional ingredients may include antiozonants, retarders, extenders, plastics or resins, other elastomers and blowing agents.

10.2.1   Curing systems for neoprene

Curatives for neoprene are limited by the ingenuity of the compounders, the processing and end user requirements and economics.

Metal oxides are an essential part of the curing system. They regulate scorch, cure rate and cure state, in addition, they serve as acid acceptors for trace amounts of hydrogen chloride that are released from the polymer during processing, curing and vulcanisate ageing. Combinations of magnesium oxide and zinc oxide produce the best balance of processability, cure rate and vulcanisate performance. With the G-types, the relationship between these properties is markedly affected by the amounts of MgO and ZnO used. Similar effects are noted in the W and T-types, but to a much lesser degree because of the pronounced influence of the accelerator.

For improved resistance to swell and deterioration by water, an oxide of lead may be used in place of the magnesia/zinc oxide combination. Red lead (Pb_3O_4) is the preferred oxide. Using lead oxide results in a sacrifice in colourability, tensile strength, compression set resistance and heat resistance compared to the characteristics offered by MgO/ZnO combination. It is more expensive usually.

The oxides of other metals are either impractical to use or exert no appreciable influence on the curing characteristics. Calcium oxide is sometimes used as a desiccant to facilitate vacuum-extrusion and with Permalux acceleration, to control scorch. Iron oxide (Fe_2O_3) and titanium dioxide (TiO_2) are often used for their tinctorial properties in coloured stocks.

The grade of magnesia used in neoprene compounds is important to the processing and properties of the finished product. The grade of magnesia used has two characteristics.

1. It is precipitated (not ground) and calcined after precipitation.

2. It is very active, having a high ratio of surface to volume.

Types of zinc oxide that have fine particle size and high surface area are preferred. This ingredient should be added late in the mix cycle to avoid scorch, with appropriate care exercised to insure good dispersion. Incomplete dispersion results in localised overcure and inferior physical properties. Because of the importance of good dispersion, of the protection of magnesia from the effect of moisture contamination and of the economics of reduced mixing cycles, predispersed forms of the various metallic oxides are available from many suppliers.

For practical cure rate and vulcanisate properties, an organic accelerator is required in addition to metal oxides for the W and T-types of neoprene. Amines, phenols, sulphenamides, thiazoles, thiurams, thioureas, guanidines and sulphur are the common accelerators and/or curing agents. The curing mechanism for organic curing agents other than thioureas has been reviewed.

Sulphur vulcanisation is probably similar to that of natural rubber or of copolymers containing butadiene. Thiourea accelerators, particularly ethylene thiourea, have for many years been the accelerators of choice for a broad range of applications.

10.2.2 Antioxidants for neoprene

Antioxidants are essential in all neoprene compounds for good ageing vulcanisates. Those commonly used in other elastomers usually behave similarly in neoprene.

10.2.3 Processing aids for neoprene

Processing aids include lubricants, tackifiers and agents for controlling viscosity and nerve. Stearic acid, microcrystalline waxes and low-molecular-weight polyethylenes make good lubricants. Hydrogenated rosin esters and coumarone-indene resins are preferred as tackifiers. Viscosity modification may be achieved in either direction. Dithiocarbamates and guanidines peptise soften the G-types. Increased viscosity is best obtained by blending the neoprene with a high-molecular-weight neoprene WHV. Decreased viscosity is achieved by using plasticisers or blending with the low-molecular-weight neoprene FB. Nerve may be reduced for smooth extrusion and calendering by blending with neoprene WB, employing high structure carbon black or hard clay as fillers.

10.2.4 Fillers for neoprene

The effects of carbon blacks and mineral fillers on the processing and vulcanisate properties of neoprene are generally similar to their effects on other elastomers, with the possible exception that reinforcing fillers are of less importance. Like natural rubber and other polymers with pronounced crystallisation tendencies, neoprene gum vulcanisates have high tensile strength. The need for reinforcement, therefore, is less for such elastomers as SBR, butyl rubber and nitrile rubber. Practical vulcanisates are obtained for the majority of uses by filling ('loading') with the soft, relatively non-reinforcing thermal type of carbon blacks (MT and FT) or with clays and fine-particle calcium carbonate.

These fillers are not only relatively inexpensive but afford extra economy because they may be added in greater amounts than more reinforcing fillers for a given hardness. If the superior physical properties ordinarily associated with higher tensile strength are required, SRF blacks are usually used. Carbon blacks of finer particle size—either furnace or channel black—are used as a rule only when the ultimate in certain properties such as abrasion and tear resistance is required.

Of the fillers other than carbon black, clays are usually preferred for good processing and overall vulcanisate quality. Whitings, especially coarse ones, provide poor vulcanisates that weather badly. Whitings of finer particle size, however, are used to provide vulcanisates with superior resistance to heat ageing. Hydrated calcium silicate and precipitated silicon-dioxide fillers of fine particle size give neoprene vulcanisates remarkably high tensile strength, hot tear resistance and abrasion resistance, but impaired dynamic properties.

10.2.5　Plasticisers for neoprene

The plasticisers and softeners most often used in neoprene are low-cost petroleum derivatives. Amount ranging from 10 to 20% by weight of the filler loading are usually required for processing reasons alone. The W-types require more than the G-types. When the total amount of petroleum plasticiser fails to exceed 20–25 parts per 100 of neoprene, naphthenic oils may be used. These have the advantage over aromatic oils of not darkening light-coloured vulcanisates or staining contacting surfaces. When plasticising oils are used in greater amounts in order to produce very soft vulcanisates or to accommodate the high amounts of filler used in low-cost compounds, aromatic oils are recommended to insure compatibility.

Petroleum plasticisers seldom improve the flexibility of a vulcanisate at low temperature. Dioctyl sebacate is excellent for this purpose, although many other organic chemicals, mostly high-molecular-weight esters, are widely used.

10.3　Processing dry neoprenes

The methods for processing dry neoprene are discussed below:

10.3.1　Mixing of dry neoprenes

With few exceptions, neoprene compounds should be mixed at as low a temperature and with as short a cycle as possible. This handling minimises the danger of scorch and is especially important with the G-types, which are likely to give soft and sticky stocks on over-mastication. Omission of both zinc oxide and accelerators until late in the mixing cycle is also necessary to avoid scorching. Magnesia should be added early, if possible, with the neoprene.

10.3.2　Calendering and extrusion of dry neoprenes

Calendering of neoprene stocks demands more critical control than other processing operations, particularly of the G-types, which are very temperature sensitive and perform best below 70°C. The W-types are usually calendered with slightly hotter rolls because they release more easily and require a higher temperature to avoid shrinkage. The three states of neoprene (elastic, granular and plastic) are most apparent in calendering, especially with stocks of low loading. In the elastic phase, the sheet will shrink and become rough on release but may be fairly smooth on the roll. The granular phase is self-descriptive and the sheet is relatively free from nerve. In the plastic phase, it is glossy smooth, weak and nerve-free. The temperature at which phase changes occur varies with the type of neoprene, the phase change from elastic to granular ranging from 60°C in the case of type WRT to 70°C for type GN. The plastic phase exists at about 93°C for all neoprenes.

For frictioning, the plasticisable, crystallisation-resistant neoprene GRT is recommended. Best results are obtained using prewarmed fabric, a top roll temperature of 93°C or above, a centre roll at 65–82°C and a bottom roll at room temperature to 65°C, depending on the weave.

Most neoprene compounds extrude best with a cool barrel and screw, a warm head and a hot die. These are the major variations in the processing of neoprene that differ from the normal routines followed with other elastomers.

10.4 Applications of neoprenes

Some of the most important uses are in adhesives, the transportation industry, the energy industry, the construction industry, wire and cable, hose, belting and consumer products.

10.4.1 Adhesives

Neoprene-based adhesives are available both in fluid and dry-film form. The fluid types are classed as:

1. Solvent adhesives.
2. Solid adhesives.
3. Dry-film adhesives.

Solvent adhesives

Solvent adhesives consist of the neoprene polymer and suitable compounding ingredients dissolved in an organic solvent or combination of solvents (toluene, ethyl acetate, naphtha, methyl ethyl ketone). Latex adhesives are composed of particles of neoprene and compounding ingredients dispersed in water. Both solvent and latex types are normally supplied in ready-to-use 'one-part' adhesive systems. When specified, 'two-part' adhesive systems can also be produced to meet specific end-use requirements. Solvent and latex neoprene adhesives achieve their bonds through evaporation of the fluid and subsequent crystallisation and curing of the elastomeric residue.

Solid adhesives

Solid adhesives, on the other hand, are based on fluid neoprene polymers and contain neither solvent nor water. Adhesives of this type are normally the choice in speciality applications where fluidity is required, yet where volatile loss or shrinkage cannot be tolerated. They possess little 'green' strength and usually require application of either heat or catalytic agent to develop an adhesive bond.

Dry-film adhesives

Dry-film adhesives based on neoprene are 100% solids materials normally supplied in the form of a tape or sheeting. This type of adhesive is compounded so that it softens when heated. Cooling then resolidifies the material and forms the adhesive bond.

Neoprene adhesives are made in a range of consistencies, i.e., from very thin liquids through heavy-bodied 'putties' to dry films. These adhesives can be controlled to offer a wide range of properties, for example, tack, 'open tack time,' storage stability, bond-development, flexibility and strength, heat and cold resistance and all of the basic characteristics of polychloroprene polymers. The most significant ingredient in a neoprene adhesive, apart from the polymer, is the resin. Various types of resin are used in order to enhance specific adhesion, improve cohesive strength and hot bond strength, or to impart better tack retention. Resins are usually compounded at a level of 25–60 phr and are rarely added above 100 phr.

The most commonly used resin types are alkyl phenolic resins, terpene phenolic resins, hydrogenated resin and rosin esters, coumarone-indene resins and hydrocarbon resins such as poly (x-methyl styrene).

10.4.2   Transportation of neoprenes

In the automotive field, neoprene is the base elastomer for a variety of components such as V-belts, timing belts, blown sponge gaskets for door, deck and trunk, spark plug boots, power brake bellows, radiator hoses, steering and suspension joint seals, tyre sidewalls, ignition wire jackets and many other items. In aviation, neoprene is used in mountings, wire and cable jackets, gaskets, seals, deicers, etc.

In railroading, it is used in track mountings, car body mountings, air brake hoses, flexible car connectors, freight-car interior linings, journal box lubricators and so forth. All of these uses require physical strength, compression set, resilience, fluid, weather and temperature degradation resistance.

10.4.3   Energy industry

A durable oil-resistant neoprene rubber has a multitude of uses in the exploration, production and distribution of petroleum, such as packers, pipeline, seals, gaskets, hose, coated fabrics, wire and cable and so forth.

10.4.4   Construction industry

This application area, while not new, is now gaining in popularity. Outstanding items are extruded window-wall sealing gaskets, calendered sheeting for membrane waterproofing, flashing and roof covering, highway joint seals, bridge mounting pads, soil pipe gaskets, sound and vibration isolation pads, etc. Neoprene-modified asphalt is gaining acceptance in roads, parking lots and airport runways. Proven serviceability and economic acceptability account for this growth.

10.4.5 Wire and cable

Practically the oldest use of neoprene is as a jacket for electrical conductors (low and high voltages). Its superior resistance to abrasion, oil, flame, weather and ageing over natural rubber accounts for its broad acceptance.

10.4.6 Hose

Neoprene hose has been made almost continuously for a long time and is used in the cover, cushion and tube compounds. All types of hose are involved.

10.4.7 Belts

Maximum durability in service climates requiring a range of temperature and flex resistance makes neoprene the polymer of choice for belting. Applications include V-belts, power-transmission belts, conveyor belts and escalator hand rails.

10.4.8 Consumer products

Sponge shoe soles made of closed-cell neoprene combine cushioned comfort with long wear. Neoprene latex foam has played a critical role in mattress and cushioning fire safety, offering an added margin of improved flame-protection and low smoke for institutional furnishings.

10.4.9 Advantages and disadvantage of neoprenes

Advantages of neoprenes
1. Good abrasion resistance.
2. Good ozone resistance.
3. Good tear strength.
4. Good oil and solvent resistance.
5. Inflammability.
6. Good adhesion to metals.
7. Increased hardness in high-temperature environments.

Disadvantages of neoprenes
High swelling in some oils, hot water, acids and some organic solvents.

10.5 Neoprene latex

Neoprene latices are aqueous, colloidal dispersions of polychloroprene or of copolymers of chloroprene and other monomers such as methacrylic acid or 2,3 dichloro-1,3-butadiene. They are available in both anionic and non-ionic surfactant systems.

A variety of articles, particularly those that are relatively thin or of complicated shape, are more easily made from latex than from dry neoprene. An example would be neoprene household and industrial gloves, which are readily made from latex. Also, the production of neoprene balloons from dry neoprene would be difficult. Other major applications include latex-based adhesives, protective coatings, binder for cellulose and other fibres and elasticising additive for concrete, mortar and asphalt.

As with dry neoprene, the neoprene latex must be compounded with other materials to be converted into useful products. Some of these materials are mandatory for optimum performance, whereas others are optional. In some cases, a more expensive type of neoprene latex may actually provide greater value-in-use to the user because it can accommodate more additives or because its inherent strength eliminates the need for curing.

End-use performance, processing and economics usually dictate the choice of neoprene latex type. Included in processing are the cost of storage and handling, coagulation and drying rates, curing time and production yield. Usually cost can be minimised by maintaining low compound viscosity at the highest practical solids content.

10.5.1 Compounding neoprene latices

All neoprene latex compounds must contain a metal oxide and an effective antioxidant. Metal oxides function in several ways. They serve to vulcanise the polymer, they make it more resistant to ageing, heat and weathering and they act as acid acceptors. Zinc oxide has proven to be the most desirable of these oxides from both processing and film-property standpoints. The presence of a good antioxidant in neoprene is as important as a metal oxide. Both are necessary to bring out the inherently good ageing properties of the neoprene and, as a general rule, it is recommended that neoprene latex compounds contain at least 5 phr of zinc oxide and 2 phr of an antioxidant.

Accelerators are used to increase the rate of cure and to enhance the physical properties of neoprene. The addition of thiocarbanilide to neoprene latex produces polymers with high modulus. Where tensile strength is of primary importance, the use of both tetraethylthiuram disulphide and a water solution of sodium dibutyldithiocarbamate is recommended. This combination imparts outstandingly high tensile strength with little increase in modulus. Other accelerators commonly used in neoprene latices are zinc dibutyldithiocarbamate and di-*o*-tolyguanidine.

Mineral oils and light process oils are used as plasticisers and to improve the 'hand' in neoprene latex. Petroleum-based plasticisers have been found most effective as crystallisation inhibitors and ester-type plasticisers are employed to improve the low-temperature serviceability of the product.

Fillers such as clay, whiting, titanium dioxide, carbon black, hydrated alumina and fine silicas can be used to impart specific properties to the neoprene or to act as low-cost diluents. Loadings vary from as little as 10 phr in dipped goods to several hundred parts in adhesives and coatings. Anionic, non-ionic and amphoteric surfactants are used as both shear and chemical stabilisers in the preparation of neoprene latex compounds. Usually 0.1–1 phr is sufficient in most applications to produce the necessary processing stability.

Thickeners are more often needed with neoprene latex compounds than with natural-rubber latex because the viscosity of neoprene latex at a given solids content is less than that of natural rubber latex. Both natural and synthetic gums are excellent thickeners for neoprene latex. If it is necessary to decrease the viscosity of neoprene latex compositions, they can be diluted with de-ionised or distilled water.

Processing neoprene latices

The manufacture of articles from both neoprene and natural rubber latex is similar and the processes used with natural rubber are adaptable in most instances for use with neoprene. Exceptions exist, however, the most important being that of cure. Neoprene has the advantage of not precuring or after curing, but in order to obtain the optimum physical properties, neoprene films must be cured at temperatures above 100°C and preferably in the range of 120–140°C.

The general techniques suggested for good rubber latex compounding procedures also hold for neoprene latex. Fine-particle-size dispersions and emulsions, compounding pH above 10.5 and the use of de-ionised or distilled water in preparing all ingredients are all recommended for processing uniformity.

10.5.2 Properties of chloroprene (neoprenes) rubber

Some of physical and mechanical properties, chemical resistance, thermal properties and environmental performance of chloroprene rubber are given below:

Physical and mechanical properties

Physical and mechanical properties	Typical values
Durometer or hardness range	20–95 Shore A
Tensile strength range	500–3,000 PSI
Elongation (range %)	100–800 %
Abrasion resistance	Very good to excellent
Adhesion to metal	Excellent
Adhesion to rigid materials	Good to excellent
Compression set	Poor to good

Flex cracking resistance	Good
Impact resistance	Good to excellent
Resilience/rebound	Fair to good
Tear resistance	Good to excellent
Vibration dampening	Good to excellent

Chemical resistance

Chemical resistance	Typical values
Acids, dilute	Excellent
Acids, concentrated	Poor
Acids, organic (dilute)	Good to excellent
Acids, organic (concentrated)	Poor to good
Acids, inorganic	Good to excellent
Alcohol's	Excellent
Aldehydes	Poor to fair
Alkalies, dilute	Good
Alkalies, concentrated	Poor
Amines	Poor to good
Animal and vegetable oils	Good
Brake fluids, non-petroleum based	Fair
Diester oils	Poor
Esters, alkyl phosphate	Poor
Esters, aryl phosphate	Poor to fair
Ethers	Poor
Fuel, aliphatic hydrocarbon	Poor to good
Fuel, aromatic hydrocarbon	Poor to fair
Fuel, extended (oxygenated)	Fair
Halogenated solvents	Poor
Hydrocarbon, halogenated	Poor
Ketones	Poor to fair
Lacquer solvents	Poor
LP gases and fuel oils	Good
Mineral oils	Fair to good
Oil resistance	Fair
Petroleum aromatic	Good
Petroleum non-aromatic	Good
Refrigerant ammonia	Excellent
Refrigerant halofluorocarbons	R-11, R-12, R-13, R-21, R-22

Refrigerant halofluorocarbons W/oil   R-11, R-12, R-22

Silicone oil   Fair to excellent

Solvent resistance   Fair

Thermal properties

Thermal properties	Typical values
Low temperature range	$-70°F$ to $-30°F$
Minimum for continuous use (static)	$-80°F$
Brittle point	$-85°F$
High temperature range	$+200°F$ to $+250°F$
Maximum for continuous use (static)	$+250°F$

Environmental performance

Environmental performance	Typical values
Colourability	Fair
Flame resistance	Fair to good
Gas permeability	Fair to good
Odour	Fair to good
Ozone resistance	Good to excellent
Oxidation resistance	Good to excellent
Radiation resistance	Fair to good
Steam resistance	Fair to good
Sunlight resistance	Good to excellent
Taste retention	Fair to good
Weather resistance	Fair to good
Water resistance	Fair to good

Product safety

For all the solid polymers, however, routine industrial hygiene practices are recommended during handling and processing to avoid such conditions as dust build-up or static charges.

Neoprene latices similarly have a very low order of oral toxicity. Since most are strongly alkaline, however, they may cause burns if they come in contact with eyes or skin. Volatile organic materials in neoprene latices include chloroprene monomers, toluene and butadiene. Compounding ingredients used with neoprene to prepare finished products may present hazards in handling and use. Before proceeding with any compounding work, consult and follow label directions and handling precautions from suppliers of all ingredients.

Chlorosulphonated polyethylene rubber

11.1 Introduction

HYPALON is a trademark for chlorosulphonated polyethylene (CSPE) synthetic rubber (CSM) noted for its resistance to chemicals, temperature extremes and ultraviolet light. It was a product of DuPont performance elastomers, a subsidiary of DuPont. Along with PVC, CSM is one of the most common materials used to make inflatable boats and folding kayaks. It is also used in roofing materials and as a surface coat material on radomes owing to its radar-transparent quality. These are special purpose synthetic rubbers.

HYPALON is also used in the construction of the decking of modern snow-shoes, replacing neoprene as a lighter, stronger alternative.

This elastomer is widely used in insulation for wire and cables, shoe soles and heels, automotive components, building products, coatings, flexible tubes and hose seals, gaskets, diaphragms. 'HYPALON' 45 can accept large amount of fillers and is used as a binder for powdered metal to produce magnetic gaskets for doors and sheet goods for X-ray burners.

The HYPALON trademark has become the common name for all kinds of CSM regardless of manufacturer. Tosoh Corporation of Japan produces CSM under the trade names Toso-CSM and extos.

These are olefin based rubbers in the form of white chips, having specific gravity 1.10–1.28. These are resistant to ozone, as well as the weather, being better than neoprene and butyl rubber in this respect. These are also resistant to oil, solvents, chemicals and abrasion. When polyethylene is treated with a mixture of chlorine and sulphur dioxide, some chlorine atoms are substituted on the chains and some sulphonyl chloride group ($—SO_2Cl$) are formed.

A typical polymer contains 25–30% chlorine (one chlorine for every seven carbon atoms) and about 1.5% sulphur (one—SO_2Cl for every 90 carbon atoms). This elastomer can be cross-linked by a large variety of compounds including many rubber accelerators. The material is poor in snap and rebound and has low elongation and some permanent set. Its abrasion resistance, flex life, low temperature brittleness and resistance to crack growth are good.

11.2 Process of manufacture of HYPALON

The process of HYPALON manufacture involves the simultaneous chlorination and chlorosulphonation of polyethylene in solution. HYPALON, as a chlorosulphonated polyethylene, is described as a CSM rubber, according to ASTM D1418. A simplified form of its chemical structure is shown below:

$$—CH_2—CH—CH_2—CH_2—CH_2—CH—CH_2—$$
$$\qquad\quad \tfrac{1}{2} \qquad\qquad\qquad\qquad\qquad \tfrac{1}{2}$$
$$\qquad\quad Cl \qquad\qquad\qquad\qquad\qquad SO_2Cl$$

The rubberiness of HYPALON is derived from the natural flexibility of the polyethylene chain in the absence of crystallinity, the introduction of the chlorine atoms along the polyethylene chains provides sufficient molecular irregularity to prevent crystallisation in the relaxed state.

The crystallinity of the polyethylene chain can also be eliminated by an alternative method of introducing molecular irregularity—by copolymerising the ethylene with propylene. The process is used in the manufacture of EPDM polyolefin elastomers.

CMS vulcanisates are seen to have better oil and flame resistance, and, as indicated by the substantial difference in their tensile strengths, better all-around mechanical toughness. The chlorine atoms on the polyethylene backbone not only provide elastomeric properties, but also give useful improvement in oil resistance and flame resistance. The sulphonyl chloride groups provide cross-linking sites for the non-peroxide curing processes.

The sulphur content of most grades of HYPALON is maintained at 1% by weight on the polymer. The chlorine content varies between the different grades and has a significant effect on the oil resistance of the polymer.

Many other properties of compounds and vulcanisates of HYPALON are affected by chlorine content, besides oil resistance. At low-chlorine levels, the polymers retain some polyethylene-like characteristics—they are harder and stiffer because they are partially crystalline, they also show good electrical properties, good heat resistance and good low-temperature flexibility. With increasing chlorine, the compounds and vulcanisates get increasingly rubbery but then at higher chlorine levels again become stiffer, because of increasing glass transition temperature.

The high-chlorine grades are characterised by their excellent oil resistance and good flame resistance.

11.2.1   Compounding ingredients of HYPALON

A typical vulcanisable compound is made up of the components listed in Table 11.1. Components of the curing system are considered to be both the acid acceptor and its activator as well as the vulcanising agent and its accelerator. These will be described first because the curing of HYPALON involves substantially different chemistry from other elastomers.

Table 11.1: A typical CSM compound.

Ingredient	Function	phr
HYPALON 40 (35% Cl)	Polymer	100
Magnesia	Acid acceptor	4
Pentaerythritol	Activator for 2	3
Hard clay	Filler	80
Aromatic oil	Plasticiser	25
Paraffin wax	Process aid	3
Sulphur	Vulcanising agent	1
Tetramethylthiuram disulphide	Accelerator for 7	2

Acid acceptor/activator of HYPALON

The choice of acid acceptor will depend on the vulcanising agent used and the functional requirements of the vulcanisate. The main functions of an acid acceptor are to act as a heat stabiliser, to absorb acid by-products of the curing reaction and to maintain sufficient alkalinity to allow effective curing reactions to proceed. Note that zinc oxide is excluded from this list because its reaction product, zinc chloride, causes polymer degradation on heat-ageing or natural weathering. For the same reason, other zinc-containing additives should also be excluded.

The activator increases the effectiveness of the acid acceptor apparently by solubilising it in the polymer. Adding an activator allows a substantial reduction in the amount of acid acceptor needed and this, in addition to a cost saving leads to lower-viscosity, safer processing stocks.

Fillers of HYPALON

Carbon black is the preferred filler for CSM vulcanisates because it gives best reinforcement of physical properties and best resistance to chemical degradation, to compression set and to water absorption. SRF carbon black gives a good balance of properties and is widely used as a general-purpose filler. Weathering of CSM with as little as three parts of carbon black as a protective pigment is outstanding.

Mineral fillers are often used to take advantage of CSM's non-discolouring characteristics. Best heat resistance is obtained with whiting or black fixe, best electrical properties and very good water resistance are obtained with calcined clay and hydrated silicas and alumina are preferred for improved flammability performance.

When compounded with suitable protective pigments, non-black CSM compositions show excellent long-term weathering performance.

Plasticisers of HYPALON

Aromatic petroleum oils are widely used as plasticisers primarily because of their low cost. Ester plasticisers give compounds that are light in colour, with good low-temperature properties, while liquid chlorinated paraffins give vulcanisates with good flammability performance and weatherability. Polymeric plasticisers are less volatile than the other types and are preferred for heat-resistant compounds.

Processing aids of HYPALON

Stearic acid and stearates are effective release agents but should not be used in litharge-containing compounds in which they reduce scorch safety. Zinc stearate should be avoided in compounds designed for heat resistance or weatherability. Stearic acid should also be avoided in maleimide cures because of cure retardation.

Paraffin wax, poly (ethylene glycol) and low-molecular-weight polyethylene, separately or in combination, are all effective process aids for CSM that do not affect scorch safety in litharge-containing compounds, or the effectiveness of maleimide cures.

Curatives of HYPALON

Many types of curing processes are available with CSM elastomers and some are effective regardless of the level of chlorine in the elastomer and are discussed below.

Ionic cures: Ionic cures of CSM are possible when the acid acceptor is a divalent metal oxide. Ionic cures are characterised by their ability to proceed at low temperatures, by the fact that they are accelerated by moisture and by the development of a high modulus. Ionic cross-linking is responsible for the limited bin stability of CSM compounds when they contain divalent metal oxide acid acceptors and are stored under humid conditions.

The ability of HYPALON to cure under mild ambient conditions is useful for both pond liners and roofing membranes. It allows a sheeting that is readily bondable to itself, by means of solvent—or heat-welding, to be put in place and then to slowly cure as a result of exposure to ambient conditions.

Covalent cures: CSM is also capable of undergoing covalent cross-linking and most applications for CSM make use of this type of curing. Three different systems are commonly used—sulphur cures, which give the widest choice of compounding ingredients, peroxide cures, which are non-discolouring and give good heat and compression-set resistance and maleimide cures, which also give good heat and compression-set resistance, but which are less sensitive to the type of compounding ingredient. Covalent curing systems rely on a cross-linking agent and one or more accelerators for their curing action.

Sulphur cures: Sulphur-based cures are very versatile and are practical for all types of molding and extrusion processes. They allow high states of cure to be achieved even in low-cost, highly extended compounds. Vulcanisates with properties of both covalent and ionic cross-linking result from the low-temperature moisture curing of a compound. Practical cures can also be achieved in hot water. Although the cure rate is slow, it is very convenient since it allows wire insulation to cure while the wire is on the reel. Typically, insulation is cured in this way by standing the reels in a steam room or immersing them in a tank of hot water. For a comparison, it would take about 20 hr in a press at 100°C to achieve the state of cure obtained with 2 hr in steam at 100°C.

Peroxide cures: CSM is readily cured with a peroxide. Frequently, a cure promotor such as triallyl cyanurate is added to improve the effectiveness of the peroxide. An acid acceptor must also be added. Compounds to be cured with peroxides can contain low levels (e.g., upto 40 phr) of fully-saturated plasticisers, higher levels will cause progressive cure retardation. Chlorinated paraffins are best and ester plasticisers are less satisfactory, but adequate. Aromatic process oils will drastically retard the cure.

Maleimide cures: (N,N'-m-phenylenedimaleimide) is a primary curing agent for HYPALON. Maleimide cures also require the presence of calcium hydroxide as an acid acceptor and the antioxidant VANOX AT (a butyraldehyde-amine condensation product) as an accelerator. Maleimide cures are very safe and characteristically produce good water resistance and excellent resistance to compression set and to heat ageing.

Unlike peroxide cures, maleimide cures are relatively insensitive to type and proportion of plasticiser. Although the maleimide cure system is deactivated by moisture, it is effective in high-pressure steam, some cure retardation may be seen in low-pressure steam cures, however.

Special additives of HYPALON

Nickel dibutyl dithiocarbamate (NBC), a dark-green powder, is widely used with CSM to improve the heat resistance of black or dark-coloured compounds. Upto 3 parts of NBC are added in sulphur-cured compounds based on mixed metal oxides. The addition of NBC reduces scorch safety. Also, NBC is not effective in peroxide cures. Other additives include low-melting thermoplastic polymers such as PVC copolymers and high styrene resins, which may be added to CSM to prevent contact-sticking and distortion of wire covering while it is being cured on the reel in steam or hot water.

11.2.2 Mixing of HYPALON

All elastomers will soften as their temperature rises while being worked on processing equipment. This softening effect is somewhat greater for CSM than for most other elastomers. However, it is entirely thermal and reversible so that a CSM stock that is warmed up on a mill and then allowed to cool back to room temperature will regain its former stiffness. This means that CSM compounds can be reworked without significant change in processing behaviour, as long as scorch is not a factor.

Mill-mixing of HYPALON

CSM compounds are easily mixed on conventional mills. Because of longer heat exposure, mill-mixed compounds are likely to be scorchier than those mixed in internal mixers. Cold water should be used on the mill except when processing the more thermoplastic grades, such as HYPALON 45 and 623, which require a mill temperature of around 70°C. A typical mill-mixing time for a normal CSM batch would be about 25 min, depending on the type and amount of fillers and plasticisers used.

Internal mixing of HYPALON

Internal mixing is the most cost-efficient volume system for producing well mixed CSM compounds. Short mix cycles are preferable due to the need to minimise the heat history of the batch. The preferred mix procedures generally use upside-down mixing without premastication of the polymer and as high a ram pressure as practical. Load factors varying from 0.55–0.80 are normal, with the higher load factors being used for the lower-viscosity stocks. Typical ram-down times will be 1½ to 3½ min.

Production compounds can generally be made on a single-pass mix cycle, but if processing safety is marginal, a two-pass mix procedure may be needed. Dump temperature should be around 121°C (250°F) if no sulphur curatives are included and around 107°C (225°F) when sulphur curatives are added.

Following, are some mixing suggestions as well as tips for achieving good dispersion:

1. Magnesia and low molecular weight polyethylene should be added with the fillers. When added alone, they tend to stick to the fast roll and take the stock with them.
2. Stearic acid and stearates should be added after the magnesia is incorporated.
3. Stocks should not be cut while loose pigments are visible in the mix.
4. Process aids should be added with the fillers whenever possible.
5. Oils can be added with soft fillers and not simultaneously with reinforcing blacks.
6. Resins used as softeners (e.g., coumarone-indene resins) should be added early in the mix. Many of these flux are at the normal mixing temperatures.

7. Add litharge and accelerators in the form of dispersions.

8. Accelerators and Vanox NBC (when used) should be added at the end of the mix. Care should be taken to flux Vanox NBC (approximate melting point 85°C [185°F]), as incomplete dispersion will result in variable heat resistance of the vulcanisates.

9. If necessary to heat the mill to incorporate high melting resins, it may be more practical to add the accelerators in a second pass on the warm-up mill than it is to cool the batch while it is on the mix mill.

10. Stocks should not be allowed to ride the mill at any time during the mix cycle or sticking will result.

11. The batch should be turned over at least once during the mix to prevent sticking. If necessary, add the fillers and plasticiser in two increments to make this possible.

Cooling and storage of HYPALON

After mixing on a mill or in an internal mixer, the stock should be cooled as quickly as possible. A water dip or spray is effective, but the slabs should be completely dry before they are stacked. Water absorbed by compounded HYPALON acts as an accelerator, causing an increase in the viscosity and possibly, subsequent scorching. For the same reason, mixed stocks should not be stored in conditions of high humidity. Mixed compounds of HYPALON should be used as quickly as possible.

If the mixed stock is likely to be stored for more than a week before it is used, it is advisable to withhold the accelerator and add it shortly before the next processing step. Blowing agents and accelerators should always be added to sponge compounds just before use.

Stock warm-up

Slabs of compounded HYPALON synthetic rubber are tough when cold and frequently will shred if fed to a tight mill. If a few slabs are passed through a more open mill until they are warmed somewhat, they will band quickly. Once a warm band is established, cold slabs can usually be fed to one end of the mill with good results. The best warmup mill conditions depend somewhat upon the subsequent operations and will be discussed under extrusion and calendering.

Compounds of HYPALON should not ride the warmup mill as they tend to stick and may scorch. If the extruder or calender operation is interrupted, the stock should be cut into the pan and spread out so its heat can dissipate. It is not advisable to feed the warmup mill unless the operation is continuous and the stock is certain to be removed in a very few minutes.

11.3 Processing of HYPALON

11.3.1 Molding of HYPALON

CSM compounds are successfully processed by compression, transfer, or injection molding. In these processes, where reproducibility of flow of the compound is important, steps should be taken to compound for good bin stability, to mix so as to minimise scorch, to store the mixed compound under cool, dry conditions and to rotate the inventory so that compounds to be molded have a similar storage history. Due to the marked thermoplasticity of CSM compounds, multicavity transfer and injection molds require carefully balanced runner systems and similar flow restrictions for the gates, so that uniform filling of all cavities is achieved.

11.3.2 Extrusion of HYPALON

CSM compounds can be processed in standard hot- or cold-feed rubber extruders. The compounds are normally fed in strip form. In general, CSM compounds require little work to attain physical uniformity and very-low compression ratio screws are usually adequate. Extruded parts can be cured in zero-pressure steam, hot-air ovens, LCM, fluidised bed, CV and by microwave. Curing with ionising radiation is also effective for CSM.

11.3.3 Calendering of HYPALON

Substantial amounts of CSM are used in calendering operations. Compounds from the intermediate chlorine grades, for example, HYPALON 40, are calendered at 60–100°C (140–212°F) onto a fabric with, normally, uneven roll speeds on the calender and usually with the top roll slightly hotter than the middle roll. The maximum practical gauge for these sheets is about 40 mils (1.02 mm) for a single pass.

The calendered sheets are normally cured in an autoclave while wrapped on a drum with a curing liner. Adhesion between the plies is usually excellent. Compounds from the low-chlorine CSM grades, as represented by HYPALON and the lower-viscosity HYPALON, are calendered without curatives for use in pond liners and roofing membranes. Because of their crystallinity and scorch safety, these compounds are calendered at higher temperatures, 100–150°C (212–302°F), with the easier-processing compounds from HYPALON 623 being calendered at the lower end of this temperature range.

11.4 Properties of CSM

The applications for CSM depend only in part on its combination of oil and heat resistance. CSM is much more resistant to corrosive or oxidising chemicals including ozone, than are neoprene and nitrile rubbers, it is tougher than silicones and EPDM rubbers and it has electrical properties that are not as good as EDPM's but exceed those of many other elastomers. CSM has excellent radiation resistance, making it well-suited for wire and cable coverings in nuclear applications. Further, it is colour-stable and highly weather-resistant.

11.4.1 Oil-resistance and low-temperature properties

The oil-resistance and low-temperature properties of CSM vulcanisates depend mainly on the chlorine content of the CSM used. Low-temperature properties can additionally be controlled by the choice of plasticiser in the compound.

11.4.2 Heat resistance

CSM elastomers, as a class, have good-to-excellent heat resistance, depending mainly on the choice of compounding ingredients, although within this class of elastomer the low-chlorine grades tend to have slightly better heat resistance than the high-chlorine grades.

11.4.3 Flame resistance

Best flame resistance is obtained with the high-chlorine grades of CSM. Chlorinated plasticisers, brominated resins and antimony trioxide synergist are commonly used additives. Mineral fillers, excluding carbonates, are preferred. Hydrated alumina is frequently used as the major filler where smoke suppression is important.

11.4.4 Water resistance

The use of magnesia acid acceptor is the major cause of water sensitivity in CSM vulcanisates. Replacing magnesia with litharge in black stocks with a sulphur cure, or with dibasic lead phthalate in all other systems, will result in improved water resistance. Where the presence of lead is undesirable, synthetic hydrotalcite will provide significantly better water resistance than magnesia. Water resistance is also affected by the choice of filler—carbon black is the best, followed by non-black fillers with calcined clay preferred.

11.4.5 Weather resistance

When suitably compounded with protective pigments, CSM vulcanisates and the unvulcanised crystalline grades are highly weather-resistant. They are also colourable with good colour stability. CSM compounds, provided they are free of nutritive compounding ingredients, are also resistant to mildew growth.

11.4.6 Oxygen/ozone resistance

CSM vulcanisates are highly resistant to attack by ozone and do not need the addition of antiozonants in their compounds. They also are very resistant to oxidation, however, the addition of an antioxidant will improve performance under severe heat-ageing conditions.

11.4.7 Electrical properties

Best insulation resistance is obtained with the low-chlorine grades of CSM. Good electrical properties may also require formulating for good water resistance in compounds containing little or no carbon black.

Compounds can be formulated so that combustion products are low in both smoke and corrosive gases.

11.4.8 Chemical resistance

CSM vulcanisates are very resistant to degradation caused by exposure to corrosive chemicals. Litharge is the preferred acid acceptor for vulcanisates for generally good chemical resistance although for certain highly concentrated chemicals such as Baume sulphuric acid, or concentrated nitric acid, magnesia will give better service. The best general-purpose filler for chemical-resistant stocks is blanc fixe or barytes. Resistance to solvents is limited by the resistance of the vulcanisates to swelling in those solvents rather than to any chemical degradation caused by the solvents.

11.4.9 Dynamic properties

The resilience of CSM vulcanisates is generally in the 55–75% range for grades having medium-and low-chlorine content. High-chlorine grades are more highly damped and the resilience of their vulcanisates are generally too low at room temperature to measure on a Yerzley Oscillograph.

11.5 Applications of CSM

11.5.1 Industrial uses of CSM

Two areas in which CSM has proven advantages are in hose and in electrical applications. In the hose industry, CSM is particularly useful because of its oil and chemical resistance, its outstanding ozone and weather resistance and because of the frequent need for bright colours.

Electrical industry

CSM is used in the electrical industry as a protective jacketing or sheathing on top of other elastomers with superior insulating properties. CSM also finds considerable usage as an integral or 'one-shot' insulation/jacket for wiring carrying upto 600 volts. The combination of insulation resistance and long-term resistance to the deteriorating effects of ozone, weather, oil and heat make CSM a clear choice for high-quality wire applications.

Automotive uses

In the auto industry, high-volume CSM usage includes hose, tubing and electrical applications. In addition, low permeability to moisture and refrigerants make high-chlorine CSM an ideal polymer for air-conditioning hoses.

Construction uses

Considerable usage of CSM in reservoir-liner construction has resulted from a combination of excellent weather resistance and the ease of fabrication of uncured liners made of CSM.

Membrane uses

For many of the same reasons, CSM use is also well-established in single-ply roofing membranes. Here, the flame resistance and moisture resistance of CSM is very desirable and the fact that weather-resistant membranes can be made in heat-reflective light colours helps reduce the cost of air conditioning.

Miscellaneous applications of CSM

Other applications for CSM include roll covers for use in strong acids and in applications where corona discharge produces ozone and electrical stress, CSM is also used for tank linings, industrial maintenance coatings and extruded sponge. CSM compositions are binders for cork in automotive gaskets and for magnetic fillers in refrigerator gaskets.

11.5.2   Product safety

When recommended handling procedures are followed, HYPALON synthetic rubber polymers and the products derived from them present no health hazards. HYPALON synthetic rubber may contain a small amount of carbon tetrachloride (CCl_4) and a lesser amount of chloroform ($CHCl_4$) as residues from its manufacturing process. Both are regulated as air contaminants. When large quantities of raw polymer are stored or processed, it is advisable for the protection of personnel to provide adequate ventilation to keep employee exposure below the regulated levels.

12.1   Introduction

Nitrile rubber, also known as Buna-N, Perbunan, or NBR, is a synthetic rubber copolymer of acrylonitrile (ACN) and butadiene. Trade names include Nipol, Krynac and Europrene. These are special purpose synthetic rubbers.

Nitrile butadiene rubber (NBR) is a family of unsaturated copolymers of 2-propenenitrile and various butadiene monomers (1,2-butadiene and 1,3-butadiene). Although its physical and chemical properties vary depending on the polymer's composition of nitrile, this form of synthetic rubber is generally resistant to oil, fuel and other chemicals (the more nitrile within the polymer, the higher the resistance to oils but the lower the flexibility of the material).

It is used in the automotive and aeronautical industry to make fuel and oil handling hoses, seals and grommets. It is used in the nuclear industry to make protective gloves. NBR's ability to withstand a range of temperatures from − 40°C to +108°C makes it an ideal material for aeronautical applications. Nitrile butadiene is also used to create molded goods, footwear, adhesives, sealants, sponges, expanded foams and floor mats.

Its resilience makes NBR a useful material for disposable lab, cleaning and examination of gloves. Nitrile rubber is more resistant than natural rubber to oils and acids, but has inferior strength and flexibility. Nitrile gloves are nonetheless three times more puncture-resistant than natural rubber gloves.

Nitrile rubber is generally resistant to aliphatic hydrocarbons. Nitrile, like natural rubber, can be attacked by ozone, ketones, esters and aldehydes.

The structure of nitrile rubber is given below:

$$+CH_2-CH=CH-CH_2+\ CH_2-CH-$$
$$\underset{\text{Butadiene}}{} \underset{\text{Acrylonitrile}}{\overset{\displaystyle |}{\underset{\displaystyle C\equiv H}{}}}$$

Nitrile rubber (NBR) is commonly considered the workhorse of the industrial and automotive rubber products industries. NBR is actually a complex family of unsaturated copolymers of acrylonitrile and butadiene. By selecting an elastomer with the appropriate acrylonitrile content in balance with other properties, the rubber compounder can use NBR in a wide variety of application areas requiring oil, fuel and chemical resistance. In the automotive area, NBR is used in fuel and oil handling hose, seals and grommets and water handling applications. On the industrial side NBR finds uses in roll covers, hydraulic hoses, conveyor belting, graphic arts, oil field packers and seals for all kinds of plumbing and appliance applications.

Like most unsaturated thermoset elastomers, NBR requires formulating with added ingredients and further processing to make useful articles. Additional ingredients typically include reinforcement fillers, plasticisers, protectants and vulcanisation packages. Processing includes mixing, pre-forming to required shape, application to substrates, extrusion and vulcanisation to make the finished rubber article. Mixing and processing are typically performed on open mills, internal mixers, extruders and calenders. Finished products are found in the market place as injection or transfer molded products (seals and grommets), extruded hose or tubing, calendered sheet goods (floor mats and industrial belting), or various sponge articles.

12.2   Chemistry and manufacturing process of NBR

NBR is produced in an emulsion polymerisation system. The water, emulsifier/soap, monomers (butadiene and acrylonitrile), radical generating activator and other ingredients are introduced into the polymerisation vessels. The emulsion process yields a polymer latex that is coagulated using various materials (e.g., calcium chloride, aluminium sulphate) to form crumb rubber that is dried and compressed into bales.

Some specialty products are packaged in the crumb form. Most NBR manufacturers make at least 20 conventional elastomer variations, with one global manufacturer now offering more than 100 grades from which to choose.

NBR producers vary polymerisation temperatures to make 'hot' and 'cold' polymers. Acrylonitrile (ACN) and butadiene (BD) ratios are varied for specific oil and fuel resistance and low temperature requirements.

Specialty NBR polymers which contain a third monomer (e.g., divinyl benzene, methacrylic acid) are also offered. Some NBR elastomers are hydrogenated to reduce the chemical reactivity of the polymer backbone, significantly improving heat resistance. Each modification contributes uniquely different properties. Figure 12.1 shows the typical NBR manufacturing process.

12.2.1   Acrylonitrile (ACN) content of NBR

The ACN content is one of two primary criteria defining each specific NBR grade. The ACN level, by reason of polarity, determines several basic properties, such as oil and solvent resistance, low-temperature flexibility/glass transition temperature and abrasion resistance.

Higher ACN content provides improved solvent, oil and abrasion resistance, along with higher glass transition temperature.

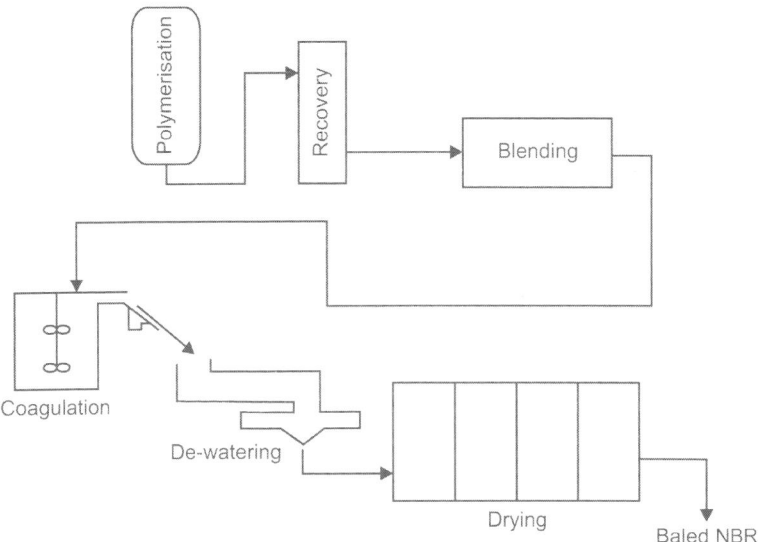

Figure 12.1: NBR manufacturing process.

12.2.2   Mooney viscosity and polymer architecture

Mooney viscosity is the other commonly cited criterion for defining NBR. The Mooney test is reported in arbitrary units and is the current standard measurement of the polymer's collective architectural and chemical composition. The Mooney viscosity provides data measured under narrowly defined conditions, with a specific instrument that is fixed at one shear rate. Mooney viscosity of polymers will normally relate to how they will be processed lower Mooney viscosity materials (30 to 50) will be used in injection molding, while higher Mooney products (60 to 80) can be more highly extended and used in extrusion and compression molding. More definitive polymer characterisation can now be achieved using newer instruments and techniques that measure properties at shear rates pertinent to specific processing requirements. Using these newer instruments, such as the RPA2000, MDR2000, Capillary Die Rheometer and the newer Mooney machines, it is now possible to rheologically measure elastic, as well as viscous characteristics. The RPA2000 and MDR2000 also measure cure rates and cure states.

12.3   General types of NBR

12.3.1   Cold NBR

The current generation of cold NBR's spans a wide variety of compositions. Acrylonitrile content ranges from 15% to 51%. Mooney values range from a very tough 110, to pourable liquids, with 20–25 as the lowest practical limit for solid material. They are made with a wide array of emulsifier systems, coagulants, stabilisers, molecular weight modifiers and chemical compositions. Third monomers are added to the polymer backbone to provide advanced performance. Each variation provides a specific function. Cold polymers are polymerised at a temperature range of 5 to 15°C, depending on the balance of linear-to-branched configuration desired.

The lower polymerisation temperatures yield more-linear polymer chains. Reactions are conducted in processes universally known as continuous, semi-continuous and batch polymerisation.

12.3.2 Hot NBR

Hot NBR polymers are polymerised at the temperature range of 30 to 40°C. This process yields highly branched polymers. Branching supports good tack and a strong bond in adhesive applications. The physically entangled structure of this kind of polymer also provides a significant improvement in hot tear strength compared with a cold-polymerised counterpart. The hot polymers natural resistance to flow makes them excellent candidates for compression molding and sponge. Other applications are thin-walled or complex extrusions where shape retention is important

12.3.3 Cross-linked hot NBR

Cross-linked hot NBR's are branched polymers that are further cross-linked by the addition of a di-functional monomer. These products are typically used in molded parts to provide sufficient molding forces, or back pressure, to eliminate trapped air. Another use is to provide increased dimensional stability or shape retention for extruded goods and calendered goods. This leads to more efficient extruding and vulcanisation of intricate shaped parts as well as improved release from calender rolls. These NBR's also add dimensional stability, impact resistance and flexibility for PVC modification.

12.3.4 Carboxylated nitrile (XNBR)

Addition of carboxylic acid groups to the NBR polymer's backbone significantly alters processing and cured properties. The result is a polymer matrix with significantly increased strength, measured by improved tensile, tear, modulus and abrasion resistance. The negative effects include reduction in compression set, water resistance, resilience and some low-temperature properties.

12.3.5 Bound antioxidant NBR

Nitrile rubbers are available with an antioxidant polymerised into the polymer chain. The purpose is to provide additional protection for the NBR during prolonged fluid service or in cyclic fluid and air exposure. When compounding with highly reinforcing furnace carbon black the chemical reactivity between the polymer and the pigment can limit hot air ageing capability. Abrasion resistance is improved when compared with conventional NBR, especially at elevated temperatures. They have also been found to exhibit excellent dynamic properties.

12.4   Compounding of NBR

Nitrile rubber can be compounded to obtain a broad range of properties. Basically, nitrile rubbers are compounded in much the same fashion as natural rubber or the styrene-butadiene rubbers. Polymer selection is important to obtain the best balance of oil resistance and of low-temperature flexibility.

The higher acrylonitrile content polymers have the highest oil and fuel resistance, but poorer low temperature properties. The lower acrylonitrile content polymers have good low temperature properties at some sacrifice in oil and fuel resistance.

As with natural rubber and styrene-butadiene rubber, zinc oxide at 3–5 phr level and stearic acid at a 1–2 part level are added for proper activation. A good antioxidant is usually added in all nitrile compounds to improve stability. Proper choice of antioxidant depends on the requirements such as high-heat resistance, extraction resistance, or staining characteristics.

Nitrile rubber is not inherently ozone-resistant, therefore protection must be added to the compound to achieve a desired degree of ozone-resistance. Protective waxes, usually at the 1–2 part level, are generally added to the compound in addition to the ozone inhibitors.

Ozone resistance can also be obtained through blending techniques with other materials such as poly (vinyl chloride), chlorosulphonated polyethylene, chlorinated polyethylene, epichlorohydrin and ethylene-propylene terpolymers.

Reinforcing fillers are necessary in order to achieve optimum properties with nitrile rubbers. Carbon black is the most widely used filler and the nitrile rubbers respond to the full range of available carbon blacks from the largest to the smallest particle sizes and range of structures. Non-black applications will require the use of reinforcing silicas of various types, calcium carbonates, hard clays, talc and other pigments.

Plasticisers are generally used in nitrile-rubber compounds to improve processing and low-temperature properties. Typically they are ester types, aromatic oils and polar derivatives and can be extractable or non-extractable depending upon the end-use requirements. Vulcanisation can be achieved with sulphur, sulphur-donor, or peroxide systems. Table 12.1 illustrates some typical vulcanisation systems in general use.

Table 12.1: Typical vulcanisation systems.

System	Level	Characteristics
Sulphur	1.5 parts	Low cost, general purpose, slow curing
MBTS	1.5 parts	
Sulphur	0.3 parts	Fast-curing, low compression set, blooms
TMTD	3.0 parts	
Sulphur	0.5 parts	Faster curing, lower compression set, non-blooming
TETD	1.0 parts	
TMTD	1.0 parts	
CBTS	1.0 parts	
DTDM	1.0 parts	EV system, excellent heat ageing
OTOS	2.0 parts	
TMTM	1.5 parts	
Dicumyl peroxide (40%)	4.0 parts	General-purpose peroxide system, low compression set

12.4.1 Mixing and processing of NBR

All of the commercially available nitrile rubbers can be mixed either on a two-roll mill or with internal mixing equipment. Compounds can be designed to be easily extruded, calendered, or molded with injection, compression, or transfer techniques.

The proper choice of polymer Mooney viscosity determines how well a compound will process. The low-Mooney-viscosity polymers lend themselves to injection molding and calendered friction compounds. The medium-Mooney-viscosity polymers will generally transfer-and compression-mold and provide good calendering characteristics. The high-Mooney-viscosity polymers provide excellent extruded compounds, particularly when high 'green' strength is required. Blends of polymers with different Mooney-viscosity ranges often provide the solution to many processing difficulties.

12.4.2 Properties of NBR

Viscosity: The viscosity of the rubber has an influence on the processibility and the comments for SBR apply here also. There is no difference in the solvent swell resistance and low temperature flexibility properties of vulcanisates from NBR's with different viscosities, as long as the nitrile content of the rubbers is the same. NBR grades with extremely low viscosities (liquid NBRs) can be used as compatible, non-volatile plasticisers in blends with other NBRs. During vulcanisation, these plasticisers are then partially chemically bound to the rubber network and therefore cannot be easily extracted.

Polymerisation temperature: NBR grades produced at low polymerisation temperatures (cold rubbers) show less chain branching than the hot NBR grades. The polymerisation temperature has an influence on the processibility of NBR's. The polymerisation temperature not only has an influence on the long chain branching, but also on the monomer sequence distribution and the *cis*-1,4, *trans*-1,4 and 1,2 micro-structure of the butadiene. For instance, a NBR polymerised at 28°C with 36% by weight of acrylonitrile in the polymer, has a butadiene component with a randomly distributed microstructure of 12.4% *cis*-1,4, 77.6% *trans*-1,4 and 10% 1,2 addition. Due to the lack of compositional uniformity along polymer chains, it is not possible for NBR, nor for SBR, to form crystallites on extension. This lack of strain crystallisation (lack of self-reinforcement) results in relatively poor tensile properties of NBR gum vulcanisates.

Pre-cross-linking: NBR grades, which have been pre-cross-linked by adding a small amount of divinyl benzene during the polymerisation, can be blended with normal NBR grades, to improve their processing behaviour, in addition to lowering the compression set and solvent swell of the final vulcanisates. However, the blending with pre-cross-linked NBRs also lowers the tensile strength, maximum elongation and the tear strength of conventional NBR vulcanisates.

Acrylonitrile content: The glass transition temperatures of polyacrylonitrile at +90°C and of polybutadiene at −90°C differ considerably and therefore with an increasing amount of acrylonitrile in the polymer, the T_g temperature of NBR rises, together with its brittleness temperature.

Stabilisers: The stabilisers added during polymerisation give NBR a good storage stability and they protect it against cyclisation reactions when subjected to high temperatures during mixing and further processing. Staining amine stabilisers generally give better protection than the non-staining phenolic types.

12.4.3   Blends of NBR with other rubbers

There is only a limited blend compatibility of NBR and non-polar rubbers, such as– NR or BR. Yet, sometimes small amounts of NR (to improve building tack) or BR (to lower the brittleness temperature) are used as blend partners. NBR and in particular those grades with a low acrylonitrile content, can be blended with SBR over the full range of concentrations, without a significant deterioration of mechanical vulcanisate properties. These blends are frequently used for economic reasons and in applications, where there is only a moderate demand on solvent swell resistance. There is a possibility to improve the ozone resistance and weatherability of NBR vulcanisates through blends with EPDM, ETER, or BIIR. Blends of NBR and ETER can at the same time be compounded for an improved swell resistance, lower temperature flexibility and higher heat resistance. Particularly well established are blends of NBR and PVC, which are compatible in all blend ratios as long as the NBR grades have a sufficiently high acrylonitrile in all blend ratios as long as the NBR grades have a sufficiently high acrylonitrile content (the compatibility limit is about 25% acrylonitrile). These blends have better ozone resistance improved swell resistance, better tensile and tear strength, but lower elasticity, poorer low temperature flexibility and higher compression set than conventional NBR vulcanisates. Phenolics are also compatible with NBR. In gum and filled NBR compounds and vulcanisates, the phenolics act as hardeners and reinforcing agents and the higher the acrylonitrile content of the NBR, the higher are these effects. Phenolic resins also impart an extremely high abrasion and swell resistance, high hardness and high tensile strength to NBR vulcanisates, but they reduce the elasticity and compression set resistance. Epoxy resins also reinforce NBR vulcanisates.

12.4.4   Properties of NBR vulcanisates

Mechanical properties: When used in combination with reinforcing fillers, vulcanisates with excellent mechanical properties can be obtained from NBR. The optimum tensile strength of up to about 25 MPa occurs at a hardness of Shore A 70 to 80, but one can formulate NBR compounds such that a wide spectrum of vulcanisate hardness can be obtained, ranging from about Shore A 20 to ebonite.

Heat and ageing resistance: NBR vulcanisates have a distinctly better heat resistance than those from NR or SBR. Therefore, it is possible to produce NBR vulcanisates which are still useful even after ageing for 6 weeks at 120°C. If oxygen is excluded (e.g., by ageing the vulcanisate submersed in oil), this ageing resistance becomes even better. The best heat resistance of NBR vulcanisates is obtained when cadmium oxide and cadmium containing accelerators are used, but these compounds are of dubious value because of their high toxicity.

The weather and ozone resistance of NBR vulcanisates is comparable with that of NR vulcanisates, but it is more difficult to improve these properties of NBR vulcanisates by means of antiozonants. In black compounds, the ozone and weather resistance can be somewhat improved by *p*-phenylene diamine and in light compounds, by non-staining enol ethers in combination with micro-crystalline waxes, but especially through blending the NBR with ozone resistant rubbers, such as EPDM, ETER or PVC. Because of the great solubility of polar *p*-phenylene diamines in NBR they show less migration and therefore their activity is reduced.

Low temperature flexibility: The same comments apply to the low temperature flexibility and the elasticity of NBR vulcanisates. An insufficient low temperature flexibility can be improved by using ester or ether based plasticisers, or by blending the NBR with other rubbers, such as BR. However, the latter worsens the solvent swell resistance, unless ETER is used as blend partner.

Swelling resistance: Vulcanisates from NBR, which is a polar rubber, have a good resistance to swelling when submersed in media, which are non-polar, or weakly polar, such as gasoline, grease, mineral oil, or animal and vegetable fats and oils. The swelling resistance depends, however, very much on the NBR grade, the compound components, filler loading, type and amount of plasticiser and the degree of vulcanisation. NBR vulcanisates also resist swelling in alcohols (methanol, ethanol). Blends of gasoline and alcohols, the so-called gasohols, which are frequently discussed as substitutes to extend the supply of gasoline, swell NBR vulcanisates more significantly than the individual components.

When compounding for NBR vulcanisates, one always has to bear in mind the oils with which the vulcanisate will be in contact. For instance, if NBR components, will be exposed to oils with sulphur and nitrogen containing additives, one has to take into account a potential post-vulcanisation from the sulphur. Therefore, one would use in this instance a TMTD concentration of less than 2 phr, to limit the formation of ZDMC.

Permeation: NBR vulcanisates are less permeable to gases than the ones from NR or SBR. As the acrylonitrile content increases, the permeation decreases and with the high acrylonitrile grades, it approaches that of IIR or epichlorohydrin (ECO), but not CO. The diffusivity in NBR vulcanisates is influenced by the amount and type of fillers in the compound and by the degree of cross-linking. Fillers with a laminar structure, such as mistron vapour, reduce the permeability considerably. The permeation of gasoline is, however, significant in NBR vulcanisates. To reduce this, one has to use NBR grades with a high acrylonitrile content, compounded with mistron vapour, or NBR blends with PVC.

Other properties: Due to their high polarity, NBR vulcanisates have a considerably higher electrical conductivity than the ones from non-polar rubbers. Therefore, NBR is hardly used at all in components which require a low electrical conductivity. The thermal conductivity and the thermal expansion co-efficient are of the same order as for NR and SBR. The dynamic properties of NBR vulcanisates are good, but NBR vulcanisates are harmed by exposure to hydrogen sulphide.

12.4.5   Applications of nitrile rubber

The uses of nitrile rubber include non-latex gloves for the healthcare industry, automotive transmission belts, hoses, O-rings, gaskets, oil seals, V-belts, synthetic leather, printer's roller and as cable jacketing. NBR latex can also be used in the preparation of adhesives and as a pigment binder. Unlike polymers meant for ingestion, where small inconsistencies in chemical composition/structure can have a pronounced effect on the body, the general properties of NBR are not altered by minor structural/compositional differences. The production process itself is not overly complex, the polymerisation, monomer recovery and coagulation processes require some additives and equipment, but they are typical of the production of most rubbers.

To sum up, nitrile rubber (Armaflex) is a versatile and flexible closed cell elastomeric insulation. It is resistant to water vapour, oil and most acids. It has excellent adhesive receptiveness and it can be easily cut and fabricated. Its maximum operating temperature is 105°C, which makes it ideal for insulating hot/cold water pipes, air-conditioning pipe work and in blanket form is used to insulate ventilation ductwork. Nitrile rubber will not contribute significantly to fire and in certain circumstances is self-extinguishing. It is available in pipe-sections, rolls and in sheet forms.

1. Elastomeric products offer excellent flexibility.
2. Resistant to water vapour.
3. Resistant to thermal transmittance properties.
4. Oil and acid resistant (refer to manufacturer's data sheets before installation).
5. Excellent adhesive and coating receptiveness.
6. Good cutting characteristics and easy to fabricate.

12.5   Hydrogenated NBR (H-NBR or ENM or HSN)

In non-aqueous solutions and by using suitable catalysts, such as pyridine-cobalt complexes, or complexes from rhodium, ruthenium, iridium and palladium, it is possible to hydrogenate NBR partially or even completely. Particularly effective is, for instance, a catalyst based on transition metals with a trivalent rhodium halide. Completely hydrogenated NBRs have a fully saturated polymer chain, which, conceptually, could have been derived from ethylene and acrylonitrile. Therefore, the code ENM has been proposed for this copolymer. Since, however, most of the hydrogenated NBRs still contain double bonds, the designation of H-NBR will be used here.

Completely saturated NBR grades can be cross-linked with peroxides. The vulcanisates give the highest resistance to hot air and hot oils that can be achieved with NBRs, a high resistance to oxidative and ozone degradation, high resistance to sulphur-containing oils, even hydrogen sulphide, sulphur and nitrogen-containing oil additives and a high resistance to industrial chemicals. In addition, the fully saturated H-NBRs have an excellent tensile strength, good low temperature flexibility and very good abrasion resistance. After exposure to high temperatures, these H-NBRs retain their mechanical properties much better than, for instance, FKM vulcanisates, so that, in spite of their lesser heat and swelling resistance, fully saturated H-NBRs can compete with FKM. Because of the excellent mechanical properties also at high application temperatures, the good low temperature flexibility and the good chemical resistance, fully saturated H-NBR is able to replace even FKM in various applications.

At a level of unsaturation of as low as 3 to 5 moles%, H-NBR can be cross-linked with sulphur. Such sulphur cross-linked H-NBR does not have the good environmental stability of a fully saturated H-NBR any longer, but it has better dynamic properties, e.g., for use in toothed drive belts working under oil.

12.6 Latest developments of NBR

Thus, nitrile rubbers containing 'bound' antioxidants, i.e., antioxidants attached to the polymer chain are less fugitive and are less likely to be soluble in fuels or oils and are less volatile, thereby improving dry-heat resistance also hydrogenated nitrile rubbers containing little or no unsaturation show promise of better heat resistance and resistance to oxidised gasoline as well as improved resistance to the harsh environments found in deep, sour wells.

Polyacrylic rubber

13.1   Introduction

Polyacrylic rubbers (ACM) are copolymers of acrylic esters and a minor proportion (1–5%) of a second monomer with reactive sites for cross-linking. These are special purpose synthetic rubbers. Examples of the esters used either singly or in combination are ethyl-, butyl- and cetyl-acrylate and methoxy- and ethoxy-acrylate. The choice depends on the balance required between low-temperature flexibility and resistance to swelling in aromatic oils, as these properties are inversely related.

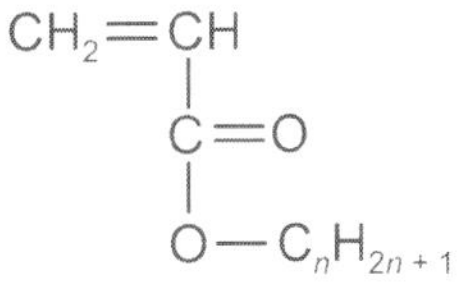

$$CH_2\!=\!CH$$
$$|$$
$$C\!=\!O$$
$$|$$
$$O\!-\!C_nH_{2n+1}$$
Alkyl acrylate

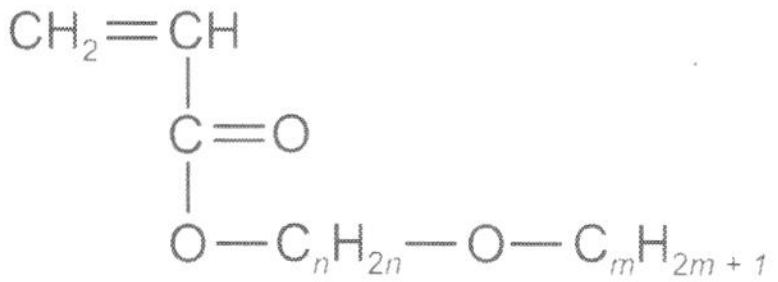

$$CH_2\!=\!CH$$
$$|$$
$$C\!=\!O$$
$$|$$
$$O\!-\!C_nH_{2n}\!-\!O\!-\!C_mH_{2m+1}$$
Alkoxyacrylate

The second monomer will contain a halogen-most commonly chlorine, epoxy or carboxyl groups.

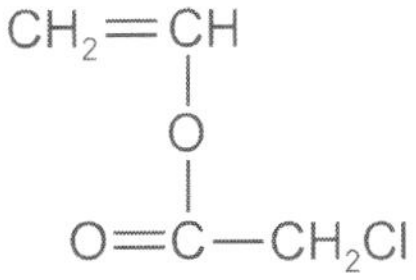

$$CH_2\!=\!CH$$
$$|$$
$$O$$
$$|$$
$$O\!=\!C\!-\!CH_2Cl$$
Halogen (vinylchloroacetate)

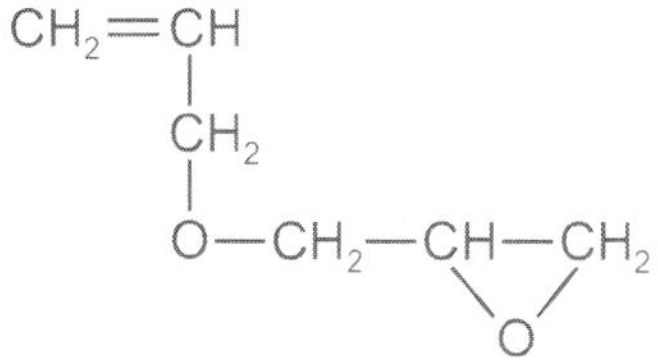

$$CH_2\!=\!CH$$
$$|$$
$$CH_2$$
$$|$$
$$O\!-\!CH_2\!-\!CH\!-\!CH_2$$
$$\diagdown O \diagup$$
Epoxy (allylglycidylether)

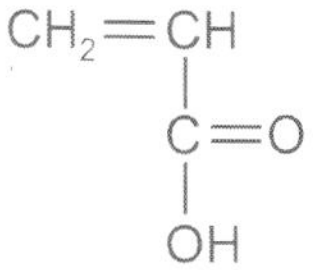

$$CH_2\!=\!CH$$
$$|$$
$$C\!=\!O$$
$$|$$
$$OH$$
Carboxyl (acrylic acid)

Polyacrylic rubber (ACM) compounds are developed to provide a rubber part that would function in applications where oils and/or temperatures as high as 204°C would be encountered. These are also very resistant to attack by sulphur-bearing chemical additives in the oil. These properties have resulted in general use of polyacrylic rubber compounds for automotive rubber parts as seals for automatic transmission fluids and extreme pressure lubricants. Polyacrylic rubber will prove most useful in fields where these special properties are used to the maximum. It is recommended for products such as automatic transmission seals, extreme pressure lubricant seals, searchlight gaskets, belting, rolls, tank linings, hose, O-rings and seals, white or pastel-coloured rubber parts, solution coatings and pigment binders on paper, textiles and fibrous glass.

13.2 Compounding and processing of ACM

Polyacrylic rubbers (ACM) can be mixed on an open mill or in an internal mixer and can be extruded and calendered. They are generally compression molded but can also be transfer-molded or injection-molded. A post-cure is usually required to develop optimum physical properties, particularly compression set.

Compounding is relatively simple. The high service temperature and oil immersion do not permit the addition of many special additives, as they are liable to be lost by extraction by hot oils or by evaporation. The choice of grade depends on the balance of low-temperature flexibility and oil swell required. Carbon black (HAF-LS grade and FEF grade blacks) provide the best overall processing and vulcanisate properties. For non-black compounds precipitated silica combined with equal parts of either aluminium silicate or silane-treated clay can be used. Wax and stearic acid are the most commonly used process aids. Synthetic gaphite may be incorporated as a lubricating agent in compounds, for rotary shaft seals.

The temperature range over which polyacrylic rubbers are serviceable varies with the grade the overall range being –40 to 175°C. These rubbers are superior to nitrile rubbers in resistance to swelling and deterioration by hot hydrocarbon oils, extreme pressure lubricants, transmission and hydraulic fluids.

Resistance to oxidation, ozone, sunlight and gas-permeation is also good. The main application is in automotive seals and gaskets.

13.2.1 Curing of polyacrylic rubber

A typical polyacrylic rubber, such as the copolymer of ethyl acrylate and chloro-ethyl vinyl ether, is supplied as a crude rubber in the form of white sheets having a specific gravity of approximately 1.1. It may be mixed and processed according to conventional rubber practice.

However, polyacrylic rubber is chemically saturated and cannot be cured in the same manner as conventional rubbers. Sulphur and sulphur-bearing materials act as retarders of cure and function as a form of age resistor in most formulations. Polyacrylic rubber is cured with amines, 'Trimene base' and triethylene tetra-mine are most widely used. Ageing properties may be altered by balancing the effect of the amine and the sulphur.

Like other rubber polymers, reinforcing agents such as– carbon black or certain white pigments are necessary to develop optimum physical properties in a polyacrylic rubber vulcanisate. Selection of pigments is more critical in that acidic materials, which would react with the basic amine curing systems, must be avoided. The SAF or FEF carbon blacks are most widely used, while hydrated silica or precipitated calcium silicate are recommended for light-coloured stocks.

Typical curing temperatures are from 143 to 166°C at cure times of 10 to 45 min depending on the thickness of the part. Polished, chromium-plated molds are recommended. For maximum overall physical properties, the cured parts should be tempered in an air oven for 24 hr at 149°C.

13.2.2 Cure systems

Because acrylic elastomers have a saturated backbone, cross-linking is accomplished via incorporation of copolymerised reactive cure sites. The nature of this cure site varies in commercial products and therefore different cure systems have been developed for specific types. Thus, acrylic elastomers from different suppliers are not generally directly interchangeable in a given recipe. Conventional domestic acrylics respond to soap/amine, activated thiol, soap/sulphur or sulphur donor, lead/thiourea, diamine and trithiocyanuric acid (TCY) cure systems. The ethylene/acrylic type responds to diamine and peroxide cures. Since the majority of acrylic cure systems are basic in nature, they are retarded by acids and accelerated by bases. Specific cure-system recommendations should be obtained directly from the polymer suppliers.

Despite significant advancements in acrylic-cure technology, all current, state-of-the-art acrylic elastomers require a relatively long cure cycle or must be subsequently post-cured (tempered) in a circulating hot-air environment to realise optimum compression-set resistance.

13.2.3 Forming

To obtain smooth extrusions, more loading and lubrication are necessary than for molded goods, because of the inherent nerve of the polymer. Temperatures of 43°C in the barrel and 77°C on the die are recommended.

Generally, those compounds that extrude well are also good calendering stocks. Suggested temperatures for calendering are in the range of 37.8 to 54°C. Higher temperatures will result in sticking of the stock to the rolls. Under optimum conditions, 15-mil films may be obtained.

Polyacrylic rubber may be coated on nylon either by calendering or from solvent solution. It also has excellent adhesion to cotton and is often used as a solvent solution applied to cotton duck to be used as belting. Solvents generally used include methylethyl ketone, toluene, xylene, or benzene.

Polyacrylic rubber is most widely used in many types of seals because of its excellent resistance to sulphur-bearing oils and lubricants.

In general, polyacrylic rubber vulcanisates are resistant to petroleum products and animal and vegetable fats and oils. They will swell in aromatic hydrocarbons, alcohols and ketones. Polyacrylic rubber is not recommended for use in water, steam, ethylene glycol, or in alkaline media.

Laboratory tests indicate that polyacrylic vulcanisates become stiff and brittle at a temperature of $-23°C$. But in actual service, these same polyacrylic rubbers have been found to provide satisfactory performance at engine start-up and operation in oil at temperatures as low as $-40°C$.

For those applications requiring improvement in low-temperature brittleness by as much as $-4.0°C$ and that can tolerate considerable sacrifice in overall chemical oil and heat resistance, a copolymer of butyl acrylate and acrylonitrile may be used.

13.2.4   Reinforcing agents of polyacrylic rubber (ACM)

Acrylic elastomers do not provide high gum strength when cured. Reinforcing agents are required to develop useful properties. Carbon black reinforcing agents provide the best overall balance of vulcanisate properties. The use of mineral reinforcing agents is primarily limited to electrical resistant and colour-coded applications.

Aluminium silicate or silica types are commonly utilised, either alone or in combination. Amino or vinyl silane coupling agents are also sometimes employed to gain improved vulcanisate properties.

Organic colourants are generally recommended since inorganic metallic-oxide colour pigments tend to have an adverse effect on the cure and heat-ageing characteristics of silica-reinforced compounds.

Synthetic graphite is also utilised in conjunction with carbon black and/or mineral reinforcing agents to promote improved surface lubricity characteristics in, for example, rotary-shaft-seal applications. Neutral-to-high-pH-alkaline pigments are recommended since acidic types will tend to retard the basic cure mechanism of most acrylic elastomers.

13.2.5   Plasticisers of polyacrylic rubber (ACM)

Plasticisers are used in acrylic-rubber compounds as process aids or to gain improved low-temperature resistance. However, type and amount is limited by their volatility and extraction characteristics, as related to post-cure conditions and service requirements. Low-volatility ester and polyester plasticisers are generally utilised.

13.2.6   Process aids of polyacrylic rubber (ACM)

Lubricating agents are essential to promote the release characteristics of acrylic compounds. Stearic acid is commonly used, usually in combination with commercial process aids, which provide both external (release) and internal (viscosity-reducing) lubricating qualities. However, it is cautioned that high lubricant levels can interfere with mold knitting and metal-bonding properties.

13.2.7 Antioxidants of polyacrylic rubber (ACM)

Although acrylic elastomers are highly resistant to oxidation, certain antioxidants can promote marginally improved dry-heat resistance. Low-volatility diphenylamines have been found useful for conventional acrylics, while hindered phenols are suggested for the ethylene/acrylic type. The use of antioxidant is specifically recommended in the ethylene/acrylic gum polymer and for all non-black reinforced compounds.

13.3 General processing of polyacrylic rubber (ACM)

13.3.1 Mixing of polyacrylic rubber (ACM)

Acrylic compounds can be Banbury or mill-mixed. However, internal Banbury mixing is preferred and generally utilised by the industry. A two-pass Banbury-mix procedure is most common, but a one-pass mix is also possible on relatively non-scorchy compounds.

An 'upside-down' mixing procedure is sometimes desirable for use with highly loaded, dry compounds. Because acrylic elastomers are somewhat thermoplastic in nature and tend to lose shear resistance fairly rapidly on mixing, it is recommended that reinforcing agents be incorporated very early in the mixing cycle to obtain good dispersion. Maximum cooling is also essential, to maintain polymer integrity during the initial phase of mixing.

13.3.2 Extrusion/calendering of polyacrylic rubber (ACM)

Most acrylic compounds extrude well enough for standard mold-preparation. However, for finish goods, special compounding may be required to obtain satisfactory green strength, size and finish characteristics. In general, the use of high-structure reinforcing agents along with increased process aid is usually effective. Compound extension through the use of softener is also sometimes necessary. The range of die temperatures is 65–107°C.

Calendering compounds are designed much the same as for finished extrusions. However, additional lubricant may be required to promote good release characteristics. The range of calender-roll temperatures is 27–107°C.

13.3.3 Compound storage stability

Typical finished acrylic compounds provide reasonably good shelf-stability characteristics in the range of one week to several weeks when stored under normal room-temperature, dry conditions. Shelf life is primarily dependent on the activity of the cure system employed. Storage under refrigerated (4°C), low-humidity conditions is also an effective means of extending shelf life. In practice, aged compounds are usually mill-freshened prior to subsequent processing.

13.3.4 Vulcanisation of polyacrylic rubber (ACM)

Acrylic elastomers lend themselves to all common cure processes. They are commonly compression-, transfer- and injection-molded, but can be steam-cured as well. Typical cure cycles (in minutes) are as follows:

Compression-mold	3′/190°C to 1′/204°C
Transfer-mold	12′/163°C to 4′/177°C
Injection-mold	3′/90°C to 1′/204°C
Open-steam cure	30–90′/163°C

In addition, acrylic elastomers respond to most continuous vulcanisation (CV) cure techniques. Special compounding may be required for the low viscosity types to gain adequate back pressure in compression molding operations. As indicated previously, a post-cure is usually necessary to develop optimum compression-set resistance. Oven post-cure cycles in the range of 4–8 hr/177°C are commonly utilised.

13.3.5 Bonding characteristics of polyacrylic rubber (ACM)

A variety of commercially available solvent-base, curing-type adhesives have been found to provide excellent bonds of acrylic compounds to metals and other substrates. Bonding is carried out during the vulcanisation process. Recommended substrate preparation and application procedures are provided by the manufacturer.

13.3.6 Solution characteristics

Certain acrylic elastomers also lend themselves to solvent-solution coatings and adhesive applications. These dissolve readily in common industrial solvents such as methyl ethyl ketone, acetone, ethyl acetate and toluene. Polymer concentrations in the 30–40% total solids range are possible.

13.3.7 Blends

Blending is a process by which one can achieve a compromise between processibility, properties and economy by suitable choice of partners. Polyacrylic rubber is a special rubber, featuring resistance to heat, oil (particularly lubricating oil) and ozone, enjoys increasing demands in automotive and related fields. The use of special heat resistant rubbers like fluroelastomers is a costly affair. Apart from the cost, because of extreme toughness, the fluroelastomers are often faced with processing difficulty. On the other hand the polyacrylic rubber is softer and easy to process. Fluoroelastomers cannot blend with general purpose oil resistant rubbers but the presence of polyacrylic rubber can act as an active interphase between the two using a common curative system.

Both the polyacrylic rubber and fluroelastomer can be cured with diamine types of curatives and a compromise can be achieved.

The blend capability of acrylic elastomers is limited. However, certain conventional types can be utilised to reduce the cost of diamine curable fluorocarbon (FKM) compounds or to improve the heat-resistance of epichlorohydrin (CO/ECO) compounds. In addition, modified low-T_g versions are believed to have potential as high-temperature resistant-impact improvers for poly (vinyl chloride), polycarbonate and poly (phenylene oxide) plastic materials. Although not reported, the ethylene/acrylic type may also have interesting blend possibilities.

13.3.8 Applications of polyacrylic elastomers

Besides the classical molding technologies, seen the innumerable potential applications, these elastomers can be designed for being turned through extrusion or calendering with LCM vulcanisation, hot air or UHF. Other more sophisticated technologies as 'roto-cure' or solvent solutions can be used lending particular care to the reological characteristic and 'shelf-life' of the compound.

The applications mainly concern the auto sector and oil drilling as:

1. O-rings.
2. Motor gaskets.
3. Shaft seals.
4. Valve cover gaskets.
5. Oil filter gaskets.
6. Pipes.
7. Spark plug boots.

13.3.9 Properties of polyacrylic rubber

Some of physical and mechanical properties, chemical resistance, thermal properties and environmental performance of polyacrylic rubber are given below:

Physical and mechanical properties

Physical and mechanical properties	Typical values
Durometer or hardness range	40–90 shore A
Tensile strength range	500–2,500 PSI
Elongation (range %)	100–450%
Abrasion resistance	Fair to good
Adhesion to metal	Fair to good

Adhesion to rigid materials	Fair to good
Compression set	Poor to good
Flex cracking resistance	Fair to good
Impact resistance	Poor
Resilience/rebound	Fair to good
Tear resistance	Poor to good
Vibration dampening	Good to excellent

Chemical resistance

Chemical resistance	Typical values
Acids, dilute	Fair
Acids, concentrated	Poor to fair
Acids, organic (dilute)	Poor
Acids, organic (concentrated)	Poor
Acids, inorganic	Fair
Alcohol's	Poor
Aldehydes	Poor
Alkalies, dilute	Fair
Alkalies, concentrated?	Fair
Amines	Poor
Animal and vegetable oils	Good
Brake fluids, non-petroleum based	Poor
Diester oils	Good
Esters, alkyl phosphate	Poor
Esters, aryl phosphate	Poor
Ethers	Poor
Fuel, aliphatic hydrocarbon	Excellent
Fuel, aromatic hydrocarbon	Poor to good
Fuel, extended (oxygenated)	Fair to good
Halogenated solvents	Poor to good
Hydrocarbon, halogenated	Poor to good
Ketones	Poor to good
Lacquer solvents	Poor to good
LP gases and fuel oils	Good
Mineral oils	Good to excellent
Oil resistance	Excellent
Petroleum aromatic	Fair
Petroleum non-aromatic	Good

Refrigerant ammonia	Fair
Refrigerant halofluorocarbons	R-11, R-12, R-13
Refrigerant halofluorocarbons w/oil	R-11, R-12, R-13, R-22
Silicone oil	Excellent
Solvent resistance	Good

Thermal properties

Thermal properties	Typical values
Low temperature range	−30°F to 0°F
Minimum for continuous use (static)	−30°F
Brittle point	−40°F
High temperature range	+350°F to + 400°F
Maximum for continuous use (static)	+400°F

Environmental performance

Environmental performance	Typical values
Colourability	Good
Flame resistance	Poor
Gas permeability	Good to excellent
Odour	Fair to good
Ozone resistance	Good to excellent
Oxidation resistance	Excellent
Radiation resistance	Poor to good
Steam resistance	Poor
Sunlight resistance	Good to excellent
Taste retention	Fair to good
Weather resistance	Excellent
Water resistance	Excellent

13.3.10 Typical properties of a polyacrylic rubber vulcanisate

Typical properties of a polyacrylic rubber vulcanisate (mold cured 4 min, 170°C, oven post-cured 8 hr, 175°C) is given Table 13.1.

Table 13.1: Typical properties of a polyacrylic rubber vulcanisate (mold cured 4 min, 170°C, oven post-cured 8 hr, 175°C).

Property	Value
Hardness (IRHD)	75
Tensile strength (MPa)	14

(Cont'd...)

Property	Value
Modulus 100% (MPa)	10
Elongation at break (%)	130
Compression set ASTM method B-70 hr 150°C (%)	17
Heat resistance 70 hr — 175°C in air	
Change in hardness (IRHD)	+3
Change in tensile strength (%)	−20
Change in elongation at break (%)	−23

13.4   Future developments of polyacrylic rubber (ACM)

Polymers available on the market are typical for oil resistance and low temperature performance. Inside these two classes there is a further differentiation according to the 'cure-site' present in the polymeric chain. The two principal families are substantially composed from an omopoliymere of the ethyl-butyl acrylate and a copolymer ethyl-acryilate.

They are differentiates for low temperature resistance. Such characteristics improve with the addition of special monomers. As for other elastomers, when polyacrylic rubbers improve their low temperature performance there is a worsening of oil resistance. Although significant advances have been made in acrylic-cure technology, continued research is aimed at the development of more highly reactive polymers and/or cure systems to completely eliminate the necessity of post-cure for short-cycle molded-goods applications.

Fluorocarbon rubber

14.1   Introduction

The term fluorocarbon elastomer includes polytetra-fluoroethylene polymer of chlorotrifluoroethylene, fluorinated ethylene, propylene polymers, polyvinylidene fluoride, hexafluoropropylene, etc. These are thermoplastics, resistant to chemical and oxidation, non-combustible, temperature range upto 287°C, high dielectric constant, resistant to moisture, weathering, ozone and ultraviolent radiation. Their structure comprises a straight backbone of carbon atoms symmetrically surrounded by fluorine atoms. These are special purpose synthetic rubbers.

It is generally available in powder, dispersion, film, sheet, tubes, rods, tapes and fibres. It is used as a high temperature wire and cable insulation material for electrical equipment, food, drug and chemical equipment, coating of cooking utensils, piping gaskets, continuous sheet, etc.

Fluorocarbon elastomer are elastomeric high polymer which contains fluorine and may be a homopolymer or copolymer. Fluorocarbon polymers include a large group of fluoroelastomers, as mentioned above including a copolymer in which the molecular skeleton is a –P=N– chain containing approximately an equal number of tri and heptafluoroethoxy side group.

Such polymers are amorphous, thermally stable, non-combustible, have low glass transition temperature (–77°C) and are generally resistant to attack by solvents and chemicals.

The fluorocarbon rubber (though very expensive) and elastomers have been of critical importance in solving urgent problems in aerospace, automotive, chemical and petroleum industries.

The first elastomeric perfluoro carbon, a TFE/perfluoro (methyl vinyl ether) copolymer, is the most stable of all known elastomers. It has found extensive use in solving problems in aggressive environments. The copolymer of vinylidene fluoride and hexafluoropropylene is vulcanisable to an elastomer which possesses a resistance to chemicals, oils and solvents and to heat, which is outstanding in comparison with any other commercial rubber. Variations in the polymerisation recipes and conditions and the use of undisclosed alternative or additional monomers have enabled the producers to supply a range of polymers varying in processing and vulcanisate properties but with the general qualities.

Fluorocarbon rubbers are very stable materials because of the strength of the bond between fluorine and carbon. The most typical grades of fluorocarbon rubbers are based on vinylidene fluoride and hexafluoropropylene (HFP) monomers, which are referred to as FKM in ASTM standards and FPM in ISO standards. There are also fluorocarbon rubbers containing chlorine in vinylidene monomers (e.g., $CFCl=CF_2$), referred to as CFM rubbers. Fluorocarbon rubbers are usually produced by emulsion radical polymerisation. Peroxide compounds act as initiators.

Monomers used in fluorocarbon rubbers are given below:

1. Vinylidene fluoride (VF_2).
2. Tetrafluoroethylene (TFE).
3. Chlorotrifluoroethylene (CTFE).
4. Hexafluoropropylene (HFP).
5. 1-hydropentafluoropropylene (HPTFP).
6. Perfluoromethylvinylether (FMVE).

Commercial names of above monomers are given in Table 14.1.

Table 14.1: Monomers and their commercial names.

Monomers	Type designation	Commercial types
VF_2 + HFP	FKM	VITON A, AHV, A-35, E-60, FLUOREL 2140, 2141, 2143, 2146 SFF-26
VF_2 + HPFP	FKM	Tecnoflon SL, SH
VF_2 + HFP + TFE	FKM	VITON B, B-50
VF_2 + HPFP + TFE	FKM	Tecnoflon T
VF_2 + TFCIE	CFM	KEL-F 3700, 5500, SKF-32
PFMVE + TFE + X	FKM	ECD 006

Structure of the fluorocarbon rubbers, VF_2/HPTFP/TFE -copolymer is given below:

$$\left[H_2C - F_2C \right]_x \left[FHC - \underset{\underset{CF_3}{|}}{CF} \right]_y \left[CF_2 - CF_2 \right]_z$$

The most commonly used FKM rubbers can be vulcanised with diamines, polyhydroxide compounds and bisphenols. The vulcanisation system has a metal oxide as acid acceptor.

14.2 Manufacturing of fluorocarbon rubbers

The major vinylidene fluoride (VF_2) and hexafluoroprene (HFP) and vinylidene fluoride (VF_2)/hexafluoroprene (HFP)/tetrafluoroethylene (TFE) elastomers are usually prepared by radical polymerisation in emulsion, using peroxy compounds such as ammonium persulphate as initiator, occasionally in redox systems with or without chain-transfer agents such as carbon tetrachloride, alkyl esters, or halogen salts. A fluorinated soap may be used but is not required. The normally continuous polymerisation is followed by centrifugation or coagulation and isolation of the dry polymer by standard manufacturing techniques. The product is sold in the form of pellets, chips, strips and slabs, which are translucent, white, or slightly coloured, depending on the presence of compounding agents.

The other fluoro-olefin copolymers are prepared by similar processes. In all cases, great care must be exercised because of the potential toxicity and explosiveness of some of the monomers or monomer mixtures, particularly TFE and HFP. Monomers must be pure since the polymerisation of fluoro-olefins is inhibited by a wide variety of substances.

14.3   Processing and curing of fluorocarbon rubbers

Fluorocarbon elastomers must be processed with care. Equipment must be clean and free of oil and grease. The usual contaminants such as sulphur, water and other polymers must not be present. They either interfere with the cure or compromise fluorocarbon elastomer properties. This need for cleanliness is especially important in Banbury mixing, where inadvertent contamination is more likely.

14.3.1   Mill mixing of fluorocarbon rubbers

This should be done on as cool a mill as possible with chilled cooling water to ensure shearing characteristics for good dispersion and minimum scorch. Raw polymer bands readily. Once the band is formed and cut a few times, compounding should be started immediately. For fast mixing, compounding ingredients should be pre-blended. This is particularly important for preventing mill-roll sticking, which commonly occurs if magnesium oxide is added separately. If polymer blends are needed, the higher-viscosity fluorocarbon rubber should be banded first, followed by the lower-viscosity variety to assure a uniform blend.

14.3.2   Internal mixing of fluorocarbon rubbers

Most fluorocarbon compounds cured with bisphenol or peroxide cure systems can be mixed in 'one-pass' in an internal mixer. Diamine cures are too scorchy for the Banbury. Properly handled with ample cooling water, compounds present less of a potential scorch problem in a Banbury than on large mills. The mix cycle should be kept to 3–4 min and the batch dropped when the temperature indicator is in the range 93–104°C (200–220°F).

14.3.3 Extruding of fluorocarbon rubbers

Extrusion of fluorocarbon rubbers (Trade name VITON and FLUOREL) is widely practised to produce finished goods. In general, lower-viscosity grades will extrude more easily. Compounds with diamine-cure systems, will generally extrude well at temperatures generated by the shear heat from the screw and need only moderate heat added. Bisphenol compounds generally require more heat. The head and die should be heated to 110°C (230°F) and 127°C (260°F) respectively. The very-high-fluorine products also require heating to assure smooth extrudates.

14.3.4 Calendering of fluorocarbon rubbers

Most of the FKM group of fluoroelastomers (VF$_2$/HFP copolymers and terpolymers) can be calendered. Quality and economy depend upon many factors, particularly viscosity at the time of calendering. Mixed stocks should be used promptly or stored cool (<18°C [65°F] and great care taken to exclude moisture. Following careful warming to 27±6°C (81±10°F) stock should be continuously strip-fed across the calender roll to maintain a uniform rolling bank.

14.3.5 Molding of fluorocarbon rubbers

Most fluoroelastomer compounds can be molded by compression, transfer, or injection processes. Higher-viscosity polymers are best-suited for compression molding and may even be required, depending on the design of the mold (vs. lower-viscosity analogues) for elimination of trapped gases. VITON is probably as low a viscosity fluoroelastomer. Transfer and injection molding generally require polymers of lower viscosity than does compression molding. Also, for transfer- and, especially, injection-molding processes, good scorch safety will be more important than in compression molding.

Injection molding is becoming very important. Injection-molding systems require proper selection of elastomeric compounds, molding conditions and mold design to insure the right combination of good flow, low scorch and optimal physical properties such as the best (or lowest) compression set for O-rings. Scorch safety is needed to allow flow through parts and throughout the mold. Shrinkage is caused by relatively low quantities of reinforcing filler and loss of volatiles generated during post-cure. Molds must be designed to correct for shrinkage to provide specification parts.

14.3.6 Curing of fluorocarbon rubbers

Although fluorocarbon elastomers are inert and slow to react, cures can be effected with surprising speed with appropriate formulations, details for which should be obtained from the respective manufacturers.

In general, full compounds are subjected to a two-stage cure cycle to achieve the best balance of vulcanisate properties. The initial cure (pre-cure) in most cases requires pressure and is usually carried out in a hydraulic press or steam autoclave at temperatures up to 204°C (400°F). Pre-cure times may be as short as 1–5 min or as long as 2–3 hr (steam). This is followed by a long oven cure at atmospheric pressure. An oven-cure cycle of 15 hr at 232°C (450°F) is required to optimise tensile and compression set properties, though a cycle as short as 10 hr may be sufficient. Higher-temperature oven-cure cycles can improve tensile strength and elongation at break for VITON E60C and compression-set resistance for VITON E60C and VITON B-910 or related FLUOREL products. In most cases, a cycle of 10–12 hr will develop over 90% of maximum physical properties. Properties suitable for special applications are obtained by careful selection of the cure system and product type.

14.3.7   Processing of other fluorocarbon rubbers

The vinylidene fluoride/chlorotrifluoroethylene copolymer, though very tough, can be formulated with ingredients on a two-roll rubber mill at 77–88°C (170–190°F). Curing is effected with amine or peroxides with metal salts or oxides as acid acceptors. The tetrafluoroethylene-propylene copolymer can be processed like the other fluorocarbon rubbers using similar fillers. Curing is effected with peroxide systems. The perfluoroelastomer in KALREZ parts is very difficult to process and is sold only as finished parts or stock.

14.3.8   Physical properties of cured fluorocarbon rubbers

The fluorocarbon rubbers are exceptionally stable, excelling all rubbers in overall resistance to combinations of heat, light, ozone, solvents and aggressive chemicals. They also have good high-temperature compression-set resistance and low-temperature flexibility.

Many products have been developed to give the particular combinations of properties required for the varied highly sophisticated applications for which very special formulations have been developed.

Heat resistance

FKM fluoroelastomer vulcanisates are considered serviceable almost indefinitely when exposed continuously up to 200°C.

Compression-set resistance

The most important market for fluorocarbon elastomers is in seals and O-rings, for which compression-set resistance has proved to be the best criterion of service performance. As a result of the discovery of the phosphonium chloride accelerator system for bisphenol and other phenol cures, excellent results are readily obtained with VITON and FLUOREL and related products. Such products are available in a variety of viscosities, both in precompounded form and uncompounded, for which accelerator masterbatches are offered. Compression-set resistance is also strongly affected by the filler.

14.3.9 Low-temperature flexibility

The commercial FKM elastomers have brittle points of about –25°C to –40°C, which are not easily improved by compounding. Plasticisers tend to harm heat stability and accelerate deterioration on ageing since they are less stable than the fluoropolymers. Peroxide-curable fluorocarbon rubbers can be blended with fluorosilicones to improve low-temperature flexibility, but at a sacrifice in high-temperature stability and solvent resistance.

14.4   Compounding of FKM

Curing agents: The first cure systems for FKM consisted of capped diamines, such as hexamethylenediamine carbamate, in combination with acid acceptors, like MgO, CaO, PbO, or other lead compounds. These cure systems offer, however, poor scorch safety, poorer vulcanisate properties and poorer compression set than the later bisphenol AF cure systems. On the other hand, the conventional amine systems produce vulcanisates with a good adhesion to metals and a relatively good resistance to amine stabilisers which are blended with motor oils. Because of the better scorch safety and better compression set, one uses now primarily cure systems based on bisphenol AF in combination with MgO and $Ca(OH)_2$ as acid acceptors. The resulting vulcanisates have good physical properties, high tensile strengths, low hot compression set and release well from molds. However, with the increased use of motor oils with amine donors as stabilisers and of methanol-blended motor fuels, the use of bisphenol cross-linking systems has to be re-evaluated, because the vulcanisates soon form cracks after exposure to these oils at higher temperatures. In addition, FKM grades with low fluorine contents, that were cured with bisphenol, have a relatively poor swelling resistance in methanol-blended motor fuels.

FKM grades with higher fluorine contents can be cross-linked with peroxides. Peroxide cured vulcanisates are more resistant to amine stabilisers in blended oils and to methanol containing motor fuels and therefore, peroxide cures are being used more extensively.

Peroxycarbamates appear to be particularly well suited to FKM cures. Acid acceptors have to be used also with peroxide cures and with high fluorine FKM grades, very good mechanical properties are obtained in addition to the already mentioned good resistance to blended oils and motor fuels. The cure systems tend to result in mold sticking of the vulcanisate, but this can be remedied to some extent by using carnauba wax.

Stabilisers, protective agents: Basic compounds or metal oxides are required as HF-acceptors. For the most demanding heat resistance specifications, the choice of the HF-acceptors assumes particular importance. PbO is recommended if the vulcanisate is exposed to hot acids and dibasic Pb-phosphite, together with ZnO, if the vulcanisate is exposed to steam or hot water, MgO and CaO give a superior vulcanisate performance in dry heat.

Fillers: To obtain good vulcanisate properties, the desired hardness, good processibility and to reduce compound cost, one uses non-reinforcing blacks and mineral fillers. MT black gives a good balance of good processibility and physical properties. Since the compounds stiffen very soon after mixing, only relatively small amounts of filler (10–30 phr) can be used. Lowest compression set values are obtained with Austin Black in the compound.

Softeners: FKM is not compatible with conventional plasticisers. To improve the processibility of FKM compounds, one can use, instead carnauba wax, pentaerythritol tetrastearate or low molecular weight polymers which are chemically similar, namely those from VF_2/HFP.

14.5   Compounding of FKM for specific properties

14.5.1   Fluids and chemical resistance

FKM fluorocarbon elastomer vulcanisates are resistant to hydrocarbons, chlorinated solvents and mineral acids. They swell excessively in many polar solvents such as ketones, some esters and ethers. They are attacked by amines, alkali and some acids, e.g., hot anhydrous HF and chlorosulphonic acid.

The main determinant of chemical resistance is the metal oxide used in the compound. For example, VITON compounded with magnesium oxide swells 61% in volume in red fuming nitric acid, but only 45% when compounded with zinc oxide-Dyphos and 24% with litharge. General-purpose FLUOREL with magnesium oxide swells 110% in concentrated hydrochloric acid at 158°C while a litharge compound will swell only 2%.

14.5.2   Adhesion

The adhesion of FKM elastomers to metals is important for many applications. Good adhesion during molding and curing is obtained if the metal surface is properly prepared and a suitable adhesive primer is used.

14.5.3   Resistance to automotive fuels

A combination of automotive regulations, higher under-hood and under-body temperatures and the use of more highly aromatic, non-leaded gasoline has focused attention on the FKM elastomers in automotive applications. As a result, they are used increasingly in automotive fuel hose because of superior resistance to 'sour' gasoline (fuel containing peroxides), to gasoline/alcohol mixtures and to permeation.

14.5.4   Resistance to exhaust gases

The largest existing application for fluorocarbon rubbers in energy-related industries is flue-duct expansion joints used in desulphurisation systems for coal-fired plants. The fluorocarbon elastomer is needed to resist the high temperatures and the wet acidic flue gas streams that together cause metal or less chemically resistant elastomeric expansion joints to corrode and/or deteriorate.

While all FKM terpolymers are more effective in this application than the copolymers, the peroxide-curable high-fluorine terpolymers exhibit the best resistance. Peroxide cures are more effective than bisphenol cures, which, in turn, surpass diamine cures.

14.5.5   Coatings and sealants

FKM elastomers are often needed in coatings and sealants for which the lower-viscosity types such as VITON and FLUOREL are usually preferred. Typical solvents used are methyl ethyl ketone, ethyl acetate, methyl isobutyl ketone and amyl acetate and related ketones and esters.

14.6   Tetrafluoroethylene/propylene-copolymer

This product has a different chemical-resistance profile and better electrical resistance properties as compared with the FKM rubbers and is of about equal thermal stability. The points of better resistance are in exposure to steam, amines and amine corrosion inhibitors, wet sour gas and oil, phosphate ester hydraulic fluids, glycol brake fluids and some other polar systems plus high- and low-pH environments. It is not as resistant to many hydrocarbons, chlorinated solvents and ethers. Because of these properties, the major areas of usage are oilfield applications, chemical processing, automotive, aerospace, wire and cable exposed to chemically aggressive media.

14.6.1   Kalrez perfluoroelastomer parts

Kalrez parts are made from a perfluoroelastomer that is in a sense a rubbery derivative of polytetrafluoroethylene plastic. In aggressive fluid or gaseous environments, no other elastomer can equal the overall performance of Kalrez parts. These products combine the resilience and sealing force of an elastomer with chemical inertness and thermal stability similar to polytetrafluoroethylene resins. They resist attack by nearly all chemical reagents including ethers, ketones, esters, amines, oxidisers, fuels, acid and alkalies. They provide long-term service in virtually all chemical and petrochemical streams, even where corrosive additives cause other elastomers to swell or degrade. They are not suitable for use with molten potassium or sodium metals.

Kalrez parts retain electric properties in long-term service as high as 288°C (550°F) and in intermittent service up to 316°C (600°F). They are usually reliable at temperatures as much as 83°C (150°F) higher than those sustainable by parts made from even other fluoroelastomers.

Consequently, Kalrez parts usually out-perform other elastomeric sealing material in difficult environments.

The polymer is extremely expensive and in fabrication requires unique tooling plus difficult and complex manufacturing techniques. Consequently, it is offered only in the form of finished parts, including O-rings, tubing, rods, sheeting and so on, plus custom-designed parts. These parts are used extensively in oil exploration and processing, in the chemical industry and in scientific instrumentation.

14.7   Properties of FKM vulcanisates

Mechanical properties: The general level of mechanical properties of FKM vulcanisates is, as with most of the other speciality rubbers, distinctly lower than for conventional diene rubbers. The tensile strength depends greatly on the temperature and it drops considerably at higher temperatures. The same applies to the hardness, which can be compounded for a range of Shore A 50 to 95, but preferably 70. A phenolic cure system results in a better hardness retention at high temperatures and this cure system is also suitable for optimum compression set. FKM vulcanisates are not very elastic.

Heat and ageing resistance: FKM has the best heat resistance of all rubbers and continuous service for 1000 hr will be 220°C and even a service life at 250°C is possible. FKM vulcanisates also resist degradation from weathering and ozone.

Low temperature flexibility: The dynamic brittleness temperature of FKM is about –18°C. More recently, FKM grades have been developed with an improved brittleness temperature of –40°C.

Resistance to swelling and chemicals: FKM vulcanisates not only resist swelling in hot oils and aliphatic compounds, but also in aromatics and chlorinated hydrocarbons. They are also very resistant to most mineral acids, even acids at high concentrations. As already mentioned, FKM vulcanisates cross-linked with bisphenol AF are relatively easily attacked by amine stabilisers of blended oils. Peroxide-cured FKM vulcanisates with high fluorine contents are more stable in this respect. FKM vulcanisates also swell more in methanol, ketones, esters and ethers, than in mineral oil and motor fuels. The resistance to swelling improves with increasing fluorine contents and, therefore, one chooses FKM grades with the highest possible fluorine content for applications with methanol-containing motor fuels. FKM vulcanisates have also a good resistant to motor fuels with hydroperoxides (sour gas), as well as to hydrogen sulphides, which are present in oil wells. Hot hydrofluoric and chlorosulphonic acids attack FKM vulcanisates.·

14.8 Advantages and disadvantages of fluorocarbon rubbers

14.8.1 Advantages of fluorocarbon rubbers

1. Excellent heat resistance (up to 200°C, temporarily 315°C).
2. Good chemical and solvent resistance.
3. Excellent oxygen, ozone and weather resistance.
4. Incombustible.
5. Good abrasion resistance.
6. Good high-temperature compression-set resistance.

14.8.2 Disadvantages of fluorocarbon rubbers

1. Low alkali resistance.
2. Relatively poor mechanical properties.
3. Limited elasticity at low temperatures.
4. The tensile strength decreases substantially at elevated temperatures.
5. Very costly.

The fluorocarbon rubbers are used for special applications that require good heat, oxygen or corrosion resistance and hot solvent and oil resistance.

14.9 Applications of fluorocarbon rubbers

1. Car and aeroplane seals and hoses.
2. Fire-resistant coverings.
3. Heat-resistant insulators.
4. O-rings, shaft seals.
5. Gaskets, fuel hoses, valve-stem seals.

14.9.1 Typical industrial uses of fluorocarbon rubbers

1. Valve seals, O-rings and special configurations.
2. Valve and pump linings.

3. Gaskets—in refineries and chemical plants.
4. V-ring packings.
5. Expansion joint.
6. Hose (for chemical resistance)—rubber-lined or rubber-covered.
7. Wire/cable cover—in steel mills and nuclear power plants.
8. Rolls—100% fluorocarbon rubber or laminated to other elastomers.

Automotive applications of fluorocarbon rubbers

1. Valve-stem seals.
2. Shaft seals.
3. Transmission seals.
4. Fuel-handling systems:
   (a) Inject or nozzle seals.
   (b) In-tank pump coupler hose.
   (c) Carburettor-pump cups, needle valves, diaphragms.
   (d) Fuel shut-off valves.
   (e) Fuel-hose or fuel-hose liner.

Aerospace applications of fluorocarbon rubbers

1. O-ring seals in jet engines.
2. Hydraulic systems.
3. Lubricating systems.
4. Fuel systems.

Oil well applications of fluorocarbon rubbers

1. Drill bit seals.
2. Packers.
3. V-ring packers.
4. Valve seals.
5. Blow-out preventers.

Other uses of fluorocarbon rubbers

1. Oil-suction and delivery hose.
2. Truck chemical-transfer hose.
3. Tank linings.
4. Chimney linings.
5. Seals for scientific instruments.

14.10 Additional developments of fluoro elastomers

14.10.1 Tetrafluoroethylene-propylene-co-and terpolymers (TFE/P)

Through alternating co-polymerisation of tetrafluoroethylene and propylene (TFE/P) one obtains the following structure:

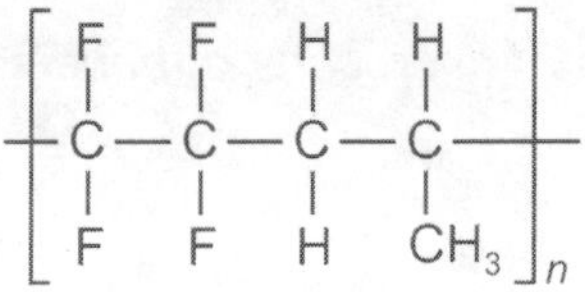

$$
\left[\begin{array}{cccc}
F & F & H & H \\
| & | & | & | \\
-C & -C & -C & -C- \\
| & | & | & | \\
F & F & H & CH_3
\end{array}\right]_n
$$

Tetrafluoroethylene

This new generation of fluoro elastomers has unique properties and it has been recently commercialised under the name of Aflas. A precise alternating arrangement of the co-monomer molecules is important, since short propylene sequence impede the cross-linking and reduce the heat resistance because of their thermoplastic nature. There are three TFE/P co-polymer grades now available—a low molecular weight grade for extrusion, a higher molecular weight for compression molding and a very high molecular grade for highest vulcanisate performance levels, such as in oil drilling applications. More recently, a TFE/P-terpolymer has also been developed, in which the termonomer is a fluorinated vinyl compound. It can be readily cross-linked with peroxides through a vinyl side-group. This terpolymer has better low temperature flexibility. Although TFE/P-grades can be cross-linked in nucleophilic reactions with dioles (e.g., bisphenol AF) and amines (e.g., hexamethylene diamine carbamate), most vulcanisates are cross-linked with peroxides in combination with TAIC as coagent and $Ca(OH)_2$, as acid acceptor.

This cure system gives a better all-round environmental resistance. As fillers one uses small quantities of blacks (N550–N990), as well as aluminium silicates and hydrophobic fumed silica. The latter gives the best long-term resistance against amine corrosion inhibitors. TFE/P vulcanisates can be obtained in hardness Shore A from 65 to 95, but mostly of 70.

14.10.2 Nitroso elastomers (AFMU)

The alternating co-polymerisation of tetrafluoroethylene with perfluoronitroso-methane, together with small amounts of perfluoronitrosobutyric acid as reactive termonomer, results in an elastomer AFMU. This elastomer has a very low glass transition temperature of $-50°C$. However, above $175°C$, AFMU is not resistant to strong bases and oxygen. Cross-linking reactions with metal oxides and epoxides produce vulcanisates with excellent resistance to strong acids and oxidising agents and the vulcanisates do not burn, even in pure oxygen. These very expensive elastomers have been especially developed for the aerospace industry.

14.10.3 Fluorotriazine elastomers

Perfluoroalkyltriazines are obtained by reacting perfluoroalkyl-dinitriles and ammonia. The resulting poly(imidoamidine) is stable and at higher temperatures, it cross-links. Like nitroso elastomers, the fluorotriazine elastomers are also resistant to acids and oxidising agents, but they are attacked by bases.

14.10.4 Poly-(fluoroalkoxyphosphazene) elastomers (PNF)

A new type of rubber based on fluorophosphonitrile derivatives (PNF) has been brought on to the market and this product too is in competition with FKM. Although PNF elastomers have a maximum service temperature which, at about 175°C (or 200°C for intermittent exposure applications), is 50 to 75°C lower than that for FKM, they have excellent low temperature flexibility, with a T_g value of –65°C. The swelling resistance is also very good, in fuels at room temperature it is appreciably better than that of FKM elastomers and at higher temperatures (fuel injection temperatures) it is considerably better. The swelling resistance in aromatic solvents and in some chlorinated hydrocarbons is also better than that of FKM elastomers. The mechanical properties and the compression set at 150°C, are also good.

15.1   Introduction

Silicone rubbers are inorganic polymers, since their main chain structure does not include carbon atoms. Structure of silicone is shown in Fig. 15.1 which shows silicone and oxygen atoms – siloxane groups - form the polymer main chain. There are typically also some pendant groups, usually methyl groups, attached to the polymer chain. The molar mass of silicone rubbers can vary over a wide range and consequently there are liquid materials as well as traditionally resinous rubbers available. These are special purpose synthetic rubbers.

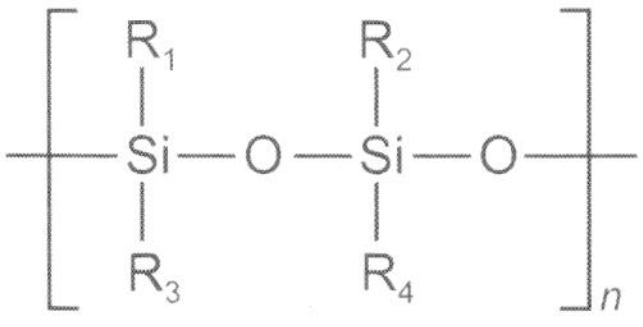

$$\left[\begin{array}{cc} R_1 & R_2 \\ | & | \\ -Si-O-Si-O- \\ | & | \\ R_3 & R_4 \end{array} \right]_n$$

Figure 15.1: Structure of silicone.

Silicone rubbers are usually polymerised from cyclic oligomers to linear macro-molecules. The vulcanisation can be carried out at room temperature or elevated temperature. Vulcanisation at room temperature occurs with cross-linking agent (e.g., ortho-silicone acid ether) or air. For high temperatures vulcanisation peroxides are used. The molar mass of silicone rubber vulcanised at elevated temperatures is higher (300000–1000000 g/mol) than in room temperature vulcanisation (10000–100000 g/mol).

Chemically, silicone rubber is a specially prepared polymer of dimethyl-siloxane. The chemical reaction is controlled to get linear chains of several thousand units of silicon-oxygen with two methyl radicals having molecular weights of the order of 500000. This elastomeric polymer in the form of a gum is mixed with suitable inorganic fillers to give mechanical strength. To make it elastic, it is heated with a suitable chemical, for a specified time (vulcanisation). This brings about elasticity through intramolecular cross-linking. The unique properties of silicone rubber are partially accounted for by the bond energy of the silicon-oxygen linkage which is about 1.5 times as strong as the carbon-carbon linkage present in organic polymers. Special rubbers are produced by modifying the dimethyl polysiloxane gum. Rubber gums with 0.1 mole per cent methyl-vinylsiloxane give low compression-set rubbers. By incorporating one boron atom per 300–400 silicone atoms, fusible rubbers have been prepared. Solvent-resistant rubbers are made by incorporating polar groups such as fluorinated alkyl and cyanoalkyl into the polymer molecule. Replacement of 5–15% of the methyl groups by phenyl groups yields rubbers which retain their elasticity even at –150°F. 'Bonding pastes' or rubbers with paste-like consistency are formulated from an elastomeric gum, calcium carbonate and benzoyl peroxide.

15.2   Vulcanisation of silicone rubber

Silicone rubber compounds are normally heat-cured in the presence of one of the organic peroxides such as: (i) *bis* (2,4-dichlorobenzoyl) peroxide, (ii) di-benzoyl peroxide, (iii) di-cumyl peroxide, (iv) 2,5-dimethyl -2,5-*bis* (*t*-butyl peroxy) hexane and (v) di-tertiary butyl peroxide.

Vulcanisation rate is conveniently studied by means of the Monsanto Rheometer and it provides a continuous measurement of complex dynamic shear modulus while a rubber is being cured in a mold under heat and pressure. This study is accomplished by measurement of the torque on a conical disk rotor that is embedded in the rubber and is being sinusoidally oscillated through a small arc. The torque is a linear function of cross-link density as determined by swelling measurements, though the proportionality constant varies with the stock. The torque readings can, therefore, be considered as relative cross-link densities.

15.3   Compounding ingredients

A typical silicone rubber formulation contains a silicone polymer, reinforcing and (or) extending fillers, process aids or softeners to plasticise and retard creep-ageing, special additives (e.g., heat-ageing and flame retardant additives), colour pigments and one or more peroxide curing agents.

Silicone polymers: Pure silicone rubber polymers, differing from one another in polymer type and molecular weight, are available from the basic suppliers.

Silicone rubber elastomers: Although pure polymer may be used, it is generally easier and more economical for the rubber fabricator to compound from silicone-reinforced gums or bases. Over the years, silicone suppliers have developed a wide variety of silicone rubber compounds to meet special requirements and specifications. However, the most recent offerings have been in bases which are mixtures of pure polymer, of process aids and of reinforcing silica fillers that have been specially processed.

Both of these product lines are made up of blendable bases plus selected additives for enhanced properties such as heat-ageing, flame retardancy and oil resistance as well as processing aids.

Colour pigments for silicone rubber are: (i) red, (ii) green, (iii) blue, (iv) orange, (v) white, (vi) yellow, (vii) buff, (viii) black and (ix) brown.

Reinforcing fillers: The fumed process silicas reinforce silicone polymer to a greater extent than any other filler. Due to the high purity of the filler, the rubbers containing it have excellent insulating properties, especially under wet conditions. Some precipitated filler may give high water absorption due to residual salts. If so, the wet electrical properties will not be very good. However, recently some precipitated silicas are available with very-low-salt content and provide better electrical properties.

Semireinforcing or extending fillers: The extenders are important for use in compounds containing reinforcing fillers in order to obtain an optimum balance of physical properties, cost and processability.

Ground silica and calcined kaolin do not provide significant reinforcement. As a consequence, they can be added to a reinforced gum or compound in relatively large quantities in order to reduce pound-volume cost. These extenders are satisfactory in either mechanical or electrical-grade rubber.

The reinforcement obtained with calcined diatomaceous silica is greater, though quite modest, than that obtained with any other extender. Therefore, as an extender, it is not as useful as ground silica. However, it is used in electrical stocks, low-compression-set stocks and in general mechanical stocks to reduce tack and modify handling properties.

Calcium carbonate and zirconium silicate are special-purpose extenders, used mainly in pastes that are coated on fabrics via solvent dispersion.

Zinc oxide is used as a colourant and as a plasticiser. It imparts tack and adhesive properties to a compound.

Additives: Organic colours and many inorganic colours, have adverse effects on the heat-ageing of silicone rubber. Usually 0.5 to 2 parts per 100 parts of compound are sufficient for tinting purposes. It is often desirable to masterbatch colour pigments in order to get good dispersion and close colour matches. Red iron oxide is used as a colour pigment and as a heat-ageing additive; 2 to 4 parts per 100 parts of gum will give improved heat stability at 600°F. Process aids are used with highly reinforcing silica fillers. These have a softening or plasticising effect and they retard the 'crepe-ageing' or 'structuring' or 'pseudocure' of the raw compound that occurs due to the high reactivity of the reinforcing filler with silicone polymer.

Curing agents: In commercial practice, it has been found that none of the six commonly used peroxides is a universal curing agent. Such curing agents are: (i) *bis* (2,4-dichloro-benzoyl) peroxide, (ii) benzoyl peroxide, (iii) tertiary butyl perbenzoate, (iv) dicumyl peroxide, (v) 2,5-dimethyl-2,5-*bis* (*t*-butyl peroxy) hexane and (vi) D-tetiary butyl peroxide.

All three 'vinyl-specific' peroxides are good for thick-section molding, with dicumyl being less preferable due to a slight tendency to 'air-inhibit' in the same manner as benzoyl peroxide. In addition, dicumyl peroxide has somewhat less volatile decomposition products (acetophenone and α-dimethylbenzyl alcohol) than do the other two. This means that external pressure during cure is less important, but still required; and that longer oven post-bakes are needed for thick sections than in the case of the other two peroxides. Just as with the diaroyl peroxides, optimum physical properties require relatively close control of dicumyl peroxide concentration.

15.4 Compounding of silicone rubber

The various steps in compounding of silicone rubber are discussed below.

15.4.1 Mixing of silicone rubber

Silicone rubber may be compounded in conventional equipment, such as Doughmixers, Banburys and two-roll mills. A Banbury works well with a dry, non-sticky stock, but is undesirable with compounds that require a relatively long heat cycle and become too tacky to unload easily.

While a two-roll mill is capable of compounding silicone rubbers, the nature of the silica fillers creates an undesirable work atmosphere. However, two-roll mills are excellent for colouring, catalysing and preforming some of the firmer stocks.

15.4.2 Fabricating of silicone rubber

Freshening

Freshening has been the first fabrication step. This is a remilling operation, to reverse the 'crepe-hardening' or 'structuring' that has taken place since the compound was made. 'Structuring' occurs more quickly and proceeds farther if the compound has been aged at higher temperatures. It is caused by the formation of hydrogen bonds between the hydroxyl groups of the filler and the hydroxyl groups of oxygen atoms of the polymer. Two factors have reduced the amount of structuring that occurs in the current silicone compounds: first, the hydroxyls on the polymer are kept to a low level; second, the hydroxyls on the filler are reduced by treating the filler with silanes or siloxane process aids. However, freshening is still often carried out by the fabricator because the material has been in inventory at a high temperature or for a long time. It is also freshened when the fabricator catalyses it.

Before the mill is loaded, the roll clearance should be set fairly loose. Usually the silicone compound will band on the slow roll first (crumbling and lacing can be, but usually are not, encountered). As the milling continues, the nip should be gradually tightened. In the case of a hard compound, the original roll settings will be somewhat tighter for a short time. Then the mill is set fairly loose and gradually tightened. This procedure, combined with the proper amount of stock on the mill (slight bank), minimises air entrapment during further milling and blending. After a smooth sheet has formed on the fast roll, compounding ingredients may be added if desired.

Milling is continued until the stock reaches the desired consistency. Under-freshened stock will flow poorly upon molding and will form parts with a rough surface upon extrusion. An over-freshened compound loses green strength and becomes sticky and hard to handle. Stock must not be allowed to warm up above 100–130°F after a curing agent has been added.

At this point, the stock is freshened and ready for use in one of the following fabrication procedures.

Molding

Compression and transfer molding are the most widely used methods for molding silicone rubber parts. However, large-volume applications, in the automotive industry for spark-plug boots and in the health-care industry for catheters, are resulting in a large growth for injection molding. Compression and transfer moldings are run at pressures of 800–3000 psi and temperatures of 220–370°F. Three mold variables that must be controlled are temperature, speed of mold closing and pressure. Temperature is influenced mainly by the choice of a peroxide curing agent. If the temperature of the mold increases after repeated molding, the temperature of platens has to be decreased to prevent scorching. The speed of mold closing has to be adjusted to allow complete filling of the mold and escape of all air, again without scorching. The pressure has to be sufficient to prevent thick flash but low enough to avoid excessive wear and tear on the mold. The molded silicone piece will vary in size if the pressure is varied, because of the great compressibility of silicones. To optimise the use of material in compression molding, the preform must be shaped and sized to minimise wasted flash.

The mold shrinkage of silicone rubber is about 2–4% and is affected by a number of factors, but primarily that the linear thermal expansion of silicone rubber is 17–20 times that of steel and about 2 times that of organic rubber. The shrinkage is, therefore, largely dependent on the temperature of the molding. It should be noted that silicone compounds with higher loadings of filler have lower shrinkage. Mold shrinkage is augmented by the release of volatiles during the cure and post-bake stages.

Injection molding involves pressures of 5000–20,000 psi and temperatures of 370–485°F. Injection times are approximately 3–10 seconds, while molding time is in the range of 25–90 seconds. This method, relative to compression molding gives less flash, better properties and greater uniformity. Blow molding has also been used on some simple parts. In this process, heavy-walled parts are usually made because of the variation in wall thickness. Pre-hardened stainless steel is recommended for the molds with chromium plating, also used where a high finish is desired and where the undercuts are minimal. Assistance in mold design is available from the basic silicone rubber suppliers. Silicone mold-release agents are unsuitable for use with silicone rubber. Household-detergent water solutions of 0.5–2.0% are recommended for spray or brush applications to the mold. A thin layer is preferred to avoid build-up on the mold.

Extrusion

Gaskets, tubing, tape, wire and cable, seals, rods, channels and hoses in a variety of shapes and sizes may be extruded. The equipment is similar to that used with organic rubber. When silicone rubber is extruded, however, the low green strength and decomposition temperature of the peroxide curing agent must be considered.

Typical extruders have screws with a length-to-diameter ratio of 10/1 to 12/1 although shorter screws are sometimes used. A single flight screw is usually used but sometimes a second flight is added near the discharge to minimise the pulsating of the extrusion. The screw has a compression ratio of 2/1 for harder stocks and up to 4/1 for stickier stocks, or when less porosity or close tolerances are needed. The compression ratio is preferably provided by a variable-pitch screw, keeping the flight depths constant from feed to discharge. If the flight depth is reduced, there is a tendency in the longer barrels to build excessive heat. Because of the abrasive nature of silica fillers, especially the larger particle sizes, both the barrel and screws should be built of abrasion-resistant alloys. However, many of the organic rubber hoses are successfully extruding tubing and wire and cable, using unhardened equipment. Both the barrel and screw should also be water-cooled to prevent scorch.

To prevent feeding problems, a roller feed is usually used. The roller is fed from a 'hat' (coiled strips of compound) or from strips removed from an in-house mill. The roller should have a higher speed than the screw. The breaker plate should contain at least two screens (40–200 mesh) to ensure clean extrusions. A fine mesh is backed by a coarser backing screen. This assembly increases the back pressure that ensures air removal and better dimensional control. The finer screens will give cleaner extrusions, but will slow production and will have to be changed more often. Usually the screens are not cleaned but are discarded after removal.

A front flange assembly is used to hold and centre both the die and the mandrel (pin) when extruding tubing. Fine adjustments are made with adjusting screws on either the die or the pin. The die should be made from pre-hardened stainless steel and designed to produce smooth flow with no dead spaces to hold up material that may start to cure. The pin should be drilled so that low-pressure air can be used to support the tubing and keep it round. The die opening is often different from both the shape and size of the extrusion because of differential flow and die swell. These differences can be compensated for to some extent by reducing the thickness of the lasts where the die dimensions are smaller. Die swell is also greater if the silicone polymer is branched or if the extrusion is speeded up. More heavily loaded compounds have less die swell. The extrusion can be considerably reduced in size by stretching it before cure. Again, the shrinkage that takes place on curing and post-bake must be taken into account. In forming wire and cable insulation, the extruder must be fitted with a cross-head. Reinforced hose may also be made by feeding tubing reinforced with high-temperature fibre or wire through a cross head and extruding a second layer of silicone rubber over it.

Several methods of vulcanisation are available for extrusions. Hot-air vulcanisation is very common and the catalyst is *bis*-(2,4-dichlorobenzoyl) peroxide. Other catalysts are too volatile and cause bubbling in the extrudate and lead to undercure due to loss of catalyst. Two types of hot-air vulcanisation are possible—vertical and horizontal. In the horizontal type, the extrudate is laid on an endless belt and passes through an oven 10 to 30 feet long. The oven is usually at 600–800°F although the first zone may be cooler to minimise loss of the catalyst. Air turbulence should be present inside the oven to improve heat transfer and to drive off the volatile by-products. A vertical hot-air vulcaniser with a variable-speed drum at the top may also be used. Cure usually takes place only on the up-cycle and speeds up to 60 feet per minute are obtained.

Although hot-air vulcanisation can be used to cure wire and cable insulation, continuous steam vulcanisation is usually used because it is much faster with speeds up to 1200 feet per minute. Benzoyl peroxide is preferred to 2,4-di-chlorobenzoyl peroxide to avoid scorching. Continuous-steam vulcanisation is more expensive, longer and more complicated. Other less common types of vulcanisation are steam autoclave vulcanisation, hot-liquid vulcanisation and fluidised-bed vulcanisation.

Calendering

Continuous thin sheets of unsupported silicone rubber are produced on a calender. One can also coat a reinforcing fabric on one side with a 3-roll calender or on both sides with a 4-roll calender. The resulting product gets its strength from the fabric and its flexibility and good electrical properties and moisture resistance from the silicone rubber. Silicone rubber should be processed at slower speeds than organic rubber. The range of 0.2 to 2 feet per minute is best for the start-up until even release from the rolls is assured and running speeds are generally around 5 to 10 feet per minute.

Silicone rubber is often calendered onto fabrics such as glass, nylon, aromatic nylons, polyester and cotton. A variety of weaves are used. Woven fibre glass has the best combination of properties and is most commonly used with silicone rubber. The high-temperature properties of aromatic nylons are making them increasingly popular.

A typical 4-roll calendering set-up is shown in Fig. 15.2. An extruder or a ram feeds silicone rubber to the first and fourth rolls. These rolls are slower and warmer than the centre rolls. Since silicone rubber moves to the faster and cooler rolls, it is transferred to the centre rolls. The centre rolls then transfer the rubber to the fabric. While the outside rolls may be warmed to facilitate transfer, the temperature must be kept below 130°F in order to prevent scorching. If the silicone rubber is unsupported, the fabric is replaced with a liner. Polyester film, holland cloth, polyethylene and release-coated paper are commonly used. The liner is stripped off after cure or just before use if the sheet is to be used green. A major use for green sheet is to cut it into strips to build hoses or to manufacture electrical tape. A 3-roll calender is actually more common than a 4-roll calender.

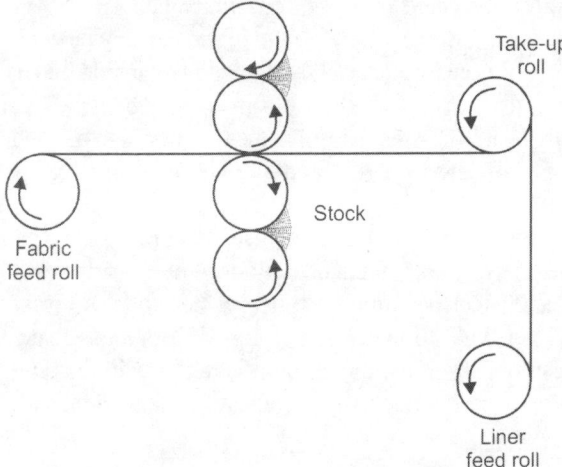

Figure 15.2: Flow diagram of set-up for calendering supported sheet.

Calendering sheet can be vulcanised in a steam autoclave, provided the roll is first pressure-taped. The thickness of the roll must be controlled and preferably heated inside and out to achieve complete cure.

The sheet can also be cured or semi-cured by eliminating the liner feed and bypassing the supported or unsupported sheet over a hot drum placed between the calender and the take-up roll.

Dispersion coating of fabric

This technique of fabric coating permits thinner coatings and provides more thorough penetration of the fabric than does calendering. A thin coat from a 5–15% silicone rubber dispersion will improve the strength and flex life of glass cloth and provide a good 'anchor coat' for calendering. Excellent high-temperature electrical insulating materials, diaphragms and gaskets can be made from glass cloth with thicker coatings. Silicone rubber can also be dispersion-coated on organic fabrics and then used in many applications such as aircraft seals, radome covers and general-purpose control diaphragms.

The silicone rubber is generally supplied by manufacturers in dispersions of soft, readily dispersed pastes that have been especially designed for cloth coating. However, any silicone rubber compound can be dispersed in solvent and probably most dispersions are made at the fabricator's plant.

Xylene, toluene and mineral spirits are common solvents, except for the fluorosilicones for which the solvent must be a ketone such as methyl ethyl or methyl isobutyl ketone. Chlorinated solvents and solvents containing anti-oxidants, rust inhibitors and similar additives should be avoided since they interact with the peroxide vulcanising agent.

The coating pastes are readily dispersed with a propeller mixer. Compounds containing reinforcing fillers should be freshened, sheeted off thin, cut into small pieces and soaked overnight in just enough solvent to cover the compound. The mixture should then be stirred with the propeller mixer until it becomes uniform. The remaining solvent is then added, with mixing, in small portions. The dispersion should be filtered through an 80 to 150-mesh screen (depending upon consistency) before use.

Benzoyl peroxide is normally used for curing because it has a high-enough decomposition temperature and a low-enough vapour pressure to permit the use of heat to remove the solvent after coating. It is best to add crystalline benzoyl peroxide to the dispersion in the form of a 5% solution in toluene or xylene. Preventing overheating during or after peroxide addition is essential.

Dip-and-flow coating is used for priming and for applying thin coatings (for example, coating a 4-mil electrical-grade glass cloth to 10 mils overall thickness). Thickness is controlled by coating speed and by dispersion solids concentration (ordinarily 5–25%). Usually, the uncoated cloth enters the dip tank at a 45° angle, passes under an idler roll at the bottom of the tank and then up into a vertical oven or coating tower. Excess dispersion flows off the cloth between the tank and the tower. The coating tower is heated by hot air and ideally, is divided into three zones. The solvent is removed in the first zone at 150–175°F. Vulcanisation takes place in the second zone at 300–400°F in the case of glass cloth and at 250–300°F in the case of organic fabrics. When glass cloth is being coated, the third zone is maintained at 480–600°F.

This last temperature range removes final traces of volatiles, including peroxide decomposition products. Maximum bond to the cloth and optimum electrical properties are developed in the third zone. Dispersion coating is somewhat faster than calendering, with speeds of 10–20 feet per minute common. Dip-and-knife coating is similar to dip-and-flow. However, thicker dispersions (35% solids is typical) are used and a knife or rod is placed between the dip tank and the tower. The coated fabric is pulled past the knife (a second knife can be placed on the opposite side of the fabric), which wipes off excess dispersion.

Thickness is controlled by dispersion solids concentration and knife position relative to the fabrics. Heavier coatings can be obtained by this method than by dip-and-flow. In reverse-roll coating, the knife is replaced by a roll which rotates in opposition to the direction of the cloth movement. This often improves penetration.

Heavy-duty hose

Heavy-duty hose for autos, trucks, aircraft and industrial use requires different techniques. Uncured and semicured, unsupported and fabric-supported silicone rubber sheets and tape may be fabricated into ducts and heavy-duty hose. This is done by wrapping a hollow mandrel, a collapsible mandrel, a core made with a low-temperature-melting alloy or a foundry sand core. Aluminium mandrels are widely used, but release of the finished part is often a problem. The mandrel should be sprayed or brushed with a dilute aqueous solution of household detergent or dusted with mica or talc.

The aluminium mandrel works best for straight sections of hose. Silicone rubber sheets are wrapped around the mandrel to the desired thickness. It is easier to get a tight, wrinkle-free construction if the mandrel is turned on a lathe. Smooth liner for hose may be made by butt-wrapping uncured tape or sheet on the mandrel, or by using extruded and cured tubing.

For more complex shapes with curves, a core made from a low-temperature-melting alloy or from foundry sand and resin is often used. The cores are spiral-wrapped with fabric-reinforced tape cut on a 45° bias. Bias-cut tapes are more stretchy and give the hose maximum flexibility.

After the hose has been built, it should be pressure-taped with wet cotton or wet nylon. This is necessary to prevent sponging during cure and to provide maximum ply adhesion.

Curing is usually accomplished by running steam into the hollow mandrel, or by placing the wrapped core in a steam autoclave. Curing time and temperature depend upon the peroxide used and the heat capacity of the hose assembly. Following cure, the pressure tape should be removed and the mandrel stripped, while the assembly is still warm. The release agent should be washed off before the part is oven post-baked. If a core of a low-temperature-melting alloy has been used, the alloy is melted to free the cured hose. If a core of foundry sand has been used, the brittle core is broken up to release the cured hose.

15.4.3 Bonding of silicone with metals

Silicone rubber can be bonded to many materials, including iron, nickel, copper, zinc, aluminium, titanium, various steels, ceramics, glass, masonry, many plastics, organic and inorganic fabrics, vulcanised and unvulcanised silicone rubber and other elastomers.

In all cases, it is essential to thoroughly clean the surface to be bonded. Metal surfaces containing loose scales, oxides, other salts and embedded dirt should be sand blasted, sand papered or acid-etched. Metal surfaces should be cleaned with a solvent such as methylene chloride or trichloroethylene. Some surfaces, such as plastics and vulcanised rubbers, should be roughened with abrasives. It helps to wash the surface with acetone before application of the primer. Primers are applied in dilute solution by brushing, dipping or spraying. If the bond is inadequate, the layer of primer may be too thick or the primer solution may be too concentrated. One may also try several primers to find the one best suited to the particular materials being bonded. The typical primer for silicone rubber has two types of reactive sites. The first reactive site reacts with the surface hydroxyls of the metals, glass, or masonry. The reactive compound is usually an alkoxy silane (e.g., SiOMe) which hydrolyses to a hydroxy silane that can then form either a hydrogen bond or a covalent bond with the surface hydroxyls. A catalyst to speed this reaction may also be present. The second reactive site in the primer is some type of unsaturation to react with the silicone rubber. The hydrolysis of the alkoxysilanes usually takes place within 60 min of the time the primer is applied. The reaction with the surface hydroxyls takes place while the primer is drying and also during vulcanisation. The reaction of the primer unsaturation with the silicone rubber takes place during vulcanisation and is catalysed by peroxide.

Bonding unvulcanised silicone rubber

A curing agent should be added to the freshened compound. After preforming, the stock should be carefully laid on the freshly primed surface. Vulcanisation must be done under heat and pressure at temperatures appropriate to the peroxide used. Approximately 15–30 min at 330–350°F is usually sufficient when using dicumyl peroxide or VAROX and 15–30 min at 260–280°F, when using powdered benzoyl peroxide. A stepped oven post-bake is usually required to develop optimum bond strength.

The self-bonding compounds may be bonded to many metals in the same manner, except that no primer is required. Unlike bonding to primed surfaces, this technique is insensitive to part geometry and to the degree of compound flow in the mold during bonding.

Bonding vulcanised silicone rubber

Cured silicone rubber may be bonded to a primed surface with a 10–40 mil interlayer of heat-curing silicone rubber adhesive. Vulcanisation is conducted under heat and pressure. The pressure must not be so great as to squeeze out the adhesive. This bonding can also be accomplished under pressure at room temperature with a room-temperature-vulcanising silicone rubber adhesive. Both techniques can be used for splicing, or for bonding cured silicone rubber to itself.

The self-bonding compounds are excellent for bonding cured silicone rubber to itself and are recommended for splicing. These bonds will usually be stronger than those obtained with adhesives.

15.4.4 Post-baking

With trimming or deflashing, oven post-baking is often the final step in the fabrication of silicone rubber parts. Post-baking removes volatile materials such as low-molecular-weight silicones and peroxide decomposition products. Removal of the low-molecular-weight silicones results in less shrinkage later and in low extractables. Removal of the peroxide decomposition products improves reversion resistance, compression set, electrical properties, chemical resistance and the bond to other substrates.

Ovens should have forced-air circulation with no dead-air pockets. Air flow should be sufficient to keep the oven atmosphere outside the explosion limits of the volatiles from the rubber. Oven vents to the outside of the building should be checked periodically to ensure that they are not plugging up and reducing air flow. Temperature control should be ± 10°F and the oven should go up to 500–600°F. The oven should be equipped with a temperature-limit switch and a safety switch should be provided to turn off the heat if the blower stops.

The rubber charge should be on stainless steel trays designed to provide maximum contact with the circulating air. The parts must not be in contact with one another, as this would slow the removal of volatiles and lead to parts sticking together.

For some applications, no post-bake is necessary. However, it is generally desirable to post-bake at least a few hours at 300–400°F. Usually, longer cures to at least 50°F above the service temperature are required. Thicker sections should be step-cured, especially if the aroyl peroxides have been used. Also, 2,4-dichlorobenzoic acid and benzoic acid will cause internal reversion if the part containing them is not step-cured.

There are no set curing schedules that will cover every situation that a fabricator may encounter. In each case, the optimum curing schedule must be worked out. The size and shape of the part, number of parts in the oven and the air flow through the oven, have major influences on the curing cycle. A possible schedule for a thicker part is 3 hr at 300°F followed by 4 hr at 350°F and finally followed by 10 hr at 400°F.

15.5 Advantages and disadvantages of silicone rubber

15.5.1 Advantages of silicone rubber

1. High temperature resistance, wide operating temperature range (even-100 to 300°C).
2. UV light, oxygen and ozone resistance (peroxides have to be used for vulcanisation).
3. Elasticity.
4. Non-toxic, odourless, tasteless.
5. Good release properties.
6. Good electrical insulation.
7. Good ageing resistance at high temperatures.
8. Good resistance to low concentrations of acids, bases and salts.

15.5.2 Disadvantages of silicone rubber

1. Weak oil resistance (exception aliphatic oils).
2. Low resistance to steam, acids and alkalis.
3. Weak mechanical properties without additives.
4. Large shrinkage in molded articles.
5. Vulcanisation to obtain good mechanical properties has to be carried out with peroxides.
6. Silicone rubber is very costly.

15.6 Properties of silicone rubber

The strong silicon-oxygen chemical structure of silicone gives the elastomer its unique performance properties. Examples include:

Temperature resistance: Silicones withstand a wider range of temperature extremes than nearly all other elastomers, remaining stable through a range of –75°F to 500°F. They may be sterilised by ethylene oxide (ETO), gamma, e-beam, steam autoclaving and various other methods.

Mechanical properties: Silicone rubbers have high tear and tensile strength, good elongation, great flexibility and a durometer range of 5 to 80 Shore A. The softest durometers available are reinforced gels.

Electrical properties: Silicones exceed all comparable materials in their insulating properties as well as flexibility in electrical applications. They are non-conductive and maintain dielectric strength in temperature extremes far higher or lower than those in which conventional insulating materials are able to perform.

Biocompatibility: In extensive tests, silicone rubbers have exhibited superior compatibility with human tissue and body fluids and an extremely low tissue response when implanted, compared to other elastomers. They do not support bacteria growth and will not stain or corrode other materials.

Silicones are odourless, tasteless and are often formulated to comply with biocompatibility guidelines for medical products.

Chemical resistance: Silicones resist water, oxidation and many chemicals, including some acids and alkali solutions. Concentrated acids, solvents, oils and fuels have a negative effect on silicone rubber and should not be used with silicone.

15.7 Applications of silicone rubber

Though the raw material, silica, from which silicone rubber is made, is cheap and plentiful, the manufacturing process is difficult and expensive. In spite of its high cost, silicone rubber is preferred in many applications as they need less maintenance and prove economical in the long run.

1. Electrical equipment and technical products in high temperatures.
2. Medical devices and hospital supplies.
3. Roll coverings.
4. Cable coverings and insulators.
5. Lining compounds.
6. Molds.
7. O-rings.
8. Seals for the aeronautical industry.

15.7.1 Applications of silicone rubber in aerospace industry

A major consumer of this speciality rubber is the aircraft and aerospace industry. It is used for gaskets and sealing rings for jet engines, in ducting for circulating hot air to the cabins and to surfaces requiring deicing, as sealing strip for doors, windows, turrets, switch boot covers, vibration dampers, for sealing cavities and junctions for environmental protection, for bonding and sealing aircraft components exposed to fuels and oils and as groove injection sealants for aircraft fuel tanks. Silicone rubber can withstand temperatures upto 9000°F (5000°C) for several minutes while still retaining good insulation properties. The superior ablation characteristics thus enable its use in the coating of rocket fuel valves, supply cables and silo doors for protection from rocket blast.

15.7.2 Applications of silicone rubber in automotive field

The automotive field is currently showing the maximum growth for the use of silicone rubber. A critical application of silicone rubber bonded to metal is in the shaft seals used in automatic transmissions. The silicone rubber not only resists the action of the hot transmission fluid and the frictional heat of the shaft on the seal lip but also prevents the corrosive pitting of the shaft during extended periods of engine idleness, which generally occurs with sulphur-containing rubbers. Silicone rubber hoses which have superior resistance to hot ethylene glycol, diesel fumes and engine-cleaning solvents, last as long as the engine itself. High ignition voltages and higher under-the-hood temperatures necessitate the use of extruded silicone rubber ignition wire insulation and molded silicone rubber spark plug boots. Molded fluorosilicone rubber check seals are used in carburettors to replace metal components. Room temperature vulcanisation (RTV) sealants are used for formed-in-place engine gaskets, repair and installation of wind shields and light covers. It is estimated that 3 to 3.5 kg of silicone rubber is used per car at present.

15.7.3 Electronics and electrical industries

In the electronics industry, silicone rubber is used for semiconductor junction passivation, protection of electronic components, general potting and encapsulation for modules, relays, power supplies, amplifiers, etc., as a-particle barrier. For calculators and adding machine keyboards it eliminates corrosion and makes bounceless contact possible, minimising electrical noise and double signals. In the electrical industry, silicone rubber is used for wire and cable insulation, encapsulation of transformers and motors, electrical connectors for industrial and household requirements, in insulating tapes for transformers, heating pads and several electrical parts. Effective microwave oven gasketing is made from conductive silicone rubber. It also helps to reduce RF (radio frequency) inter-ference and is used in oven sealing. Special silicone rubbers which are high-loss dielectrics are used in microwave transmission lines. The largest growth area in the coming years will be for wire and cable insulation with emphasis on fire safety in nuclear plants.

15.7.4 Applications of silicone rubber in biomedical industry

Silicone rubber is largely used in biomedical and prosthetic devices to replace a missing ear, to restore movement to arthritic hands and for many other body parts. Their characteristic inertness, resistance to ageing, flexibility, softness, dependability, sterilisability and biocompatibility make them ideal for this purpose. As a biomaterial in extracorporal devices, polymethylsiloxane tubing is used with heart-lung machines, dialysers and related devices, after incorporating heparin as an anti-coagulant for the blood. By reacting poly-urethane with polyorganosiloxanes having reactive end groups, a tough material called Avcothane has been developed by AVCO Everett Research Laboratory in the U.S. This tough elastomer is able to withstand millions of flexings without failure and as such is used widely in heart assist devices like interaortic balloons, artificial hearts and artificial blood vessels.

Other uses of silicone rubber are in surgical tubings, heart valves, catheters, hydrocephalus, shunts, nerve cuffs, replacement of bladders, contact lenses and prosthetic parts to correct deformation due to heredity, disease or accident. In the prosthetic applications, it is used widely for reconstructing breast, ear, eye, bone, finger joint, limb, tendon and ligament.

The most popular of silicone implants is the mammary implant which consists of a strong, thin but very stretchy silicone rubber envelope filled with a clear silicone gel of the same general weight and texture as the breast tissue. Other gel or liquid-filled silicone rubber devices include testicle implants for cosmetic replacement, vaginal forms for the reconstruction of vagina, penile implants for males who have erectile incompetence.

15.7.5   Miscellaneous applications of silicone rubber

Silicone rubber is also used for many other applications. Flexible mold materials of silicone rubber allow excellent reproduction of details of casting of metals, plastics, waxes and plaster. The non-stick characteristic leads to its use in rolls for handling hot plastics in the production of films, embossing, laminating and smoothing of hot tin over tin plate steel. Its release properties find application in coatings on release paper and parts of machinery or rolls handling sticky materials such as hot polyethylene, candy and adhesives.

Constancy in transmissibility or resonant frequency in the temperature range of -65–$300°F$ and the retention of dynamic absorption characteristics in spite of ageing makes it an ideal material for effective noise and vibration control.

Thin films of silicone rubber are selectively permeable to different gases and as such may be used for separating gases to obtain relative enrichment of one gaseous component over another. Its outstanding weatherability in the hottest and flexibility in the coldest of climates makes it ideal for the building industry to caulk joints of masonry, metal, wood or plastics, to seal windows, as protective coatings for plywood, metal or masonry building materials indoors as well as outdoors.

Thermoplastic polyurethane

16.1   Introduction

The first commercially available thermoplastic elastomers were polyurethanes (TPUs) formed of long flexible polyether or polyester chains linked by polar polyurethane units which associate into microdomains by hydrogen bonding. These are specialty rubbers. These segmented copolymers have the general formula $(AB)_x$, whereas a triblock copolymers has the general formula ABA. Polyurethanes are generally manufactured from an aromatic disocyanate, an oligomeric diol and a low molecular weight diol.

The low molecular weight diol is typically called a chain extender because it links AB segments together. A typical thermoplastic polyurethane based on diphenyl methylene-4,4/diisocyanate (MDI), poly (tetramethylene oxide) and butane diol is given below:

where, n = 1–5, m = 20 and x = 20

Thermoplastic polyurethanes have excellent strength, wear and oil resistance and are used in fibres, footwear, automotive bumpers, snowmobile treads, adhesives, etc., and in high performance structural applications. Approximately 15% of the thermoplastic elastomer market is claimed by polyurethanes.

The best-known polyurethane rubbers are the flexible polyurethane foams (often illogically classified as plastic foams), but in addition solid vulcanisable materials have become of use as a result of their oil resistance and excellent abrasion resistance, whilst the thermoplastic polyurethane rubbers exhibit these properties together with the ability to be processed as thermoplastics.

Polyurethanes, as a class of materials, are one of the most versatile available. By varying the reactants, their amounts and the reaction conditions, one can obtain millable elastomeric gums, hard rigid plastics, reactive liquids and foams.

16.2   Basic reactions in polyurethane chemistry

The basic reactions in polyurethane chemistry are:

RNCO	+	R'OH	$\rightarrow$	RNHCOOR'		...(16.1)
Isocyanate		Alcohol		Urethane		
RNCO	+	R'NH$_2$	$\rightarrow$	RNH – CONHR'		...(16.2)
Isocyanate		Amine		Urea		
RNCO	+	R'NHCOOR"	$\rightarrow$	R'NCOOR"		...(16.3)
					CONHR	
Isocyanate		Urethane		Allophonate		
RNCO	+	R'NHCONHR"	$\rightarrow$	R'NCONHR"		...(16.4)
					CONHR	
Isocyanate		Urea		Biuret		

In reactions (16.3) and (16.4) the isocyanate is capable of reacting with the active hydrogen in a urethane or urea groups, to give branching or cross-linking by the formation of an allophonate or biuret group. The most important reactions for the production of elastomers, however, utilise diisocyanates and polyols and the elastomeric products formed can be castable polyurethanes, millable gums, thermoplastic polyurethanes and polyurethanes of other types.

Castable polyurethanes: These liquid systems can be produced either in a one-shot system (i.e., the diisocyanate, polyol and chain extender reacted in a single stage) or more usually, as a pre-polymer, which is chain extended and cross-linked at a later stage. In the pre-polymer system, the diisocyanate and polyol (either a polyether or a polyester) are reacted to give a pre-polymer, which may be either a liquid or a waxy solid. The reactant ratios used ensure the pre-polymer contains isocyanate groups at the chain ends.

By altering the foregoing components, it is possible to make products elastomeric or rigid and by the inclusion of water or a blowing agent, cellular products can be obtained. Clearly, so many possible variations make general comments difficult, but the following are generally accepted.

1. Polyester polyurethanes generally give superior mechanical properties and chemical resistance, but inferior hydrolytic stability.
2. Polyether polyurethanes give superior low temperature properties and hydrolytic stability.
3. Diamine chain extenders give superior properties to diol-cured elastomers.
4. Mechanical properties are generally improved as the hardness increases.

Processing of liquid systems proceeds as follows:

Pre-polymer

Mix → Degas → Cast → Cure (Solid)
↓

Chain extender   Post-cure

The mixing can be done by hand, or in low pressure mixer/dispensers and in reaction injection molding (RIM) machines. In the latter operation, no degassing is required.

16.3   Polyurethane components

Urethane polymer formation requires di- and poly functional reactants to enable the building of large linear chains and three-dimensional networks. The actual components used to prepare polyurethane elastomers are usually difunctional and the classes are listed in Table 16.1. However, the formulations used to make some polyurethane systems such as foams, reaction-injection molding (RIM) products, coatings, caulks, adhesives and so on. Almost all polyurethane components are room-temperature liquids or low-melting solids for process advantages such as transport, metering and mixing.

Table 16.1: Polyurethane components.

Component	*General structure*
Diisocyanate	OCN — R — NCO
Macroglycol (polyol)	HO — R' — OH
Chain extender (cross-linker)	HO (or NH_2) — R'' — OH (or NH_2)

Diisocyanates: The diisocyanate component may be any of a multitude of relatively small molecules. These have achieved commercial importance in the polyurethane industry.

Polyols: The polyol component of polyurethane elastomers is usually a macroglycol, of molecular weight about 500 to 4000 (usually 1000 to 2000), whose bifunctionality permits the formation of long, strong, linear chains. In some elastomer formulations, the polyol may be a triol or even of higher functionality.

Chain extender (cross-linker): The chain extender component of polyurethane elastomers is a relatively small, usually difunctional molecule, about the same size as the diisocyanate. In the chain extending reaction with isocyanate, its small size result in structures rich in urethanes groups and/or urea groups, both of which have the dual ability to form vinyl chloride (VC) through hydrogen bonding and CC or chain branching through allophanate formation and biuret formation by further reaction with isocyanate.

16.4   Polymerisation of polyurethane

Basically, there are two commercial methods of polyurethane polymerisation – the 'pre-polymer' (two step) process and the 'one-shot' process.

Pre-polymer process: This process involves the preparation of a low-molecular-weight, isocyanate-terminated pre-polymer, then its chain extension to a high polymer.

In the first step the macroglycol units with excess of the diisocyanate through urethane link formation in an end-capping reaction to produce linear chains that terminate in isocyanate groups, remain reactively low in molecular weight and remain low melt viscosity, thus enabling liquid processing of the pre-polymer. The diisocyanate-macroglycol chain segments (singly under-lined) comprise the urethane-sparse 'soft-segment' structure in polyurethane chains.

In the second step the terminal isocyanate groups of the pre-polymer react with the added chain extender through urethane like formation to couple the pre-polymer molecules and produce a high molecular weight polyurethane elastomer. In the process, a new structure, the diisocyanate-chain extender product (urethane-rich 'hard segment,' doubly underlined) has been built into the polyurethane chains to alternate with the soft segments.

One-short process: This is single-step process in which all of the polyurethane components are mixed together at one time. Here the polymerisation proceeds to completion again yielding a high polymer with alternating soft and hard segments. The one-shot process is used to make thermoplastic polyurethane elastomer resins, RIM (reaction injection molding) products, foams and so forth. As in the pre-polymer process, polymer hardness and modulus can be increased by increasing the diisocyanate and (balancing) chain extender levels charged in the polymerisation, which, in effect, produces more and longer diisocyanate-chain extender hard segments in the polyurethane.

16.4.1   Chemical cross-linking

The polyurethane chemical cross-linking can be part of the polymerisation process in the liquid processing systems (casting, RIM), accomplished by proper adjustment of the polymerisation recipe. But in the case of the millable gum and thermoplastic resin urethane elastomer, conventional rubber-processing and curing techniques (mill/Banbury mixing, compression/injection molding) are applied.

16.5   Properties of polyurethane (AU)

The properties of polyurethane depend greatly on the type of cross-links.

Viscosity: AU grades which can be cross-linked with isocyanates (AU-I grades) are relatively soft, with Mooney viscosities between 14 and 25, while peroxide-cross-linkable grades (AU-P grades) have conventional Mooney viscosities of about 55.

Storage life: This is good in AU. AU-I grades should be stored in a cool and dry environment.

Solubility: AU can only by dissolved in very few solvents. Only dimethyl-formamide completely dissolves AU-I grades, while AU-P grades also dissolve in ketones and tetrahydrofuran.

Hydrolysis resistance: AU grades differ in their resistance to hydrolysis and resistant or non-resistant grades are commercially available. EU grades are more resistant in this respect.

16.6   Compounding of polyurethane (AU)

16.6.1   Vulcanising agents

Isocyanates and peroxides are the main vulcanising agents for AU, while sulphur vulcanisation, which is possible with some AU grades, plays only a minor role. AU-I grades are commonly cross-linked with toluylene diisocyanate (TDI) and for products with a high hardness one uses higher TDI concentrations together with hydroquinonedioxyethyl ether. Organic lead salts acts as accelerators. Although, in principle, it is also possible to use other diisocyanates, they do not, as a rule, offer any advantage over TDI as cross-linkers.

Instead, most of them have serious shortcomings. TDI can be readily dispersed in compounds, it has a good storage life in these compounds after mixing and yet, it gives fast cure rates.

The physical properties of TDI vulcanisates are also much better than vulcanisates from other diisocyanates. Only stable peroxides qualify for a peroxide cure of AU-P grades, because a sufficient scorch safety required for the mixing and subsequently processing steps.

The choice of peroxides is also governed by the well known rules. If particularly high cure states are required, special co-activators, like triallyl-cyanurate (TAC) have to be used. They do not affect the potlife and processing safety of compounds significantly. Peroxide cures cannot be accelerated and the compounds must not contain sulphur or sulphur compounds, since these interfere with the peroxide cure.

16.6.2   Protective agents against hydrolysis

Hydrolysis inhibitors are important, particularly for hydrolysis-prone AU grades. Polycarbodiimides are successful for this purpose and they also improve the resistance of AU vulcanisates to lubricants, hot air and weathering, particularly in tropical climates.

16.6.3 Fillers

AU-I grades are frequently processed without filler and in this instance, the hardness of the vulcanisates has to be adjusted through the amount of TDI and the co-reagent, hydroquinonedioxyethyl ether. The hard isocyanate segments by themselves give reinforcement, a high hardness and a good tensile strength. And yet, fillers with varying degrees of reinforcement can be used with AU-I grades. Already small amounts of reinforcing blacks or fumed silica can improve considerably the hardness and tear resistance of vulcanisates. Although fillers can be readily incorporated, the amount of filler that can be used is limited, because the mixing temperature has to be kept low due to the presence of free diisocyanate. In this way, premature scorch is avoided. Semi- or non-reinforcing fillers are used to modify the processibility of compounds and to reduce compound cost. Moisture in fillers can adversely affect the isocyanate cure and also the ageing resistance of vulcanisates. AU-P grades are compounded mostly with reinforcing or semi-reinforcing fillers to obtain optimum vulcanisate properties. Although N330 and N550 blacks are most frequently used, their choice regarding compound processibility and vulcanisate properties is governed by the same rules as for other rubbers. Silica fillers can also be used.

16.6.4 Softeners

The low swelling of AU-I grades in many solvents also means that the rubber is incompatible with most plasticisers. Therefore, they are hardly used at all in lower processing temperatures and thus improves the scorch life in injection molding or calendering processes. AU-P grades can be plasticised with small amounts of phthalates and polyadipates to reduce the vulcanisate hardness.

16.7 Processing of polyurethane (AU)

In general, AU compounds are mixed and processed using conventional rubber machinery and following generally accepted rules. With AU-I grades, however, there can be processing problems and they should be treated like any other scorchy compound. Moisture and water should not come into contact with AU-I compounds, nor can open steam vulcanisations be used. Humidity, cure temperature and cure time have an influence on the vulcanisate properties and bimodal cures also change the property spectrum.

16.8 Properties and uses of polyurethane (AU) vulcanisates

16.8.1 Mechanical properties

The tensile strength of AU-I vulcanisates reaches as high as 40 MPa and it is thus usually higher than that of other rubber vulcanisates. The hardness is generally high and it ranges for AU-I grades from 70 to 99 Shore A. And yet, the AU-I vulcanisates exhibit a relatively high degree of elasticity at every hardness level.

The abrasion resistance is excellent and better than that of other rubber vulcanisates, in spite of the high hardness of the AU-I vulcanisates.

16.8.2   Heat resistance, ageing resistance

AU vulcanisates withstand a long-term exposure at 75–90°C or even higher. In keeping with their chemical nature, AU vulcanisates can be attacked by hydrolysis due to hot water, steam, acids or bases, as well as by some lubricants, by heat and particularly by longer exposure to tropical climates. Vulcanisates from the hydrolysis-resistant AU-P grades can be used for one year at 65°C. The ageing, weathering and ozone resistance of AU vulcanisates is excellent. Products from AU are not attacked by oxygen nor ozone.

16.8.3   Low temperature flexibility

AU vulcanisates have reactively good low temperature properties. The dynamic brittleness temperatures are mostly between –22 and –35°C.

16.8.4   Swelling resistance

AU swells only very little in aliphatic and many other solvents. Highly polar solvents, like chlorinated hydrocarbons, aromatics, esters and ketones swell AU vulcanisates to a greater extent, but in contact with these solvents and with high-octane motor fuels, AU vulcanisates perform much better than many other rubber vulcanisates.

If high vulcanisate hardnesses can be tolerated in specific applications, AU-I vulcanisates from high TDI and hydroquinone-dioxyethyl ether can be used, since the degree of swelling is inversely proportional to the modulus of the vulcanisates.

16.8.5   Uses of AU

Products made from AU are particularly used in the automotive and mechanical engineering industries for seals, elastic-, shock-absorbing or damping members, for power transmission elements, for flexible joints, for suspensions and for supports with high abrasion resistance.

In these applications, the chemical and physical performance characteristics are exploited, namely a combination of good weather and solvent swell resistance, high abrasion resistance and elasticity even at high hardness levels and good low temperature properties. However, in all applications, one has to consider the potential hydrolysis, the limited heat resistance and the relatively high price of the rubbers.

16.9 Selection criteria for cast polyurethane elastomers

Selecting the right material for a given application requires detailed knowledge about the various materials available in the market, their properties, their performance and their economics. Sometimes steel, aluminium or other metals are the materials of choice, in other cases rubber or plastics such as ABS, polystyrene, PVC or phenol-formaldehyde resins are best.

In a growing number of needs, cast polyurethane (PU) elastomers offer superior design and performance characteristics. A wide variety of elastomers are now available. Hence it is essential to understand the types of cast PU elastomers that are available, the kinds of physical and environmental resistance properties they have and typical applications that each will fulfil.

16.9.1 Need for using castable polyurethane elastomers

Cast polyurethanes are made by mixing a pre-polymer and a curative and pouring the mixture into a mold. These two liquids are blended either by hand mixing or by a processing machine, poured into a mold and allowed to cure into a final shape. The major characteristic of these materials is that they have extraordinary physical properties. They are actually engineering materials and are chosen for use on the basis of these properties.

Some of the reasons for using cast polyurethanes are: (i) performance, (ii) abrasion resistance, (iii) toughness, (iv) tear resistance, (v) load-bearing ability, (vi) cost effectiveness, (vii) reduced down time in process operations and (viii) lower tooling and equipment costs for small production runs.

The two most important reasons are performance and cost effectiveness. In some cases, performance characteristics of these materials allow them to be used in applications where other materials simply cannot be used satisfactorily due to the tough demands. In other cases, end uses select polyurethanes because they can outperform other materials by a large margin. This is usually a result of their particular properties, abrasion resistance and toughness, that is, resistance to breakage on impact or in rough handling, very high tear resistance and high load-bearing ability.

These four properties, although certainly not the only outstanding properties of urethanes, are the ones that usually make them stand out far above other materials in many applications.

Cost effectiveness is the second reason. Even though polyurethanes are often more expensive than other materials including various rubbers the extra cost is frequently justified in terms of less downtime in actual service. This is particularly critical, for instance, in mining and paper mills. Downtime in these operations is very expensive.

16.10 Advantages and limitations of polyurethane

16.10.1 Advantages of polyurethane versus metal

(i) Lighter weight, (ii) less noise, (iii) better wear, (iv) cheaper fabrication and (v) corrosion resistance.

One of the advantages of urethanes versus metal is lighter weight. Pans fabricated from polyurethane weigh far less than metal pans and are much easier to handle and, typically, result in having to move less mass in machinery. In addition, metal pans tend to generate noise while polyurethane absorbs noise.

The reduction of this 'noise pollution' in the workplace when polyurethanes replace metal can often be dramatic. Polyurethanes will also outwear metals in many applications and can be easily cast in inexpensive tooling as discussed earlier. In contrast, making metal pans requires foundry operations, welding and machining and, as a result, can be very costly, particularly with high hardness alloys.

Polyurethanes are also corrosion resistant. For example, in many mining operations, highly corrosive solutions cause rapid deterioration of steel. Wherever there is a combination of abrasion and corrosion at the same time such as in copper mining railings and pipelines – the lifetime of metal parts can be remarkably short. Urethanes, because they have high resistance to abrasion and to corrosion, outlast metals by a large margin.

16.10.2 Advantages of polyurethane versus plastics

(i) Non-brittle, (ii) elastomeric memory and (iii) abrasion resistance.

One advantage of urethane elastomers over plastics is that they are not brittle. Many plastics, particularly in the higher hardnesses, tend to crack and break under impact and shock loading. Polyurethanes remain true elastomers maintaining their high impact resistance even up to very high hardnesses. Polyurethanes also have elastomeric memory, that is, they can be stretched even at high hardnesses to substantial elongations and will return to their original dimensions. Most plastics, once they have been stretched beyond a certain point, remain permanently stretched. Finally, plastics, as a class, do not have the high abrasion resistance of polyurethanes.

16.10.3 Advantages of polyurethane versus rubber

1. Abrasion resistance.
2. Cut and tear resistance.
3. Higher load bearing.
4. Clarity and translucence.
5. Ozone resistance.

6. Harder durometer range.

7. Resistance to mold, fungi, etc.

8. Non-marking.

The main advantages of urethanes versus rubber are the higher abrasion resistance, greater cut and tear resistance and higher load bearing ability. In addition, most cast polyurethanes have natural colours ranging from completely clear to opaque white or amber. Their ready acceptance of a wide variety of pigments and dyes also permits colouring ranging from black to brilliant fluorescent oranges, reds or greens. This is especially useful in colour coding of parts. A good example is in business machines where rolls and belts are colour coded so that correct replacement can be easily made.

Rubber is subject to ozone cracking, particularly around electrical equipment where ozone concentrations can be high. Polyurethanes have no ozone-cracking problem. The fact that polyurethanes are pourable and castable, which was mentioned earlier, makes for cheaper tooling and makes possible the fabrication of complicated parts. Moreover, most rubber compounds when compounded up to 90 or 95A durometer, have sacrificed a good deal of their physical properties. On the other hand, polyurethanes in the 80 to 95A durometer range are approaching the peak of their properties and give extremely good performance at these hardnesses.

16.10.4 Limitations of polyurethanes

(i) High temperature service, (ii) moist hot environments and (iii) certain chemical environments.

There are three main limitations of polyurethanes. Polyurethanes are not high temperature materials. Owing to a certain amount of thermoplasticity in their nature, properties tend to fall off at elevated temperatures. Generally speaking, polyurethanes are not useful materials under heavy service loads at temperatures above approximately 105–107°C.

Another limitation is that all polyurethanes are subject to hydrolysis in the presence of moisture and elevated temperatures. The combination of the two factors creates a problem. While at low temperatures, most polyurethanes can withstand continual contact with water for many years, no polyurethane can stand prolonged contact with live steam. In between, there is a wide range of temperature and moisture conditions under which polyurethanes may, or may not be suitable for use. Newer developments in polyurethane chemistry, however, show promise to push these limits further. Lastly, there are certain chemical environments that are unsuitable for cast polyurethanes. Very strong acids and bases generally are detrimental as are certain solvents, specifically the aromatic solvents such as toluene or ketones such as methyl ethyl ketone

(MEK) or acetone and esters such as ethyl acetate. (There are many solvents, on the other hand, which urethanes resist very well and are suited for in-contact service. These include many oils and petroleum-based materials).

16.11 Polyester and polyether

There are two main types of polyurethane: polyester and polyether. Both are highly effective in a diverse range of industries. Polyester urethane and polyether urethane are elastomers, meaning that they possess elastic properties and both offer unique performance properties. What are the main differences between polyester vs polyether urethanes? Below is a breakdown of each type's physical properties and what kind of projects they are suited for.

16.11.1 Abrasion resistance properties

There are two types of abrasion resistance sliding and impingement. Sliding refers to scraping and rubbing abrasion while impingement refers to particles or objects striking the urethane surface at a high angle.

1. Polyester has a high capacity for sliding abrasion resistance which makes it ideal for applications like scraper blades and chute liners.
2. Polyether offers excellent impingement abrasion resistance which makes it the choice for sandblast curtains and bumpers that get hit head-on.

16.11.2 Heat resistance properties

Both polyester and polyether urethanes perform well at elevated temperatures however:

1. Polyesters withstand high temperatures longer and are more resistant to heat ageing.
2. Polyethers are much less susceptible to dynamic heat build up. That is why they are the choice for high speed rollers where the rapid flexing creates heat.

16.11.3 Low temperature flexibility

All elastomers (rubbers, urethanes, silicones, etc.), get harder and less flexible as temperatures drop and will eventually reach a temperature at which they become brittle. Urethanes brittle point is between $-40°F$ and $-100°F$ depending on formulation. Polyether are less effected by cold temperatures.

16.11.4 Hardness properties

Both polyesters and polyethers can be made to any hardness from marshmallow soft to bowling ball hard.

16.11.5 Cut and tear resistance

Urethanes are known for their toughness however, polyesters have higher tensile strength and higher cut and tear resistance than polyethers.

16.11.6 Water and moisture resistance

If the product is to be submerged or used in a very humid condition it should be a polyether because polyether have excellent hydrolytic stability.

16.11.7 Oil, fuel and solvent resistance

If the product will come in contact with oils, fuel or solvents (even vapours) your choice should be a polyester as they are more resistant to attack from these harsh chemicals.

16.11.8 Rebound properties

Products such as skate wheels need to return the energy they absorb when your foot pushes them into the pavement (rebound), otherwise you would feel like you are running in soft sand. Sand absorbs most of the energy from your foot and makes you run sluggishly. Polyether provide much higher rebound and therefore is the choice for skate wheels and high speed rollers.

16.11.9 Shock absorption properties

Sometimes you want the product to absorb the energy it receives (opposite of rebound). Polyester is the choice for shock absorption and is used widely in vibration dampening applications.

PEVA, chlorinated polyethylene and ethylene acrylic elastomers

17.1   Introduction

This chapters discussed Poly(ethylene-co-vinyl acetate) (PEVA), chlorinated polyethylene and ethylene acrylic elastomers. These are specialty rubbers and widely used in various industries.

Poly(ethylene-co-vinyl acetate) (PEVA) is a random copolymer of ethylene and vinyl acetate, which is commonly referred to as PEVA copolymer or simply PEVA. The character of PEVA varies strikingly with change in vinyl acetate (VA) content. At very low proportions of VA, the copolymer resembles low density polyethylene (LDPE) but exhibits the physical properties and processing behaviour of a tough thermoplastic. PEVA with high VA content (20 to 40% by weight) are less crystalline and hence have more transparency and flexibility. The wide range of properties, depending on the VA content, extends the application profile from plastics to rubber like elastic products.

The method of preparing ethyl-vinyl acetate depends on the desired vinylacetate content. Mass polymerisation gives 45 weight per cent content at the most, emulsion polymerisation gives over 50 weight per cent content and solution polymerisation 30–90 weight per cent content. PEVA can be vulcanised using peroxides or ionising radiation. Sulphur cannot be used because of the saturated main chain. PEVA is often blended with NR and SBR to improve the ozone resistance.

17.2   Chemical structure of PEVA

PEVA is a random copolymer with a long ethylene chain and pendant acetate groups. It is represented by the following structure (Fig. 17.1).

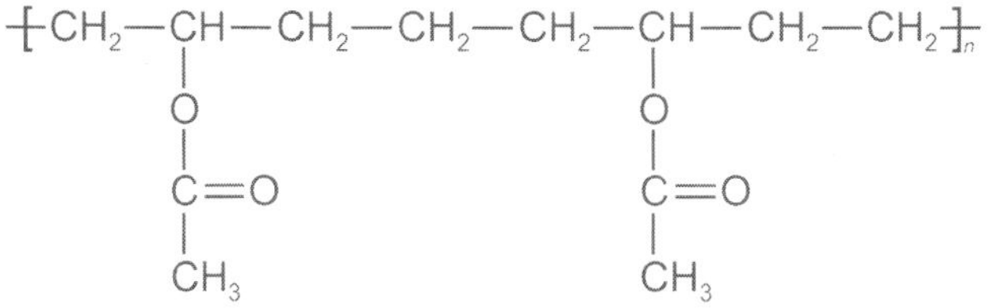

$$\begin{array}{c} \overline{}CH_2\!-\!CH\!-\!CH_2\!-\!CH_2\!-\!CH_2\!-\!CH\!-\!CH_2\!-\!CH_2\overline{_n} \\ \quad\qquad | \qqu\qquad\qquad\qquad\qquad | \\ \quad\qquad O \qquad\qquad\qquad\qquad\qquad\; O \\ \quad\qquad | \qquad\qquad\qquad\qquad\qquad\; | \\ \quad\qquad C\!=\!O \qquad\qquad\qquad\qquad C\!=\!O \\ \quad\qquad | \qquad\qquad\qquad\qquad\qquad\; | \\ \quad\qquad CH_3 \qquad\qquad\qquad\qquad\; CH_3 \end{array}$$

Figure 17.1: Structure of poly(ethylene-co-vinyl acetate).

The amount of VA in the copolymer controls its character. At low VA content it has a plastic nature. Increase in the VA content results in increased flexibility. The crystallinity decreases and it becomes more rubber like or elastic in nature. The presence of acetate groups also gives polarity to the polymer that reflects in its solubility, adhesion and filler intake properties.

17.2.1 General properties of PEVA

1. Excellent oxygen, ozone and light resistance.
2. Extremely good water and oil resistance.
3. Good heat resistance.
4. No resistance to organic solvents.
5. Fire resistance.
6. Good tack to other materials.
7. Low price.
8. Poor tear resistance.
9. Low abrasion resistance.
10. Low elasticity due to the thermoplastic character.
11. With reinforcements, high tensile strength can be obtained.

17.2.2 Physical properties

Vinyl acetate content: The Vinyl acetate (VA) content has a very important effect on the crystallinity and polarity of the copolymer. As the VA content increases, due to the relative bulkiness of the acetoxy side chain, the close packing of polymer chains becomes difficult and the ability of the polymer to form crystalline regions, often referred to as crystallites, decreases. Hence as the VA content increases, the crystallinity decreases.

Another overriding effect of VA content results from the polar nature of the acetoxy side chain. Thus, as the VA content increases so does the polarity of the copolymer. An increase in polarity gives rise to a number of interesting properties such as compatibility with other polymers, increased adhesion, printability and electrical properties.

An increasing VA content can increase the flexibility, transparency, tackiness, filler retention, shock resistance, etc., and decrease the hardness, stiffness, melting point, chemical resistance, etc.

Melt flow index: Melt flow index (MFI) is a direct indication of the average molecular weight of the PEVA copolymer. As the melt flow index increases the flowability of the melt increases but viscosity of the melt, toughness, impact strength and stress cracking strength decreases.

Crystalline melting point: The crystalline melting point (T_m) of the PEVA ranges between 70 and 150°C. The value decreases with increase in VA content. Short term heating above service temperature without mechanical strength is possible right upto the crystalline melting point. Physical or chemical cross-linking can bring about an increase in thermal stability under load.

Thermal conductivity: In comparison with their homopolymers, LDPE or high density polyethylene (HDPE), PEVA shows lower thermal conductivity.

Electrical properties: PEVA copolymer does not have the same dielectric properties compared to LDPE due to the polarity of the VA group.

Creep properties: At low VA content, say upto 20%, the creep or cold flow behaviour of PEVA is similar to that of LDPE. But as VA content increases, resistance to creep decreases.

17.2.3   Chemical properties

Water absorption: PEVA absorbs small quantities of moisture due to its polar nature. The water absorption increases with increase in VA content.

Chemical resistance: Unstressed samples of PEVA show excellent resistance to strong alkalis, brine solution, detergents and non-oxidising media but have poor resistance to hydrocarbons and chlorinated solvents due to the low crystallinity (higher degree of branching) and its polar nature. The resistance deteriorates with increase in VA content.

Solubility: PEVA copolymer is soluble in aliphatic, aromatic and chlorinated solvents. The solubility in all solvents is minimal at room temperature but higher VA content grades can be dissolved in warm solvents.

Resistance to stress cracking: PEVA copolymers are highly resistant to stress cracking even in the presence of polar organic solvents or surface active agents like detergents. The resistance to stress cracking increases with increase in VA content.

Gas and vapour permeability: PEVA is permeable to gases and vapours such as CO_2, O_2, water vapour etc. Because of its lower crystallinity, PEVA is more permeable to these gases than LDPE. Permeability increases with increase in VA content.

Weathering: Exposure to UV radiations, in the presence of atmospheric oxygen slowly degrades unprotected PEVA.

17.2.4   Applications of PEVA

Due to great versatility of PEVA copolymers it could find a wide range of applications. The properties of PEVA such as their intrinsic flexibility, toughness chemical resistance, environmental stress cracking resistance (ESCR), easier processing and ozone resistance give them an advantage over other polymers. PEVA offers a cost advantage over rubber and polyvinyl chloride (PVC) compounds due to low density. PEVA is superior to LDPE due to its properties like ESCR, low softening temperature, better UV resistance, etc. PEVA has advantage over PVC and elastomers with outstanding low temperature flexibility, retention of flexibility over long periods, better chemical resistance, better UV and ozone resistance, better ESCR, high processability, etc.

PEVA is compatible with a wide range of fillers, waxes, resins and other thermoplastics and hence widely used in hot melt and other blend applications. It is also ideal for packaging and storing food stuffs since they are devoid of plasticisers. The products that require greater flexibility at low temperature such as freezer doors, gaskets, boots, ice cube trays can be made. Medico-pharmaceutical applications such as anesthetic facemask, contact lens holders, orthodontia products and disposable feeding bottle teats can be made.

PEVA copolymers are widely used in the manufacture of various types of films which require flexibility, toughness and optical properties, films for fresh meat packaging, horticultural films, cling films and disposable surgical gloves can be manufactured.

Other uses of PEVA are given below:

1. Cable and wire coverings.
2. Seals.
3. Floor materials.
4. Some medical extrusions.
5. Hoses.

17.3   Chlorinated polyethylene (CPE)

CPE is the standardised acronym for thermoplastic chlorinated polyethylene elastomer, which is produced by chlorination of polyethylene. The chlorine content could be as high as 70% by weight, but the current grades have a chlorine content in the range of 25% up to 42%.

Chlorinated polyethylene elastomers (CPE) and resins have excellent physical and mechanical properties, such as resistance to oils, temperature, chemicals and weather. They can also exhibit superior compression set resistance, flame retardancy, tensile strength and abrasion resistance.

CPE polymers can range from rigid thermoplastics to a flexible elastomer, making them highly versatile. These polymers are used in a variety of end-use applications such as wire and cable jacketing, roofing, automotive and industrial hose and tubing, molding and extrusion and as a base polymer.

Chlorinated polyethylene is a heat, weathering and oil resistant elastomer which is used in many wire/cable, hose, automotive and industrial parts for many years. Its unique combination of heat and oil resistance makes it an ideal choice for many under the hood applications. It is also gaining increasing use in many industrial hoses, replacing some well-established materials such as chloroprene rubber (CR) due to concerns over high cost and availability. It is increasingly used to replace some traditional materials to upgrade the end products performance.

By using HPDE of different molecular weight and molecular weight distribution, adjusting the production process, many types of chlorinated polyethylene (CPE) elastomers can be produced.

17.3.1   Characteristics of chlorinated polyethylene (CPE)

As special synthetic rubber, chlorinated polyethylene (CPE) has superior comprehensive performances and good performance-price ratio.

Some of the important characteristics of chlorinated polyethylene (CPE) are given below:

1. Excellent chemical, heat and ignition resistance.
2. Performance over a broad temperature range.
3. Outstanding heat-ageing characteristics.
4. Unique crumb form, permitting easy blending, formulating and processing.
5. Polymer shelf stability measured in years.
6. Good processability on equipment common to the rubber industry.

17.3.2   Application of chlorinated polyethylene (CPE)

1. Automobile hoses and industrial hoses: CPE has outstanding temperature resistance, oil resistance, chemicals resistance, ozone resistance and weatherability, CPE has become good materials for automobile industry, including applications in oil tubes, delivery tubes for cooling fluid, power steering tubes and outer layers of many other rubber hoses. CPE is also used for automobile vacuum tubes and vent tubes. Hoses for oil delivery and chemical delivery in industries are another quickly increasing market.

2. Wires and cables: CPE is widely used in fields of wires and cables, including isolation layers of various types of oil resistant parallel flexible wires (such as HPN cords), sheath for flexible wires or flexible cables for electrical appliances (like electric heater, cooking utensils, air conditioners and refrigerators), sheath for various types of light, middle-weight and heavy cables, sheath for mining cables, marine cables and locomotive cables, isolation layer or sheath for various types of power cables, instrument cables and control cables. Due to CPE superior performance-price ratio and its vulcanised compound's long time storage

at room temperature, it has become the first choice for customers for substitution of CR.

3. Various types of specialised rubber products: Due to CPE superior heat resistance, oil resistance, chemicals resistance and ageing resistance, it can be used for production of special rubber products like sealing rings, pads and liners.

4. Combined use of CPE with other rubbers: Like EPDM, NBR and BR will not only reduce production cost, but compensate performance deficiencies of other compounds.

5. Application in magnetic materials: CPE has very high filling capacity for ferrite magnetic powders. Magnetic rubber products produced from CPE has superior low temperature performances and is widely used in application like refrigerator sealing strips and magnetic cards. Magnetic components used in electromechanical products can also be made from CPE by using its thermal resistance. (Refrigerator sealing strip; magnetic rubber rings in computers and micro motors).

6. Miscellaneous applications: Furthermore, it is used for softening PVC foils, without risking the migrate of plasticisers. Chlorinated polyethylene can be cross-linked peroxidically to form an elastomer which is used in cable and rubber industry. When chlorinated polyethylene is added to other polyolefins, it reduces the flammability.

CPE blends well with many types of plastics such as polyethylene, PEVA and PVC. Such blends can be formed into final products with adequate dimensional stability without the need of vulcanisation. The excellent additive/filler acceptability characteristics of CPE can provide a benefit in blends where compound performance and economics are critical.

1. Blends of CPE and vinyl, which produce very robust compounds with exceptional oil resistance, physical and flame characteristics. If the environments contain oil, acids or alkalis, CPE jackets give more protection than PVC.

2. TPO based CPE, with very low halogen content, superior low temperature properties and good flame characteristics.

3. CPE systems suitable for radiation cross-linking. They are oil resistant, flame retardant, E-beam curable.

17.4   Ethylene acrylic elastomers

Ethylene acrylic elastomers can be made into cured compounds that have excellent resistance to high temperatures and good resistance to automotive fluids such as transmission fluids and engine oils.

Low temperature performance: The low temperature performance of surpasses that of most other heat- and oil-resistant polymers. Typical compounds meet OEM specifications for performance at –40°C.

Resistance to fluids: End products based have excellent resistance to hot oils and hydrocarbon- or glycol-based lubricants, transmission fluids and power steering fluids. Low oil swell can be obtained with proper grade selection and compounding.

It is not recommended for use in components immersed in gasoline or highly aromatic fluids, but can be used as gasket for air intake manifold and as cover material in fuel line applications to reduce costs.

Good resistance to blow-by and exhaust gas acid. Condensates makes the · material of choice in many automotive applications. This property becomes more important as more exhaust gas is recycled.

Excellent vibration damping: The high vibrational damping characteristic of compounds remains nearly constant over broad ranges of temperature, frequency and amplitude.

NHFR compounds: Ethylene acrylic elastomers are not inherently resistant to burning. These NHFR compounds exhibit a combination of good oil resistance, good heat resistance and good low temperature properties.

High-temperature durability: Parts made with this elastomer retain elasticity and remain functional after continuous air oven exposures. Conventional filled compounds can meet heat requirements of six weeks at 165°C, 18 months at 121°C or five days at 204°C.

Compressive stress relaxation (CSR): These compounds perform exceptionally well in seal and gasket applications and have good CSR performance in engine oils out to 5000 hr at 150°C.

17.4.1 Properties of ethylene acrylic elastomers

Some of physical and mechanical properties, chemical resistance, thermal properties and environmental performance of ethylene acrylic elastomers are given below:

Physical and mechanical properties

Physical and mechanical properties	Typical values
Durometer or hardness range	35–95 shore A
Tensile strength range	500–3,000 PSI
Elongation (range %)	200–850%
Abrasion resistance	Good to excellent
Adhesion to metal	Good
Adhesion to rigid materials	Good

Compression set	Poor to good
Flex cracking resistance	Good
Impact resistance	Good to very good
Resilience/rebound	Poor to fair
Tear resistance	Good to excellent
Vibration dampening	Good

Chemical resistance

Chemical resistance	Typical values
Acids, dilute	Good
Acids, concentrated	Poor to fair
Acids, organic (dilute)	Good to excellent
Acids, organic (concentrated)	Poor to excellent
Acids, inorganic	Fair to good
Alcohol's	Good to excellent
Aldehydes	Fair to good
Alkalies, dilute	Good to excellent
Alkalies, concentrated	Poor
Amines	Good
Animal and vegetable oils	Good
Brake fluids, non-petroleum based	Poor
Diester oils	Poor
Esters, alkyl phosphate	Poor
Esters, Aryl phosphate	Poor
Ethers	Poor
Fuel, aliphatic hydrocarbon	Good
Fuel, aromatic hydrocarbon	Poor to fair
Fuel, extended (oxygenated)	Fair
Halogenated solvents	Poor to good
Hydrocarbon, halogenated	Poor
Ketones	Poor
Lacquer solvents	Poor
LP gases and fuel oils	Poor
Mineral oils	Poor
Oil resistance	Poor
Petroleum aromatic	Poor
Petroleum non-aromatic	Poor
Refrigerant ammonia	Poor to good

Refrigerant halofluorocarbons — Poor to good
Refrigerant halofluorocarbons W/oil — Poor
Silicone oil — Good to excellent
Solvent resistance — Poor

Thermal properties

Thermal properties	Typical values
Low temperature range	−55°F to −30°F
Minimum for continuous use (static)	− 50°F
Brittle point	− 75°F
High temperature range	+ 250°F to + 350°F
Maximum for continuous use (static)	+ 350°F

Environmental performance

Environmental performance	Typical values
Colourability	Good
Flame resistance	Poor
Gas permeability	Excellent
Odour	Good
Ozone resistance	Excellent
Oxidation resistance	Excellent
Radiation resistance	Good
Steam resistance	Poor to fair
Sunlight resistance	Excellent
Taste retention	Fair to good
Weather resistance	Excellent
Water resistance	Good to excellent

Polysulphide, norbornene and polyphosphazene rubbers

18.1   Introduction

This chapter discusses polysulphide, norbornene and polyphosphazene rubbers. These are specialty rubbers and are widely used in various industries. The important applications include – pickling tanks, water sewers, oil pipes, drying printing inks. They are also exclusively used in sealing. They are also used as binders for rocket propellants other applications include insulating glass industry due to their excellent adhesion to aluminium and glass. They are used as sealants in aircraft industry, marine and construction applications.

18.2   Polysulphide rubber

Since the commercial introduction in 1929 of the polysulphide polymers, they have been utilised in specialty applications due to their excellent oil and solvent resistance as well as good ageing properties. Although the original polymers were solid rubbery materials, today the predominant product, discovered some 20 years later, is the mercaptan terminated liquid polymer (LP).

Polysulphide rubbers are formed when dihalide reacts with sodium polysulphide. Polysulphide rubbers can be divided into four different groups: Thiokol A, FA, ST and LP rubbers. A-type polysulphide rubbers have ethylene dichloride as a dihalide, FA rubbers are produced from the blend of ethylene dichloride and dichloroethylene form. ST-rubbers are produced from dichloro-ethyenel form and trichloropropane. LP types are liquid polymers. They are formed by breaking down a high molecular weight polymer in a controlled manner.

The sulphur content of type A is high (84%), the sulphur content FA types is 49% and that of ST types 37%. It can be transformed *in situ* from a liquid state into a solid elastomer, even at low temperatures, which makes its use convenient for adhesives, coatings and sealants.

Polysulphide rubbers are produced by the condensation of sodium poly-sulphide with dichloroalkanes:

$$R\ Cl_2 + Na_2\ S_x \rightarrow R - S_x -$$

The polymer varies both in characters of R and x and in the length of polysulphide chain. In the year 1929, Thiokol Chemical Corporation, New

Jersey first introduced a polysulphide rubber (Thiokol A) based on the reaction product of ethylene dichloride and sodium tetrasulphide.

The different polymers produced by the Thiokol chemical corporation are given in Table 18.1.

Table 18.1: Various grades of polysulphide rubbers (Thiokol).

Polymer	Dihalide, R	X	% Sulphur
Thiokol A	$ClCH_2\ CH_2\ Cl$	4	84
Thiokol B	$Cl\ (CH_2)_2\ OCH_2O\ (CH_2)_2\ Cl$	4	64
Thiokol FA	$ClCH_2\ CH_2\ Cl\ CH_2\ (OCH_2\ CH_2\ Cl)_2$	2	47
Thiokol ST	$CH_2\ (OCH_2\ CH_2\ Cl)_2$ 2% Trichloropropane	2.2	37

Manufacture of polysulphide rubber: The general method of preparation of polysulphides is to add the dihalide slowly to an aqueous solution of sodium polysulphide. Magnesium hydroxide is often employed to facilitate the reaction, which takes 2–6 hr at 70°C. Sodium polysulphide is usually produced directly from sodium hydroxide and sulphur at elevated temperature.

$$6\ NaOH + 2\ (x + 1)\ S \xrightarrow{\ 100°-150°C\ } 2\ Na_2S_x + Na_2S_2O_3 + 3\ H_2O$$

where, $x = 1 - 4$.

Properties of polysulphide rubber: The solid polymers are used almost exclusively in applications where good resistance to solvents is required. This depends on the amount of sulphur in the molecule. Thiokol A is resistant to every type of organic solvent. However, its odour, processing characteristics and mechanical properties are very poor and the other types which have moderate physical properties and better all-round solvent resistance than neoprene or nitrile rubbers are more widely employed.

Curing agents for thiokols are diverse, but it is customary to use an organic accelerator (e.g., MBTS or TMTD with zinc oxide and stearic acid for Thiokol FA). The thiol terminated polymers (Thiokol ST) can be cross-linked by metal oxides, metal peroxides, inorganic oxidising agents, peroxides and *p*-quionone-dioxime. Carbon black, usually SRF or FEF in 40–60 phr loading, is essential for adequate strength.

The polysulphide rubbers have very good resistance to oils, fuels, solvents, oxygen and ozone, impermeable to gases but have poor mechanical properties and poor heat resistance. They are however, not recommended for use against strong oxidising acids in any concentration. They are blended with other synthetic rubbers for improved processing.

Applications of polysulphide rubber: Because of their excellent oil and solvent resistance and impermeability to gases, polysulphides find applications in specialty areas. Thiokol FA is used in the manufacture of rollers for can lacquering, quick drying printing ink application and grain coating of paint on metals. Another major application of Thiokol FA is in solvent hose liner.

Type ST is used in the gas metal diaphragms. Primary use for type A is as flexibiliser for sulphur. 2–5 parts of Thiokol A dissolved in molten sulphur prevents it from crystallisation so that it can be used as a mortar for acid pickling tanks, water sewers and oil pipes. Several applications for polysulphide rubber are, as linings and sealants in airplane fuel tanks, concrete fuel storage tank linings, tank car linings, self-sealing aircraft tanks and deicer on wings.

18.2.1   Liquid polysulphide rubber

A series of liquid thiol terminated polymers (Thiokol LP 2,3,4,31, 32 and 33) are available based on the diethylene formal disulphide structure but containing some branching and Thiokol LP 205 based on dibuthylene formal disulphide. These low molecular weight polymers are formed by reductive cleavage of disulphide linkage in solid rubber by means of a mixture of sodium hydrosulphide and sodium sulphite. The reaction is carried out in water dispersion and the relative amount of the hydrosulphide and sulphite controls the extent of cleavage and liquid polymers of varying molecular weights can be readily prepared. The sodium hydrosulphide splits a disulphide link to form a thiol and a sodium salt of thiol.

The extra sulphur atom is taken up by sodium sulphide.

$$- R - S - S - R- + NaSH + NaSO_3 \rightarrow - RSNa + HSR - + NaS_2O_3$$

The sodium salt of the polysulphide is converted back to the free thiol on coagulation with acid. While the commercial liquid polymers contain terminal thiol groups produced by the above method, liquid polymers have been prepared experimentally with terminal alkyl, aryl, hydroxyl, allyl and carboxyl groups.

These materials can be produced by using a mixture of dihalide with the appropriate monohalide in the initial reaction with sodium polysulphide. The molecular weight of the product is easily controlled by the mole ratio of monohalide to dihalide.

These liquid polymers have molecular weights in the range 600–7500 and viscosities 2.5–1400 poise at 24°C. The most useful reaction for conversion of the liquid polymers to the high polymer state is that of direct oxidation. This reaction results in a linking of the two thiols to form the polymeric disulphide with liberation of water as a by-product.

Typical reactions are:

$2 - RSH + PbO_2$	$\rightarrow$	$- R-S-S-R- + H_2O + PbO$
$2 - RSH + ZnO$	$\rightarrow$	$- R-S-Zn-S-R- + H_2O$
$2 - RSH +$ organic peroxides	$\rightarrow$	$-R-S-S-R- + H_2O$

It is customary to incorporate carbon black or white fillers and plasticisers, such as dibutyl phthalate for enhancing properties. They are almost exclusively used in sealing, casting and impregnation applications. Although initially employed as binders for rocket propellants, at present, their largest single use is in the insulating glass industry due to their excellent adhesion to aluminium and glass and inherent resistance to UV radiation and moisture transmission.

They are used as sealants in aircraft industry, marine and construction applications. Other uses include dental molding compound, cold molding compound, formed -in-place gaskets, concrete coatings and bounding, as epoxy flexibiliser for indoor applications and filled molding compounds.

18.3   Polynorbornenes rubber

Norbornenes are important monomers in ring-opening metathesis polymerisations (ROMP) with for instance the Grubbs catalyst. Polynorbornenes are polymers with high glass transition temperatures and high optical clarity.

In addition to ROMP polymerisation, norbornene monomers also undergo vinyl-addition polymerisation.

18.3.1   Uses polynorbornenes rubber

Polynorbornene is used mainly in the rubber industry for anti-vibration (rail, building, industry), anti-impact (personal protective equipment, shoe parts, bumpers) and grip improvement (toy tyres, racing tyres, transmission systems, transports systems for copiers, feeders, etc.).

Reachable performances:

1. Loss factors (tan delta) larger than 3.
2. Rebounds of less than 1%.
3. Tear strengths of 50 N/mm^2.
4. Friction coefficients of 2 and more.
5. Shore hardness between 4 and 90 Shore A.

Second main application: Oil-binding system with absorption capability of hydrocarbons, 10 times of own weight

Ethylidene norbornene is a related monomer derived from cyclopentadiene and butadiene.

18.3.2   Properties polynorbornenes rubber

Mol. wt.	Average M_w >2,000,000
Hardness	18–80 (Shore A)
Particle size	800 µm
Density	0.96 g/mL at 25°C (L)

18.4   Polyphosphazene rubber

Polyphosphazenes include a wide range of hybrid inorganic-organic polymers with a number of different skeletal architectures that contain alternating phosphorus and nitrogen atoms. Nearly all of these molecules contain two organic or organometallic side groups attached to each phosphorus atom. These include linear polymers with the formula $(N = PR^1R^2)_n$, where R^1 and R^2 are organic or organometallic side groups. The linear polymers are the largest group, with the general structure shown in Fig. 18.1. Other known architectures are cyclo-linear and cyclomatrix polymers in which small phosphazene rings are connected together by organic chain units. Other architectures are available, such as block copolymer, star, dendritic, or comb-type structures. More than 700 different polyphosphazenes are known, with different side groups (R) and different molecular architectures.

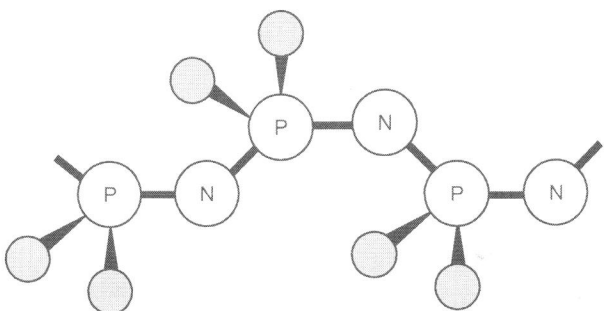

Figure 18.1: General structure of polyphosphazenes. Grey spheres represent any organic or inorganic group.

18.4.1   Synthesis polyphosphazene rubber

The method of synthesis depends on the type of polyphosphazene. The most widely used method for linear polymers is based on a two-step process. In the first step a cyclic small molecule phosphazene, known as hexachlorocyclotri-phosphazene, with the formula $(NPCl_2)_3$, is heated in a sealed system at 250°C to convert it to a long chain linear polymer with typically 15,000 or more repeating units. In the second step the chlorine atoms linked to phosphorus in the polymer are replaced by organic groups through reactions with alkoxides, aryloxides, amines or organometallic reagents. Because many different reagents can participate in this macro-molecular substitution reaction and because two or more different reagents may be used, a large number of different polymers can be produced, each with a different combination of properties. Variations to this process are possible using poly(dichlorophosphazene) made by condensation reactions.

Another synthetic process uses a living cationic polymerisation that allows the formation of block copolymers or comb, star, or dendritic architectures. Other synthetic methods include the condensation reactions of organic-substituted phosphoranimines.

Cyclomatrix type polymers made by linking small molecule phosphazene rings together employ difunctional organic reagents to replace the chlorine atoms in $(NPCl_2)_3$, or the introduction of allyl or vinyl substituents, which are then polymerised by free-radical methods. Such polymers may be useful as coatings or thermosetting resins, often prized for their thermal stability.

18.4.2 Properties and uses of polyphosphazene rubber

The linear high polymers are of more than 700 different macro-molecules which are known with different side groups or combinations of different side groups. In these polymers the properties are controlled partly by the high flexibility of the backbone, its radiation resistance, high refractive index, ultraviolet and visible transparency and its fire resistance.

However, the side groups exert an equal or even greater influence on the properties since they impart properties such as hydrophobicity, hydrophilicity, colour, useful biological properties such as bioerodibility, or ion transport properties to the polymers.

Thermoplastics

The first stable thermoplastic poly(organophosphazenes), isolated in the mid 1960s by Allcock, Kugel and Valan, were macro-molecules with trifluoroethoxy, phenoxy, methoxy, ethoxy, or various amino side groups. Of these early species, poly[*bis*(trifluoroethoxyphosphazene], $[NP(OCH_2CF_3)_2]_n$, has proved to be the subject of intense research due to its crystallinity, high hydrophobicity, biological compatibility, fire resistance, general radiation stability and ease of fabrication into films, microfibres and nanofibres. It has also been a substrate for various surface reactions to immobilise biological agents.

Phosphazene elastomers

The first large-scale commercial uses for linear polyphosphazenes were in the field of high technology elastomers, with a typical example containing a combination of trifluoroethoxy and longer chain fluoroalkoxy groups. The mixture of two different side groups eliminates the crystallinity found in single-substituent polymers and allows the inherent flexibility and elasticity to become manifest. Glass transition temperatures as low as $-60°C$ are attainable and properties such as oil-resistance and hydrophobicity are responsible for their utility in land vehicles and aerospace components. They have also been used in biostable biomedical devices.

Other side groups, such as non-fluorinated alkoxy or oligo-alkyl ether units, yield hydrophilic or hydrophobic elastomers with glass transitions over a broad range from $-100°C$ to $+ 100°C$. Polymers with two different aryloxy side groups have also been developed as elastomers for fire-resistance as well as thermal and sound insulation applications.

Polymer electrolytes

Linear polyphosphazenes with oligo-ethyleneoxy side chains are gums that are good solvents for salts such as lithium triflate. These solutions function as electrolytes for lithium ion transport and they have been the focus of much research designed to incorporate them into fire-resistant rechargeable lithium-ion polymer battery.

The same polymers are also of interest as the electrolyte in experimental dye-sensitized solar cells. Other polyphosphazenes with sulphonated aryloxy side groups are proton conductors of interest for use in the membranes of proton exchange membrane fuel cells.

Hydrogels

Water-soluble poly(organophosphazenes) with oligo-ethyleneoxy side chains can be cross-linked by gamma-radiation techniques. The cross-linked polymers absorb water to form hydrogels which are responsive to temperature changes, expanding to a limit defined by the cross-link density below a critical solution temperature, but contracting above that temperature. This is the basis of controlled permeability membranes. Other polymers with both oligo-ethyleneoxy and carboxyphenoxy side groups expand in the presence of monovalent cations but contract in the presence of di- or tri-valent cations, which form ionic cross-links. Phosphazene hydrogels have been utilised for controlled drug release and other medical applications.

Bioerodible polyphosphazenes

The ease with which properties can be controlled and fine-tuned by the linkage of different side groups to polyphosphazene chains has prompted major efforts to address biomedical materials challenges using these polymers. Different polymers have been studied as macro-molecular drug carriers, as membranes for the controlled delivery of drugs, as biostable elastomers and especially as tailored bioerodible materials for the regeneration of living bone. An advantage for this last application is that poly(dichlorophosphazene) reacts with amino acid ethyl esters (such as ethyl glycinate or the corresponding ethyl esters of numerous other amino acids) through the amino terminus to form poly-phosphazenes with amino acid ester side groups. These polymers hydrolyse slowly to a near-neutral, pH-buffered solution of the amino acid, ethanol, phosphate and ammonium ion. The speed of hydrolysis depends on the amino acid ester, with half-lives that vary from weeks to months depending on the structure of the amino acid ester. Nanofibres and porous constructs of these polymers assist osteoblast replication and accelerate the repair of bone in animal model studies.

18.4.3   Commercial aspects

The cyclic trimer, $(NPCl_2)_3$, is commercially available and has formed the starting point for most commercial developments. Prominent among these developments has been the high performance elastomers known as PN-F or Eypel-F, which have been manufactured for seals, O-rings and dental devices. An aryloxy-substituted polymer has also been developed as a fire resistant expanded foam for thermal and sound insulation.

The patent literature contains many references to cyclomatrix polymers derived from cyclic trimeric phosphazenes incorporated into cross-linked resins for fire resistant circuit boards and related applications.

Section III

Vulcanisation mixing and calendering of rubber

Chemistry and technology of vulcanisation

19.1   Introduction

Vulcanisation of rubber is a process of improvement of the rubber elasticity and strength by heating it in the presence of sulphur, which results in three-dimensional cross-linking of the chain rubber molecules (polyisoprene) bonded to each other by sulphur atoms.

Vulcanised rubber is a material that undergoes a chemical process known as vulcanisation. This process involves mixing natural rubber with additives such as sulphur and other curatives. Vulcanisation makes rubber much stronger, more flexible and more resistant to heat and other environmental conditions. Vulcanised rubber makes both soft and hard objects, ranging from rubber bands to bowling balls. In fact, almost any rubber object consists of vulcanised rubber.

It is necessary to use vulcanisation to make commercial-grade rubber because natural rubber is not stable enough to produce goods with. In fact, natural rubber melts when warm, breaks apart when cold and is very sticky. This is because natural rubber consists of independent polymer chains that allow the rubber to be deformed. Vulcanisation creates bridges between these polymer chains, allowing the rubber to be deformed when stress is applied and to return to its original position when the stress is removed.

Applications of vulcanised rubber: Vulcanised rubber has many applications and is used to make a vast assortment of objects. For example, vulcanised rubber is used to make rubber hoses, shoe soles, tyres, bowling balls, bouncing balls, hockey pucks, toys, erasers and instrument mouthpieces. Most rubber products in the world are vulcanised, whether the rubber is natural or synthetic.

Advantages of vulcanised rubber: Vulcanised rubber is stronger than non-vulcanised rubber because its bonds are made of cross-links at an atomic level. This allows vulcanised rubber to stand up to more stress and damage. Vulcanised rubber is also more rigid than non-vulcanised rubber, so it is more stretch-resistant in the first place. However, vulcanised rubber has the same elasticity level as non-vulcanised rubber.

Disadvantages of vulcanised rubber: Vulcanised rubber does not have many disadvantages but does exhibit several negative attributes. Rubber is mildly toxic when burned and cools quickly after being melted, making it some what hazardous to those handling it.

19.2   Vulcanisation systems

There are four types of vulcanisation systems.

1. Conventional cure system.
2. Efficient vulcanising (EV cure) system.
3. Semi-EV vulcanising (semi-EV cure) system which also includes sulphurless (sulphur donor) cure.
4. Non-sulphur vulcanisation system.

Conventional, EV and semi-EV systems are based on sulphur/accelerator ratio and applicable only to NR as well as isoprene and butadiene based synthetic rubbers such as SBR, NBR, IIR and EPDM (unsaturated rubbers).

19.2.1   Conventional sulphur cure system

1. It contains high proportions of sulphur (2.0–3.5 phr).
2. Low proportions of accelerator (0.4–1.2 phr).
3. Accelerator to sulphur ratio is 0.1–0.6.
4. Sulphur to accelerator ratio >1.0.
5. (Low accelerator to sulphur ratio = high sulphur to accelerator ratio).

Characteristics of conventional cure system

1. Network will be mainly polysulphidic (above 65%), which are thermally unstable.
2. Fair degree of wasted sulphides and main chain modification.
3. Excellent mechanical strength.
4. High set.
5. Poor heat and ageing resistance.

19.2.2   EV sulphur cure system

1. Contains very little sulphur (0.4–0.8 phr).
2. High proportion of accelerator (2–5 phr).
3. Accelerator to sulphur ratio is 2.5–12.
4. Efficient use of sulphur.
5. Mainly monosulphidic bonds (75% mono and 25% disulphidic).

19.2.3   Semi-EV sulphur cure system

1. Sulphur levels are intermediate between conventional system and EV system.
2. Used for a compromise in cost and/or performance.

3. Particular application in NR where a compromise between heat ageing and fatigue life is sought after.

19.2.4 Sulphurless cure in EV system

1. Can employ sulphurless cure where there is no elemental sulphur present.
2. In sulphurless cure the sulphur available for cross-linking is donated by partial decomposition of sulphur containing accelerators (sulphur donors).

Properties of EV cure vulcanisate

1. Low set and slightly lower mechanical strength.
2. Excellent oxidative ageing resistance.
3. Excellent heat and reversion resistance.
4. Poor flex-fatigue properties–not suitable for dynamic applications.

19.3 Mechanism of rubber vulcanisation

Vulcanisation is the cross-linking process that prevents permanent deformation under load and ensures elastic recovery on removal of the load on the product. Since vulcanisation was first discovered, a major focus on elastomer systems has been to characterise the network structures formed. Complex mechanisms are involved in vulcanisation processes rather than a simple chemical reaction. A series of consecutive and competing reactions occur during the sulphuration of rubber under vulcanising conditions and hence no single mechanism can be appropriate.

Further, the network structures are complex and rich in types of structure. Many of the traditional analytical techniques are not useful, as the concentration of the chemically modified structures induced by the vulcanisation is extremely low. Even then many approaches have been attempted in an effort to relate the chemical microstructure to the physical properties of both raw and cured elastomers. NMR spectroscopy is one of the powerful spectroscopic methods used to directly evaluate the chemical structures of polymeric materials. Solid state C-13 NMR has also been widely applied for the characterisation of vulcanised rubber systems.

For the study of reaction mechanism, sulphur vulcanisation reactions can be broadly classified into two, the unaccelerated and the accelerated types. Unaccelerated sulphur formulation consists of rubber and sulphur while the accelerated systems contain rubber, accelerator and sulphur. In addition to this, both the types include zinc oxide - stearic acid activator system also. There are also accelerator systems in which elemental sulphur is not present instead, the accelerator provides sulphur for vulcanisation. This sulphur free vulcanisation can also be referred to as sulphur donor systems. The most widely used accelerator in this type is tetra methyl thiuram disulphide (TMTD), although other accelerators such as 4-morpholinyl 2-benzothiazyl disulphide (MBDS) are also used. If a full understanding of the relationship between vulcanisation chemistry and network structure is possible, one can tailor formulations to produce the desired mechanical and chemical properties.

Unaccelerated sulphur only vulcanisates while alleviating many of the disadvantages of uncross-linked elastomers, does not provide an optimum product. Vulcanisation with sulphur but without accelerators is an extremely slow process. Relatively large amount of sulphur and long vulcanisation time are necessary and the vulcanisates are not of high quality. They have strong tendency to revert and their resistance to ageing is poor. A problem of sulphur blooming is also found to occur.

Vulcanisation with sulphur alone is therefore of no technical importance. The yield of cross-linked polymer is low when sulphur is used alone, which may be due to the formation of multivalent polysulphidic bridges, cyclic sulphidic and vicinal bridge links. It is known that several reactions by different mechanisms (of a radical or ionic nature) may take place simultaneously or consecutively during vulcanisation.

These reactions range from double bond migration, isomerisation, chain cleavage, cyclisation and formation of vicinal cross-links. Several techniques including the use of radical scavengers and electron paramagnetic resonance (EPR) analysis have been used to study the reaction mechanism involved in sulphur vulcanisation. According to Shelton and McDonel the unaccelerated sulphur vulcanisation is a polar process. Blokh also concluded from his EPR studies that unaccelerated sulphur vulcanisation proceeds through a polar mechanisms.

There is the possibility for a free radical mechanism for sulphuration where sulphur radicals are formed, via., a homolytic fission of the S_8 ring. Although a radical process would explain certain experimental results, the general agreement is that a polar mechanism operates during the sulphur vulcanisation. A general version allowing for either proton or hydride transfer for the unaccelerated vulcanisation can be represented as:

Initiation: S_8 $\rightarrow$ $S_x^+ + S_y^-$

Propagation: $S_x^+ + RH$ $\rightarrow$ S_xRH^+ (RH = Rubber)

$$S_xRH^+ \begin{cases} \xrightarrow{\text{H}^- \text{ transfer}} S_xR + RH_2 \\ \xrightarrow{\text{H}^+ \text{ transfer}} S_xRH_2 + R^+ \end{cases}$$

$$RH_2^{\cdot} \left.\begin{array}{c} \\ \\ \\ R^{\cdot} \end{array}\right\} + \ S_8 \ \longrightarrow \quad \begin{array}{c} RH_2S + S_x^{\cdot} \\ \\ RS_x \end{array}$$

Termination:

$$\left.\begin{array}{c} RH_2^{\cdot} \\ \\ R^{\cdot} \\ \\ S_x^{\cdot} \\ \\ S_xRH^{\cdot} \end{array}\right\} + \ S_y^{-} \ \longrightarrow \quad \text{Non-chain carriers}$$

According to Dogadkin and Shershnev the differences in the points of view regarding the mechanism of vulcanisation are so much a matter of approach to the interpretation of experimental factors, as the fact that for such a complicated phenomenon as vulcanisation, it is improper to support a single mechanism. Even though the use of sulphur alone in rubber vulcanisation is typically ineffective requiring 45–55 sulphur atoms per cross-link and tends to produce a large proportion of intramolecular (cyclic) cross-links, such ineffective cross-link structures are of interest in the understanding of complex nature of vulcanisation reactions. Spectroscopic studies of unaccelerated sulphur vulcanisation point to the formation of polysuphidic, monosulphidic and also cyclic sulphidic linkages.

19.3.1   Accelerated sulphur vulcanisation

By far the common vulcanisation systems used in industrial applications are the accelerated sulphur formulations. The accelerated sulphur systems can be classified into single and binary accelerator combinations. Almost all accelerators need metal oxides for the development of their full activity. Zinc oxide is being used as the best additive.

The mechanism under which accelerated sulphur vulcanisation occurs is a function of the class of accelerators/activators.

A generally accepted scheme of the reactions is as follows:

1. Accelerator (Ac) and activator interacting with sulphur to form the active sulphurating agent.

   $Ac + S_8 \rightarrow Ac - S_x - Ac$ (Active sulphurating agent)

2. The rubber chains interact with the sulphurating agent to form polysulphidic pendant groups terminated by accelerator groups.

$Ac - S_x - Ac + RH \rightarrow AcH + R - S_x - Ac$ (Pendent sulphurating agent)

   where, RH is the rubber chain.

3. Polysulphidic cross-links are formed.

$$R - S_x - Ac + RH \rightarrow AcH + R - S_x - R \text{ (Cross-links)}$$

4. Network maturing and competing side reactions and thermal decomposition leads to the following reactions:

$$R - S_x - Ac \rightarrow \text{Cyclic sulphides} + \text{Dienes} + \text{ZnS (Degradation)}$$

$$R - S_x - Ac \rightarrow S_{x-1} + R - S - Ac \text{ (Desulphuration)}$$

$$R - S_y - R \rightarrow S_{y-1} + R - S - R \text{ (Monosulphidic cross-links)}$$

$$R - S_{x+y} - R + Ac - S_2 - Ac \rightarrow Ac - S_{y+2} - Ac + R - S_x - R \text{ (Sulphur exchange)}$$

There are at least three competing reactions (cross linking, desulphuration and degradation) that occur during the cure and network maturing period. The first step is the formation of active sulphurating agent. The accelerator and activator first interact to form a species, which then reacts with sulphur to form active sulphurating agent. The active sulphurating agent reacts directly with the rubber molecule to give a rubber bound pendent group. In this pendent group a fragment derived from the accelerator or sulphur donor is linked through two or more sulphur atoms to the rubber chains. These form cross-links either by direct reaction with another rubber molecule or by disproportionation with a second pendent group of a neighbouring rubber chain. Polysulphide cross-links formed undergo further transformation by two competing reactions, desulphuration or decomposition. Progressive shortening of the polysulphide producing finally monosulphide links by desulphuration. Then a number of competing reactions, which are termed network maturing lead to the final cross-linked structure. The ultimate network structure formed depends on the temperature, accelerator types and concentration.

The ratio of types of cross-links formed (*poly*, *di* or *mono* sulphidic) depends on the ratio of sulphur to accelerator in the formulation. Changes may continue to occur in the network structure, especially if the vulcanisate is in service under elevated temperature, after the formal vulcanisation is over. Layer proposed that sulphur determines the overall amount of reaction but the accelerator determines the length of the sulphur chains. Although mechanism of accelerated vulcanisation has been extensively studied, there is still much disagreement to its exact mechanism. Craig, Dogadkins, Bevilacquas, Scheele, Blokhs Tsurugi and Fukudas all have advanced free radical mechanism to explain the results of accelerated sulphur vulcanisation, where as Baternarr Porter and Allen and others suggested polar mechanism as a logical extension of their proposed mechanism for unaccelerated sulphur vulcanisation. On the other hand Shelton and McDonel and Coran proposed mechanisms involving both free radical and ionic species.

In radical mechanism of accelerated sulphur vulcanisation, the accelerator cleaves to form persulphenyl radicals, which then abstract protons. The rubber radical reacts with another intermediate to form rubber bound intermediates. Two such rubber bound intermediates then form the actual cross-link.

$$XS_aX \xrightarrow{\text{Thermolysis}} XS_b^+ + XS_c^+$$

$$XS_b + R\text{--}H \longrightarrow R^+ + XS_b\,H$$

$$R^+\,XS_aX \longrightarrow RS_xX + XS_d^+$$

$$R^+ + XS_c^+ \longrightarrow RS_cX$$

$$2RS_xX \longrightarrow RS_yR + XS_aX$$

The proposed polar mechanism is shown below:

$$\begin{array}{l} \text{1. } 2RS_aSX \;\rightarrow\; RS_yR + XS_zX \\ \text{2. } RS_aSX + R'\text{--}H \;\rightarrow\; RS_yR' + XS_wH \end{array}$$

The active sulphurating agent is assumed to be the zinc accelerator perthiolate complex. The co-ordination of electron donating ligands to the zinc atom will increase the electron density on the sulphur atom of the perthiolate groups. In the transition state C–S bond formation will be enhanced by the increased nucleophilicity of the XSS_a group while C–H bond fission will be limited by the reduced electrophilicity of the XS_b group, which increases the SN_2 character of the process. The rubber bound pendent group RS_aSX is converted to cross-links by the process (1) and (2) mentioned above.

19.3.2 Binary accelerator systems

A binary accelerator system refers to the use of two or more accelerators in a given formulation. The use of such systems finds wide technological applications. One of the motivations for the use of binary systems is the synergistic behaviour of accelerator combinations in that the final properties are better than those produced by either accelerator separately. These systems are widely used in industry and are becoming increasingly popular due to the fact that such mixed systems provide better acceleration, better control of processing safety and improvement in physical and chemical properties of the vulcanisates. Though the practice of using binary accelerators is quite old, the mechanism of the combined action of these accelerators are not studied adequately and only recently scientists began to fully probe the complicated mechanism of binary systems. Accelerator present at a relatively higher concentration is called primary and that present in smaller amounts is called secondary accelerator.

A lot of binary systems have been developed for practical applications. These include thiuram disulphides, sulphenamides, benzothiazyl disulphides, some derivatives of piperidine, pyrrole, piperazine, phthalmides etc. Amines like diphenyl guanidine (DPG) are used in combination with other accelerators such as MBT or sulphenamides to activate the vulcanisation reaction.

Usual binary systems consist of thiocarbamate derivatives and benzothiozoles. Thiuram systems generally show very little scorch safety. In order to increase scorch delay period, often sulphenamides and MBT are added. Thiourea and its derivatives are also known to be good secondary accelerators in rubber vulcanisation.

Even though a good deal of work has been reported to elucidate the mechanism of vulcanisation by single accelerator, little attention has been paid to the chemistry of vulcanisation of binary systems. Higher accelerating action of binary system is assumed to be through the formation of either a eutectic mixture or salt forming compound having greater chemical reactivity and better solubility. Dogadkin and collaborators investigated a number of popular accelerator combinations and found mutual activation with many of them.

Based on the experimental observations in the vulcanisation by using various combinations of accelerators, they classified the various binary systems into three different groups:

1. Systems, which show synergism.
2. Systems with a pair of accelerators in which the mutual activity of the pair does not exceed the activity of the most activated accelerator.
3. Systems with only additive action of accelerators.

The first group consists of disulphides (e.g., MBTS, TMTD, etc.), or mercaptans (e.g., MBT) with nitrogen containing organic bases or disulphides with sulphenamides. Sulphenamides with nitrogen containing organic bases belong to the second group. The third type exhibiting additional effect include systems containing sulphenamides (and some disulphides) with TMTD and those containing combinations of the same chemical class. In the case of systems with mutual activity such as MBTS with DPG or CBS, the reaction between the accelerators were observed higher under vulcanisation conditions than when they are reacted alone. It was suggested that in the initial stage of vulcanisation, the accelerators interact with one another to form an *active* complex. This complex then disintegrates with the formation of active free radicals responsible for initiating the interaction of rubber with sulphur.

$$RSSR \;+\; R'R''NH \;\rightarrow\; \frac{RSSR}{R'R''NH} \;\rightleftharpoons\; RSH \;+\; RS^{\cdot} \;+\; R'R''N^{\cdot}$$

| Disulphide | Organic amine or sulphenamide | Intermediate complex | Free radicals |

The reaction is believed to propagate as follows:

$$RS^{\boldsymbol{\cdot}} + R'R''NH \quad \rightarrow \quad RSH + R'R''N^{\boldsymbol{\cdot}}$$

$$R'R''N^{\boldsymbol{\cdot}} + RSSR \quad \rightarrow \quad RSNR'R'' + RS^{\boldsymbol{\cdot}}$$

RSH and RSNR'R'' are considered as highly active accelerators.

The accelerator activity of sulphur bearing accelerators (XSSX, XSX, etc.) depends partly on the nature of X and partly on the mode of attachment of the functional sulphur to other atomic grouping in accelerator molecule. In a report by Moore and others which presented the investigations on the TMTD-TU binary accelerator systems, a novel probable mechanism for the synergistic activity of TU was suggested.

This theory recognises the importance of the polysulphidic intermediates (I) formed during the vulcanisation process, which subsequently reacts with the rubber chain to yield further intermediates (II). These intermediates finally react to yield sulphurated cross-links.

The process is schematised as follows:

$$(m-1)\,XSSX \xrightarrow{\ ZnO\ } XS_mX + \frac{(m-2)}{2}\ (XO)_2\,Zn + (XS)_2\,Zn$$

$$(I)$$

$$...(19.1)$$

$$RH + XS_mX \xrightarrow{\ ZnO\ } RS_{m-1}X + \frac{1}{2}\ (XS)_2\,Zn + H_2O$$

$$(II)$$

$$...(19.2a)$$

$$RH + RS_{m-1}X \xrightarrow{\ ZnO\ } RS_{m-2}R + \frac{1}{2}\ (XS)_2\,Zn + H_2O$$

$$...(19.2b)$$

where, $X = Me_2N\cdot C{:}S$ and RH = Rubber hydrocarbon.

Since the cross-linking reaction 19.2b follows from the products of the reaction 19.2a which itself requires the thiuram polysulphides (I) produced in the reaction 19.1 it follows that any acceleration of the latter must also lead to a corresponding increase in overall vulcanisation rate. Studies on the basic oxyanion type nucleophiles suggests that oxygen atoms of ZnO prefer to attack the thiocarbamyl carbon atoms of TMTD causing the following polar substitution

$$Zn^{++} \ldots O^{\frown\frown} XSSX \longrightarrow Zn^{++} \ldots \overline{O}X + \overline{S}SX \qquad ...(19.3)$$

Reaction 19.3 yields a perthioanion ($XS\overline{S}$) which will rapidly effect the heterolysis of S–S bond in another TMTD molecule to give the trisulphide (III).

$$XS\overline{S} + \overset{\frown}{S} - \overset{\frown}{S} \longrightarrow XSSSX + \overline{S}X$$
$$\quad\quad\quad |\quad| \quad\quad\quad\quad (III) \quad\quad\quad\quad\quad\quad ...(19.4)$$
$$\quad\quad\quad X\quad X$$

Participation of (III) in processes similar to 19.3 and 19.4 will lead to the formation of higher polysulphide of the type I (m ≥ 4). However when thiourea is present, it is assumed that it will interact with TMTD under the prevailing basic conditions providing an easier and faster route for the formation of polysulphide (I) replacing the slow processes 19.3 and 19.4 mentioned above. Thus addition of thiourea to TMTD causes acceleration in the production of polysulphide.

Binary accelerator systems are of great importance because of their wide-spread use in the industry. Smith and Collin also investigated the technology of binary systems and their conclusions also points to the formation of complex compounds. In binary accelerator systems also evidences are available for either polar or radical mechanism as in the case of single accelerator systems, even though no conclusive evidences have been reported.

Researchers first tried APT as a secondary accelerator along with TMTD/MBTS/CBS as primary accelerator in NR gum compounds. Sulphur vulcanisation of natural rubber was carried out using standard compounding recipes and their properties were compared with control mixes containing thiourea as secondary accelerator and also different other binary combinations. Considering the fact that favourable cure characteristics were obtained in these systems, detailed investigations of these mixes were carried out.

Tensile and other physical properties of the vulcanisates were studied and the network structures were estimated by using swelling studies. Although reasonable in physical strength, gum natural rubber vulcanisates are suitable only for very few commercial applications. In this context the effect of APT on filled natural rubber systems were also investigated. Considering the effectiveness of amidino phenyl thiourea in these various binary systems, the effect of change in concentration of APT on the cure characteristics of these systems was also investigated.

Rubber latex is a colloidal system and latex compounding is different from that of dry rubber. Here the ingredients are added either as solutions or dispersions. There is no problem of scorching for latex systems and thus processing problems are less. But the stability of the compounded latex is to be considered. Usually ultra accelerators like dithiocarbamates and xanthates are used in latex processes. Philpott studied several accelerator combinations such as TMTD, MBTS and CBS in various latex vulcanisation systems. The vulcanisation of latex using TMTD or CBS alone proceeds only at relatively higher temperature. At low temperatures (100°C or below) the reaction is very slow. Philpott showed that certain sulphur containing nucleophiles like thiourea, are able to activate vulcanisation by TMTD and CBS so that well cured vulcanisates may be prepared rapidly at or around 100°C. In present study binary NR latex systems were prepared using TMTD and CBS as primary accelerators along with APT and latex films prepared were vulcanised at two different temperatures, viz., 120 and 100°C. Various mixes were prepared with different concentrations of APT. The mixes requiring optimum concentrations of APT were studied in detail. Further the rheological behaviour of the compounded lattices were also investigated with a view to study the stability of the compounds in various systems at various temperatures.

It is to be noted that the mechanism of rubber vulcanisation depends on the type of elastomer used also. Based on the favourable results obtained with NR, the researchers tried APT in a synthetic rubber of the diene type, viz., styrene butadiene rubber. SBR is slower curing than NR and needs either more accelerators or more active accelerator combinations. Scorch problems are less likely in SBR than in NR. Lack of crystallisation is responsible for the lower green strength and lower gum tensile strength of SBR. To overcome this type of behaviour incorporation of fine reinforcing fillers are necessary for synthetic rubbers. Yet in another study SBR vulcanisates were prepared with carbon black/precipitated silica as filler. Different mixes with varying concentrations of APT were tried in standard recipes in both gum and filled compounds. Gum formulations were prepared as investigations on reaction mechanisms are easier in such systems.

The effect of change in concentration of APT on the cure characteristics was investigated and the results showed that as the concentration of the secondary accelerator increases the optimum cure time is seen to decrease accordingly. Considering the cure characteristics, practical cure systems with optimum concentration of APT have been developed for these SBR systems. Cure and vulcanisate properties of the experimental mixes were compared with those of control. To understand the variations in the physical properties of the various vulcanisates, chemical cross-links were also estimated using the equilibrium swelling method. Compounding, curing, etc., of polar rubbers are different from other hydrocarbon rubbers. The effectiveness of amidino phenyl thiourea in the compounding and vulcanisation of a typical polar rubber like polychloroprene is also tried in this study. Thiourea derivatives are popular in neoprene compounding. For example ethylene thiourea (NA22) is a common ingredient of compounding of this rubber. Metal oxides are also an important part of the curing system in CR. Considering the carbon backbone structure with double bonds, the possibility of sulphur cross-linking is also sometimes followed. In another study sulphur vulcanisation of Neoprene-W along with MgO and APT. TMTD-NA22 combination was used as control. Investigations were carried out to find out the effect of this amidinothiourea derivative on filled CR systems also. Carbon black and precipitated silica were used as fillers. It is also to be emphasised that conventional accelerator used in CR vulcanisation, viz., ethylenethiourea (NA22) is reported to be toxic while the amidino thiourea derivative is a non-toxic additive.

All rubbers have shortcomings in one or more properties. There are therefore technical reasons for blending, as it makes possible to obtain the right compromise in properties by blending different elastomers. The difficulties encountered in the processing and vulcanisation of some rubbers also emphasise the need for blending. Economic reasons can also be given for blending since appreciable price differences exist between different rubbers. For example, the resistance of polychloroprene to ozone is outstandingly good but its price is high and accordingly blending of CR with cheaper rubbers is normally beneficial for various applications.

The mechanism of vulcanisation reaction in a single elastomer is bound to be different from that of a blend. In this context attempt was made to use APT as a secondary accelerator for the vulcanisation of NR-CR blends. A 50:50 blend of NR and CR was used and compounding ingredients were added after making a uniform mix of the elastomers. Microheterogenity may diminish when carbon black is added to the preblended elastomers. Systems with varying concentrations of APT along with TMTD and other compounding ingredients were prepared in gum and carbon black filled blends. TMTD-TU and TMTD-NA22 binary combinations were taken as controls. Different cure characteristics and various physical properties of the experimental as well as control systems were investigated.

Materials for compounding and reinforcement

20.1   Introduction

Substances that bring about the actual cross-linking process are called vulcanising agents. Numerous and varied vulcanising agents are now used in rubber industry. In addition to sulphur they include various organic peroxides, quinones, metal oxides, bifunctional oligorners, resins, amine derivatives, etc. Vulcanisation can also be achieved by using high-energy radiation without any vulcanisation chemicals. The cross-links formed by peroxides are purely carbon-carbon linkages. The importance of peroxides is their ability to cross-link saturated elastomers such as ethylene propylene rubber, silicone rubber, etc., which cannot be cross-linked by other vulcanising agents. Chloroprene rubbers are generally vulcanised by the action of metal oxides along with other chemicals. Sulphur and nonsulphur systems have advantages and disadvantages of their own, but sulphur systems still remain versatile.

There are several advantages for sulphur as the vulcanising agent, viz., (i) higher flexibility during compounding, (ii) easier adjustment of the balance between the vulcanisation stages, (iii) possibility of air heating, (iv) possibility to control the length of the cross-links, (v) better mechanical properties to the vulcanisates and (vi) economic reasons.

However compared to peroxide curing, sulphur systems show lower heat and reversion resistance, higher compression set and higher possibility of corrosion in cable metals. Vulcanisation reaction is determined in large measure by the type of vulcanising agents (curatives), the type of process, temperature and time of cure. The number of cross-links formed, also referred to as degree of vulcanisation or state of cure, has an influence on the elastic and other properties of the vulcanisate. Therefore the type of vulcanisation process is the important connecting link between the raw material and the finished product.

Vulcanisation of rubber with sulphur alone is a very slow process and it takes several hours or even days to reach optimum cure depending on the temperature of vulcanisation and the nature of rubber used. The vulcanisates formed are of very low physical strength and mechanical properties also. Further they have a strong tendency for reversion and their resistance to ageing is poor. The use of sulphur alone is ineffective and requires 45 to 55 sulphur atoms per cross-link and tends to produce a large portion of intramolecular (cyclic) cross-linkss. Sulphur bloom is also very common. Vulcanisation with sulphur alone is therefore of no technological importance at all. A major break-through came with the discovery of organic nitrogen compounds known as accelerators. Bases like aniline, thiocarbanilide, etc., were the first organic vulcanisation accelerators of rubber. Now a large number of compounds have been suggested as vulcanisation accelerators.

In the vulcanisation network sulphur is combined in a number of ways. In the form of cross-links it remains as monosulphide, disulphide or polysulphide. It may also be present as pendent sulphides or cyclic monosulphides or disulphides as shown below:

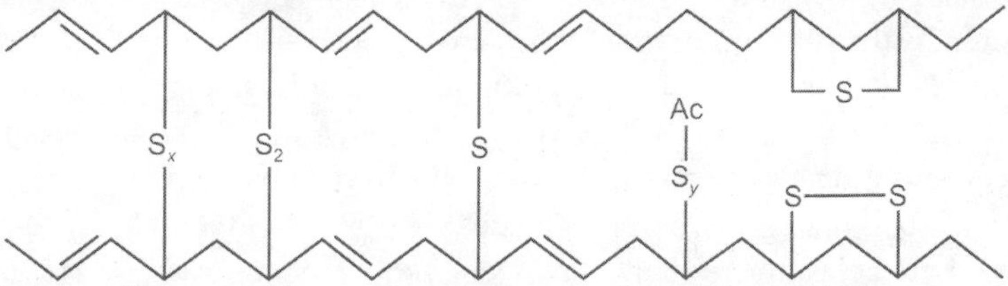

| Poly sulphidic | Di-sulphidic | Mono sulphidic | Accelerator terminated pendent group | Cyclic sulphides |

The accelerator sulphur ratio determines the efficiency by which sulphur is converted into cross-links, the nature of cross-links and the extent of main chain modification. Depending on the sulphur accelerator ratio the sulphur vulcanising systems can be categorised as: (i) the conventional or high sulphur vulcanisation system (CV) where sulphur is added in the range of 2–3.5 parts per hundred rubber (phr.) and the accelerator in the range 1–0.4 phr., (ii) the efficient vulcanising (EV) system where sulphur is added in the range of 0.3–0.8 phr. and accelerator in the range of 6.0–2.5 phr. and (iii) the semi efficient (SEVI system where sulphur is added in the range 1–1.8 phr. and accelerator in the range 2.5–1 phr. As the CV system has got greater amount of sulphur compared to the accelerator the possibility of polysulphidic linkage formation is higher. At higher temperatures the polysulphidic linkage may break to mono and disulphidic. This explains the reversion at higher temperatures, which leads to low strength and modulus. Properties like compression set and thermal stability are better for EV systems primarily due to the lower amount of polysulphidic linkages.

20.2 Metal oxides

Carboxylated nitrile, butadiene and styrene - butadiene rubbers may be cross-linked by the reaction of zinc oxide with the carboxylated groups on the polymer chains. This involves the formation of zinc salts. Other metal oxides are also capable of reacting in the same manner.

Magnesia is included in the formulation to act as a scavenger for the chlorine atom also. More rapid vulcanisation is achieved by the use of organic accelerator, e.g., ethylene thiourea (ETU).

Difunctional compounds: Certain difunctional compounds form cross-links with rubbers by reacting to bridge polymer chains into three dimensional networks. Phenolic resins are used with NR, BR and SBR, quinone dioximes with butyl and diamines with fluororubbers.

Diene rubbers such as NR, butyl and BR can be vulcanised by action of phenolic compounds.

The cross-linking of butyl with *p*-quinone dioxime proceeds through an oxidation step by an oxidising agent such as red lead and the reaction has been proposed to take place via the formation of dinitrosobenzene as an intermediate, followed by addition of this intermediate to two molecules of rubber with removal of two hydrogen radicals, which react with the polymer or with more nitroso-benzene when quinone dioxime will be regenerated.

The basic ingredients for vulcanisation of fluorocarbon rubbers comprise a metal oxide (ZnO, MgO or PbO) with a diamine such as hexamethylenediamine carbamate. The vulcanisation probably occurs in three stages, hydrogen fluoride is eliminated in the presence of basic materials to form regions of unsaturation, difunctional agents then react through additions to double bonds or through substitution of an allylic fluoride atom to form chemical cross-linkages. Finally, during the high temperature post cure, conjugated double bonds are formed which undergo Diels – Alder condensation and subsequent aromatisation leads to very stable aromatic cross-links.

20.2.1   Organic peroxides

Organic peroxides are useful for cross-lining saturated as well as unsaturated polymers or those which contains no sites for attack by other vulcanising agents. This type of vulcanising agent does not enter into polymer chains but produces radicals which form carbon – carbon linkages between adjacent polymer chains.

They are useful for ethylene-propylene rubber (EPR) and silicone rubber. They are not generally useful for vulcanising butyl rubber, poly (isobutylene co-isoprene), because of a tendency towards chain scission rather than cross-linking when the polymer is subjected to the action of a peroxide.

Elastomer derived from isoprene and butadiene are readily cross-linked by peroxides. However, many of the vulcanisate properties are inferior to those of accelerated sulphur vulcanisates. Peroxide vulcanisates of these diene rubbers may be essential when improved thermal ageing and compression set properties are required.

Some widely used peroxides for cross-linking of elastomers are given in Table 20.1.

Table 20.1: Organic peroxides for vulcanisation.

Chemical name	Remarks
Benzoyl peroxide	Used for silicone rubber Not suitable for olefin rubbers
2,4 Dichlorobenzoyl peroxide	Used for silicone rubber Not suitable for olefin rubbers
Dicumyl peroxide	Used for natural and synthetic olefin rubbers Also used for silicone rubber
2,5-Di-(*t*-butyl peroxy)-2,5-dimethyl hexan	Used for natural and synthetic olefin rubbers Also used for silicone rubber
Di-(*t*-butyl peroxy)-*p*-diisopropyl benzene	Used for natural and synthetic olefin rubbers Also used for silicone rubbers

The initiation step in peroxide induced vulcanisation is the decomposition of the peroxide to give free radicals:

$$\text{Peroxide} \rightarrow 2\ R\bullet$$

where, R is an alkoxy, alkyl or acyloxy radical, depending on the type of peroxide used, e.g., benzoyl peroxide gives benzoyloxyl radicals and dicumyl peroxide gives cumyloxyl radicals. In case of unsaturated hydrocarbon elastomers such as butadiene or isoprene, the next step is the abstraction of a hydrogen atom from an allylic position on the polymer molecule or the addition of the peroxide – derived radical to a double bond of the polymer molecule.

20.3 Accelerators

Substances that are added in small amounts during compounding to accelerate the vulcanisation reaction and to improve the physical and service properties of the finished products are called accelerators. These substances can reduce the cure time from days or hours to minutes or seconds at the vulcanisation temperature. The decrease in vulcanisation time is of tremendous economic importance because of increased turnover and consequent reduction in cost of production. Further, the amount of sulphur required can be reduced considerably in presence of an accelerator. Generally 0.1 to 3 phr. is sufficient to give a vulcanisate of desired properties. The first accelerators used in rubber vulcanisation were in fact inorganic compounds. Magnesium oxide, litharge and zinc oxide were the most widely used among them. A major breakthrough came with the discovery of organic nitrogen compounds acting as accelerators for the vulcanisation process. An intense search for the vulcanisation accelerators started around 1906 by Oenslager. He introduced organic base aniline as accelerator into rubber compounds to improve the quality of low-grade rubber and to accelerate the vulcanisation reaction. Though aniline was proved to be a powerful accelerator, it was unacceptable because of its toxic nature. A number of aniline derivatives were then investigated. Among these thiocarbanilide was found to be effective in combination with zinc oxide. Later several other organic compounds were shown to have accelerating activity and majority of them was nitrogen containing organic bases.

First, it was believed that it is the basicity of these substances rather than the chemical constitution that is responsible for the accelerator activity. Later it was established that the activity of organic bases is not proportional to their basicity. With the discovery of nitrogen free accelerators like zinc alkyl xanthate and zinc thiophenols the theory that the element nitrogen was responsible for the accelerator activity was rejected. A large variety of accelerators were developed during the first two decades of the 20th century. By this time dithiocarbamates and alkyl xanthates were widely used as accelerators. In the early 1920's it was discovered that thiocarbanilide reacts with sulphur to yield 2-mercapto-benzothiazole. This and its derivatives are still the most important accelerators used in rubber industry, particularly because they impart outstanding properties to the vulcanisates. Sebrell and others and Bruni and others discovered independently that 2 mercaptobenzothiazole and its homologues, its disulphide and its metal salts are very effective accelerators and these yield vulcanisates of improved physical properties.

A reaction product of amines from beet molasses with CS_2 was introduced by Molony, which was later identified as tetramethylthiuram disulphide (TMTD). Around 1920 it was discovered that thiuramdisulphides enable vulcanisation to proceed without sulphur. Investigations in the field of accelerators aimed at development of those, which reacted slowly and safely at the processing temperature but rapidly at the vulcanisation temperature resulted in the introduction of dibenzothiazyl disulphide (MBTS) which gave greater scorch safety at higher processing temperatures.

Later more delayed action and yet fast curing vulcanisation systems were made possible from thiazole derivatives of sulphenamides. Thiocarbamyl sulphenamides are reported to be more productive than the corresponding benzothiazole derivative, due to the combined scorch delay of sulphenamides and the fast acceleration activity of thiocarbamate both being present in their structure. With the discovery of ultra accelerators vulcanisation can be achieved even at room temperature. Thus there are different classes of compounds, which can serve as accelerators in sulphur vulcanisation. Table 20.2 shows main classes of organic compounds that are commercially useful either as primary or secondary accelerators in sulphur vulcanisation of diene rubbers.

Table 20.2: Organic compounds as accelerators with examples.

Class	Speed	Examples
Thiourea derivatives	Slow	DPTU, DBTU
Guanidines	Medium	DPG, OOTG
Benzothiazoles	Semi fast	MBT, MBTS
Sulphenamides, Sulphenimides	Fast, delayed action	CBS, TBBS, MBS
Oithiophosphates, Xanthates	Fast	ZOBP
Thiurams	Very fast	TMTO, TMTM, TETO
Dithiocarbamates	Very fast	ZDC, ABDC

Noting the structure of these vulcanisation accelerators a feature common is some form of a tautomerisable double bond and many of them contain the $-N = C-S-H$ functionality. The time to the onset of cure varies with the class of the accelerator used. Usually a long delay period before the onset of sulphur cross-linking occurs with sulphenamide and sulphenimide accelerators. Prior to cross-linking reaction, it is the role of accelerators to react with elemental sulphur, metal oxide and/or the rubber.

Accelerators offer many advantages such as lowering the cure temperature and shortening of the cure time, thus reducing thermal and oxidative degradation. Also, optimum physical properties could also be obtained with lower sulphur content.

20.3.1   Accelerator activators

In order to achieve the full potential of vulcanisation accelerators it is necessary to use organic or inorganic 'activators'. Inorganic activators are metallic oxides such as– ZnO, PbO, MgO, etc. Zinc oxide is the most important of these additives. Originally ZnO was used as an extender for cost reduction and then it was found to have a reinforcing effect and was later found to reduce vulcanisation time. Usually an activator system, a combination of zinc oxide and a long chain fatty acid such as stearic acid that act as a co-activator is used.

Generally it can be stated that increasing the pH leads to activation of the vulcanisation. The basic activators mentioned lead to improved strength properties of the vulcanisates and come to a shortening of the vulcanisation time. Better processing and improvement in dispersion of fillers and other chemicals can also be achieved by the use of fatty acids and fatty acid salts as co-activators. 5 phr. zinc oxide with 1–3 phr. stearic acid is the commonly accepted combination.

20.4 Retarders

These ingredients are used to reduce the accelerator activity during processing and storage, i.e., to prevent scorch during processing and prevulcanisation during storage. They should either decompose or not interfere with the accelerator during normal curing at elevated temperature. In general, these materials are organic acids such as salicylic acid, phthalic anhydride (PA) which function by lowering the pH of the mixture thus retarding vulcanisation.

Presently, though N-(cyclohexylthio) phthalimide (CTP) is the largest tonnage retarder used in rubber industry, N-nitrosodiphenylamine (NDPA) and thio-sulphonamides also constitute special class of retarders. The use of retarders should be avoided, if possible, by the proper selection of accelerator – sulphur combination and careful control of processing conditions.

20.5 Fillers

Fillers are usually inorganic powders of small particle size incorporated during compounding for various purposes like improvement in strength, cheapening the product, etc. Choice of the type and amount of the fillers to be used depends on the hardness, tensile strength and other properties required in the product. Some fillers are incorporated primarily to reinforce the product and they are termed as reinforcing fillers. Carbon blacks, silicas, silicates, etc., are in this class. Others are included mainly to cheapen and stiffen the final product. China clay, barytes, etc., come under this type.

Reinforcement by filler is the enhancement of one or more properties of an elastomer by the incorporation of that filler, thus making it more suitable for a given application. It is generally agreed that strong links exist between rubber chain and reinforcing filler particles. The effect of filler on rubber vulcanisates depends on its physical properties such as particle size, surface area, surface reactivity, electrical charge on the particle and chemical properties such as pH and reactivity with accelerators. Reinforcing fillers substantially improves the mechanical and dynamic properties of the rubber. As the filler dose increases the properties increase progressively and then decreases. This also depends on the type of filler and rubber used.

The most common and effective reinforcing filler is carbon black. There are varieties of blacks characterised by the particle size, method of manufacture, etc. They are essentially elemental carbon and are composed of aggregated particles. During vulcanisation carbon blacks enter into chemical reaction with sulphur, accelerator, etc., participating in the formation of vulcanised network. Thus the filler will influence the degree of cross-linking also. Carbon black also interacts with the unsaturated hydrocarbon rubbers during milling and the rubber is adsorbed on to the filler. This alters the stress - strain properties and reduces the extend of swelling of the product in solvents. Porter reported that the cross-link density of a black reinforced vulcanisation system increased by about 25% compared to the corresponding unfilled ones. Carbon black generally increases the rate of vulcanisation and improves the reversion resistance. However, carbon blacks can be used in dark coloured products only.

Precipitated silica is the best non-black reinforcing filler so far developed and come closer to carbon black in its reinforcing properties. They have particle size as fine as that of carbon black and have an extremely reactive surface. Precipitated silica is highly adsorptive and hence in formulations containing them, it is necessary to use more than the normal quantity of accelerator or a combination of accelerator system, which is more reactive. Proper choice of the accelerator and activator are done to obtain appropriate scorch and cure times in silica and silicate filled mixes. One distinct advantage imparted by silica to many rubbers is the increased resistance to air ageing at elevated temperature.

Contrary to most types of synthetic rubbers, natural rubber (NR) does not require the use of fillers to obtain high tensile strength by virtue of its higher stress crystallisation. However, the use of fillers is necessary in order to achieve the level and range of properties that are required for technical reasons. Reinforcing fillers enhance the already high tensile properties of gum natural rubber and they improve in particular the abrasion and tear resistance. It must be stated that hardly any filler will enhance all properties to the same optimal degree. The reinforcing effect of active filler as well as the dosage required could be quite different for different elastomers. The amount and type of fillers required in different rubbers or their blends are also different. For example, the activity of fillers in SBR, BR and NBR is often quite more pronounced because of their lack of strain crystallisation than in NR and partially also in polychloroprene rubber (CR). The variation in the effectiveness of NR and synthetic rubbers with regard to fillers can be explained with the theory of overstressed molecules. Other compounding ingredients include antidegredents and other special additives.

20.5.1 Carbon black

The carbon blacks constitute the most important class of reinforcing fillers for rubbers. They are prepared by incomplete combustion of hydrocarbons or by thermal cracking. Most of the carbon blacks used today is made by oil furnace process. A fuel, either gas or oil, is burnt in an excess of air, producing a turbulent mass of hot gases into which the feed stock (generally a residual oil of high aromatic and low asphaltene content) is injected. Reaction to form finely divided carbon is completed within milliseconds and the black is separated from the combustion gases by means of cyclones and filter bags. It is mixed with water in pin-type mixers and then dried in horizontal rotating drums. The yield from the furnace process varies from 25% to 75% of the available carbon, depending upon the particle size, particle diameter ranges from 20 nm to 80 nm. The main types of carbon blacks are: FEF (fast extrusion furnace), HAF (high abrasion furnace), SAF (super abrasion furnace) and ISAF (intermediate super abrasion furnace) blacks. ISAF Black is the one used mostly for treads of passenger car and truck tyres.

In the thermal process, oil, or more frequently, natural gas is cracked at 1300°C in absence of oxygen in a hot refractory surface. The recovery is about 40–50% of the available carbon and blacks thus obtained range in particle diameter from 120 nm to 500 nm. Main types of these blacks are: FT (fine thermal) and MT (medium thermal).

In the now obsolete and virtually extinct channel process, natural gas is burnt in small burners with a sooty diffusion flame and the carbon is deposited by the impingement of the flame on a cool surface such as a large rolling drum or on to slowly reciprocating channel irons. The deposited black is scraped and collected. The yield from this very inefficient process is 5% or less, but very fine blacks can be made, the particle diameters ranging from 9 nm to 30 nm.

The five most important properties of carbon black are: (i) Particle size, (ii) structure, (iii) physical nature of the surface, (iv) chemical nature of surface and (v) particle porosity.

Particle size: The particles of carbon black are not discrete but are fused 'clusters' of individual particles. The fusion is especially pronounced with very fine carbon blacks.

However, the reinforcement imparted by the black is not influenced by the size of the clusters but greatly by the size of the particles within it. The particle size of various grades of carbon blacks is given in Table 20.3.

Table 20.3: Particles size of various types of carbon black fillers.

Type name	Type code	Average particle diameter (nm)
Super abrasion furnace	SAF	20 (Oil furnace)
Intermediate super abrasion furnace	ISAF	23 (Oil furnace)
High abrasion furnace	HAF	23 (Oil furnace)
Fast extrusion furnace	FEF	40 (Oil furnace)
General purpose furnace	GPF	50 (Oil furnace)
Semi-reinforcing furnace	SRF	60 (Oil furnace)
		80 (Gas furnace)
High modulus furnace	HMF	60 (Gas furnace)
Fine thermal	FT	180
Medium thermal	MT	470

Structure: The term 'structure' refers to the joining together of carbon particles into long chains and tangled three-dimensional aggregates. This aggregation of the particles takes place in the flame during carbon black manufacture. By virtue of their irregular morphology, the aggregates are bulky and occupy an effective volume considerably larger than that of the carbon itself. 'High structure' refers to a high degree of bulkiness, manifested in low bulk density and a high capacity to absorb oil. In rubber technology it is customary to associate high structure of filler with high modulus of vulcanisate. The thermal process produces blacks with little or no structure. On the other hand, the oil furnace process, using highly aromatic raw materials gives blacks of high structure. Structure is normally measured by determining the total volume of the air spaces between aggregates per unit weight of black. The test is done by measuring the volume of a liquid dibulyl phthalate (DBP), required to fill the voids.

The gross morphology of reinforcing silicas is similar to that of carbon blacks, except that fusion tends to be more extensive with precipitated silicas.

Physical nature of particle surface: The carbon atoms in a carbon black particle are present in layer planes. Diffracted-beam electron micrographs have shown that in the blacks with low reinforcing potential (thermal blacks), the layers are highly oriented. They are mainly parallel to the surface, have regular spacing and are quite large with very few defects in their network structure. On the contrary, blacks with high reinforcing potential, show less crystallite orientation. Their particles are more irregular in shape, the layers are much smaller, less frequently parallel and have more defect, this may indicate the presence of significant amounts of 'non-graphitic' carbon.

It is possible that the nature of the carbon atoms at the surface of a particle may affect rubber reinforcement. The carbon atoms can be relatively unreactive if they are an integral part of the layer plane, more reactive if attached to a hydrogen atom and very reactive if present as a resonance stabilised free radical.

Chemical nature of particle surface: Carbon blacks consist of 90–99% elemental carbon. The other major constituents are (combined) hydrogen and oxygen. The hydrogen comes from original hydrocarbon and is distributed throughout the carbon black particles, oxygen is confined to the surface. The principal groups present are, phenolic, ketonic and carboxylic together with lactones which are chemically combined.

Particle porosity: The surface of carbon black particles are not smooth owing to the attack on them by high-temperature oxidising gases immediately after their formation. Oxidation takes place at the 'non-graphitic' atoms and can progress into the particle to give pores.

Effects on rubber properties: In general terms, the smaller the 'particle size', the poorer the processability and the higher the reinforcement. The effect of 'structure' is more noticeable on processing properties than on the properties of the vulcanisate. In general, the higher the structure, the stiffer and less 'nervy' the unvulcanised compound and the harder the vulcanised material. The term 'nerve' relates to the elastic recovery from deformation of a raw rubber when a stress is removed from it.

The role of the physico-chemical nature of the surface in the rubber reinforcement is still not fully understood. It has been postulated that a black with high structure gives a high modulus rubber, not because the carbon black agglomerate restrict the cross-linked network but because the high shear forces during mixing break these agglomerates down to give active free radicals capable of reacting with rubber. High structure, however, does not increase either tensile strength or tear resistance – the two properties usually associated with reinforcement. On the other hand, it has been shown that, as a carbon black is progressively graphitised by heat treatment, tensile strength and tear resistance progressively decrease, indicating that the physico-chemical nature of the surface is important.

As far as the chemical nature of the surface is concerned, experimental evidences overwhelmingly point to the existence of chemical bonds or chemisorptive linkages, i.e., interactions of higher energy physical adsorption. First, carbon black surfaces contain functional groups capable of reacting with polymer molecules to form grafts during processing and vulcanisation and second, numerous reactions of hydrocarbon polymers with carbon black have been demonstrated. In fact, several possible mechanisms exist by which grafts may be formed. For example, carbon blacks chemisorb olefins at vulcanisation temperatures. The degree of chemisorption is increased in the presence of sulphur. Shear generated polymeric free radicals have been shown to graft to the carbon black surface during mixing. The cross-linking rate of a rubber is affected by phenolic and carboxylic groups, the black slows the rate of cure in proportion to its total acidity. Channel black for example, is slower curing than furnace black. However, on the whole, the physical adsorption activity of the filler surface is of much more importance than its chemical nature for the mechanical properties of general purpose rubber.

Carbon black particles with 'pores and cracks' have higher surface areas than blacks of similar particle size without such features. This can result in cure retardation owing to the increased adsorption and inactivation of rubber curatives.

Main effects of filler characteristics on vulcanisate properties: Although rubber properties are interconnected and relate to the combination of all filler properties, a brief summary of the main influence of each of the four filler characteristics is given below:

1. Smaller particle size (larger external surface area) results in higher tensile strength, higher hysteresis, higher abrasion resistance, higher electrical conductivity and higher mooney viscosity, with minor effects on extrusion shrinkage and modulus.

2. An increase in surface activity (physical adsorption) results in modulus at the higher strain ($\geq 300\%$), higher abrasion resistance, higher adsorption properties, higher 'bound rubber' and lower hysteresis.

3. An increase in persistent structure (bulkiness) results in lower extrusion shrinkage, higher modulus at low and medium strains (upto 300%), higher mooney viscosity, higher hysteresis and longer incorporation time. Higher electrical conductivity and heat conductivity are found for higher structure blacks. This property is interrelated with surface activity, structure changes on fillers without surface activity (graphitised black) showing the effects indicated above only rather faintly. At constant high activity, the structure effects are most pronounced.

4. Porosity results in higher viscosity and higher electrical conductivity in the case of carbon blacks.

Fillers are known to influence the cross-linking reaction during vulcanisation, e.g., the retardation of cure by channel blacks compared to furnace blacks, hard clay as compared to whiting, or some silicas as compared to silicates, all having corresponding particle size. In most cases, the cause of this retardation may be attributed to acidity of the filler indicated by the pH of its aqueous slurry, influencing the kinetics of the cross-linking reaction. The slurry pH of a channel black is 4–4.5, of a furnace black 7–9, for clay and whiting the values are approximately 4.5–5.5 and 8–10, for silica it varies from 3.5–7, whereas silicates approach 10. The mixing stage in the course of the production of a vulcanisate introduces more variance in the mechanical properties than any other step. The primary process in mixing of carbon black and rubber is the penetration of the voids between the aggregates by the rubber, the primary products are concentrated agglomerates held together by the rubber vehicle. When all voids are filled with rubber, the black is considered 'incorporated' but not yet dispersed. Immediately after and even during mixing of their formation, these concentrated agglomerates are subjected to high shear forces that tend to break them down again into smaller and smaller units until the final dispersion is reached. However, this effect will not be so evident with coarse fillers. High structure fillers incorporate more slowly than low structure fillers, but once incorporated, the former disperses more easily and rapidly than their low structure counter part.

During the milling process, rubber chain molecules become attached to reinforcing fillers, so they are no longer soluble in the usual rubber solvents, this process is the basis for the formation of 'bound rubber'. It continues after mixing and eventually a system of interconnecting chain and particle results, which appears as an insoluble fragile black gel containing all the black and part of the rubber (the bound rubber). The process may continue for weeks at room temperature and is accelerated by increasing the temperature. Bound rubber is usually expressed quantitatively as the percentage of the rubber originally present, i.e., a compound of 50 parts of black and 100 parts of rubber with a bound rubber of 35% has 35 parts of rubber bound to 50 parts of black.

Bound rubber can be considered as a measure of surface activity and because of this, the percentage of bound rubber may run parallel to rubber properties related to surface activity such as modulus, abrasion resistance and hysteresis. The formation of bound rubber is usually explained by assuming that mechanical breakdown of polymer chain molecules results in the appearance of free radicals at the newly formed chain ends. Reactive sites on the filler surface then combine with these free radicals to form the bound rubber. Since there are many sites on a filler particle, it can act as a giant cross-link. The amount of bound rubber first increases with milling and then goes down, as the polymer (natural rubber) breakdown becomes the dominating factor.

Application of carbon black in rubber compounding: Between 90% and 95% of the total carbon black produced is used in rubber industry and approximately 80% of this is used in the manufacture of tyres and related products such as inner tubes and retreading compounds.

By far the largest amount of carbon black used today is of the furnace type. Furnace blacks are practically used in every type of black-filled rubber article. The high structure, high reinforcing furnace grades are used in synthetic rubber treads to give increased tread life. The low reinforcing furnace grades are used in carcasses, the high reinforcing ones in treads. The fine particle blacks are used where high strength and resistance to abrasion are required, i.e., in conveyor belt-covers and certain types of footwear. The coarse particle blacks are used in such articles as hose, cables, footwear uppers, mechanical goods and extrusions. The thermal blacks are used in inner linings and inner tubes. They provide a low degree of reinforcement and can be used at high loading. They are used in V-belts because of their low heat build-up and other applications are in mats, sealing compounds and mechanical goods.

20.5.2   Non black fillers

Addition of silica to a rubber compound offers a number of advantages such as improvement in tear strength, reduction in heat build up and increase in compound adhesion in multi component products, such as tyres. Two fundamental properties of silica and silicates influence their use in rubber compounds, ultimate particle size and extent of hydration. Other physical properties such as pH, chemical composition and oil absorption are of secondary importance.

Silicas, when compared to carbon blacks of the same particle size, do not provide the same level of reinforcement, though the deficiency of silica largely disappears when coupling agents are used with it. In general, silicas produce relatively greater reinforcement in more polar elastomers such as NBR and CR than in non polar polymers such as SBR and NR. The lack of reinforcement properties of silica in NR and SBR can be corrected through the use of silane coupling agents.

Some other filler systems are worthy of mention, not because of their reinforcement qualities but because of their high consumption. These include kaolin clay (hydrous aluminium silicate), mica (potassium aluminium silicate), talc (magnesium silicate), limestone (calcium carbonate) and titanium dioxide. As with silica, the properties of clay can also be enhanced through treatment of surface with silane coupling agents. Such clays show improved tear strength, an increase in modulus, improved adhesion of components in multi component products and improved ageing properties.

Calcium carbonate is used as a low-cost filler in rubber products for static applications such as carpet underlay. Titanium dioxide finds extensive use in white products such as white sidewalls of tyres where appearance is important.

20.6 Antidegradants

The presence of carbon-carbon double bonds renders elastomers susceptible to attack by oxygen, ozone and also thermal degradation. The oxidation of elastomers is accelerated by a number of factors including heat, heavy metal contamination, light, weather, fatigue, oxygen, ozone and atomic radiation.

The loss in physical properties associated with natural and accelerated ageing processes, is normally caused by either chain scission resulting in reduction of chain length and average molecular weight, cross-linking resulting in a three dimensional structure and higher molecular weight and chemical alteration of the molecule by introduction of new chemical groups.

Natural rubber, polyisoprene and butyl rubbers degrade predominantly by chain scission resulting in a weak, softened stock often showing surface tackiness. Chemical analysis shows the presence of aldehyde, ketone, alcohol and ether groups resulting from oxidative attack and mostly at α-hydrogens and double bonds. SBR, neoprene, EPDM, polybutadiene and acrynitrile degrade by cross-linking, giving boardy and brittle compounds with poor flexibility and elongation. Since all rubber like materials, whether natural or synthetic, cured or uncured, contain a certain amount of chemical unsaturation, they are subject to chemical attack by oxygen. Generally, when 1–2 weight % oxygen is combined with most polymers, the product is no longer useful.

By speeding up the effect of oxygen, metallic salts are some of the most powerful catalysts of oxidation, manganese, copper, iron, nickel and cobalt being the worst, however, vulcanisation helps to diminish their harmful effects. Consequently antidegradants are added to retard or prevent the polymer break down, to improve the ageing qualities and to extend the service life of the product involved.

20.7   Antioxidants

A proper stabiliser or antioxidant has to function by interrupting the degradation reaction sequence either by: (a) capturing the free radicals formed and/or (b) by ensuring that the peroxides and hydroperoxides produced decompose into harmless fragments without degrading the polymer and without initiating new free radicals capable of propagating the reaction chain.

The majority of commercially available inhibitors belong to two main chemical classes: amines and phenolics, which represent respectively staining and non-staining types. In general the amines are used only where colour is not important. Phosphites are mainly used as stabilisers for SBR.

The non- staining antioxidants are subdivided in four groups: phosphites, hindered phenols, hindered bisphenols and hydroquinones.

Hindering the phenolic hydroxyl group with at least one bulky alkyl group in the *ortho* position appears necessary for high antioxidant activity. Nearly all-commercial phenolic anti-oxidants are hindered in this manner (examples 2,6, di - tert - butyl-*p*-cresol). Steric hindrance decreases the ability of a phenoxy radical to abstract a hydrogen atom from the substrate and produce an alkyl radical capable of initiating oxidation. However, because of their low molecular weight hindered phenols tend to be volatile and hence hindered bisphenols such as 4,4' - thiobis (6-*t*-butyl -*m*- cresol) are the most persistent of four classes of materials because of their low volatility.

Hindered phenols are employed as stabilisers for uncured, unsaturated elastomers during processing and storage in relatively low concentrations, i.e., ca 0.1%–0.5%. According to use requirements, up to 3% of a phenolic or amine antioxidant is added before vulcanisation.

20.7.1   Staining antioxidants

Antioxidants derived from *p*-phenylenediamine and di-phenylamine are highly effective scavenger of peroxy radicals. They are more efective than the phenolic antioxidants for the stabilisation of easily oxidisable unsaturated elastomers. *p*-phenylene diamine derivatives are used primarily for elastomers containing carbon black because of their intense staining effects.

N,N'-disubstituted-p-phenylenediamines: These products protect unsaturated elastomers against oxidation and degradation by ozone.

Substituted diphenylamines: They are rarely used alone, usually found as a constituent of proprietary antiflex cracking - antioxidant blends. They tend to show a directional improvement in compound fatigue resistance.

20.7.2 Antiozonants

When diene rubbers are exposed to ozone under stressed conditions, cracks develop which are perpendicular to the direction of stress. The mechanisms of ozone attack and its inhibitions are not well understood. It is believed that an antiozonant reacts with either the zwitterion or the ozonide to form an inert protective film. Every time this film is broken, is repaired by the formation of fresh film produced from the three reactants (rubber, ozone and antiozonant) *in situ. para* phenylenediamines (PPDs) are the only class of antiozonants used in significant quantities in combating the ozone – initiated degradation.

They not only serve to protect rubbers from ozone but also improve resistance to fatigue, oxygen, heat and metal ions.

There are three general categories of *para* phenylenediamines, which are: (i) dialkyl PPDs, (ii) alkyl–aryl PPDs and (iii) diaryl PPDs.

Waxes are an additional class of materials used to improve ozone protection of rubber primarily under static conditions. Wax protects rubber against static ozonolysis by forming a barrier on the surface. Wax migrates from the bulk of the rubber continuously, maintaining an equilibrium concentration at the surface. Microcystalline wax migrate to the rubber surface at a slower rate than paraffin wax and performs best at high service temperature whereas paraffin waxes protect best at low temperatures. It can be noted that under dynamic conditions, the protective wax film breaks down, after which the antiozonant system in the rubber formulation will take over as the primary stabiliser system for ozone protection. Waxes are used to ensure protection against ozone for products in storage such as tyres in a warehouse.

Selection criteria for antidegradants

The selection criteria governing the use of antidegradants are as follows:
1. Discolouration and staining: For elastomers containing carbon black, more active amine antioxidants are preferred, phenolics are mainly used in light coloured goods, where colour retention is imported.
2. Volatility: As a rule, the higher the molecular weight of the antioxidant the less volatile it will be, though hindered phenols tend to be highly volatile compared with amines of equivalent molecular weight. Thus correct addition of antioxidants in the compound is critical for avoiding any loss of material.

3. Solubility: Low solubility of an antioxidant cause blooming to the surface with consequent loss of protection of the product. Therefore solubility of antidegradants, particularly antiozonants, controls their effectiveness. They should not be extracted by water or other solvents, such as hydraulic fluid, during their service life.

4. Chemical stability: Stability of antidegradants against heat, light, oxygen and solvents is required for durability.

5. Concentration: Most antidegradants have an optimum concentration for maximum effectiveness after which the material solubility becomes a limiting factor, *para* phenylenediamine offer good oxidation resistance at 0.5–1.0 phr and antiozonant protection in the range 2.0–5.0 phr. Above 5.0 phr *para*-phenylenediamine tends to bloom.

20.8   Softeners

Softeners include a wide variety of oils and synthetic organic materials, which do not react chemically with rubbers but serve primarily as processing aid. They are used for a number of reasons. Some of these reasons are:

1. To decrease the viscosity and thereby improve the workability of the compound.
2. To reduce mixing temperature and power consumption.
3. To reduce hardness.
4. To reduce low temperature brittle point.
5. To aid in the dispersion of fillers.
6. To reduce mill and calender shrinkage.
7. To provide lubrication to aid in extrusion and molding.

The most important class of the softeners are hydrocarbon oils which fall into one of the three primary categories, paraffinic, naphthenic and aromatic. All the three classes of oils used at 2–10 phr, contain high levels of cyclic carbon structures, difference are in the number of saturated and unsaturated rings. For general all round properties, the naphthenics are preferred. The proper selection of the oils for inclusion in a formulation is important. The oil must be compatible with the rubber and the other compounding ingredients used in the recipe. Incompatibility will result in poor processing characteristics and/or bleeding in the final products. Certain esters of organic acids or phosphoric acid are used as plasticisers in the circumstances where petroleum oils may be unsuitable, e.g., because of incompatibility with the polymer. They are used particularly in NBR and CR polymers as processing aids. Examples are, dibutyl phthalate (DBP), dioctyl phthalate (DOP), poly (propylene adipate) (PPA) and trixylyl phosphate (TXP), etc.

20.9 Peptisers

Peptisers are generally used to lower the viscosity of uncured compounds. They function in the thermo-mechanical and thermo-oxidative breakdown of rubbers, as oxidation catalysts at high temperatures but as radical acceptors at low temperatures. They have little effect on vulcanisation properties. Examples of peptisers are, pentachlorothiophenol, phenyl hydrazine, certain diphenyl sulphides and xylyl mercaptan. Peptisers are most effective in natural rubber, SBR and polyisoprene, but they are relatively ineffective in other synthetic rubbers. Each peptiser has an optimum loading in a compound for maximum efficiency. Peptisers such as pentachlorothiophenol are generally used at 0.1–0.25 phr level. This enables significant improvement in compound processability, reduction in energy consumption during mixing and improvement in compound uniformity. High levels can, however, adversely affect the compound properties, as excess peptiser continues to catalyse polymer break-down during the service life of the product.

20.10 Plasticisers

Although plasticisers represent a separate large group of compounding materials, they can also be considered as processing additives. They do not only modify the physical properties of the compound and the vulcanisate but can also improve processing as shown in Table 20.4.

Table 20.4: Influence of plasticisers.

On physical properties	On processing
Lower hardness	Lower viscosity
Higher elongation	Faster filler incorporation
Improved flex life	Easier dispersion
Better low temperature performance	Lower power demand and less heat generation during processing
Swelling tendency	Better flow
Flame resistance	Improved release
Antistatic performance	Enhanced building tack

As a property modifier in rubber compounds, plasticisers can reduce the second order transition point (glass transition point) and the elasticity modulus. As a result cold flexibility is improved. The static modulus and tensile strength are lowered in most cases and correspondingly a higher elongation at break results. Speciality plasticisers provide improved flame retardance, antistatic properties, building tack or permanence. The softening effect of plasticisers leads mostly to improved processing through easier filler incorporation and dispersion, lower processing temperatures and better flow properties. Plasticisers act on elastomers through their solvent or swelling power. They can be split up into two groups: Primary or true plasticisers which have a solvating effect and secondary plasticisers or extenders which are non-solvating and act as a diluent. It is common practice to divide plasticisers into mineral oils and synthetic plasticisers. Mineral oils, by-products of the lubricating oil industry, have the largest market share as relatively inexpensive plasticisers which are used on a large scale in tyre compounds and general rubber goods to reduce costs. At high dosage levels they allow for higher filler loadings. The mineral oils are split into paraffinic, naphthenic and aromatic types. They all exhibit a high compatibility with the low polar or non-polar diene rubbers.

Compatibility of plasticisers with the elastomer is of major importance for their optimum effectiveness. It is largely determined by the relative polarity of both polymer and plasticiser. A homogeneous and stable mixture of plasticiser and elastomer is achieved when their polarities are nearly the same. In any case sufficient compatibility is required to achieve the processability and physical properties intended without separation problems which are observed as as exudation or bloom or volatility or fuming during processing.

Table 20.5 lists different elastomers and ester plasticisers according to their polarity and facilitates the selection of suitable plasticisers.

Table 20.5: Different elastomers and ester plasticisers.

Elastomer	*Plasticiser*
NBR, very high ACN	Phosphate
AU, EU	Dialkylether aromatic esters
NBR, high ACN	Dialkylether diesters
NBR, medium ACN	Tricarboxylic esters
ACM, AEM	Polymeric plasticisers
CO, ECO	Polyglycol diesters
CSM	Alkyl alkylether diesters
CR	Aromatic diesters
NBR, low ACN	Aromatic triesters
CM	Aliphatic diesters
HNBR	Epoxidised esters
SBR	Alkylether monoesters
BR	Alkyl monoesters
NR	
Halo-IIR	
EPDM	
EPM	
IIR	
FKM	
Q	

Mineral oils are not included. Among them the high aromatic products have a higher polarity while the paraffinic ones are practically non-polar. Liquid elastomers are plasticisers which can be viewed as processing additives. They co-cross-link during vulcanisation and cannot be extracted. The vulcanisate properties are insignificantly changed.

Among the synthetic plasticisers esters are the most widely used types. For cost and compatibility reasons they are mainly used in polar rubbers. Their main function is to modify properties rather than to improve processing. In many cases they enhance low temperature flexiblity and the elasticity of vulcanisates. They are preferably used in NBR, CR and CSM.

The ester plasticisers can be split up into general purpose plasticisers and speciality plasticisers with the latter the property modification is given as under:

1. Cold flexibility.
2. Heat resistance.
3. Resistance to extraction.
4. Flame retardance.
5. Antistatic behaviour.

From the monomeric ester plasticisers the phthalic acid esters represent the largest group as they are relatively inexpensive. The carbon chain length of the alcohol components range from C4 to C11 and often mixed alcohols are used in the esterification process. The number of C-atoms on the chains and the degree of branching determine the properties of the esters. An increasing number of C-atoms reduces compatibility, volatility and the solubility in water. It worsens processability and enhances oil solubility, viscosity and cold flexiblity. A higher degree of branching leads to poor low temperature performance, higher volatility, easier oxidation and higher resistivity.

Plasticisers which improve in particular low temperature performance and elasticity of the vulcanisates are aliphatic diesters of glutaric, adipic, azelaic and sebacic acid. They are mostly esterified with alcohols having branched chains, such as 2-Ethylhexanol or isodecanol. Oleates and thioesters are often used in CR. Esters based on triethylene glycol and tetraethylene glycol or glycol ethers of adipic and sebacic acid and thioethers are used as low temperature plasticisers in NBR and CR. A wide variety of low temperature plasticisers is available while differences in effectiveness are often marginal. The choice is finally determined by properties such as volatility or compatibility.

Heat resistant vulcanisates require plasticisers having a low volatility. It should be noted that not the volatility of the pure product is decisive but the volatility from the vulcanisate which depends on compatibility and migration. Particularly suitable plasticisers for polar elastomers are, for example, trimellitates or pentaerythritol esters, polymeric esters and aromatic polyethers which also act as tackifiers. In comparison with common ester plasticisers their processability is more difficult. Especially the polymeric esters exhibit a remarkable resistance to extraction by oils and aliphatic solvents. This group of plasticisers has proven to be of use in heat resistant vulcanisates based on thermally stable elastomers such as HNBR, ACM and CSM.

Flame retardant ester plasticisers play a relatively important role since halogen containing products, such as chlorinated paraffins, are not now generally permitted in use. Phosphate esters are often used. Several types are commercially available permitting a proper choice regarding heat resistance or low temperature performance. They are alkyl, aryl and mixed esters.

Antistatic plasticisers are another important group. Having a limited compatibility they accumulate at the vulcanisate surface and reduce the surface resistance. The best known representatives of this group are polyglycol esters and ethers.

20.10.1 Processing of plasticisers

The incorporation of plasticisers, at moderate dosage levels, on two-roll mills or in the internal mixer is relatively easy. They act dispersingly during filler incorporation and at the same time the compound viscosity and consequently the processing temperature is reduced. Plasticiser containing compounds generally have enhanced building tack and better extrusion performance. In general the synthetic plasticisers have very little influence on shelf life or scorch safety of the compounds.

20.11   Colours and pigments

Colourants are the materials used for colouring non-black rubber goods. They must be stable to curing conditions, colour fast, should be free from staining or bleeding and reasonably priced.

Colourants are commonly chemically classified as inorganic white and coloured pigments, polymer – soluble organic colourants (dyes), carbon black pigments and organic pigments (e.g., metal flakes). Pigments are virtually insoluble in rubbers whereas dyes are soluble. The term pigment is linked with a specific range of particle size ($\sim$0.01 to $\sim$ 1.0 μm). Colouration with pigments requires dispersion, colouration with dyes involves a dissolving process. Dyes can only absorb light and do not scatter it and are therefore transparent, whereas the optical effect of pigments is based on reflectance resulting in opaque and coloured materials. With all coloured pigments that selectively absorb and reflect, the shade is influenced by particle size. To be efficient, they should have a strong covering power. No reflectance occurs when the particle sizes are very small. Intensive research on organic pigments has produced a variety of such materials which are stable to curing conditions and to light and are non-bleeding either to adjacent rubber compounds or to other finishes. Many of these pigments are available as pastes or as master-batches in rubber, which greatly assist in dispersion and lead to appreciable economics in the amount needed to produce the colour required in the final compounds. Certain rubber soluble dyestuffs are also used to produce delicate shades in translucent materials. Inorganic pigments are generally insoluble in the rubber in which crystal or particulate structure is retained to some degree, colour results from the dispersion of fine particles throughout the rubber.

Some of the inorganic pigments, which are used for colouring rubbers, are:

Titanium dioxide: A white pigment, mainly used for its whitening power in tyre sidewalls, hospital accessories, floor tiles, etc.

Zinc carbonates: A white powder, cheaper than titanium dioxide, used when maximum whiteness is not required.

Antimony sulphides: Antimony trisulphide gives strong crimson colour (crimson antimony) or red-orange colour (golden antimony) depending on presence of sulphur and other chemicals like antimony tetrasulphide or calcium sulphate.

Cadmium sulphide: Gives colours ranging from yellow through orange to deep red.

Chromium oxide: A dull green pigment.

Iron oxide: Mainly ferric oxides, with a range of colours from deep red, through orange to yellow depending on the method of preparation, are available.

Mercuric sulphide: Gives strong bright red colour (vermilion).

Nickel titanate: Highly stable yellow pigment.

Ultramarine blue: It is sulphur – containing complex of silicates. The colours vary from deep blue to greenish shades.

The inorganic materials usually give rather dark colours in rubber compounds, for brightly coloured materials more expensive organic pigments are used which are available in a wide range of colours and shades.

20.12 Tackyfying agents

Natural rubber displays a phenomenon known as 'natural tack'. When two clean surfaces of masticated rubber (rubber whose molecular weight has been reduced by mechanical shearing) are brought into contact, the two surfaces become strongly attached to each other. This is a consequence of interpenetration of molecular chain ends followed by crystallisation. Amorphous rubbers such as SBR do not exhibit such tack and it is necessary to add tackyfying agents such as rosin derivatives and polyterpenes to impart tack for efficient compounding. Other tackifiers include pine tar, coumarone resins, petroleum resins and non-reactive phenolic resins.

20.13 Blowing agents

Many polymers are used in cellular form in which the polymer matrix contains gas filled cells which may or may not be intercommunicating. Over the years many methods have been devised for preparing cellular polymers such as sponge and microporous rubber, the most important ones are the following:

1. Incorporation of a chemical compound which decomposes at some stage of the processing operation to yield volatile reaction products such as N_2, CO_2, H_2O, NH_3, etc. These are known as chemical blowing agents. Most widely used blowing agent is azodicarbonamide. However hydrazide derivatives such as benzene sulphonohydrazide, hydrazine derivatives, e.g., trihydrazinotriazine, nitrosoamine derivatives such as dinitrosopentamethylene tetramine, terephthalamide derivatives such as N,N'-dimethyl-N,N'-dinitrosoterephthalamide and in a few cases inorganic bicarbonates such as sodium bicarbonate, are also used as chemical blowing agents.

2. Physical blowing agents liberate gases as a result of physical processes such as evaporation or desorption at elevated temperatures or reduced pressures. This class includes mostly volatile liquids, e.g., Freons, aliphatic hydrocarbons, powdered solid carbon dioxide.

3. Diffusion of gases into the polymer under pressure with subsequent expansion of the composition at elevated temperatures after decompression.

Mixing and curing of rubber compounds

21.1   Introduction

Mixing is the first and most critical process. If each of the various components, rubber, fillers, oils and chemicals is not thoroughly distributed and dispersed through the mass of the compounds, then problems cascade down through the subsequent processes of shaping and curing and result in less than optimum physical properties in the end product. There are three general shaping processes extrusion, calendering and molding. The first two are followed by a curing stage, often an inline continuous system, whereas in molding whether compression, transfer or injection, curing takes place in the mold.

It is important to differentiate between compounding and mixing. Compounding requires deciding which base elastomer, or elastomers, to use, together with the quantities and types of other ingredients, to achieve, after mixing and curing the properties required for the specific application. In a well-designed compound each ingredient and its level, are such as to achieve specific properties in the compound, either to affect its behaviour in the subsequent processes of shaping and curing, or to give certain physical or chemical properties in the end product.

The compounder has primarily to consider the needs of the end application, above those of the processing plant. In other words, the formulation is dictated by the properties required by the end application. However, at the same time, it has to be possible to mix, shape and cure the compound in standard equipment. In addition, the unit cost of the final product, which depends on the cost per unit volume of the cured compound and the cost of processing, has to be considered. End-users, especially the automotive industry, have become increasingly insistent on high quality and low cost. Thus, the compounder has to balance all these factors in designing the compound.

Mixing, on the other hand, requires deciding for a given formulation what equipment to use and the times, speeds, pressures, temperatures and procedures that are required to blend those ingredients into an adequately mixed compound. This also requires consideration of costs. The aim should be to minimise the labour, energy and equipment costs per unit volume of product.

Thus in summary:

1. Compounding involves determining what to mix.
2. Mixing involves determining how to mix.

Mixing is usually carried out in a series of batches in a high-speed internal mixer, in combination with a dump mill, which cools and shapes the product for the next process. Optimisation of the mixing process requires an understanding of the chemistry, physics and engineering involved. In this chapter, the most recent developments in procedures, equipment and control systems for rubber processing will be described. Unfortunately, few plants in practice have all these in place. Processors have to do the best they can with the equipment they have. However, the better understanding the processor has of the mixing process, the more readily he or she can optimise the plant's operations. Optimisation means producing compound that is adequately mixed at the minimum cost per unit volume.

The three stages in the mixing process are:

1. Feeding the ingredients to the mixer in the correct quantities, at the correct times and at the correct temperatures.

2. The actual mixing of the ingredients.

3. Discharge of the mixed compound from the mixer and its shaping, cooling and packaging for the next process.

The single most important aspect of mixing is consistency. Consistency in type, in quality and in quantity of ingredients, in temperatures, speeds and pressures, in times of addition of ingredients, of mixing steps and of dumping: and in the input of work, in shear rates and shear stresses are all required. Only if these things are all consistent, batch after batch, will the product of mixing be consistent, batch after batch, in its behaviour in subsequent processes of shaping and curing and in finished product properties.

21.2   Material flow to the mixer

There are several aspects to be considered here. These are receipt and storage of the raw materials, weighing the ingredients for each batch, conveying the weighed quantities to the mixer and finally, charging these to the mixer. How each of these steps is managed for any specific material in any particular plant depends on the form in which the material comes, the quantity of the material used and of course, the material handling equipment in place. The forms in which the various materials arrive at the plant are dictated by the nature of the material and its manufacturing process. For example, most elastomers arrive in the form of bales, most oils and plasticisers as liquids, carbon black as pellets, white fillers as powders, curing system ingredients and other small chemicals mostly as powders. The material handling system and the quantities used are interrelated. If the plant has one small mixer and produces only a few thousand kilograms of mixed compound a day, ingredients will arrive in bags and drums and be weighed and charged manually. If, at the other extreme, there are several mixers and a daily production of tens or hundreds of thousands of kilograms of compound, as many as possible of the ingredients will arrive in bulk containers, such as road or rail cars and be pumped or pneumatically conveyed to bulk storage hoppers and tanks. From these the materials will be transferred to day-bins or tanks and the required quantities for each batch will be automatically weighed and charged to the mixer, probably under the control of a computerised process control system. In either case, the cure system and other small chemicals may be weighed and charged manually, or by a precision automatic system, or as is often preferred, they are prepared by an outside supplier specialising in producing chemical mixtures in pre-weighed packages.

21.2.1 Receipt and storage of raw materials

As mentioned above, to achieve consistency in mixing, the first requirement is that each component of the mix should arrive at the mixer in the correct amount, at the correct time in the mixing sequence and at the correct temperature, uncontaminated by other materials. This can only be achieved if the plant has systematic procedures in place for acceptance, handling, storage, weighing and charging. These procedures should enable the processor to subsequently identify for any mixed batch, the exact shipment identity of each raw material used.

Reliance on the suppliers quality assurance (Q.A.) reports is normally sufficient to confirm that incoming raw materials meet the agreed specifications. This requires firstly that there be in place a procedure to check the Q.A. reports before the material is accepted into the plant system. Such reports, of course, guarantee the quality and properties of the material as it left the supplier's plant. It is necessary to consider what adverse changes may have taken place in transit. The most obvious item here is to check that no contamination has taken place. Initially the shipment should be checked to see that the packaging has not been damaged and that containers have no leaks. One important item that should be checked by a laboratory test is that bulk carbon black has not developed an excessive level of fines.

It is important that all raw materials should be stored under conditions that ensure that no contamination or deterioration occurs. No raw materials should be stored in the open exposed to the weather. Identification labelling should be such that no confusion arises about the exact identity and receipt date of each shipment of each raw material. The supplier's literature should clearly indicate appropriate storage conditions and the material safety data sheet should indicate any potential hazard in handling the material. Expected shelf life should be considered when ordering, so that usage rate of the material and quantity ordered are balanced in order to avoid overage material in the system.

The storage conditions for each material depend on its nature. However, as will be described later, all major ingredients should ideally be added to the mixer at 35°C or even above and so should not be exposed to low temperatures in storage. Thus bales of elastomer, which are poor conductors of heat, should be stored in a temperature-conditioned warehouse. This is of especial importance with elastomers such as natural rubber, polyisoprene, butyl rubber and EPDM, which can form regions of ordered structure, pseudocrystallisation, at low temperatures. Such partially ordered regions are very difficult to remove once formed and can result in undispersed polymer phase in the material after mixing. Oils and plasticisers should be stored in temperature-controlled tanks in order to keep their viscosity relatively low to facilitate pumping. This also ensures that the temperature of the material, when added to the mixer, is consistent from batch to batch.

Carbon blacks and some white fillers, are hygroscopic. Some high-structure, high-surface-area blacks will pick up as much as 7% water if exposed to moist air. Thus in handling and storage, whether in bag or bulk, the materials should not be exposed to air.

21.2.2 Feeding, weighing and charging of raw materials

The procedure for feeding, weighing and charging any raw material to the mixer depends on several factors:

1. The physical nature of the material. That is, does it come as a solid, a free-flowing liquid, a paste, powder or granules.
2. The packaging and/or container in which it comes. This can be as a wrapped bale, a bag, a drum, or a bulk container, road or rail car.
3. The conveying system. This could be forklift, manual, conveyor belt, screw conveyor, pneumatic, or gravity feed or some combination of these.
4. The type of metering equipment. This could be weigh scales, volume displacement, metering pump, or tare weighing.
5. The quantity of the material required for a batch also affects the procedure. Various ingredients in a batch could be required in quantities from 100 g to 250 kg.
6. The procedure to be followed depends on whether, at one extreme, feeding, weighing and charging of ingredients are all done manually, or at the other extreme, virtually everything is carried out under computer control.

If one is designing a completely new mixing plant, then all these points can be considered and equipment, system and procedures put in place to best suit the throughput, complexity and variety of raw material and compounds envisaged for the facility. However, this is rarely the case. The processor has to design a procedure that takes into account all the above factors as they apply to his or her particular plant, equipment, materials and compounds.

It is preferable that the weighing of major ingredients be performed under computer control and prompting, either directly onto a feed conveyor, or if it can be pumped or conveyed, directly into the mixer. Minor ingredients can either be weighed, once again preferably in an automatic system, or bought in from a supplier who specialises in producing multiple ingredient mixtures. Most commercially available computer control systems for rubber mixing have two main components:

1. First, the control of the weighing of each of the ingredients as required by the compound formulation or recipe.
2. Second, control of the addition of the ingredients to the mixer and of the mixing procedure itself, including determining when to dump the batch.

Such control systems must be capable of handling elastomers in bale form, carbon black and other fillers in bag or bulk, oils and plasticisers in bulk or drum and a variety of minor ingredients in solid, liquid and paste form. They also control the raising and lowering of the ram, the opening of the hopper, the mixer temperature at the start of mixing and the discharge of the batch at the end of the mixing cycle. Most mixing plants produce a wide variety of compounds in varying quantities. Thus, consideration has to be given to production scheduling in order to achieve maximum productivity, preventing cross-contamination of batches and maintaining high quality of mixed product.

Each ingredient has to be transported to daybins or short-term storage, weighed, conveyed to the mixer and charged. How each of these steps is carried out depends on the list of considerations given above.

The two most important factors from that list are the physical form of the material and the quantity used by the plant. Most dry ingredients are delivered in bags, on pallets. Liquids such as– oils, extenders and process aids, can arrive in drums on pallets or in bulk.

Weighing major ingredients: The accuracy of weighing is the prime consideration and that this requires correct sizing of scales. Too large a scale for the quantity being weighed reduces accuracy.

Weighing of small components: In addition to the major ingredients, elastomers, fillers and oils, most rubber compounds require up to seven or eight other ingredients in small amounts. These usually include a vulcanising agent, a cure activator, cure accelerators, the combination being termed a cure system. Other ingredients such as plasticisers, softeners, process aids, antioxidants, antiozonants, UV-stabilisers and blowing agents may also be needed depending on the base polymer, the process system and the service conditions of the end product. As mentioned earlier, these minor ingredients may be obtained from an external supplier, pre-weighed and packaged, or some or all of them may be weighed on site.

21.3 Mixing process

The aim of the rubber mixing process is to produce a product that has the ingredients dispersed and distributed sufficiently thoroughly that it will shape readily in the next process, cure efficiently and give the required properties for the end application, all with the minimum expenditure of machine time and energy. Rubber mixing is usually carried out in closed high-intensity mixers, though some, mixing is still carried out on open mills. The mix when dumped from a mixer is a large misshapen mass, which has to be both cooled and converted into sheet or strips for feeding to the next process. This cooling and shaping is normally done on a mill situated directly under the mixer, though in semi-continuous systems a large-mouthed dump extruder is used to produce sheet. An extruder also provides the opportunity to screen the material through a mesh to remove undispersed fillers and any possible impurities. This is important if the end application is a membrane, aerosol seal, inner tube, or tyre inner liner. The four main components of the mixing process are incorporation, dispersion, distribution and plasticisation are discussed in detail.

1. Incorporation, sometimes called wetting, is the first stage in mixing, during which the previously separate ingredients form a coherent mass.

2. Dispersion is the process during which filler agglomerates are fractured and reduced to their ultimate size.

3. Distribution is simple homogenisation, during which the various ingredients are randomly distributed throughout the mass of the mix.

4. During plasticisation, the mix reaches its final viscosity as plasticisers effectively internally lubricate the mix.

These four processes are not entirely distinct. They all, especially distributive mixing, take place throughout the mixing cycle. However incorporation predominates in the early stages, dispersion in the middle and plasticisation towards the end.

It is important to avoid overmixing. It wastes time and energy and reduces throughput and profitability dramatically. In addition, exposure to shearing at high temperatures can result in excessive interaction between black and rubber. Also, cross-linking and viscosity may increase due to premature curing. In the case of natural rubber, the reverse. Scission of carbon-carbon bonds resulting in a reduction of viscosity, may occur. This latter process is called reversion.

The four components of the mixing process mentioned above, incorporation, dispersion, distribution and plasticisation, are discussed in detail here.

21.3.1 Incorporation

Incorporation is the preliminary step in mixing, in which the initially separate ingredients form a coherent mass, which, although still far from homogeneous, has a consistency such that the mixer rotors or mill rolls can effectively work it. In effect, the incorporation process is the wetting of carbon black with rubber and the squeezing out of entrapped air. In the mixing process, whether on a mill or in a high shear internal mixer, the rubber experiences a large deformation, which exceeds the breaking strain. Thus material failure, rather than flow, is the operative process, especially in the incorporation stage.

The incorporation step has two parallel mechanisms. In the first, as the elastomer undergoes deformation, it provides an increased surface area for accepting filler agglomerates and then seals them inside. This can be readily visualised in mill mixing, where cutting and blending operations by the mill operator directly produce this effect. In the second mechanism, the elastomer is deformed beyond its breaking strain and fractures. In this fractured state, it mixes with the filler agglomerates and as it passes out of the high strain region, once again it seals the agglomerates inside. On an open mill, this sequence takes place as the material passes through the nip. In an internal mixer, depending on the rotor design, it can take place both between the rotors and between the rotor tip and the chamber wall.

Crumbling and tearing in the internal mixer can be caused by failure of the rubber in the extensional flow region, immediately in front of the rotor. With a material having a low elongation at break, especially if the mixer is under-loaded, such tearing and crumbling may occur mostly in the void formed behind the rotor. What is clear is that any model of the mixing process must take account of both extensional and shear flows and of the limited extensibility of the rubber. Extensional flow is more effective in incorporation and in breaking up of rigid aggregates than shear flow at the same deformation rate. However, because internal mixers are designed to provide a region of very high shear, in practice extensional flow accounts for only a minor amount of total energy dissipated. During the initial stages of carbon black incorporation, the black is compressed by the polymer and the agglomerates are crushed by the shearing and compressive forces. At this stage, the interstices within the agglomerates are still filled with air, giving a very weak, crumbly composite. As mixing proceeds, rubber is forced into the voids in the carbon black and the air is expelled.

Thus, the progressive wetting of carbon black by polymer may be followed by measuring the density of the compound as mixing proceeds. When all the air has been expelled and replaced by polymer the mix reaches its final density. Further distributive or dispersive mixing does not affect this and density does not change further once incorporation, or wetting, is complete. However, polarity, as indicated by solubility parameters, is the most significant factor in the mutual wetting of ingredients. Ease of wetting increases as the difference between solubility parameters decreases.

21.3.2 Dispersion

In the incorporation stage of mixing, the carbon black forms relatively large (10 to 100 μm) agglomerates. This can be shown by examination of compounds under the microscope. During dispersion, these agglomerates are broken down to a size of less than 1 μm. The final degree of dispersion of carbon black depends not only on the specific characteristics of the carbon black (morphology, surface activity) but also on the mixing conditions (time, temperature, total shear strain) and on the properties of the rubber (molecular weight, molecular weight distribution, chemical structure).

The dispersion stage of mixing requires higher shear stress and energy input than the incorporation stage and the physical properties of the mix alter as the dispersion proceeds. The changes in power consumption in a typical rubber mix are indicative of stages in the process and can be related to the development of end-product properties.

It has to be emphasised that the four processes are not distinct sequential steps. Incorporation predominates initially, then dispersion and finally viscosity change, with distributive mixing, or general homogenisation, taking place throughout the mixing cycle.

The first power peak, which usually occurs within 1 to 1.5 min of the start of the mix, signals that the ram has reached the bottom of its travel. The second power peak, results from the balance between wetting of the polymer (incorporation) and breakdown of the carbon black agglomerates (dispersion). This is because, as previously stated, as incorporation proceeds, most of the black is present as large agglomerates with part of the rubber occluded within each agglomerate (between aggregates). These rubber/black agglomerates act as large filler particles, whose effective volume is higher than that of carbon black alone because of both the rubber inside the particles and the rubber immobilised between them. This decrease in effective volume fraction of rubber tends to raise the viscosity of the mix. As the carbon black becomes dispersed, the occluded rubber is gradually released and this results in lower viscosity.

Other properties such as die swell, which, like viscosity, are related to the effective volume fraction of rubber in the mix, approach a steady state value as dispersion tends to completion. A well-dispersed compound will always have a lower viscosity and a higher die swell than a comparable compound that is less well dispersed.

High structure blacks, such as N347 and N765, need higher peak power and thus more energy, to mix than lower structure blacks, such as N327 and N762. The surface area of the blacks has no obvious effect. White fillers, clays, talc, etc., behave similarly to low structure blacks. These results are congruent with the above discussion of the physical processes taking place in dispersion.

21.3.3 Distribution

In addition to breakdown of filler agglomerates, dispersive mixing, it is important that at the end of the mixing cycle the entire batch should be homogeneous. That is, a sample taken from any part of the batch should contain each of the ingredients in the proportions designated by the formula or compound recipe. This requires a low-shear stirring action, in contrast to the high shear nip action needed to achieve dispersive mixing.

21.3.4 Plasticisation

Plasticisers, or internal lubricants, are added to rubber compounds to modify their bulk viscosities. As these become more closely incorporated into the mix, their effect is felt as a slight reduction in viscosity. This is not always seen as a change in Mooney Viscosity as this is measured at a very low shear rate compared to those experienced by the compound in the subsequent processing operations of milling, calendering, extrusion, or molding.

21.3.5 Natural rubber mastication

Natural rubber after coagulation contains up to 5% by weight of non-polymeric constituents. At the temperatures the rubber experiences in storage and shipping, some of these constituents act to produce cross-links between the polymer chains and thus increase the molecular weight and viscosity. Before this high viscosity material can be effectively mixed, it is necessary to lower its viscosity. A combination of heat and shear and sometimes the addition of peptisers, can produce physical and chemical changes that result in a reduction in viscosity. The maximum effect can be obtained by mechanical breakdown on an open mill at temperatures below 55°C, or by thermo-chemical breakdown by a free-radical mechanism involving oxygen in an internal mixer at temperatures above 130°C.

It should be noted that normally, natural rubber producers have been supplying rubber in controlled viscosity (CV) grades that avoid the need for such mastication. These are produced by adding a free-radical scavenger at the coagulation stage. This inhibits the cross-linking reactions, which operate by a free-radical mechanism.

Flow visualisation

The earliest work on flow visualisation was that of Freakley and Wan Idris. They fitted a transparent endplate to a small laboratory mixer of the Banbury design and filmed the flow patterns of a silicone rubber with coloured markers. This and later work enabled a detailed analysis to be made of flow and mixing characteristics in the region of the rotor wing. One interesting finding was that dispersive mixing, which depends on the stress levels generated, takes place throughout the entire mass of material swept in front of the rotor wing, not simply at the rotor tip. In addition, they found that the stress levels generated depend more strongly on the batch temperature than on rotor speed.

At about the same time, Asai and others reported flow visualisation studies in a laboratory mixer in which the whole chamber was transparent and the rotors were built up of thin plates splined onto a shaft. This design enabled them to easily vary the shape of the rotors.

One significant point was that different elastomers behave differently in the mixer chamber. For example, natural rubber formed a tight band around the rotors, whereas others showed tearing to different extents. They studied the effect of different rotor designs and the effect of starved, under-filled, conditions. All these studies, have led to a better understanding of the effect of rotor design and flow behaviour in internal mixers of the tangential design.

Modelling

Most reported work on modelling the mixing process has focused on dispersive mixing of carbon black because it is more difficult to achieve and more complex than incorporation, distribution, or plasticisation. Valsamis and others point out that although many improvements have been made in internal mixer design in the 90 years that they have been used for mixing rubber, the underlying principle of their operation has not changed.

This is the repeated passage of compound through regions of high shear stress, followed by intimate distributive mixing in the other regions of the mixer. Only a small portion of the material is experiencing high stress at any moment, but the rotor configuration ensures that the entire compound passes repeatedly through the high stress region. The chamber, gate and ram designs ensure that there are no dead spots where material can sit unmixed and no hot spots where it can develop a high temperature.

McKelvey developed a model of dispersive mixing involving two-particle agglomerates suspended in a viscous fluid under shear. This predicted that some minimum critical shear stress must be exceeded before agglomerates will breakdown. In addition, it showed that only agglomerates oriented properly to the direction of the shearing force will be ruptured, even when this minimum stress is exceeded. Tadmor and others fitted their model to actual mixing in an internal mixer. They showed that the rate of agglomerate rupture is independent of agglomerate size, but instead depends on the size of the initial aggregates. This is because the rate of agglomerate rupture depends on the number of particle-particle contacts. Cotten's experimental finding that the rate of dispersive mixing increases with decreasing surface area of carbon black and increasing structure of the aggregates supports this analysis. Valsamis comments that in practical systems, that is internal mixers mixing commercial compounds, the parameters involved in models are rarely known with any accuracy and so a number of simplifying approximations are made. In other words, theory and practice are still some way apart and likely to remain so.

Another significant point is that the shear stress experienced by the material passing over the rotor tip is not homogeneous. Material near the moving rotor tip experiences a minimum stress for only a short time, while the small amount of material near the stationary wall is subjected to a much higher shear stress and for a much longer time.

However, if the number of passes is high and if the material is thoroughly redistributed between passes, a significant portion of the material experiences high stresses, even if the fraction of the material that experiences high stress is low for any particular pass.

Valsamis and coworkers then introduce the concept of maximum shear distribution (MSD) and derive an expression for the fraction of the material that experiences shear rates above a given value at least once during the mixing time. They used this model in developing the synchronous technology (ST) rotor design.

21.3.6   Flow behaviour of rubber on mills

The flow behaviour of rubber in internal mixers and the processes of dispersion and distribution, need for both a high shear nip action and a lower shear homogenising or blending action. These two actions occur when rubber is mixed on a mill. In fact of course, for many years, until the development of internal mixers, mills were the primary piece of mixing equipment in the industry. They are still used in certain specialised situations for mixing and of course, for cooling and shaping -batches of compound dumped from an internal mixer.

Tokita and White classified the mill behaviour of raw elastomers as falling into four regions. A schematic illustration of this in Fig. 21.1 shows the effect of temperature on milling behaviour as temperature is increased. In addition to temperature they showed that mill behaviour depends on the nip gap between the rolls, the ratio of roll speeds, the actual roll speeds, and, of course the viscoelastic properties of the elastomer.

They found that only in Region 2 did elastomers give good processing behaviour and accept fillers, oils and chemicals readily. Tokita illustrated the effect of altering nip gap on whether the material would band on the front (slow) roll or on the back (faster) one (Fig. 21.2). Nakajima interpreted the four regions of milling behaviour in terms of modulus and elongation at break, based on his concept that elastomers in the nip are deformed beyond their breaking strain. Modulus and elongation at break are rate and temperature dependent as is mill processability. The deformation rate is primarily controlled by the size of the mill gap or nip, when roll speed is constant. The initial response of an elastomer

Regions of mill behaviour

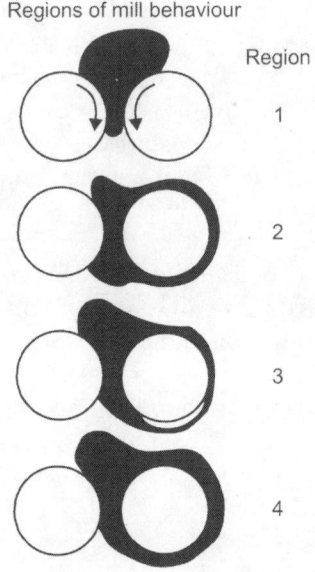

Figure 21.1: Effect of temperature on behaviour of rubber on a mill.

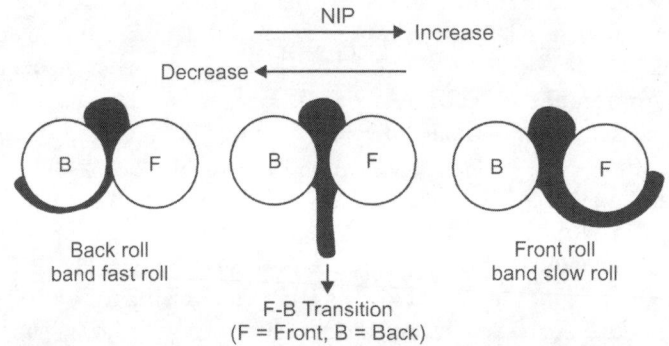

Figure 21.2: Effect of nip distance on behaviour of rubber on a mill.

placed on a mill is governed by its behaviour at low shear and slow deformation. As the material approaches the nip, it is rapidly stretched, so that large elongational deformation behaviour is also important. Nakajima developed this analysis to show how mill behaviour of gum elastomers is determined by their non-linear viscoelastic properties.

21.4   Internal mixers

A rubber internal mixer is a heavy duty batch mixer designed and made for mixing raw rubber with ingredients in rubber compound preparation.

There are several companies that manufacture mixers for the rubber industry. Such mixers all have certain common features. These are:

1. A chamber with a feed hopper and discharge door.
2. Rotors which can exert both a high localised shear stress (a pip action) and a lower shear rate (a stirring or homogenisation action) to the material being mixed. It is the combination of these two effects, high shear stress plus large shear deformation, that is effective in producing both good dispersive and good distributive mixing.
3. A ram which exerts pressure on the mix in the chamber.
4. A heating/cooling system, which controls the temperature of the chamber walls, drop door, ram and rotors.

Figure 21.3 shows section of a Banbury mixer. There are two basic designs of rotors for internal mixers, the non-intermeshing or tangential type and the intermeshing type. Tangential rotors, like the rolls of a mill, can rotate independently and therefore, if required, at different speeds. Intermeshing rotors, like gear wheels of the same diameter, have to rotate at the same speed.

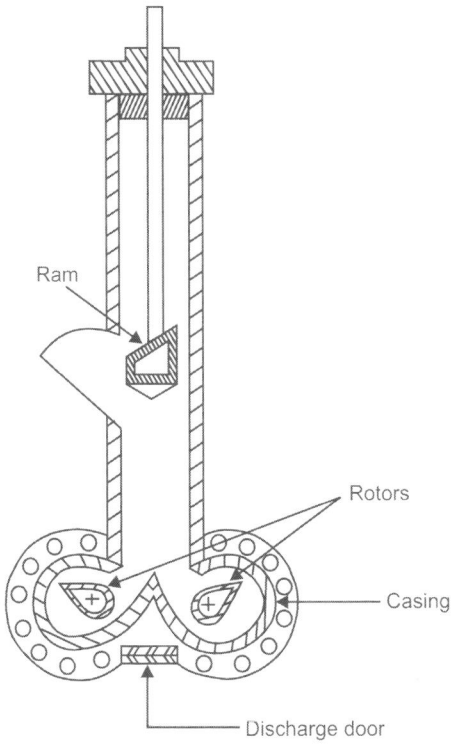

Figure 21.3: Diagrammatic section of a Banbury mixer.

The mixing action of a mill, which has three components:

1. High intensity shearing action in the nip.
2. Rolling action in the bank and elongational shearing as the material enters the nip.
3. Cutting and folding of the sheet on the mill roll by the operator aided usually by a travelling stockblender.

'The action within the nip and approach to the nip, is simulated in the tangential rotor by the rotor shape acting on compound being squeezed and sheared against the mixer side. The cutting and folding is simulated by the transfer of compound from one rotor to the other and the movement of compound along each rotor due to rotor shape. In an intermeshing process, emphasis is placed on the cutting and folding, enforced by the impingement of the radius of rotation of one rotor within the other and the archimedian screw shape of each rotor driving the material in opposite directions.

The nip of the mill and the friction ratio of the rolls is simulated by the nip between the rotors being offset from the centerline of the machine, causing a difference in the surface speeds of the top of the wings and the root of the rotor and hence the simulation of the friction ratio'.

The specific mixing action also depends on rotor design as stated above. The following general points have to be considered:

1. Tangential rotor machines accept ingredients faster and discharge mixed compound faster. As total cycle time is from dump to dump, this means that unproductive time in the cycle is less.
2. Intermeshing rotors generally have a larger volume and take up more of the chamber space. That is, for two mixers of the same chamber volume, the batch size is greater for a tangential mixer.
3. Intermeshing machines, because of the more intensive mixing action, can achieve a better degree of dispersion in a given mixing time.
4. Intermeshing rotor machines can control the batch temperature better because of the mixing action which produces more contact between rubber and cooled metal surfaces.

The above points mean that tangential rotor machines are preferred in situations, such as most tyre mixes, where high productivity is important. With difficult to mix compounds, typical of the industrial goods sector, inter-meshing machines can perform better. Also, since temperature can be controlled better, single-pass mixing can more readily be achieved with intermeshing machines.

21.4.1 Choosing a mixer

The preceding sections do not completely answer the question of what design of mixer best suits a particular operation. It obviously depends on how varied a selection of compounds are to be mixed. At one extreme, there are dedicated mixers in tyre plants, which mix one compound for one application. At the other extreme and more common, many mixers in the technical goods sector produce as many as 50 or 100 different compounds in the course of a month. In the first case, the design can be exactly specified to optimise the mixing of that one compound. In the second case, flexibility is the obvious need and productivity is of secondary importance. Another consideration is whether downstream equipment, such as mixing mills, can finish dispersive mixing, or allow the addition of curatives.

It is difficult and expensive, to change an existing mixing room configuration and thus careful planning and analysis of future needs is important. Each mixer manufacturer offers technical help in making this decision, however they each have a bias towards their own equipment. They also offer to run trials for the prospective customer in their development laboratories to show that their equipment will cope with his or her compounds.

However, such trials are expensive and are rarely offered free of charge. In most cases, the choice is made based on past experience together with the considerations given in this chapter. That is why in Europe, where intermeshing rotor machines, manufactured by Francis Shaw in England and Werner and Pfleiderer in Germany, have been available for decades, they have established a major position in the mixing of technical goods compounds. In the European tyre industry, tangential rotor machines predominate. In most of the rest of the world, tangential machines are most common in both sectors, in the tyre industry for reasons of throughput, in the rest of industry due to historical availability. Past experience, therefore maintains this distribution, even though all manufacturers now provide both tangential and intermeshing machines in all world markets.

21.4.2   Preventative maintenance of mixers

An internal mixer for rubber, like any piece of mechanical equipment, has to be properly maintained if it is to give a long and satisfactory service life. Each manufacturer provides guidance in this area as part of sales service. This, ideally, lists the inspections, lubrication and maintenance recommended daily, weekly, monthly, etc. One area for which the processor probably does not have either the experience or the equipment to check is wear. The rate of wear of rotor tips, chamber walls, throat, etc., depends on the compounds which the mixer processes, specifically whether they contain any particularly abrasive materials. As wear progresses, the clearances through which the compound passes widen and thus, the shear rates, shear stresses and total strain per unit of time experienced are reduced. This reduces mixing efficiency, especially dispersive mixing and requires lengthening of the mixing cycle. Eventually, when the hard surfaces are abraded and the underlying softer steel is exposed, the quality of the mixed product becomes unacceptable, no matter how long it is mixed.

21.4.3 Operation of internal mixer

There are a number of process and operating variables, which have to be considered in any mixing operation. To what extent the variables are under the control of the operator depends on the equipment in place. Of course, some of these variables are interdependent. For example, as rotor speed is increased, energy input rate goes up, heat generated is greater and temperature rise is faster. In order to produce a mixed product that has the ingredients well dispersed and distributed, it is necessary to design a mixing procedure for each specific compound. These procedures have to take into account the limitations of the equipment in terms of power, temperature control, available rotor speeds, etc. Thus, it is necessary to understand the mixing process and the importance of incorporation, dispersion, distribution and viscosity modification and of flow behaviour, shear rate and shear stress. Even if the majority of these variables are controlled by a computerised process control system, such an understanding is necessary in setting up the control criteria. In other words, whatever equipment you have and however sophisticated the control system is, it is still necessary to have a clear process strategy.

Mixing procedures

Two of the most important aspects of mixing, which are incidentally almost completely under the operator's control, are the order in which ingredients are added to the mixer and the times at which they are added. Such questions as to whether the polymer should be added first, followed by fillers, then oil, or the reverse, or should the oil and filler be added together are part of the process strategy controlled by the operator. There are a number of techniques used in the industry. The decision as to which is best depends on the formulation. It has to be stressed that the choice of material input sequence to the mixer profoundly affects the efficiency of mixing:

1. In general, the number of additions requiring the ram to be raised during the mixing cycle should be kept to a minimum. With the ram up and no pressure on the mix, little effective mixing occurs.
2. It is also necessary, in general, to add particulate fillers early in the mixing cycle. This helps to achieve good dispersion as a result of the high viscosity at the lower temperatures prevailing at this time. A higher viscosity gives a higher shear stress at a given shear rate/rotor speed.
3. Oils and plasticisers which reduce viscosity should be added later for the same reason.

4. Oils and plasticisers may coat the rotors and chamber wall and cause slippage, which reduces mixing efficiency. They are therefore often added together with fillers to absorb them and reduce this action.

Points such as the above, which are admittedly to some extent contradictory, are the basis of standard mixing procedures found throughout the industry.

The curative package, or mixture, which at the appropriate stage causes the polymer to cross-link, must not be exposed to elevated temperature until that stage is reached. In many mixing cycles, the temperature reached exceeds that at which it is safe to add curatives. Therefore, it has been standard practice to mix the polymer, oil and fillers in a first pass through the mixer. The batch is then dumped, cooled on the mill and taken off as sheet. This sheeted material is then returned to the mixer, fluxed and the curative package added. At this stage, the only requirement is for distributive mixing to ensure that the curatives are spread equally throughout the batch.

This can usually be done in a short time, less than two minutes usually and so the temperature reached is safely below the threshold temperature at which the curatives begin to react. Such a two-pass system reduces throughput as a significant portion of mixer time is taken up mixing the curatives in the second pass. It also, of course, adds a further increment to the cost of mixing and processors, in order to reduce costs, have developed systems that enable addition of curatives in one pass. The most common technique is to add everything except sulphur or sulphur-donor to the first pass. After dumping and cooling on a mill under the mixer, it is transferred to another cool mill, which has a relatively high friction-ratio and the sulphur is added. To avoid handling powdered sulphur in the open, sulphur dispersed in a polymer is normally used. Other equipment, which is also used downstream of the mixer to cool the batch and allow addition of curatives, are dump extruders, dispersion mixers and tandem mixers.

However, there are five situations in which unaccelerated compound, that is a first pass without curatives, may be required, rather than a single pass:

1. Very intensive mixing required to disperse fillers.
2. Downstream process requires fast cure activity.
3. First pass must be screened, transported, or stored.
4. First pass is the basis of several final compounds.
5. Blends of several first pass mixes used in final compound.

There are four general ways of mixing rubber in an internal mixer, namely the so-called conventional method, the late oil addition method, the upside-down method and the sandwich method. Just about any mixing method used in the industry, when examined, can be seen to fit into one or other of these four categories.

Conventional mixing method: This method consists of adding all the elastomers initially and fluxing them. Fillers and liquid ingredients are then added, starting with those most difficult to incorporate and distribute and those, such as carbon black, which require high shear stress to disperse. This technique gives a long mixing cycle, especially if multiple incremental additions are made. A benefit of this is a slow generation of heat and a lower temperature rise, which may allow a single-pass mix. This technique can achieve good distribution and dispersion of all ingredients, even fillers of very small particle size. With fillers of low bulk density, or with fillers that cake when dry, a variation of this technique is to add part of the liquid ingredients at the same time as the dry ingredients. The oil is absorbed onto the filler surface and into the interstices between the agglomerated particles. This increases the bulk density and reduces the tendency to cake.

Late oil addition method: This involves adding the elastomer first, followed immediately by the fillers and mixing at a rotor speed fast enough to incorporate, disperse and distribute in less than two minutes. At this point, if the oil loading is relatively low, it is all added at once and the mixing continued at a lower speed, until it is all incorporated. This method gives good dispersion as the filler is subjected to high shear stress because of the absence of oil and relatively low temperatures, in the first two minutes. If the recipe calls for a large volume of liquid plasticisers, they cannot all be added at once because they will coat the metal surfaces and cause slippage and reduce mixing efficiency considerably. In this case, the plasticisers are added slowly so as to moderate but not slow down mixing. In this situation, the ability to inject oil into the chamber with the ram down is obviously an advantage.

Upside-down mixing method

This method involves adding all the dry ingredients, other than the elastomer, to the mixer first, then all the liquids and finally the elastomers on top.

1. It is often the fastest and simplest way of mixing.
2. It is commonly used with polymers having poor self-adhesion, such as EPDM.
3. It is used if the polymer content of the mix is less than 25%. In this case, with polymer alone in the mixer as it would be for a conventional or late oil addition method, it is difficult for the mixer to start mixing because the chamber has so little material in it.
4. It is especially effective for compounds having a large volume of liquid plasticisers and large particle-size fillers.
5. Low structure, small particle size carbon blacks cannot be effectively mixed by this technique, as it does not provide a high level of dispersion.

6. With very high viscosity polymers, upside-down mixes give a rapid build-up of temperature, which can also lead to poor dispersion.

7. Clays which are often difficult to wet due to low surface energy, do not incorporate well in an upside-down mix

Sandwich mixes: When a compound requires a blend of two polymers of different solubility parameters, it is often an advantage to add one first together with fillers and oils, mix for a short while and then add the other polymer on top. The compound, having a broader solubility parameter than the polymer alone, accepts the second polymer more readily. An example of this is the blending of SBR and polychloroprene.

Collins found that when blends of polymers having different polarity are mixed, it is important to consider how other ingredients will distribute. Polar materials, such as carbon black, will prefer to associate with the more polar polymer. In such a case, the choice of which elastomer to add first will have a large effect on the properties of the compound.

Thus, for any one compound in a given mixer, there are several mixing procedures that will work. If the further factors of temperature, rotor speed and fill factor are added and the possible choice of one pass or two passes, the number of possible procedures is increased further. Therefore, it is not unusual for two plants with similar equipment to employ different procedures to mix the same compound both of which work. It is often impossible to determine which is better measured by the two criteria of cost and quality. In most situations, there is neither time nor personnel to experiment with the procedures in order to optimise them. The compounder or plant supervisor uses past experience to establish an initial procedure. Only if the compound is produced in sufficient volume to require long runs is it economical to spend the time improving the procedure.

Temperature control in internal mixers

This is a major concern in any mixing operation. The energy that produces the shearing action is largely converted into heat and this results in temperature rise during mixing. Rubber is an inherently poor conductor of heat and heat can be removed from the mass of rubber only if fresh surfaces are generated and brought into contact with cooler metal surfaces, that is, the surface of the rotors and the inside of the chamber. Modern internal mixers have a high heat exchange capacity and when the mixer chamber is empty, these metal surfaces cool rapidly to the temperature of the cooling medium. If, at the charging of the next batch this temperature is too low, the fresh rubber in the mixer may slip on the cold metal surfaces and delay the start of mixing. Thus, it is preferable to control the cooling medium temperature to ensure that at the start of each mix the temperature is above the softening point of the elastomer.

Later in the mixing cycle when heat removal is the major concern, colder water may be needed to effectively control the mixer temperature. The use of temperature-controlled coolant results in more consistent mixing and a reduction in mixing time, due to a significant reduction in the time required for compaction and thus in ram seating at the beginning of the mix cycle.

Intermeshing rotor machines are less dependent on adhesion between the rubber and metal and are therefore less affected by initial temperature.

Rotor speed

This directly affects the total shear strain and deformation rate and thus the speed of mixing, which, in practice, is limited by the maximum allowable temperature: that is, the balance between heat generation and heat removal. It is also limited by the reduction in viscosity and thus shear stress, which results from temperature rise.

Since there is a minimum shear force necessary to disperse aggregates, temperature rise reduces the rate of dispersive mixing. Thus, there is a tradeoff between increased speed of mixing and less well dispersed compound, though up to a limit, higher rotor speed leads to faster mixing.

It is especially useful to have a DC motor, which allows variable speed. For example, when adding large amounts of dry fillers, the batch loses cohesiveness until the powders are incorporated and the addition of oils or plasticisers can slow down incorporation because of their lubricant effect on metal surfaces. In both cases, dead time can be reduced by increasing the rotor speed until the temperature rise indicates that incorporation is complete. At that point, rotor speed can be reduced again to control heat build up.

Ram pressure

The main function of the ram is to keep ingredients in the mixing chamber. As ram pressure is increased, voids within the mixture are reduced and slippage between the rubber and rotor surface is reduced. This is because as pressure increases, the contact force between the rubber and the rotor surface increases and the flow begins at a lower temperature. This is especially true for high viscosity polymers, which require a higher effective batch pressure. There is a practical limit with any machine, which is the maximum pressure the seals will withstand.

Batch size

Batch size must also be optimised. If the mixer is underloaded, or overloaded, poor mixing and longer mixing time may result. If the batch is too small, the ram can seat right down and acts merely as part of the chamber wall, exerting no pressure on the materials in the mixer. The flow patterns in the mixer are altered and in a tangential rotor machine, the compound does not experience the required number of passes through the high shear stress region. If the batch is too large, then the ram never seats down and material, especially powders, accumulates in the throat and never enters the mixing chamber and is not incorporated into the mix. This has two adverse effects. Firstly, the batch is not homogeneous, nor does its composition agree with the formulation. Secondly, when the unincorporated powders are dumped with the batch they can cause a hygiene problem.

The optimum weight for any particular batch depends on the type and level of rubber, filler and plasticiser. The specific gravity of the finished compound is only a rough guide to the required volume. There always has to be some free volume, which means that at the end of the mix, the compound must not fill the mixer completely. This allows the vigorous cross-flow and turbulence necessary for distributive mixing. The more elastic the compound is, the more this action occurs. The level of the three major constituents and their nature determine the elasticity of the compound.

In determining the optimum volume for a batch, the action of the ram has to be observed. If the ram seats down quickly and completely and shows no oscillation, it indicates that the mixer is underloaded and that the ram is exerting no pressure on the contents, but is merely acting as a section of the chamber wall. If the ram remains ten or more centimeters above the stops and oscillates vigorously, then it is clear that there is material in the throat. What is required is for the ram to reach a position two or three centimeters above the stops and to oscillate gently. This shows that the ram is exerting pressure on the batch as it surges against the bottom of the rain.

Dumping the compound for cooling

One important point for any compound when formulation, loading sequence, coolant temperature settings, rotor speed and batch size have all been determined, is the decision on when to dump the compound for cooling, sheeting out and transfer to the next process.

This reduces to determining` when the compound is adequately mixed for the subsequent processes. Overmixing wastes machine time and may actually adversely affect the properties of the compound. Whatever criterion is used to determine when to end the mix, the aims are to guarantee the quality of the product, avoid overmixing and to reduce variation between batches. At one time the standard procedure was to allow the batch to mix for a given time, or until a specific temperature was reached.

The determination of these two criteria for a batch was based on experience, plus, the knowledge of the temperature sensitivity of the cure package, if present in the mix.

In the earlier discussion of the mixing process it was noted that the changes in power consumption in a rubber mix are indicative of stages in the process and can be related to the development of end-product properties. A precise and reproducible control of the mixing cycle for any compound can be achieved using energy input, as long as the mixer and major ingredients were at pre-determined temperatures at the beginning of the mixing cycle. The process control systems available all use energy usage as the main criterion determining the dump point.

Time, temperature and instantaneous power input are also monitored. Temperature is often a second, or override, control criterion to ensure that an excessive temperature is not reached if some process malfunction occurs.

21.4.4 Control of the mixing process

At one extreme, the mixing process may be completely under manual control. In this case every action from weighing, feeding, temperature settings, charging, ram up, ram down, sweeping and final dump are performed by the operators prompted by a written set of instructions. Each mixer manufacturer and some materials handling companies, now supply computerised control systems that take many of the process steps out of the operators hands, either performing them automatically, or prompting the operators via a computer screen. Most mixers in the industry have control systems somewhere in the range between completely manual and mostly automatic. In either case, as was stressed earlier, optimisation of any mixing process requires an understanding of the chemistry, physics and engineering involved and a precise and clear process control strategy. Sophisticated control equipment is no substitute for a good under-standing of the process. It is also necessary to understand the capabilities and limitations of the specific equipment available and to adjust the process, if necessary, to cope with them.

21.4.5 Scale-up

Most formulations are developed in the laboratory, using a laboratory mill, or a small internal mixer with adequate capacity. However, a mixing procedure, which works well in a small mixer, cannot necessarily be directly transposed to an industrial scale mixer. The clearances between rotor tip and chamber wall in a laboratory tangential mixer are not very different to those in an industrial mixer, but in order to generate the same relative speed and thus shear rate, the rotor of the small mixer has to rotate at a much higher rate. This produces much more extensional shear and also turbulent mixing. In addition, the ratio of metal weight to compound weight is much higher for a laboratory mixer. This provides a relatively large heat sink and so temperature rise is less for a similar energy input per unit weight of compound.

21.5 Take-off systems

After mixing, a batch of rubber has to be cooled and converted into strips or sheet, suitable for feeding to the next process. This is normally done by dumping the batch through the drop door at the bottom of the mixer onto a cool mill, capable of handling the entire batch. An alternative to this, especially if it is desired to turn the batch process into a continuous one, or if the compound must be strained, is to discharge the batch into a wide mouthed dump extruder.

21.5.1 Dump mills

The dump mill under the mixer has several functions:
1. Reduce the temperature of the mix.
2. Further homogenise the mix in both material distribution and temperature.
3. Continue dispersive mixing (the lower temperature and consequent higher viscosity helps).
4. Further modify viscosity.
5. Allow (in some cases) addition of curatives and accelerators.

When a hot batch of rubber is dumped onto a mill, it is hoped that it will band quickly and easily on the front roll, be easy to cut and handle and allow easy removal from the mill. Unfortunately, this is not always the case:
1. The rubber may fail to band.
2. It may pass through the nip and fall into the pan.
3. It may tear and crumble like dry pastry or clay.
4. It may be so tacky or sticky that it cannot be removed from the mill.
5. It may be so nervy that it won't enter the nip at all but will bounce on the mill like a baby elephant on a trampoline.

The actual behaviour of a compound depends on its viscoelastic properties, the nip gap of the mill, the roll speeds and the operating temperatures, as described earlier. Cold rubber on a cold mill simply will not pass through the nip. As the temperature of both metal and rubber is increased, significant flow through the nip begins and a tight elastic band clings to the slow roll. In this region, effective mixing occurs and curatives, accelerators and other ingredients can be added if needed. With some compounds, especially highly filled ones, as the temperature is raised the band sags or bags away from the roll surface. The material may crumble or tear and even fall off the roll. The rubber passing through the nip is extended so far that it tears, if the stresses built up have insufficient time to relax as the rubber is transported during one revolution of the roll, they accumulate until the tears stretch across the roll and the rubber crumbles into the pan. The other phenomena referred to above (nerviness, stickiness, banding on the fast roll rather than the slow roll, failure to adhere to either roll, etc.), can be explained from the viscoelastic properties of the compound, the nip gap and the actual and relative speeds of the rolls. Unless the mill has independent DC drive on the two rolls, hydraulic adjustment of the nip gap and accurate temperature control of the mill rolls, it is difficult to handle a wide variety of materials on the one mill.

21.5.2 Packaging

Once removed from the mill, as strips of the required size for the next process, the material is coated with an anti-tack material, usually in a dip tank containing a slurry of the coating material. Then it is carried on a conveyor through a cooling system using either or both water spray and blown air cooling. Next, strips are packed in wire mesh cages or hardboard cartons using a wig-wag system. It is of utmost importance that the strips be cooled sufficiently before packaging, for two reasons.

First, they may cold flow and stick together, making their removal and feeding to the next process difficult and second, if the material being packaged is finished stock containing curatives, curing (vulcanising) reactions may take place in storage, leading to problems in extrusion or molding.

21.5.3 Single pass mixing

Many processors avoid the problems and costs of double-pass mixing by transferring the material from the dump mill to an independent mixing mill with DC drive, on which the curatives are added. Others dump from the high shear mixer into a low shear blender for this step. Continental A.G. developed a tandem mixing process, which has been licensed for manufacture to Francis Shaw Ltd. In tandem mixing, the masterbatch is discharged from the top mixer into the tandem mixer (a large capacity ramless internal mixer), whose operation is synchronised with the masterbatch unit. In this second mixer, the masterbatch is cooled and mixing is completed by, the addition of the curatives package. Simultaneously, the next masterbatch is being produced in the top machine. When the second mix cycle is finished, the completed mix is discharged to a mill or extruder for working up.

21.6   Other mixing equipment

21.6.1 Mill mixing

Open mills were the original equipment used in mixing rubber compounds and are still used for specialty compounds, for small batches, in small shops where the production volume does not warrant expenditure on an internal mixer and for curative addition in a second pass, or directly after the dump mill in a single pass mixing.

Temperature control is very important in mill mixing. Cooling is usually accomplished by flooding or spraying the inside of the rolls with water, or by circulating water through channels drilled in the roll walls. Drilled rolls provide the best heat exchange and thus the best temperature control. Compound temperature is adjusted by regulating the rate of flow and temperature of the water through the rolls. Steam heating or hot water is used where temperature increase is required.

Roll temperatures suitable for mixing a given compound depend upon the nature of the polymer and such factors as the types and quantities of fillers and plasticisers that are to be incorporated. Nitrile rubber and lightly loaded EPDM compounds are usually milled with roll temperatures in the range of 10 to 30°C, chloroprene compounds at 20 to 40°C and butyl and highly loaded EPDM compounds at 60 to 80°C. For butyl and EPDM compounds, the front roll should be approximately 10°C cooler than the back roll because these polymers tend to release from the hotter roll.

Mixing is caused by the shearing action in the nip between the rolls. Originally, mills had a roll speed ratio of 1.4: 1.0, the faster roll at the back, to provide efficient breakdown of natural rubber. With the advent of synthetic rubbers and standard controlled viscosity natural rubber, this ratio has been reduced to 1.2 : 1.0 and often the fast roll is at the front because many synthetic rubbers band more easily on the faster roll.

In mill mixing, the elastomer is first added to the nip, open to about 6 mm, at the top of the rolls. A band of elastomer then passes through the nip and is formed preferably around the front roll. Depending on the elastomer used, varying degrees of difficulty may be encountered in forming the band initially. However, after a few passes, a band will form and can be fed back into the nip continuously. The elastomer is then cut back and forth twice to assure proper blending and to allow the elastomer in the bank to go through the nip. It is important for efficient mixing to maintain a rolling bank on the mill during incorporation of ingredients.

All dry ingredients, except the fillers and the cure system, are then added to the nip. The compound is then cut back and forth twice to assure good dispersion of these dry ingredients throughout the batch.

The next step is to open the mill slightly and add the fillers slowly to the batch. In order to prevent excessive loading of fillers at the center of the mill, strips of compound are cut from the end of the roll several times during this operation and thrown back into the bank. When most of the fillers have been mixed into the compound, the liquid plasticisers and the rest of the fillers are added slowly and alternately to the batch. When no loose filler is visible, the batch is cut back and forth twice more to assure good distribution of fillers and plasticisers. It is often a useful procedure to 'pig' roll the batch and feed the 'pigs' back into the nip at right-angles as a part of the cross-blending process. The next step is to open the mill more and add vulcanising agents to the batch. When the vulcanising agents are well distributed, the entire batch is cut back and forth at least five times to assure thorough cross-blending, before being sheeted off the mill.

Mill mixing is a slow process, it takes up to 30 min to mix a batch. It requires constant physical effort from a mill operator, whose skill is the determining factor of the quality of the product. Another problem is that some of the ingredients may create an environmental hazard and to avoid this they have to be added in costly masterbatch form.

21.6.2 Continuous mixing of rubber compounds

Continuous mixing of rubber compounds requires that the elastomer be available in free-flowing granular or crumb form. Unfortunately, most raw rubbers if produced in this form have to be treated with a surface coating to prevent adhesion between the particles on standing. Therefore, it is more usual to granulate the rubber on site and use it immediately. Rubber granulation usually requires a temperature below the glass-transition point and a high energy input for the granulation itself. These limitations result in increase in cost of the rubber component in the mix and in an industry as price-sensitive as the rubber industry, this is a disadvantage. Another problem is that each ingredient has to be metered and fed to the process in a constant, controlled, rate and this makes continuous mixing suitable only for relatively simple mixes, that is one raw rubber, one black, one oil and premixed chemicals.

Single screw extruders cannot normally generate the requisite critical shear stress to achieve good dispersion of carbon black, however, it has been shown that a co-rotating twin-screw extruder can. There are some problems to be overcome in controlling temperature in the extruder, due to the high viscosity of the mix, but this approach, subject to the above limitations, shows promise. However, equipment costs, for equivalent output, are of the same order as for a conventional mixing line.

21.7   Custom compounding

One alternative to installing a mixing plant is to buy compound from one of the many companies, called custom mixers or custom compounders, whose business is to supply mixed compound. If the amount of compound the plant requires is insufficient to justify an in-house mixer and ancillary equipment, then buying compound is a sensible alternative. Before doing so it is wise to visit the company and assure yourself of their competence.

21.8 Troubleshooting the mixing process

Often, problems in mixing do not become apparent until the final product is produced. In such cases, it is useful, in effect, to retrace the process shown in Table 21.1. Useful and often essential, end properties are dependent on microscopic properties. These, in their turn, are determined by the conditions experienced in the mixer and these depend on the mixer, its capabilities and limitations. Corrective action is often simple, once the cause of the problem has been identified. Problems usually are due to inadequate dispersion, too active a cure (usually called scorchiness), contamination, poor processability on the dump mill and batch-to-batch variation.

Table 21.1: The mixing process.

Machine parameters	Process conditions	Material properties	Product properties
Speed	Shear rate	Dispersion	Tensile strength
Clearances	Shear stress	Homogeneity	Modulus
Rotor geometry	Total strain	Flow properties	Elongation
Cooling efficiency	Time	Cure characteristics	Flex
	Chamber loading	Die swell	Tear strength
	Temperature		Abrasion

The lists below suggest possible causes of such problems. The connections between causes and effects of such problems should be clear from the earlier discussions of the mixing process and mixer operation and control. It is assumed that no major sins of either omission or commission are the cause of the problems. That is, the formulation has been developed and tested in the laboratory and no weighing or compounding errors have been made.

21.8.1 Inadequate dispersion or distribution

1. Insufficient work input, mixing time.
2. Order of ingredient addition not suitable.
3. Batch size too large or too small.
4. Wrong rotor speed.
5. Insufficient ram pressure.
6. Excessive wear of rotors and chamber wall.
7. Cold polymer (this applies especially to natural rubber, EPDM and butyl).
8. Excessive moisture in fillers.
9. Oils added at temperature below pour point.

21.8.2 Scorchy compound

1. Too high a heat history after addition of curatives.
2. Accelerator added too soon.
3. Inadequate distribution of curatives.
4. Too high a rotor speed.
5. Materials added at too high a temperature.
6. Inadequate cooling of compound after take-off.

21.8.3 Contamination

1. Physical contamination of one or more ingredients.
2. Insufficient clean-out between batches of different base polymers.
3. Oil-seal leak

21.8.4 Poor handling on dump mill

1. Incorrect roll temperatures, speeds, friction ratio.
2. Too high a loading of clay fillers, viscous plasticiser.
3. Poor distribution and/or dispersion.
4. Scorchy compound.
5. Compound left on mill too long.

21.8.5 Batch-to-batch variation

1. Variation in initial loading temperatures, ram pressure, cooling waterflow, or temperature.
2. Variation in compounding ingredients.
3. Variation in dump time, temperature, or energy input.
4. Variation in milling time or mill settings.
5. Variation in amount of cross-blending on mill.

Thus, to be a good mixer requires an understanding of compounding and vice versa and both require an understanding of the conditions that the product will have to meet in further processing and end product application.

21.9 Curing process

Rubber compounds are viscoelastic fluids after mixing. They comprise long polymer chains, which although intermingled, are distinct from one another. The viscoelastic behaviour of any particular compound depends on the properties of the base polymer and on the types and amounts of fillers, oils, plasticisers and other additives in its formulation. Such mixed but uncured compounds have poor physical properties, they are readily soluble in suitable solvents and are easily deformed. In order to develop properties that are required in end applications, these separate chains have to be linked by chemical bonds to give an elastic three-dimensional network. The process of generating these links is termed vulcanisation, cross-linking, or, more commonly, curing. Uncured compounds have little use or commercial value except as precursors of cured compounds. After curing, the compound becomes an elastic solid, no longer readily deformed.

Therefore, obviously, curing is normally the final stage in processing, after the final shape has been reached. Curing is a chemical process in which links are formed between previously separate chains and is largely irreversible. After choice of the base elastomer for a particular application, deciding on the cure system is the most important step in compounding. Sulphur based curing systems are the most common because of their ability to generate cross-links of one or several sulphur atoms, especially with diene polymers. Other systems used are based on phenolic resins, metal oxides, peroxides, or silanes.

Sulphur or sulphur-donor systems are the most complex, in that they contain, in addition to the main cross-linking agent, other auxiliary chemicals.

These are typically:

1. Accelerators, which increase the rate of the cross-linking reaction.
2. Activators, which speed up the reaction by forming complexes with the accelerators, which are rubber soluble.
3. Retarders, which delay the initiation of cross-linking, reducing the tendency to premature vulcanisation or scorch.

Balancing these multi-component systems to give the exact balance of cure properties for the process and the required properties is obviously a complex and tricky business.

From the point of view of the processor the important points about any cure system in addition to, of course, giving the required end properties, are as follows:

1. It should prevent cross-linking during mixing, milling and extrusion. That is it should have good scorch safety. Obviously, keeping heat history to a minimum prior to the curing stage is a good policy.
2. It should have a sharp trigger, or threshold, temperature below which no appreciable curing takes place. Above that temperature, rapid and complete curing should occur. It is difficult to obtain both of these in a system.

The processor also needs to be aware of the behaviour of the compound in the curing process. In particular, the behaviour of some elastomers, such as natural rubber, which can begin to, in effect, reverse the curing reaction, if kept at the curing temperature for too long.

This can result in chain scission and a less densely cross-linked network. The process described above is known as reversion and can result in inadequate physical properties of the end product. As with all chemical reactions, the rate of curing reactions increases with temperature.

Thus determining the effect of temperature on cure rate and scorch, especially when a new compound is being developed, is a necessary step. It helps establish the sensitivity of the compound to temperature changes during processing, seasonal temperature changes, or changes due to drafts, which affect heat losses from equipment and thus the temperature of the material.

21.10   Scorch or premature vulcanisation

If a small percentage, as low as 5%, of polymer molecules has undergone one cross-link, flow properties of the compound are seriously impaired. This makes further processing difficult and will affect the physical properties and surface appearance of the end product. Thus, as stated above, in developing a new compound, the compounder needs to determine how sensitive to premature vulcanisation it is.

This obviously is an important piece of information for the processor. It determines how much care is needed in controlling the heat history of the compound, whether it needs more thorough cooling after mixing, or storage in a refrigerated area.

The two most common tests for scorchiness involve the Mooney viscometer and a Curemeter/Rheometer.

ASTM D 1646 describes the method of measuring scorch of rubber compounds using a Mooney Viscometer. The test is similar to the standard Mooney test except that the test temperature is 120°C. After a minimum is reached in the torque curve, the test is continued until a five-point rise above that minimum is reached.

The 'scorch time' as determined by this test is rather arbitrary, in that it does not necessarily correlate with performance in processing, even at 120°C. Like so many other processing tests in the rubber industry, its value is as a comparative measure between materials, or between batches of the same compound.

The oscillating disc rheometer (ODR) and moving die rheometer (MDR) curves have been described earlier. These instruments measure the resistance to deformation during the complete curing cycle. However, it is not possible to determine precisely when the compound changes from a viscoelastic fluid to an elastic solid. Therefore, conventionally, scorch is defined as the time for the measured torque to rise either one or two units above the minimum torque measured, depending on whether an ODR or an MDR is employed. Curemeter tests are performed at temperatures at which the compound is expected to cure fairly rapidly. In the interests of standardisation the respective ASTM tests, D1646 and D2084, specify temperatures of 120°C for Mooney Scorch and 160°C for curemeters. For some curing systems, higher curemeter temperatures are used.

Therefore, inevitably, curemeter scorch times are shorter than Mooney scorch times and the two often do not correlate. Once again, we have a test, which is useful for process control and comparative purposes, but the processor has to determine the relevance to each specific compound and process.

Calendering, extrusion and molding of rubber compounds

22.1   Introduction

A calender is similar to a mill in its operation, in that the rubber and other components pass between two or more metal rolls. However, in order to produce product of controlled thickness, significantly higher pressures are needed and therefore, calender rolls, journals and frames have to be more carefully designed and more strongly constructed than those of mills.

Calendering machines are used to produce continuous sheets from rubber compounds, sometimes incorporating reinforcing materials such as textile or wire cord and for impregnating or coating fabrics with compound. Calendering is one of the oldest processes in the rubber industry, having been in use for over 200 years.

The rolling bank formed in the nip during the calendering process readily entraps air, which needs high compression in the nip to be squeezed out again. As a result, standard calender lines cannot produce sheet free of air inclusions thicker than 2 to 3 mm. Thicker sheets can only be produced by laminating these, or using a combination of a roller-die and calender, which can produce sheets up to 15 mm thick.

22.2   Equipment for rubber calendering

Rubber calenders are differentiated by the number of rolls, their arrangement and their size (diameter and width). They can have two, three, or four rolls in a variety of configurations. However, for the production of tyre stock, belting and sheeting the three-roll vertical calender with 24″ diameter, 68″ width rolls and four-roll Z and L calenders with 28 × 78″ rolls are standard. Four-roll calenders are used for applying compound to both sides of tyre cord fabrics in one operation.

The rolls, which carry out the actual process, have high quality requirements. They have to be concentric, not distort at the operating temperature and have high surface hardness and finish. Because the high separating forces tend to deflect the rolls during operation, the rolls are given a roll crown, with a mid-roll diameter being up to 0.1 mm larger (or in some applications smaller) than the ends. Modern calenders have drilled heating/cooling passages under the roll surface and the surface temperature of the rolls is controlled to ±1°C.

The required thickness and therefore the gap between the rolls, may vary from one product to the next and thus there has to he a robust mechanism to allow accurate, controllable roll-gap adjustment. Other devices compensate for deflection under load.

22.3   Process of calendering

22.3.1   Feeding

To ensure steady operation of the calender and to control shrinkage, the compound has to be pre-heated to around 93°C and thoroughly fluxed and plasticised before being fed to the calender. This is often done on an open mill. The disadvantages of mills are that they take up a lot of space, use a lot of energy, require an extra operator and the quality and consistency of the feedstock to the calender depends on the skill and care of the mill operator. This is countered in some cases by feeding the finished compound after mixing in an internal mixer to an intermediate mill, which acts as a buffer and from which it is fed to the calender. This eliminates one cooling and subsequent reheating step. Also, cold-feed pinbarrel extruders are increasingly used for the pre-heating step, as they require less energy per unit weight of compound to produce strip feed suitable for a calender. And as mentioned above, in thick sheet production an extruder and sheet die can be used.

If a cord or fabric is also to be fed to the calender, the spools are fed from equipment, which is pneumatically controlled to ensure constant tension in the feed. Accumulators are also required to allow constant feed to the calender compensating for splicing and feed spool changes.

22.3.2   Sheeting

This process is normally carried out on a three-roll calender with thickness control by a feedback system from the product. Usually, the rolls are crowned to compensate for deflection under load and so to maintain a constant roll gap across the width of the sheet.

22.3.3   Frictioning

This is impregnating a textile or metallic fabric between two rolls running at different speeds so that the rubber compound is forced into the interstices of the substrate.

22.3.4   Coating

This differs from frictioning in that the rolls run at the same speed and as a result, the compound is just laid down and pressed onto the substrate, at the required thickness.

22.3.5   Roller dies

This system ensures that an extruder conveys evenly plasticised rubber compound into a wide sheet die. Runners within the sheet die spread the compound over the entire working width. From the die, with a thickness of 30 to 60 mm, the compound enters the nip of a two-roll calender, where it is formed into the desired sheet thickness. As the die-lips reach deep into the nip, no rolling bank develops and air inclusions, normal when calendering thick rubber sheets, are avoided.

For the production of rubber sheets of even thickness within close tolerances, it is necessary to adjust the output of the calender to that of the extruder. For this purpose, there is a sensitive stock pressure sensor positioned in the foremost part of the lips, which in conjunction with a control device, maintains a constant stock pressure through the alteration of the calender roll speed. During operation, the extruder is locked to the calender so that they form one unit. For stock changes and maintenance work, the extruder, positioned on a mobile trolley, is moved backwards.

A device locking the extruder between the calender frames ensures short down-times. For quick cleaning, the sheet die is split horizontally and is hydraulically opened and closed.

Roller die systems are used to produce tyre innerliners, conveyor belt cover sheets and V-belts.

22.3.6   Downstream processes

An important requirement, after calendering, is to measure product thickness. This can be done using electromechanical sensors, beta gauges, or lasers, which feed closed-loop control systems that feed back adjustments to roll speed and other roll controls as necessary. Drum-cooling equipment is normally used to remove heat from the product and is arranged so that the cooling takes place from both sides of the sheet.

Pull-off speed can be adjusted to compensate for shrinkage as the product cools. Rubber sheets, which do not contain reinforcement, have to be supported on a belt as they are cooled. This is often done in a multi-layer unit spraying water onto both the top and the bottom surface. Often in order to prevent adhesion in storage, the surfaces are coated with zinc stearate or talc, applied by brush rolls.

Besides above there are various methods available to calculate such variables as flow rate, shear rate and pressure in the nip-gap when calendering a rubber compound. These usually assume a combination of internal viscous flow and free surface flow, but none gives more than an approximate picture of the process.

22.4   Troubleshooting problems in calendering

22.4.1   Scorch

1. Poor temperature control.
2. Bank too large, leading to excessive heat history.
3. Running speed too fast, leading to excessive shear heating.
4. Stock warmed up on mill too long.

22.4.2   Blistering

1. Roll temperature too high.
2. Feed bank too large, resulting in entrapped air.
3. Sheet thickness too high.

22.4.3   Rough or holed sheet

1. Inadequate stock warm-up.
2. Amount of material in bank too small, or too large, to form rolling bank.
3. Varying stock temperature.

22.4.4   Tack

1. Temperatures of rolls too high.
2. Incorrect stock feed temperature.

22.4.5   Bloom

Low solubility of some ingredients in formulation. Thus, calendering is an efficient process for producing flat rubber sheets and for laminating, frictioning, coating and embossing operations. It is an important primary process in the rubber industry.

22.5   Extrusion of rubber

Extrusion of profiles, tubing and hose is one of the most important fabrication operations in the rubber industry. Extrusion is used to form the final product in the case of hoses, tubes and profiles and many tyre components are extruded before being built into a tyre. The extruder itself has two main functions: to pump the rubber compound through the barrel and to generate enough pressure in the process to force the material through a die to give the required cross-sectional shape. The total process, of course, is more involved than that. The compound, usually in the form of a strip, is fed at a controlled rate into the extruder (Fig. 22.1), which imparts heat and some distributive mixing, moves

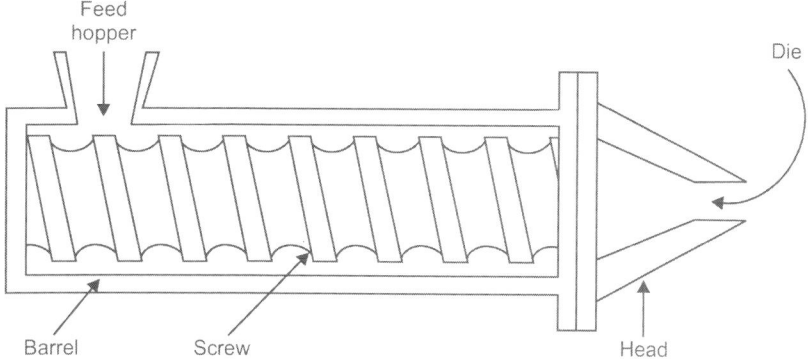

Figure 22.1: Section of a screw extruder.

the material along the machine and generates the necessary pressure. After leaving the die, the product has to be cured, cooled, often cut to length and packaged or coiled for transport to the customer. The flow behaviour of rubber compounds is complex. Process analysis, modelling and developments in equipment and controls have usually concentrated on the various subsections of the process.

22.5.1   Feeding

For high precision extrusion, the feed strip has to be of consistent width and thickness and be fed at a constant rate to the feed port of the extruder. It is important to keep the feed section of the screw full, without either flooding or starving. For this, a feed roll co-operating with the screw is necessary to help control feed of the strip to the machine's feed port. This is usually supplied as an integral part of the extruder. Designs of roll vary considerably but the diameter is normally from 1 to 2 screw diameters and length between 1 and 1.5 screw diameters. They are usually driven by a direct gear drive from the reducer output shaft at a peripheral speed about 1.1 times that of the screw. The limitation of the fixed ratio roll drive is that it requires minimal variation in feed strip dimensions if starving and flooding are to be avoided.

There are more complex drive systems that permit speed adjustment to handle varying feed conditions. One approach involves a slipping clutch, which limits the torque. Another is to have independent speed and torque control. Each of these allows a high rate of feed when necessary, but the torque limit slows the roll down when a bank develops in the feed port. In operation, the torque setting is adjusted to give the optimum bank size. Another technique to control the bank size is to feed the strip through a back tension system, which reacts to the size of the bank. As the bank increases in size, increased tension is applied to reduce it.

One useful modification is the torque feeder, which is a feeding device comprising two rolls, each driven separately by small hydro-motors. It can be easily mounted on most extruders and provides regular filling of the screw and uniform pressure build-up, free of pulsations. The device works on the principle that the pressure of the stock on the two rollers controls the hydrostatic drives and thus, automatically regulates the compound feed to the screw's requirements. One system, which uses a pressure transducer in the feed pocket to control the hydraulically driven rollers, can be run in the slipping clutch mode by setting the hydraulic pressure release valve at an appropriately low level. In many cases, an adjustable knife blade is positioned to scrape the roll surface and form a running seal that keeps rubber in the feed opening. In addition to this, a combination of a spirally-undercut barrel feed section plus a spirally-overcut screw feed section is often used to stabilise feed to the screw and maintain high feed rates without flooding. On large machines internal water cooling is required on feed rolls to counteract the heat buildup due to shearing of the rubber on the rolls.

22.5.2   Cold-feed versus hot-feed extruders

Originally, all rubber extruders were hot-feed. That is they were fed with pre-warmed strips from a mill directly into the feed section of the screw. Hot-feed extruders usually have a relatively large screw depth and short length/diameter (LA) ratio, ranging from 3 : 1 to 8 : 1 depending on the pressure required for the particular operation. Because the stock has been pre-warmed, there is little requirement for heat generation in the barrel. In fact, the major goal in hot-feed extruder design is to keep temperature rise to a minimum, the function of the screw is simply to compact and pump the compound. Their main disadvantage is their limited ability to control temperature. If the temperature and thus viscosity, of the stock being fed to the extruder varies, then this affects extrusion rate, die swell and extrudate dimensions. Hot feed extruders are still used in some applications, for example truck inner tubes, where it is vital to be able to screen the compound immediately before the extruder. Cold-feed extrusion first found wide application in the insulation of wire and cable.

This was due to:

1. Rheological properties of the more easily plasticised wire and cable compounds.
2. The high head pressures caused by the relatively restrictive insulating heads, which produced the necessary retention time within the extruder for compound fluxing.
3. The steam vulcanising tube, the limiting factor for line speed, which made the machine's low productivity tolerable. These restrictions have

largely been removed by improvements in screw design and in continuous vulcanisation techniques and by the use of more powerful drives and transmissions.

Screw design is obviously more important for cold-feed extruders because, in addition to compacting and pumping the stock, it also has to plasticise it. This leads to the main limitation of the cold-feed extruder, the output is limited by the maximum permissible extrusion temperature and most of the energy input must be removed, as heat, by cooling the screw and barrel. With cold-feed extruders, the short time from ambient temperature to extrusion from the die keeps heat history to a minimum. This is a definite advantage for most compounds and allows higher extrudate temperatures without premature scorching.

It is possible to vary the temperature profile along the barrel of a cold feed extruder by having separately controlled zones. The zones separately controlled are usually die, head, two or more barrel zones and the vacuum zone. Zone temperatures are not the same as the temperature of the compound passing through that particular zone, due to the comparatively small amount of surface area, relative to volume, exposed to contact with the barrel in each zone and to the short time of exposure when running the extruder at production speeds. In addition, a temperature gradient is created between the screw and the various barrel zones by the differences in the screw and zone temperatures and by heat generation in the compound.

As long as sufficient power is available, quality and temperature are the two factors that limit cold-feed extruder output. If both are not at their limit, then the extruder is not operating optimally. The main limitation, in practice, is the maximum permissible extrusion temperature. Most of the energy input is dissipated as heat, by cooling the screw and barrel. Thus, the main goals in cold-feed extruder design are to reduce the amount of frictional heat generated and to maximise the overall heat transfer rate by generating complex flow patterns so that fresh rubber surfaces are constantly exposed to both barrel and screw. Rubber is inherently a poor conductor of heat and heat is only removed where it directly contacts a cooled surface.

When the screw is cooled, pressure development is reduced, thus reducing the pumping capacity. However, this results in longer residence times and increased distributive mixing and also produces better thermal homogeneity, which is necessary for control of extrudate dimensions.

22.5.3   Flow mechanism

The primary function of the screw is to convey material along the barrel. Polymers are non-newtonian, that is their viscosity decreases at higher shear rates and so theoretical models, which have mostly assumed Newtonian behaviour, provide, at best, only a qualitative prediction of actual behaviour. An experimental investigation of flow mechanisms in extruders used the technique of feeding the extruder with rubber strips, which were comprised of two layers, one white the other black. After steady state extrusion had been achieved, the screw was stopped and the barrel was heated to 150°C and kept at this temperature for 30 to 40 min to complete vulcanisation. The screw, together with the vulcanised rubber in its flights, was then pushed out of the barrel with a specially designed hydraulic assembly. First, the overall distribution of rubber in the screw was investigated. Then, the rubber was peeled off the screw and some of the ribbons of rubber were sectioned. The cross sections were examined to define the flow patterns of the rubber compound in the screw. White and coworkers found that the phenomena observed during the extrusion of rubber are different from those observed for thermoplastic materials, which exhibit three consecutive stages: solid-conveying, melting and melt-conveying. In these studies, a transverse circulating flow was seen in all cross-sections along the length of the screw channel. That is, rubber behaves as a viscous fluid all the way along the screw. They suggest that this results because, although the cold rubber compound feed is above its T_g, it behaves as a solid in that there is a threshold yield stress, which has to be reached before shearing flow will begin. They also suggested that since the yield value of the rubber compound is low (approximately 10 kPa), the transition from solid to melt flow takes place in the entrance region as the rubber is subjected to the influence of the screw. As a result, the only zone clearly distinguishable is the metering or melt-conveying zone. This was true for both starved feeding and fully-filled channel feeding. They concluded that hot feed rubber extruders operate with a metering mechanism and that in a cold feed extruder this is preceded by plastication must be wrong because, in both cases, the rubber is above its glass transition temperature.

They also found that circulating flows occur in both starved and fully-filled extruder screws and that the level of circulating flows in the extruder depends upon back pressure. The flow mechanism of the metering region represents the sum of a forward drag flow and the backward pressure flow.

The starvation behaviour is related to four factors:

1. Die resistance.
2. Operating parameters.
3. Dimensions of the feeding strip.
4. Feeding system.

Die resistance and operating parameters influence the head pressure and the length of the fully-filled zone. At some level of head pressure, the screw becomes fully filled and the extruder is no longer feed controlled, but operates in a metering mode. Fully-filled screws can also be achieved by assuring better feed. This can be clone by increasing the dimensions of the rubber strip and probably by increasing the undercut near the feed throat and certainly by using a power feed system. Starvation feed tends to reduce the distributive mixing action of the extruder and to reduce viscous heat generation.

22.5.4   Extruder designs

A well-designed screw provides high output rates at low extrudate temperatures, high physical and thermal homogeneity at both high and low speed, good self-cleaning properties to facilitate compound changes and the ability to handle compounds with a wide range of viscoelastic properties. It is not easy to achieve all these in a single screw, thus, most manufacturers provide a choice of designs or interchangeable sections to enable changes from a standard to either a high-shear or high-output configuration. The main improvements required in the early cold-feed extruders were to ensure physical and thermal homogeneity of the rubber when it reached the die. Another requirement was to increase output for a given power usage by reducing the pressure drop along the screw, but without reducing homogeneity. Extruder manufacturers have developed mixing elements and screws whose aim is to introduce complex flow patterns in the melt, preferably without increasing pressure drop.

Mixing sections can only perform local homogenisation or distribution, macro-homogenisation and dispersive mixing are not achieved by passage through low-shear extruder sections.

Vacuum (or vented) cold-feed extruders are used when it is necessary to remove air, water, or other volatiles so that profiles can be vulcanised in continuous vulcanisation systems at atmospheric pressure without porosity problems. Such an extruder has a longer screw length (up to 22:1) than a standard cold-feed extruder and is basically two extruders in one, a cold-feed in series with a hot-feed. Therefore, extrudate temperatures are generally higher (up to 135°C) and outputs lower. Most developments in screw design have been based on an intuitive understanding of screw behaviour and had to be evaluated empirically after each modification. This is an expensive method of development, as designs which didn't work had to be scrapped.

22.5.5   Cavity transfer mixer (CTM)

The screw designs produce improvements in the efficiency of blending, but laminar flow, although interrupted, is only slightly diverted. It is not turned through a large enough angle to double-up the blending effect at each interruption. True turbulent flow cannot occur because the Reynolds number for rubber compounds is generally less than one, a value far below that required for turbulent flow. What is required to produce significant blending is a system that simulates the cutting and folding action used on rubber mills to blend compounds and distribute material.

Applications of the CTM in the rubber industry are:

1. Temperature and viscosity blending.
2. Workaway of off-specification material without temperature effects on viscosity or cure-rate.
3. Mixing of masterbatch with base stock.
4. Improved electrical consistency.
5. Reduction of nerve and increased output with natural rubber (NR) stock
   It is not self-cleaning and it generates some heat and back pressure with high viscosity materials.

Vented extruders: Vented extruders are used to remove volatiles present in the compound. The screw has two distinct zones separated by a dam, which has small axial grooves, acting like a die.

Dump extruders: In some large throughput, semi-continuous mixing operations, especially in the tyre industry, the batch is dumped into an extruder, rather than onto a mill.

Strainers: The function of a strainer is to pass the stock through gauze pack and screen plate to remove all traces of solid contaminants. These are often hot-feed extruders, though cold-feed machines can also be used. After the strainer, the final extruder may not need a restrictive filter, this reduces back pressure and improves output.

Extruder barrels: Barrels have to be of a hard, high quality steel so that they are not readily abraded by compounds.

22.6 Extruder operation and control

Extrusion variability may be caused by variations in the compound fed to the extruder, inconsistent feed flow and variations in extruder conditions. The quality and consistency of elastomers as received by the processor have improved significantly as control of the mixing process has improved. Even so, there will still be some feedstock variation to adjust to. Feed rate consistency is important. A rolling bank of a consistent size must be maintained in the feed throat. If its size varies then the heat input per unit volume will change. This will alter the feed temperature as it enters the extruder and thus, ultimately, affects output rate and extrudate dimensions. Variations in extruder conditions can occur in screw speed and zone set temperatures, though modern control techniques minimise these.

Extrudate temperature has to be kept below the threshold level at which significant scorch occurs. At start-up and low screw speeds the main determinants of compound temperature are the zone set temperatures. At higher speeds, more heat is generated and zone temperatures cease to have any significant effect on the extrudate temperature. Heat flow is now in the reverse direction. The head pressure required to extrude the compound through the die is also important. The higher the pressure, the longer the material stays in the extruder at any given screw speed and the temperature limitation is more quickly reached. With any combination of screw design and compound there is a trade-off between output rate and the pressure that can be generated before the temperature ceiling is reached.

Screw and barrel temperatures have a significant effect on extruder performance. Although the bulk temperature of the stock is not determined by the set temperatures of the screw or the barrel zones, the rubber actually in contact with the metal does experience those temperatures. Thus, the adhesion between the rubber and the screw and barrel does depend on the set temperatures. The transfer of rubber through the extruder is determined by the relative grip on the rubber between the screw and the barrel, the better the adhesion to the barrel the higher the output. Cold rubber will not adhere to a cold metal surface. At some temperature, depending on the type and grade of rubber, but usually in the range of 50 to 70°C, adhesion between rubber and metal is developed. Above that temperature adhesion decreases again as the rubber softens. Thus, for maximum output the feed zone should be at that temperature and the screw temperature should remain high and the barrel kept at a lower temperature.

In practice, setting-up and maintaining an efficient extrusion line often depends on the, skill of the operator. He adjusts the various controls using his experience and intuition rather than well-defined mathematical relations between product output and quality. The conditions under which an extruder will operate depend on screw and die characteristics.

22.7   Shaping of rubber

When it reaches the end of the screw, the rubber compound fills the entire cross-section of the barrel and is at the pressure generated by the screw. When it has passed through the die, it obviously has the cross-sectional shape of the die, modified by any die swell that may occur and is at atmospheric pressure. The two extruder components, which cause this change in shape and accommodate the drop in pressure, are the head and the die.

22.7.1   Extruder heads

The purpose of the head is to equalise pressure across the profile, to guide compound into the die and to hold other items, such as screen pack, pressure and temperature sensors and spider. The flow paths for the rubber have to be evenly balanced and streamlined to ensure that there are no dead areas where rubber can be held up and scorch. The volumetric flow does not change as the material passes through the head, but the cross-sectional area does change, often dramatically. As a result, the linear velocity increases as the rubber undergoes elongational flow, or stretching. The die head should have a hinge and lock, which automatically locate it in place relative to the extruder barrel. Sieve pack and die changes must be simple and quick. The head should be long enough to eliminate any pulsating flow from the screw's drive mechanism.

In the case of severe compound contamination and the resultant frequent screen changes, the resulting down-time can be reduced by means of two hinge-mounted heads or front head sections. These can be opened or closed quickly by means of a bayonet locking ring, either manually via a reduction gear or by means of a hydraulic ram. In this way it is possible to continue working with one head section while the screen pack and the breaker plate on the other one are being changed.

In order to enable screen changing to take place without interrupting production, automatic screen-changing devices are used.

Coextrusion

Many extruded products have two or more separate compounds that perform different functions in the final product. For example, many automotive seals have a hard section to grip the sheet metal, a soft section to form a lip-seal against moving glass and a sponge section to form a metal-to-metal seal. All three have to be coextruded. That is the products of three extruders have to meet and mate, in the die. For steady performance, the rates of flow, temperatures and especially pressures, must balance. Each set of flow channels must be simple and extensional flow, pressure drop and die swell controlled. The extruder for the minor component should not be oversized, or it will give surging flow at different orientations of the screw flights.

Cross-heading

This is required for reinforced hose, wire and cable and weather strip where a layer of rubber has to be put on an existing core. The extruder is usually at right angles to the extrusion line to allow space for upstream and downstream equipment. Flow channels have to be designed so that the rubber will pass easily around the core and join, seamlessly, on the other side.

Shear heads

For continuous vulcanisation applications, a shear extrusion head is a useful device. This fits between the extruder and the die and raises the compound temperature to the required prevulcanisation temperature. The material is sheared in an annular shearing gap between a rotating mandrel and a stationary cylinder and at the same time is forced by the material behind it to flow past the mandrel. This double shearing action generates heat, which raises the temperature of the material entering the die. The intensity of the energy conversion due to friction depends on the relative motion of both shearing surfaces (i.e., on the speed of the mandrel, which is infinitely variable) and on the viscosity of the rubber compound. In practice, the rubber compound can be heated to 150 to 190°C in the shearing gap immediately before the die.

These devices are cheaper than a microwave heating system of the same capacity. It is possible to extrude very complex shapes without collapse of thin sections or undercuring of thick sections. Dimensional stability is maintained because curing starts before the exit from the die.

22.7.2 Dies

The shape of the die obviously depends on the required shape of the product. There are two phenomena of importance in the flow of rubber through dies, pressure drop and die swell.

Practical extrusion dies are normally short relative to their aperture size and often have to produce a profile of complex shape. Possible extruder output rates are very much dependent on the extensional flow characteristics in the die entry region. Extensional flow occurs whenever there is a convergence of flow lines such as at a die entrance. An abrupt lead-in taper to a die can lead to extrudate distortion at low output rates. Another factor in producing extrudate distortion is velocity differences across the profile at the die exit.

Pressure drop

The pressure indicated in the die-head, just before the die, is completely dissipated in the passage through the die, if the profile exits the die to atmospheric pressure. There are three components to this pressure drop: entrance losses, die losses and exit losses. The entrance pressure loss is related to extensional flow and is usually larger than exit loss, while pressure loss in the die is dependent mainly on its length.

The effect of added lubricants can be considerable. Whether they act as internal or external lubricants depends on their degree of compatibility with the rubber compound. An incompatible material (e.g., oil in NR) acts to increase wall slippage, giving a greater entrance pressure drop. Compatible materials (e.g., fatty acid derivatives) decrease entrance pressure losses.

Die swell

After a polymer is forced through an orifice, it expands in cross section (die swell) and correspondingly shrinks in length (extrusion shrinkage). Die swell is usually defined as the ratio of the cross-sectional area of the extrudate to the cross-sectional area of the die. For a given compound, extrusion shrinkage at constant shear stress is independent of die geometry and temperature of extrusion and is, in fact, related to the molecular orientation imposed on the polymer during extrusion, this orientation being dependent on the shear stress developed in the die. In elastomeric materials, the relaxation of internal stresses starts immediately after exit from the die and continues as the material cools down. The three-dimensional entanglement network gives the polymer an additional ability to remember its previous shape for a short time after extrusion. Thus, the extrusion shrinkage is mainly determined by molecular orientation. The rate of molecular relaxation is extremely rapid, 80% of limiting extrusion shrinkage in less than ten seconds. Two different relaxation processes take place in the post-extrusion relaxation of SBR compounds containing various carbon blacks.

Die swell obviously has to be taken into account in die design. It cannot be eliminated, so the approach is to minimise and control it. If there are long flow channels in the head, stress build-up will be lower. Variations in die swell can be countered by a feedback control from one or more extrudate dimension, altering the speed of the take-off conveyor. However, the main concern is to control die swell and allow for it in the die exit shape. This is obviously a complex problem, not easily amenable to mathematical analysis. Therefore, die design is still to a large extent an art, especially when the required profile cross-section is complex. It is important to ensure streamlined flow, no dead spots and minimum pressure drop.

The more complex the profile, the more important is thermal homogeneity in the stock as it enters the die. Modern CAD/CAM techniques enable the final die design to be arrived at with less metal cutting and fewer plant trials, with a concomitant reduction in development costs.

22.8   Take-off and curing

After passing through the die, most extrudates are ideally suited for immediate curing. They are already at or near the curing temperature, they have attained their final form and in essence, all that is required to cure them is to hold them at the curing temperature for a few minutes. Thus, profiles, hoses and tubes are usually cured in a continuous process immediately after shaping. Tyre components, of course, have to be built into a tyre in the green, or uncured, state and subsequently cured in a bladder.

22.8.1   Continuous vulcanisation systems

As a rubber compound containing a curative system is held at the curing temperature the production of cross-links causes it to change from a viscoelastic fluid to an elastic solid. As it leaves the die the compound is still a fluid and as the stresses built up in the passage through the head and die relax, the dimensions of the profile, originally those of the die, change. This dimensional change is termed die-swell or extrusion shrinkage. As it is a fluid still, the shape of the profile may also be distorted by gravity, unless it is supported by internal reinforcement, by tear-off strips, or by immersion in a fluid medium. Curing systems, which provide heat to the outside surface, do not necessarily raise the temperature of the internal cross-section because of the poor heat conductivity of rubber. The various continuous curing systems described below all attempt to cope with these problems.

Pressurised steam systems

These are commonly used for products having a core or other reinforcement and profiles that are easy to seal, such as wire, cable and hose. The time required for heat to penetrate to the center of the cross-section depends on the diameter. The weight of rubber per unit length is proportional to the square of the diameter and as line speed has to be reduced in proportion to the increase in weight per unit length, because output is approximately constant, the time in the system is proportional to the square of the diameter, too. The reduced line speed with larger cables (limited by the output of the extruder) automatically results in the required higher residence time. This applies equally to the other continuous systems.

Hot air curing systems

These consist, basically, of an insulated tunnel, a metal mesh conveyor to support and move the profile through and a countercurrent of air, heated up to 300°C. With compounds (e.g., EPDM), which are not readily susceptible to oxidation ultrahigh temperature, shorter ovens can be used.

Hot air fluidised bed systems

Hot air fluidised bed systems use ballotini, 0.2 mm diameter glass beads and produce much higher heat transfer coefficients (1500) than hot air alone. Thus, shorter lengths of oven and lower temperatures can be used. Another advantage is that the bed itself provides support to the extrudate. A disadvantage is that beads, which adhere to the surface, have to be brushed or washed off the product at the end of the line.

Liquid salt bath systems

These bath systems use a molten eutectic salt mixture (usually sodium nitrate and sodium nitrite). The heat transfer coefficient is 2800 and the salt prevents air oxidation of the rubber surface, in addition, the molten salt supports the profile. To avoid air oxidation of the top surface and to ensure good heat transfer for the floating section, a film of the molten salt is poured over the surface in a cascade, or alternatively the profile is kept submerged by a belt and roller system.

Any adhering salt readily drips from the profile after leaving the bath and final cleaning is carried out by air jets. A final water wash is often used in the cooling section, but this produces a waste stream of salt water, possibly containing nitrosamines, which needs to be disposed of. This latter problem has become more acute as environmental regulations have tightened and many companies, while not eliminating salt bath systems, no longer choose them for new installations.

Pressurised liquid salt bath systems are used with cross-head systems, introducing wire or reinforcement in cable and hose manufacture. This avoids the need to use a vented extruder to remove volatiles. Because of the support provided by the molten salts, the effective weight of the cable is significantly lowered. The high temperatures and heat transfer reduces cure time and line length. Within limits, the temperature and pressure can be varied independently. Temperatures can be raised to whatever the limit is for the material being cured and sufficient pressure to eliminate porosity can be applied.

Microwave systems

Microwave systems provide quick and uniform heating throughout the profile, which is especially useful for thick profiles, profiles of varying thickness and for sponge. The compound has to be microwave receptive (i.e., polar), which most polymers are not. However, many carbon blacks are and if necessary, it is also possible to add other chemicals specifically to increase microwave receptiveness. Usually, a short microwave section is used immediately after the die to boost the extrudate temperature to curing temperature, followed by a hot air tunnel to maintain temperature until the profile is cured. There is much less heat loss with a microwave system than with the systems described previously because the heat is generated in the rubber itself. This high energy-efficiency makes the use of electricity, a more expensive source, economically feasible. Microwave systems can easily be used with inert gas atmospheres or pressurised systems. If the profile contains a metal component, precautions have to be taken to prevent arcing.

Shear head systems

Shear head systems have the same advantage as microwave systems, in that heat is generated throughout the mass of the rubber, as it moves through the shearing gap. The relative speed of mandrel and sleeve can be very high and the rubber temperature rises very rapidly. The process obviously operates to very fine tolerances, care has to be taken that the rubber is heated uniformly and that only minimal cure occurs before the material passes through the die. If not, the product will have an uneven surface and possibly poor physical properties. Due to the sensitivity of the process, temperature sensors and microprocessor control are usually needed to optimise it.

Steel belt presses

Steel belt presses are used to cure rubber sheets. The sheet is fed to a heating drum and is pressed against it by a continuous pressure belt. It then passes through a post-heating channel, cooling unit and winding unit. This type of system is used for rubber sheets for roofing, shoe soles and sheets with fabric plies, printing blankets, floor covering and rubberised fabrics, etc.

22.9   Troubleshooting the extrusion process

Despite the many feedback microprocessor control systems available on the market, it is still often the skill, experience and understanding of the extruder operator that determines the success or failure of an extrusion operation. Success or failure has to be measured in economic terms, that is the hourly production rate of the process. This depends on minimising scrap and downtime both in start-up and normal running. One important aspect of this is the ability to diagnose problems and to take effective corrective action. This section briefly lists some of the more common problems in extrusion and suggestions on diagnosis and for corrective action. Obviously, a good understanding of the extrusion process in general and of the specific process and equipment in use, is a necessary basis for such troubleshooting.

Poor dimensional stability of extrudate: This may be caused by variation in takeaway speed, screw speed, feed rate, temperature, compound viscosity and poor temperature homogeneity.

Excessive heat buildup in compound: Possible causes are screw design, screw speed, head pressure and compound formulation.

Rough surface on extrudate: Possible causes are premature scorch, volatiles and insufficient mastication.

Contamination: If the nature of the contamination is known, its source may be obvious. If the contamination is obvious in the feed strip, it may have come in with one of the compounding ingredients, or from the mixing system. Check that the screen pack is working effectively.

Porosity in extrudate: Possible causes are volatiles in the stock or, in a thick section, undercure at the center.

Strip difficult to feed or feeds irregularly: Possible causes are incorrect or varying size of strip or excessive slab dip on the surface.

Surging output: Possible causes are wrong screw design for the process such as a short metering section, worn screw or barrel, varying and inadequate feed rate, poor temperature control.

Thus, extrusion is a very important operation in rubber processing because the majority of compounds are extruded either as an intermediate or final step in the manufacturing process. An understanding of the principles of extrusion is essential to profitable operation of the process. It is important to optimise compound, equipment and processing conditions.

In most plants, the extruder operators run the system from feed to final packaging of product without direct supervision. Therefore, it is essential that they, in addition to their experience, have a basic understanding of the process, as discussed above.

22.10 Molding of rubber

Many rubber articles are produced by molding, a process in which uncured rubber, sometimes with an insert of textile, plastic, or metal, is cured under pressure in a mold. There are three general molding techniques: compression, transfer and injection molding.

In compression molding, a pre-weighed, generally preformed piece is placed in the mold, the mold is closed, with the sample under pressure, as it vulcanises. Cavity pressure is maintained by slightly overfilling the mold and holding it closed in a hydraulic press. Heat is provided by electricity, hot fluid, or steam.

Transfer molding is a form of injection molding. The compound to be cured is held in a heated reservoir, the amount of rubber to fill the mold is forced through sprues or a runner system and into the mold cavities by a piston. Pressure is maintained by the piston and a mold closure system. The mold has to be hot enough to ensure curing of the rubber, but the reservoir has to be at a lower temperature.

Injection molding combines an extruder, which heats and fluxes the rubber, with a reservoir and mold. While the rubber in the mold is curing, fresh rubber is prepared for the next cycle so that molding is a semi-continuous process. The compounds used have to be optimised for injection molding. They must flow through the nozzle and runner system and fill the mold under the pressure available. They should not cure before filling the mold, but cure quickly once in the mold.

A high viscosity compound may not fill the mold properly or may generate too much shear heat, leading to overheating and scorch. A low viscosity compound may not generate enough heat, leading to undercure. These considerations mean that there is a process window within which a specific material can be satisfactorily molded in a given process.

22.11 Compression and transfer molding

Both compression molding and transfer molding are still widely used in the industry worldwide, even though injection molding has a number of advantages. This is often because the fabricator has an existing press in good working order. When starting from scratch both economic and technical factors have to be considered.

The main economic factor is that the capital cost of machines and molds increases in the order compression molding, through transfer molding, to injection molding. A production volume may be too small to justify a high initial capital cost. In compression molding, Fig. 22.2a, the weight, dimensions and positioning of the charge have to be closely controlled, or the dimensions of the product can vary widely. This is often due to variation in the amount of material lost from the cavity as flash. This can be controlled, rather than be eliminated, by employing a shallow plunger, which, because of close clearances, means that excess rubber can only escape when high pressure is applied, after the mold is completely filled.

In transfer molding, Fig. 22.2b, rubber flows from a separate reservoir through a narrow flow channel, called a sprue, into the mold. There are three advantages over compression molding.

1. As the mold is closed before the rubber charge is forced into it, closer dimensional control is achievable.

2. In the transfer process, fresh rubber surfaces are produced and this allows development of a strong rubber-metal bond with any insert in the mold.

3. Unit production costs are lower due to shorter cure times as a result of heating the rubber due to flow through sprue, runner and gates and shorter downtime between runs, as only one charge blank is necessary, even if a multicavity mold is used.

22.12 Injection molding of rubber

Injection molding (Fig. 22.2c) is now a well-established fabrication process in the rubber industry. Its advantages comprise reduced labour costs, shorter cure times, better dimensional control and more consistent mechanical properties of the product.

The operation of an injection-molding machine requires: feeding, fluxing and injection of a measured volume of compound, at a temperature close to the vulcanisation temperature, into a closed and heated mold, a curing period, demolding and if necessary, mold cleaning and/or metal insertion, before the cycle starts again. For maximum efficiency, as many of the above operations as possible should be automatic.

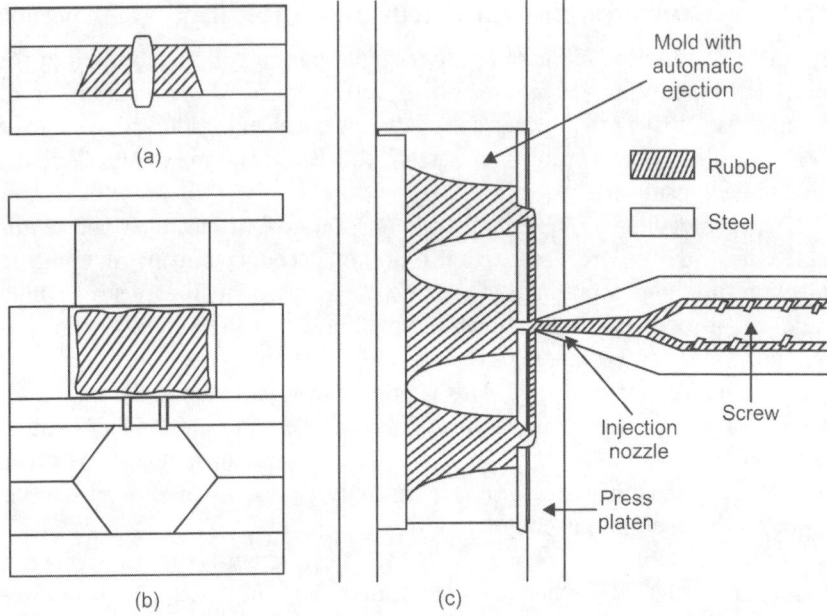

Mold with
automatic
ejection

▨ Rubber

☐ Steel

Injection
nozzle

Screw

Press
platen

(a)

(b)   (c)

Figure 22.2: Moulding methods: (a) compression (b) transfer and (c) injection.

There are three main types of injection machine: the ram type, the in-line reciprocating screw type and the out-of-line non-reciprocating screw type.

Simple ram machines cost less than screw machines and because the ram can be made to fit very tightly in the cylinder, they can develop very high injection pressures. However, as the mix receives heat only by thermal conduction from the barrel, high injection temperatures and thermal homogeneity are difficult to achieve and they are not widely used.

The screw in the in-line reciprocating screw type acts both as an extruder and a ram. In this type of machine, the mix is heated and plasticised as it progresses along a retractable screw. When the necessary shot volume has accumulated in front of the screw it is injected by a forward ramming action of the screw. With this system, a more uniformly controlled feeding of the material can be achieved, together with more rapid heating of the stock from mechanical shearing, additional distributive mixing from the rotational screw action, a greater degree of thermal homogeneity and a temperature 20 to 30°C higher than the jacket temperature. However, during the injection stage, when the screw is acting as a ram, there is inevitably some leakage back past the flights and this limits the achievable injection pressure.

The out-of-line non-reciprocating screw machines have separate screw and injection chambers and combine the advantages of the above two types. The screw plasticises the compound and delivers it through a non-return valve into a separate injection chamber. Machines of this type can generate injection pressures of up to 200 Mpa and can efficiently mold high viscosity mixes and effectively fill large volume molds.

In the standard injection process described above, the compound is injected into a closed mold, there are two variations on this:

1. Injection-compression molding: The mold is partially opened and a vacuum applied to the cavity area which is sealed by a compressible silicone gasket. A measured amount of rubber is injected into the partially opened mold. The mold is then closed and the excess rubber is forced outward to flow off channels. This process is used for articles such as precision O-rings, where runner marks are unacceptable.

2. Injection-transfer molding: The rubber is injected into a transfer chamber and then forced from the transfer chamber into the mold. This combination process uses the plasticisation and heat generation advantages of the injection unit with the controlled flash pad and cavity layout advantages of the transfer press.

22.12.1 Injection molding equipment

In specifying an injection molding system it is necessary to decide on the type of injection machine, the size of the machine in terms of shot capacity, platen size, pressure, the number of mold cavities per platen and the layout and size of runners and gates.

Delivery systems

The three main types of injection molding operations were described in Section 22.12. They differ mainly in the way in which they heat, plasticise and deliver the mix, using a simple ram, reciprocating screw, or separate screw and ram. Screw extruders provide better temperature control and homogeneity and are now generally preferred to the simpler systems.

Simple ram machines: The simple ram machines cost less than other types and deliver mix efficiently to the mold as a result of the ram fitting very tightly in its cylinder. One disadvantage is that, depending on the ratio of the shot volume to the barrel volume there may be three, four or five shots in the barrel, all of which need to be pressed forward to deliver one shot. This causes an undesirable pressure drop and heat generation in the nozzle region of the barrel where it cannot be used to advantage.

Another disadvantage is that the mix is heated slowly, by the thermal convection or conduction from the barrel of the machine. It cannot become hotter than the barrel until it is forced through the nozzle into the mold. For products having thin section, these operations may be adequate.

Reciprocating screw machines: Most of the early screw machines were of the reciprocating type but it is now generally accepted that the use of the pre-plasticising screw as the injection ram is only possible for low shot volumes (<500 ccs) or very soft compounds. The advantage of screw machines in feeding material efficiently plasticised and thermally homogeneous, can only be realised if there is constant intake of feed compound into the feed port. In other words, it is important to ensure that the feedstock is not severed when the screw, acting as a ram, moves forward to inject the shot into the mold. This can be achieved by having a positive drive arranged so that the feed strip hangs freely over the barrel opening. Another method is to machine a short tangential tunnel undercut at the front of the base of the feed throat where the material feeds into the barrel bore. This acts as a reservoir, or buffer and smoothes out variations in the feed flow. It eliminates choking or balling up of the material in the feed throat.

Screw machines with separate injection chambers and rams: These are the preferred basic design for rubber injection molding machines because they combine the advantages of both screw and ram machine. In the standard 'V' configuration the plasticated compound is fed through a check valve into an accumulation chamber. One disadvantage of this is that the first rubber fed through the check valve is actually the last rubber to be injected. This can lead to adhesion and build-up of rubber on the face of the injection piston, which can cure, break off and cause rejects and molding problems. Two modifications have been developed to circumvent this. In the 'first-in-first-out' system, the screw and ram, although separate, are in line. Initially, the injection ram, is in the forward position and the injection chamber is empty. As compound enters through the ram, it is forced backwards by the incoming material until a limit switch controlling shot volume is activated. Injection then takes place through a special ball-type torpedo, which completes the plastication and thermal homogenisation. As the material does not reach the final injection temperature until it reaches the nozzle, temperatures earlier in the system can be relatively low (–70°C). Such inline systems have gained in popularity in recent years.

Shuttle press and multi-station rotary press: One disadvantage of the single-station machines is that the injection system is idle during the curing and demolding stages. Shuttle presses are designed for applications where the time to inject and cure is approximately the same as the time required to unload, clean and service the mold. It is especially useful for the production of items containing inserts of tall moldings and of moldings with a long stripping time. The press is equipped with two sliding bottom platens, which are individually heated. After the press has opened, both bottom heating platens and molds shuttle, positioning the mold that was inside the press into the unloading station while simultaneously positioning the other mold inside the press.

The main advantage to this format is that unloading takes place on one mold while injection and curing takes place on another mold. Therefore, the time needed for stripping and loading inserts remains entirely outside the cycle time. Such machines may have automatic ejection of parts and runner system, cleaning and spraying of the molds and automatic loading of metal inserts.

Nozzles, runners and gates

Nozzles: The rubber is injected into the mold through the nozzle and obviously, with multi-station systems, its shape and outside dimensions have to be such that it mates positively with each mold. As the inside nozzle diameter is increased, the pressure drop, injection time and temperature rise, decrease.

Runners: These are the channels provided to convey the rubber from the injection point, the sprue, to the mold cavities, in cross section, they are usually trapezoidal, half-round, or round. They should be as short and direct as possible and streamlined to reduce pressure loss. The thickness of the molding at the gating point governs the cross-section size of the runner.

Gates: These are restrictive passageways from the runner to cavity which are sized to allow easy flow of the rubber into the cavity and at the same time raise its temperature to the final desired level. Generally they are full-round, fanned, with a bore one-half of that of the feed runner and as short as possible. The final sizing of gates is usually done by carrying out a practical molding test. The gate position is often more important than the size or type of gate. It should be positioned so that it feeds the thickest section of the molding, preferably in the non-stressed area of the molding. If weld lines are likely to be a problem, the gate position is critical.

Molds

Obviously, the major design consideration with a mold is the shape and size of the required molding. However, within this restriction, there are a number of points that have to be considered in determining the optimum mold design for a particular product.

Automatic ejection

To take full advantage of the short cure cycles obtained with injection molding, an automatic ejection system is required. Thin sections can be stripped from the mold by a compressed air jet. Simply shaped and flat parts can be removed by a rotating brush. Delicate parts can be removed by a robot pulling device, which places each part on a turntable, rather than ejecting it into a tote box.

Deflashing: The ideal mold design would produce either no flash or would make flash removal simple, as pointed out above. However, some parts, especially multi-cavity molds, will need deflashing. There are a number of techniques for deflashing. The most efficient involves chilling to $-150°C$ and then tumbling or shot blasting. The latter will handle complex parts, even items with inaccessible internal flash.

22.12.2 Injection molding process

During injection molding, rubber compounds are subjected to more severe processing conditions than during compression or transfer molding. Temperatures, pressures and shear stresses are higher, though cure times are shorter. Control over process variables can be more precise. The cycle time can be minimised by independently controlling barrel temperature, screw speed, mold temperature, cure time and injection pressure. Compounds with widely differing flow and cure characteristics can be molded into a variety of complex shapes. The skill lies in the optimisation of the process, this depends on close interaction and understanding between rubber compounder, mold designer and processor.

The highest productivity is achieved when the compound is injected at a temperature close to the curing temperature, into a mold at a slightly higher temperature, because these conditions minimise both injection and cure time. Scorch safety of the compound is the limiting factor in the process and the effect of machine variables on compound temperature needs to be understood if high injection temperatures and short injection times are to be achieved without scorch. Equally, it allows adjustments for batch-to-batch variations in viscosity, cure rate and scorch safety to be made, if necessary.

Thus, to a large extent, controlling the injection molding process reduces to a question of homogeneity of temperature and heat history of the compound at each stage in the process. The temperature of the mix in the injection chamber, prior to injection is determined by the temperatures of the extruder and injection chamber, by the screw speed, screw design and by the applied back pressure. The injection, or mold filling, time and temperature depend on the temperature of the mix, as determined by the above factors, the injection pressure, the dimensions of nozzle, runners and gates and the viscosity response of the compound. The cure time depends on the mold temperature and the temperature of the compound as it enters the mold.

Injection temperature

Cold feed extruder designs have been optimised to enable control of the amount of frictional heat generated and to maximise the overall heat-transfer rate by constantly exposing fresh rubber surfaces to both barrel and screw. Proper flow also ensures thermal homogeneity of the entire volume of compound in the injection chamber, whose temperature is kept as high as possible without scorching.

Screw speed

Screw speed and design can both significantly affect heat generation in the extruder. In a given machine, screw design is fixed and thus screw speed is the primary control factor. Plasticisation time also depends on screw speed and a high screw speed can be used to minimise heat history by having the screw rotate only for the time required to fill the injection chamber.

Back pressure

Back pressure is controlled by the pressure against which the screw must work whilst filling the injection chamber. Elevated back pressure raises the temperature of the mix and is only normally needed for low viscosity compounds that might not otherwise generate sufficient shear heat in their passage through the extruder.

Injection pressure

The work done by the ram in injecting the compound through the nozzle and runners is dissipated as heat and this can boost the material temperature by over 40°C above that in the injection chamber. Thus as high an injection pressure is used as possible, consistent with freedom from scorch.

22.12.3 Monitoring and modelling the injection molding process

In injection molding, the two most critical parameters are flow at the high shear rates that occur and scorch time. Most of the work has concentrated on these and ignored other less significant factors.

The approach taken by many operators of injection molding machines is to determine the effect of machine controls on mix temperature at each stage and to determine how to achieve high injection temperatures and short injection times without scorching or underfilling.

Equally, this approach allows adjustments for batch-to-batch variations in viscosity, scorch time and cure rate. This technique forms the basis of most trouble-shooting aspects published by equipment manufacturers.

At the other extreme, in terms of sophistication of approach, is to measure various properties of the compound that are relevant to flow, temperature rise and curing of the material at various stages in the process. For example, one study assessed the:

1. Behaviour during heating in the screw section by use of a defometer.
2. The cure time using a Brabender plasticorder.
3. The change of viscosity with time at different temperatures and the resistance to scorch at low shear using a Mooney viscometer.

4. The behaviour during injection using a capillary viscometer.

5. The duration of induction period and the rate of cure using a MonSanto Rheometer.

Such a complete evaluation would be time consuming and uneconomical for a molding shop to undertake for each compound, machine and mold.

Another approach is to generate an 'operating window' for the injection molding process, using a capillary rheometer. This defines those combinations of inject time and inject temperature at a given mold temperature, which should produce completely filled unscorched parts.

22.12.4 Control of the injection molding process

Equipment manufacturers offer microprocessor control systems and software packages with their injection molding machines. These can be very simple, such as an individual control on a press to monitor and control cure time/cure temperature/mold pressure.

22.12.5 Compounds for injection molding

Quality control for injection molding compounds has to be tighter than for those that are compression molded because the injection-molding process itself can be more precisely controlled and is therefore more sensitive to variations in compound properties. Compounds must have sufficient processing safety (scorch safety) to flow through the nozzle, runners and gates without scorching, but still cure rapidly in the mold. Thus, the balance of viscoelastic and curing characteristics of the compounds are extremely important.

Designing rubber compounds for injection molding has often been a trial and error process because the processor usually does not have data on the behaviour of such compounds at the temperatures, pressures and shear rates involved in injection molding. In fact, most rubber compounds that will compression mold can be satisfactorily injection molded provided that they flow well enough and are not too scorch sensitive and that the machine controls are properly adjusted. However, for maximum productivity, the compound has to be injected rapidly into the mold at near vulcanisation temperature. It is in this area of optimisation that laboratory tests, properly interpreted, can be of value in delineating an operating window of time, temperature and pressure, within which a particular material will flow well and cure effectively without danger of scorching.

The three major areas in which data on a compound are required are theological behaviour, rate of vulcanisation and heat flow into and through the compound.

22.12.6 Problems in injection molding of rubber

In addition to the operating problems discussed earlier, there are a number of problems that do not become apparent until the mold is opened and it and the product are examined. Some of these are briefly described below.

Shrinkage and part dimensions: On cooling, both the mold cavity and the molded part contract, usually by a differential amount because the metal and rubber have different coefficients of thermal contraction. Shrinkage is usually defined as the difference between the dimensions of the mold cavity and those of a molded part, when both are measured at room temperature. It might be more exact to refer to the difference between the hot mold and the cold part but the above definition is easier to use in practice.

The amount of shrinkage has to be allowed for in mold design, but it is itself variable, being affected by batch-to-batch variance in the compound, cure time, temperature and pressure and so there must be adequate dimensional tolerance allowed in the specifications for the part.

Adhesion: Adhesion in injection molding has two aspects – adhesion to the mold surface, which is not wanted and adhesion (bonding) to a metal or plastic insert in the part, which is wanted. Mold release agents are used to prevent the one and adhesion promoters to ensure the other.

Release agents, often silicone based, are sprayed onto the mold surfaces between shots. With complex molds this can be time consuming and it is difficult to ensure thin, even coating on all surfaces. There is also a problem of pollution from the spray carrier. One improvement on this is to deposit a layer of diamond-like carbon on the surfaces by a plasma vapour deposition system. This leaves a semi-permanent easy release surface, which does not wear readily.

Rubber-to-metal bonding agents have been used in the industry for many years and there are many types available. Most of these depend on a two-coat system. The first coat, or primer, applied to the clean metal surface is usually halogenated polymer with heat-reactive resin, either dispersed or dissolved in organic solvents. The metal bond is the result of van der Waals forces and is described by standard adhesion theory. Second layers, or cover cements, depend on the base polymer being used. They are usually chlorinated, brominated, or chlorosulphonated polymers, with appropriate cross-linking systems, which form cross-links with the rubber part and with the primer layer. Concern about environmental pollution due to the solvents used in the traditional systems and in the U.S., government regulations on volatile organic compounds (VOCs) and chlorine containing solvents, has led to the development of water-based adhesive systems, which are replacing solvent ones.

Backrinding: This is the term applied to the torn or gouged look that occurs at the mold parting line of compression molded parts and at the gates of transfer and injection molds. It is caused by thermal expansion of the rubber after cross-linking, which can force the cross-linked rubber into the space at the parting line or gate, causing it to rupture. The smaller the surface area/mass ratio of the part, the worse it is (i.e., a sphere is the most severe case). The best way to minimise backrinding is to minimise the shot weight commensurate with filling the cavity, of course. Increasing scorch time can also help because it ensures that the mold is filled before curing begins, as the injection temperature can be raised.

Mold fouling: The build-up of material in a mold, especially in corners, is a major problem. The more severe it is, the fewer the number of cycles before the surface or sharp edges of the part are affected and the mold has to be taken out of service and cleaned. The cause is usually deposition of chemicals and perhaps their subsequent oxidation or degradation. These agents may originate in the rubber, in fillers, curatives, waxes, etc., or from release agents. Thus, there are a wide variety of deposits whose severity varies from compound to compound and also depends on injection rate and mold temperature. One of the major concerns in compounding for injection molding is to minimise mold fouling. Costs of downtime and mold cleaning can be considerable.

Mold cleaning is often done by blasting with some abrasive particulate material such as glass beads, plastic, or metal beads. One effective, clean, but expensive and noisy process is blasts with solid carbon dioxide. Other techniques use liquids such as detergent solutions or hydrocarbons, often with ultrasonics. When the deposit is volatile at higher than curing temperatures, heating in a salt bath, fluidised bed, oven, or by induction can be used.

Orange peeling: This is usually caused by the initial layer of rubber in contact with the heated mold surface having cross-linked before succeeding layers have filled the mold. The cure is usually to increase the scorch time of the compound.

Porosity: This is due to undercure and the presence of volatiles, especially water, in the compound. Higher injection and mold temperature or longer mold closed time should resolve this.

Blisters: Air entrapped in the rubber is the usual cause. This can be eliminated by a higher back pressure, slower injection rate, or effective venting of the mold.

Rubber Technology

Volume II

Rubber Technology

Volume II

S. C. Bhatia

BE (Chemical), MBA

Avishek Goel

(B.E. Hons.) in Mechanical Engineering from BITS Pilani, Dubai Campus.
Research Associate at The Energy and Resources Institute (TERI)
in the Renewable Energy Technologies Division

WOODHEAD PUBLISHING INDIA PVT LTD

New Delhi

Published by Woodhead Publishing India Pvt. Ltd.
Woodhead Publishing India Pvt. Ltd.,
303, Vardaan House, 7/28, Ansari Road,
Daryaganj, New Delhi - 110002, India
www.woodheadpublishingindia.com

First published 2019, Woodhead Publishing India Pvt. Ltd.
© Woodhead Publishing India Pvt. Ltd., 2019

Woodhead Publishing India Pvt. Ltd. ISBN: 978-93-88320-00-9
Woodhead Publishing India Pvt. Ltd. e-ISBN: 978-93-88320-01-6

Typeset by Asian Enterprises, New Delhi
Printed and bound by Replika Press Pvt. Ltd.

Contents

Section IV

Manufacturing techniques of rubber products

23.1   Introduction

Tyre is a rubber ring placed over the rim of a wheel of a road vehicle to provide traction and reduce road shocks. Most tyres, such as those for automobiles and bicycles, provide traction between the vehicle and the road while providing a flexible cushion that absorbs shock.

The materials of modern pneumatic tyres are synthetic rubber, natural rubber, fabric and wire, along with carbon black and other chemical compounds. They consist of a tread and a body. The tread provides traction while the body provides containment for a quantity of compressed air. Before rubber was developed, the first versions of tyres were simply bands of metal fitted around wooden wheels to prevent wear and tear. Early rubber tyres were solid (not pneumatic). Today, the majority of tyres are pneumatic inflatable structures, comprising a doughnut-shaped body of cords and wires encased in rubber and generally filled with compressed air to form an inflatable cushion.

Pneumatic tyres are used on many types of vehicles, including cars, bicycles, motorcycles, buses, trucks, heavy equipment and aircraft. Metal tyres are still used on locomotives and railcars and solid rubber (or other polymer) tyres are still used in various non-automotive applications, such as some casters, carts, lawnmowers and wheelbarrows.

23.2   Tyre manufacturing process

Process flow diagram (Fig. 23.1) represents general tyre manufacturing processes, giving the flow of work ranging from the preparation of intermediate products (members) from various raw materials to the completion of tyres by a combination of those members, along with the outlines of all the equipment used for respective processes. The processes consist mainly of mix formulation, compounding, molding, vulcanisation and finishing, accompanied further by the processes of manufacturing tyre cords and bead wires.

23.3   Characteristics of manufacturing tyres

The tyre manufacturing processes consist of:

1. Preparing intermediate products (members) utilising the fluidity and plasticity of crude rubber.

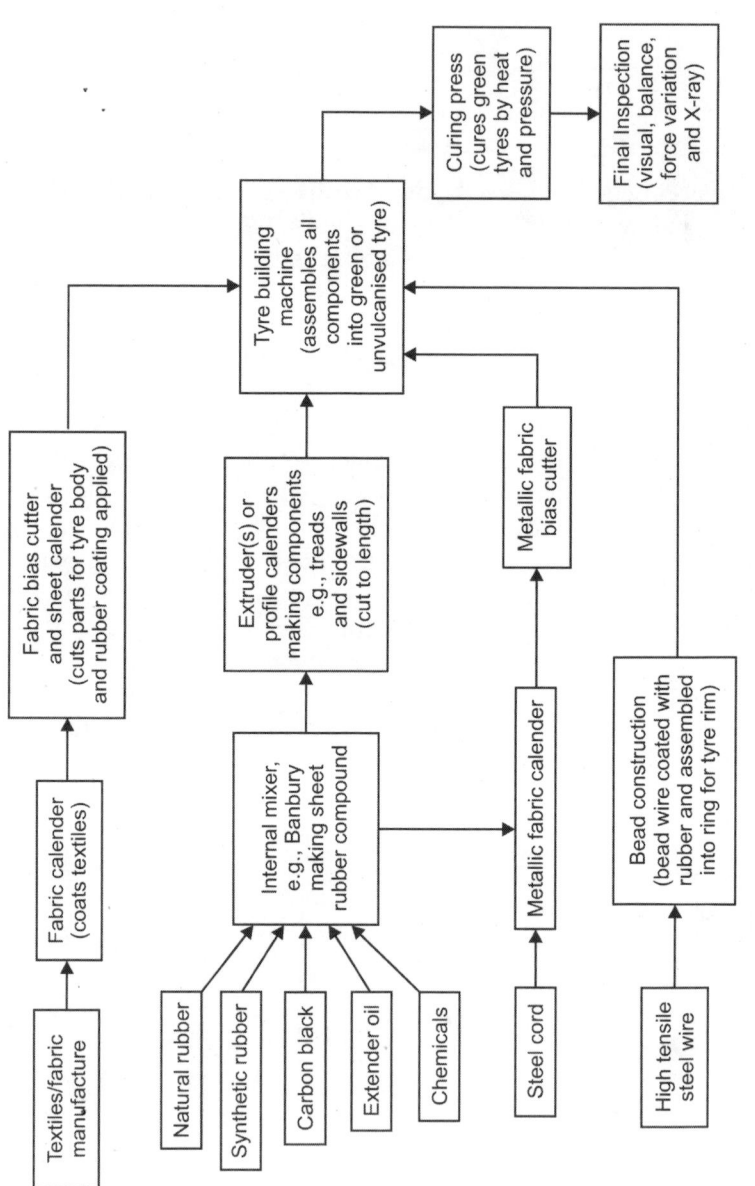

Figure 23.1: Tyre manufacturing process flow diagram.

2. Laminating the members covered with crude rubber utilising the tackiness of the covering crude rubber.

3. Assembling the members to make raw tyres.

4. Vulcanising them at the final stage to produce chemically stable and elastic tyres.

Since green rubber is unexpectedly stiff, it is generally sheeted or extruded with a large-capacity motor after it has been softened by heating. The tyre manufacturing industry is therefore classified as a large energy consuming industry.

The plasticity of rubber is greatly influenced by the quality of raw material elastomers, the methods of compounding and hysteresis of rubber and working conditions of respective processes. Tackiness also varies widely depending on the kinds of rubber and the formation of a thin coating on the surface of rubber, called blooming.

The degree of blooming depends largely on the kinds of compounding ingredients, moisture when rubber was compounded, length of time and temperature after compounding and stimulation by rubbing. In an extreme case, rubber does not adhere at all. It is impossible to eliminate the variance resulting from the fluidity of rubber and accompanying movement of cords even in the process of thermal vulcanisation.

The variance in plasticity and tackiness makes it inevitable to largely rely on manual work even today when automation is highly advanced in the molding process in which raw tyres are assembled. Given this situation, elimination of variance can contribute a great deal to energy conservation, to say nothing of the improvements in quality and productivity.

23.4   Preparation of materials for tyre manufacturing

The manufacture of tyres begins with preparing materials for rubber compounding and for pre-treatment and subsequent rubber-coating of cords (to wrap them in rubber) so that various raw materials processed can be used for the later processes of making intermediate products (members).

23.4.1   Rubber compounding (mix formulation and compounding)

A variety of raw material elastomers and various compounding ingredients are used for tyres by mixing and compounding them for use in respective members. In former days, this compounding was carried out with open rolls and naturally the working site was made terribly dirty due to the scattering of carbon black and other chemicals. Today, intensive mixers, including internal and Banbury mixers, are widely used. This intensive mixer is of enclosed type and computer controlled so that raw material elastomers, various compounding ingredients and oil are automatically fed and compounded. This has resulted in reducing dirt to a considerable degree.

Since the properties of rubber, uncured and cured, vary greatly depending on various factors as described below, attention has been focused on producing rubber compounds to specifications with a slight variance by computer control. The various factors include the kind, quantity, order and time of feeding, the extent to which ingredients are mixed evenly, compounding time and temperature of raw material elastomers and compounding ingredients.

In this process, high-capacity motors are used. This inevitably involves large power consumption, which accounts usually for 35 to 55% of the total power consumption of the factory. It is a common practice to recycle cooling water used in large quantities.

23.4.2 Pre-treatment of cords

Pre-treatment of tyre cords and canvas is a very important process aiming not only at processing fibrous materials and rubber for good adhesion, but also at modifying the properties of fibres, particularly synthetic fibres, including nylon, polyester and rayon, into ones fit for tyre cords. It is a process in which cords are dipped in adhesives and at the same time, subjected to a great tension at high temperature so that they are made not to be readily stretched out and thermally stable to be best fit for tyre cords. This process requires a large equipment in which dipping and drying can be carried out simultaneously.

23.4.3 Calendering

Calendering, also called rolling, is a process in which coating operation is carried out by covering treated textile fabric or steel cords with thin rubber layers on both sides so that materials to be used for sandwiched plies and belts can be made. The quality and thickness of rubber layers depends on respective applications.

The important point in this process lies in the accuracy of thickness in both directions of length and width. Inaccurate thickness leads to the poor performance of tyres and further to vibration due to increased imbalance. Accuracy is therefore required to 1/100 mm.

A calender with three or four rolls is generally used. The temperature of rolls and gauges are computer-controlled. In addition to the use for coating cords and canvas, a calender is also used for preparing various kinds of rubber sheets, squeegees (belt-like rubber sheets for reinforcing plies) and strips (strings) and one of the important equipment at the rubber factory. It also consumes much electric power.

23.5 Preparation of members

This section discusses the processes of preparing necessary members according to respective sizes of various tyres, the forming of a bead, cutting of rubberised cords including plies and belts and extrusion of a tread and a sidewall.

23.5.1 Rubber coated cord cutting

The operation is called 'cutting', in which rubber coated cords and canvas are cut to the angle and width according to respective kinds and applications of tyres. The machine used is called 'a bias cutter', classified into two types: one is a cutter as used for press-cutting paper or thin steel sheets in principle, while the other is a ring cutter that runs at high speed along the beam used for checking the cutter and cords. Two types are available for cutting devices: one is a vertical one with which coated materials, wound off the roll and suspended vertically, are cut after they have been held down, while the other is a horizontal one with which coated materials, wound off the roll onto the conveyor for horizontal lamination, are cut while they are held down. Since the width and the angle of cutting are required to be accurate at present, a horizontal type is more widely used.

23.5.2 Extrusion of treads and other moldings

An extruder is used for preparing rubber members with definite cross-sections, such as treads and sidewalls. Rubber for treads, compounded in the Banbury mixer, is softened by kneading through the hot rolls, fed to the extruder, forced through the tread die with a given cross-section and cut to a desired length after cooling.

The process of extruding treads is one of the most important ones in the manufacture of tyres, the uniformity of which is strictly required, because the tread, accounting for nearly half of the total weight of a tyre, tends to cause trouble when imbalanced. It is therefore important that the extruded and cooled tread is cut to a correct and uniform length, thickness, shape and weight.

Various extruders are used for tread extrusion depending on respective discharges and multi-layer extruders aiming at the simultaneous extrusion of the combination of various kinds of rubbers have come to be used widely in recent years.

The multi-layer extruder extrudes the compound, fed from more than two extruders in respective quantities required, to a desired shape. Two types of extruders are available, hot and cold types, as described earlier. The former requires heating, whereas the latter does not. Electric power is required in large quantity for softening rubber with a screw.

23.5.3 Bead molding

Bead wires, arranged at a given interval and in given number, are rubber coated and extruded to a flat bead, which, in turn, is wound around the core drum, with an inside diameter given according to respective kinds and sizes of tyres, by the number of steps required.

This is a process most commonly used. Usually, thin rubberised fabric tape, called bead covering tape, is wound further with apex rubber attached there on. This process is carried out by another equipment.

23.6 Manufacture of green tyres

Green tyres are made by the molding process, in which various members, prepared through the processes described earlier, are laminated. Two methods are available for molding: one is used for bias tyres and the other for radial tyres, each differing in carcass structure.

The flat former lamination is a method by which all members, including a ply, a bead, a breaker and a tread, are laminated on a cylindrical former, also called a drum, with a diameter slightly larger than that of a tyre rim, to make a cylindrical green tyre. A carcass is inflated, a process called shaving, during vulcanisation to make a doughnut-shaped tyre. Then, the interval between cords widens to lead to a decrease in pressure resistance. The flat lamination is therefore a method of molding tyres for low pressure use to be used for small automobiles, light trucks and agricultural machines.

The crown former lamination process is a method generally used for molding green tyres, for which resistance to high pressure is required, to be used for trucks, buses, construction machinery and aircraft, because in this process the degree of inflating a green tyre to a finished one is comparatively low thus enabling to secure a required strength in addition to the ease of molding a bead part even if the number of plies increases when necessary. In this process, the so-called crown drum is used with a large space for laminating cords within a range allowable for the operation of narrowing plies down toward the bead part. When the crown drum is used, several plies are pre-laminated in a ring form, called a band, in consideration of the operation of narrowing plies down and fitted to the drum so that the plies can be narrowed down toward the bead part. Since another machine, called a band builder, is used for making a band, a combination of a molding machine and a band builder is commonly employed. Needless to say, a breaker, a tread and a sidewall are also laminated in process.

When canvas is used for a tyre ply, it requires much labour in the core lamination in process when molding a tyre. A tyre is not formed then unless pleats, made from a sidewall through to a bead, are folded. The invention of cords, however, has led to a higher productivity because it has enabled to make green tyres almost cylindrical in shape by molding processes.

On the other hand, in the process of molding a radial tyre, in which the belt part is so made as not to extend in the circumferential direction, it is impossible to inflate a tyre after it has been molded flatly unlike a bias tyre.

It therefore becomes inevitable to apply a belt and a tread after inflating a laminate close to a finished tyre in shapes described later. The molding operation is therefore carried out usually in two stages: an inner liner, a carcass ply, a bead assembly and a sidewall are laminated on the easy-to-operate flat drum in first stage because those rubber products can be easily shaved. And a belt and a tread are applied after the resulting laminate has been inflated close to a finished tyre in size in second stage. In the tyre molding factory, green tyres cylindrical in shape are bias tyres, laminated on the flat drum, for use under low pressure, while radial tyres are close to finish tyres in shape.

23.7   Vulcanisation of tyres

Molded tyres are fed to a mold (a metal mold with a tread pattern, a side pattern, a marking and a trademark carved there on) of the specified vulcaniser, pressed against the inside of the mold from the inside and heated simultaneously from both sides, internal and external, with heating media, such as steam and hot water, so that, after a given period, vulcanisation proceeds throughout the entire tyre. Thus, a finished tyre with a vulcanised rubber structure is elastic and stable.

Automatic vulcanisers, such as BAG-O-MATIC® and AUTOFORM®, are widely used. With these machines, insertion of green tyres and taking out and transfer of cured tyres are carried out completely automatically with no one attending. Operators have only to prepare green tyres and watch the process.

Since synthetic fibres shrink by nature if left standing when hot, hot tyres after vulcanisation diminish in size when left standing. A device (a post-cure inflator) is therefore provided, with which bias tyres in which synthetic fibres are used are inflated by applying air pressure immediately after vulcanisation and cooled in an inflated state.

Two types of molds are used for molding tyres: one is a full mold that splits into upper and lower parts and mainly used for molding bias tyres, while the other is a split mold widely used for molding radial tyres. The split mold is one that splits into 6 to 9 segments along its perimeter.

The process of vulcanisation consumes more steam than any other processes for making tyres. Since fuel consumption of the factory is greatly influenced by this process, it is very important as to, how to save energy for this process, which usually accounts for 60 to 90% of the total steam consumption of the factory.

23.8   Finishing of tyres operation

When finishing tyres, a vent hole, drilled right through a metal mold, is used for discharging air from the space between the tyre and the mold. An excess rubber is forced out and forms hair-like vent spew in the vent holes and other shape of spews at the split parts of upper and lower molds and joint parts between mold segments. These spews should be removed in terms of good appearance. Automatic finishing machines have recently been introduced for this purpose in the most of factories. The finished tyres are subjected to 100% inspection including that of appearance (a sensory test by an inspector) for rejection of defective units. Those for use in passenger cars, trucks, buses and aircraft are subjected further to a balancing test to screen unbalanced units. A uniformity machine is also incorporated in the production line for measurement of uniformity of tyres for use in passenger cars, trucks and buses.

23.8.1   Names of tyre parts

Figure 23.2 is a cross-section of the steel-belt radial tubeless tyre for passenger cars and Fig. 23.3 shows the appearance and the internal structure of both bias and radial tyres for better understanding.

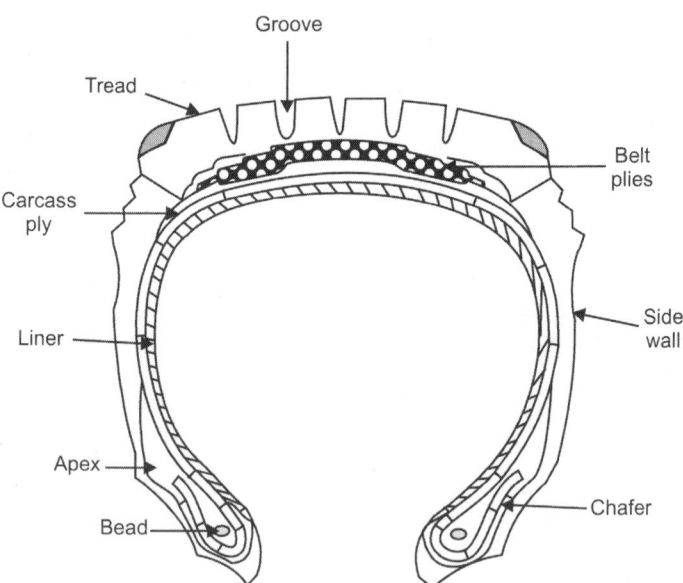

Figure 23.2: Cross-section of passenger car radial tyre.

Tread: That portion of a rubber layer which contacts the road surface. The tread pattern is so engraved on the surface as to give the property of a nonskid.

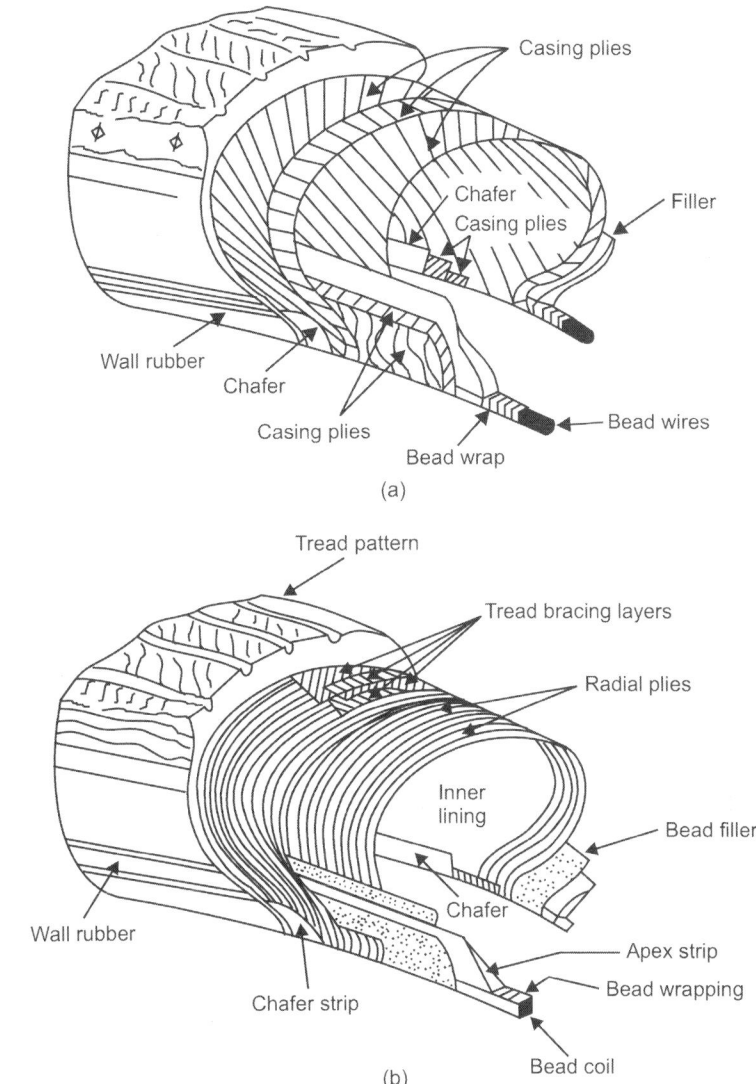

Figure 23.3: (a) Bias-belted tyre construction and (b) radial tyre construction.

Side: A portion between the tread and the bead. The surface rubber layer only of this portion is sometimes called a sidewall.

Bead: A portion made to suit the rim, with a circular assembly of steel wires wrapped with plies.

Shoulder: An interval between the tread part and the side part. The boundary is not definitely defined.

Carcass: A portion constituting the structure of a tyre composed mainly of ply and bead parts, including a belt. In some case, it includes a breaker. Ply is a thin layer of rubberised fabric.

Breaker: One to several layers of textile material, inserted between the tread and the carcass of a bias tyre to protect the carcass from a road shock or from external damage.

Belt: In a radial tyre, it is placed in the same position with a breaker, with textile material, durable and hard to stretch out, arranged almost circumferentially around the tyre to give a hoop effect.

Inter liner: In a tubeless tyre, it is a thin butyl rubber layer, made gas tight to hold air pressure and used for lining a carcass. In a tyre with an inner tube, the tyre is lined only with an ordinary rubber layer since the tube itself is made of highly gas tight butyl rubber.

Apex: Apex stands for the uppermost peak of a triangle. It is a hard rubber member with a triangle-shaped cross-section, sometimes added onto bead wires to put the bead part in shape as well as to give rigidity. It is called 'Apex' and sometimes 'bead filler' or 'stiffener'. This member is an essential for supporting a radial structure with a low degree of carcass rigidity since a ply cord has no angle and used for truck tyres of bias type that carry heavy loads.

23.9   Manufacturing of rubber products and energy consumption

Among rubber products manufacturing processes, the rubber materials milling process, the extruding process and the rolling process have a relatively higher electric power consumption which is more than 50% of the total consumption and the vulcanising process have 80% more or less of the total consumption.

It is most important to take counter measures for energy conservation, but it should be also important to improve the yield in manufacturing processes. The yield rate is the ratio of the amount of a rubber product completed to the amount of rubber materials consumed when the rubber materials are processed.

The yield rate would be decreased due to a loss of weight during processing, an occurrence of defective units and so. Since any occurrence of the defective units consumes their ration of materials, energy and labour expense, it results in directly increasing in the cost. Therefore, attention should be paid to the following points for managements.

1. Management of rubber materials: Checking of the raw rubber, compounding ingredients and so, checking of characteristics of the milling rubber.
2. Management of materials other than the rubber: As regards the tyre, checking of the strength of tyre cord, checking of the binding wire.

3. Management of manufacturing conditions of equipments in each process: Checking of operating conditions, checking of process specifications.

4. Management of inspection: Checking whether or not various conditions are correspond to the drawings and standards for the manufacturing factory.

Energy conservation in rubber industry is also discussed in detail in Chapter 34.

23.10 Tyre manufacturing and energy conservation

First of all, the amounts of both materials and energy to be consumed which are most important items in manufacturing products should be soundly seized. In the case of the fuel, items to be controlled are as follows:

Amount of fuel consumption/production amount (weight of mixing rubber) $\rightarrow$ Fuel consumption per unit weight of mixing rubber.

(Coal) weight of coal consumption (T)/(T), per 1 T of mixing rubber.

(Heavy oil) weight of heavy oil consumption (kL)/(T), per 1 T of mixing rubber.

(Gas) amount of gas consumption (Nm^3)/(T), per 1 T of mixing rubber.

Since the calorific value of each fuel is already known, the unit of kcal/T can be also controlled. In the case of the electric power, it should be as follows:

Electric power consumption/weight of mixing rubber $\rightarrow$ Electric power consumption per the unit weight of mixing rubber.

$$(kWh)/(T) \text{ per 1 T of mixing rubber}$$
$$\downarrow$$
$$860 \text{ (kcal/kWh)} \times \text{Electric power consumption (kWh) kcal/T}$$

Such a 'unit requirement' or 'energy consumption rate' is the basic numerical value in executing counter measures for the energy conservation. When carrying out the energy conservation activity until finishing it up to its index 100, there will go by some stages having a respective index value. Therefore, by confirming at each stage to what degree the index has been reduced, the effectiveness of the energy conservation could be ascertained. And as the result of it, the effectiveness in terms of Mooney viscosity is calculated.

The numerical values of the unit consumption should be hopefully confirmed once in a month at least, also from the viewpoint of factory management.

Further, in order to seize the present situation, if the specifications of main equipment in factories and the wiring and piping installations are ready to be clearly shown, this makes it convenient to select the energy conservation activity and to analyse the present situation.

Carrying out the energy conservation activity means reducing in the unit consumption, in other words. However, there are another measures to reduce in the energy cost, that is, utilising the fuel and electric power of a lower cost, or making all employees present any proposals for a remedy of the energy conservation through levelling up their consciousness on it. In the case of the rubber manufacturing factory, the electric power and the fuel are the main items to be controlled.

23.11 Reduction in the energy cost

23.11.1 Reduction in the energy consumption rate (energy conservation)

Reduction in losses

1. Optimisation of the equipment capacity and optimisation of the manufacturing capacity of equipment.
2. Prevention of the idle running and leaks.
3. Reduction in failures of the equipment.
4. Reduction in defectives.
5. Recovery of the waste heat utilising the waste heat to a heating and so.

Improvement in efficiency

1. Improvement and renewal of the equipment.
2. Rationalisation of operation processes.
3. Establishment of advanced technologies improving in the establishment of advanced technologies.

23.11.2 Utilisation of the fuel and electric power of a lower price

1. Cutting of the contract demand.
2. Effective utilisation of the power.

23.11.3 Improvement in employee's consciousness on the energy conservation

1. Levelling up of consciousness on the energy conservation.
2. Activities for the energy conservation.

23.12 Energy conservation in the tyre manufacturing factories

23.12.1 Energy consumption at tyre manufacturing factories

Table 23.1 shows an example of the state of energy usage in the tyre manufacturing factories.

Table 23.1: State of the energy usage by processes.

Process	Fuel (heavy oil)	Electric power
Mixing	9.7%	32.8%
Extruding	0.8	17.0
Sheeting	0.5	4.9
Cutting	0.2	5.7
Beading	0.1	0.1
Forming	0.1	4.7
Vulcanisation	81.1	4.3
Finishing and testing	0.9	2.2
Driving	1.9	17.0
Others	4.0	10.4

Table 23.1 shows the process which consumes lots of fuel is the curing process and both the refining section and the driving section consume the electric power prominently.

The key points to improve the manufacturing processes from the viewpoint of carrying out the energy conservation are as follows:

Mixing: Heating up the crude rubber, investigating on the peptiser, investigating on milling conditions, circulating the warm water, exhausting fan.

Extruding: Temperature of the warming sheets, investigating on the roll size (face length and number of roll), heating the mouth rings, controlling the remolding amount, narrowing the width of cooling conveyer.

Sheeting: Same with extruding.

Vulcanisation: Controlling the outgoing radiation, improving the curing method, pre-heating the die assembly, shortening the time for exchanging the die assembly, investigation on the blowing air, improving the ventilation fan.

Driving: Rising the temperature of boiler feeding water, drain recovery, jointing the steam and the warm water systems, miniaturising the boiler, withdrawing the spoil pipings, reducing in the number of air compressor.

Others: Natural illumination, controlling the steam and air leaks, reducing in the idling time, inspecting the optimum capacities of equipment and motors, installing the instruments and gauges.

In the case of the mixing process among the above said processes, the relationship between the internal mixer and the power load plays an important role. In order to reduce in the electric power for milling process, followings are the items to be noted, as enumerated also in the key points for improving processes.

Reducing in the idling time should be important, too:

1. Heating up the green rubber (resulting in minimising the peak electric power).
2. Investigating on the peptiser (resulting in shortening the masticating time).
3. Investigating on the milling conditions (resulting in shortening the cycling time and so).

Further, almost all amount of the fuel or steam used is consumed in the curing process. Table 23.2 shows problems and measures for improving them.

Table 23.2: Problems and measures for improving them.

Problems	Loss	Reasons	Measures for improving
Big loss of boiler heat	10%	High temperature of the exhaust gas	Recovery of exhaust heat
Big loss of outgoing heat in the vulcanisation process	55%	Insufficient heat insulation	Complete thermal insulation
Big loss of the waste heat in the vulcanisation process	20%	Big un-recuperated heat	Recovery of waste gas
Big amount of indistinct parts (unable to measure)	25%	Many leaks	Prevention of leaks
Small amount of recovered heat from the pre-process	18%	Big un-recuperated outgoing heat	–

From the Table 23.2, it should be noted that thermal insulation or adiabatic measures, recovering the waste gas and preventing the leaks, should be further paid attention too. Especially main energy saving point' are in the vulcanisation process.

Compressors and auxiliaries: With regard to the compressors which consumed the biggest amount of electric power, some counter measures should be taken, such as automating the operation by a pressure switching, controlling the number of compressors, reducing in the blasting pressure, reducing in the air leaks and so. Therefore, the activity should be directed to the equipment for cooling water to be supplied to factories, the nitrogen gas equipment and the boiler equipment in which any reduction in the electric consumption seemed to be possibly attained with fitting to the amount of production.

23.12.2 Improvement of nitrogen gas generators

1. Role of the nitrogen gas in the tyre manufacturing processes:
   (a) The nitrogen gas is used in the tyre curing process as a pressing source for pushing a crude tyre to a die from the inside of the tyre in the curing machine through a rubber bag.

(b) The nitrogen gas absorbs air and separates it. The nitrogen gas is generated by the nitrogen generating equipment and pressured by the high pressure compressor to send it to the curing machine.

(c) In order to prevent a deterioration of the rubber bag, the residual concentration of oxygen gas existing in the nitrogen gas should be made as lower as possible.

23.13 Role of cooling tower in tyre factories

Fans, pumps, blowers, cooling towers and many other applications are subject to varying loads. The variation may occur due to various factors, e.g. in cooling tower applications the load variation occurs may be due to utilisation of the installed capacity, variation of process conditions, etc.

Conventionally, control valve throttling, pump discharge bypass recirculation control, etc., are used in order to match the requirements of varying loads in cooling tower outlet temperature control.

Controlling varying load can be significantly reduced by the use of variable frequency drive. The importance of maintaining optimum cooling tower cold water return temperature need not be over emphasised. It is very important for the economic reasons, i.e. from the point of view of desired product yield and quality. Cooling towers are designed to take care of maximum adverse conditions. But normally both of these conditions do not occur simultaneously. Hence, there is lots of scope to improve the performance of operating cooling towers. The design cold water inlet temperature in the secondary circuit of the cooling system, to various systems is approximately between 31–32°C and the maintenance of this design temperature is very important from the point of view of process performance.

However, due to the wide, fluctuation of weather conditions and also due to the partial loading of cooling tower, it had virtually become impossible to maintain optimum cooling tower cold water return temperature, i.e., 28–29°C and operating cold water inlet temperature to various system, i.e., 31–32°C.

23.13.1 Improvement of cooling water pumps at factories

The factory cooling water are used for cooling the rolling machines, various hydraulic units and compressors, for cooling the extruding process and so. The cooling water is generally supplied at 1.5 kg/cm² for directly cooling the rubber in a shower condition in which pressure is maintained 3 kg/cm² for the other usage.

23.14 Energy conservation opportunities in boiler

The various energy efficiency opportunities in boiler system can be related to combustion, heat transfer, avoidable losses, high auxiliary power consumption, water quality and blow down. Examining the following factors can indicate if a boiler is being run to maximise its efficiency:

Stack temperature: The stack temperature should be as low as possible. However, it should not be so low that water vapour in the exhaust condenses on the stack walls. This is important in fuels containing significant sulphur as low temperature can lead to sulphur dew point corrosion. Stack temperatures greater than 200°C indicates potential for recovery of waste heat. It also indicate the scaling of heat transfer/recovery equipment and hence the urgency of taking an early shut down for water/flue side cleaning.

Feed water pre-heating using economiser: Typically, the flue gases leaving a modern 3-pass shell boiler are at temperatures of 200–300°C. Thus, there is a potential to recover heat from these gases.

The potential for energy saving depends on the type of boiler installed and the fuel used. For a typically older model shell boiler, with a flue gas exit temperature of 260°C, an economiser could be used to reduce it to 200°C, increasing the feed water temperature by 15°C. Increase in overall thermal efficiency would be in the order of 3%. For a modern 3-pass shell boiler firing natural gas with a flue gas exit temperature of 140°C a condensing economiser would reduce the exit temperature to 65°C increasing thermal efficiency by 5%.

Combustion air pre-heat: Combustion air pre-heating is an alternative to feed water heating. In order to improve thermal efficiency by 1%, the combustion air temperature must be raised by 20°C. With coal firing, unburned carbon can comprise a big loss. It occurs as grit carry-over or carbon-in-ash and may amount to more than 2% of the heat supplied to the boiler.

Effect of boiler loading on efficiency: The maximum efficiency of the boiler does not occur at full load, but at about two-thirds of the full load. If the load on the boiler decreases further, efficiency also tends to decrease. At zero output, the efficiency of the boiler is zero and any fuel fired is used only to supply the losses.

The factors affecting boiler efficiency are:

1. As the load falls, so does the value of the mass flow rate of the flue gases through the tubes. This reduction in flow rate for the same heat transfer area, reduced the exit flue gas temperatures by a small extent, reducing the sensible heat loss.

2. Below half load, most combustion appliances need more excess air to burn the fuel completely. This increases the sensible heat loss.

In general, efficiency of the boiler reduces significantly below 25% of the rated load and as far as possible, operation of boilers below this level should be avoided.

Proper boiler scheduling: Since the optimum efficiency of boilers occurs at 65–85% of full load, it is usually more efficient, on the whole, to operate a fewer number of boilers at higher loads, than to operate a large number at low loads.

Boiler replacement: The potential savings from replacing a boiler depends on the anticipated change in overall efficiency.

23.14.1 Reduction of electric power for the boiler equipment

The boiler with a capacity of 50 T/h are being operated, but the actual loading should be reduced by about 30% owing to the counter measures of the energy conservation for the secondary system.

The fuel unit consumption can make progress in good shape and improvements are as follows:

1. A thorough heat insulation with 35 to 70 mm in thickness on the curing machine, the hot water piping installment, the hot water equipment, the steam piping installation and so on.
2. Recovery of the drain from both the curing machine and the hot water piping installation.
3. Adjustment and integration of both the steam piping and the hot water piping installations.
4. Rising the temperature of the boiler supplying water.
5. Improvement in the effectiveness of the hot water equipment by changing its type.
6. Rationalisation of operating conditions for the hot water circulating installment.
7. Reduction in the cycle time through rationalising the curing conditions.

Besides above some of the auxiliaries of boiler also need attention.

Piping: There are many complex piping parts and naked pipes by installations.

Traps: There are some traps having an excessive discharging capacity against the loading, resulting in generating an energy loss. Further, such traps are located at a place where the inspection work is hard to carry out and their actuation is sometimes difficult to be judged.

Heat insulation: The main pipes are fairly heat-insulated, but some of flange valves, pressure reducers and covers of the curing cans are not.

Loading equipment:

1. The loading might be light in the case of operating the one machine, but its fluctuation became violent in the case of operating the multi-equipment

in parallel at the same time. Therefore, the boiler loading sometimes got 18 T/h at the maximum against its capacity of 10 T/h, resulting in decreasing the boiler pressure and in interfering with the production.

2. There are some curing equipment which threw away the steam each time when the curing processing is finished.

3. There are leaks of the steam.

Valves: There are marked leaks from the ground parts of valves for a high pressure and from the parts of bonnets.

Problems and counter measures for energy source

1. Heat recovery from the boiler exhaust gas temperature of the exhaust gas is generally 250°C at the exit of the economiser, the gas thus being exhausted at a higher temperature. In order to lower this temperature, a pre-heater for the supplying water can be installed at a place between the economiser and the chimney as shown in Fig. 23.4, to supply water to the feed tank and to devise to recover the heat.

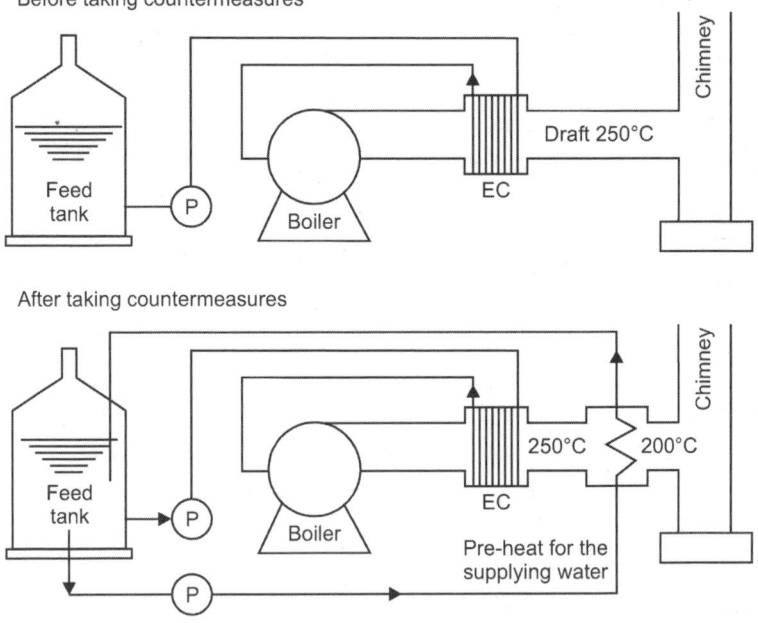

Figure 23.4: Heat recovery from the boiler.

2. The boiler is operated with the oxygen content of about 3.7% in the exhaust gas from the boiler. However, in order to reduce in the holding

heat of the exhaust gas, a low oxygen combusting equipment is installed to enable to make the boiler operated at 2.5% oxygen content in the exhaust gas.

3. Rationalisation of piping installation for the tyre curing machine. The piping installations for jacketing the tyre curing machine are much complex and their total length amounted to 7 m. The complexity is because the piping had a structure in which two pipes enabled to cure the tyre at the same time and therefore even if one of them got out of order, the other alone enabled to cure it through closing the opposite valve. However, since there had not been actually such a one-pipe curing operation, the piping installation is altered, in which the piping was shortened to 4 m and was thoroughly heat insulated to reduce in the heat loss due to an outgoing radiation.

4. Recovery of the exhaust heat from the pot heater. The steam is thrown away from the pot heater each time when the curing processing is finished. This steam was made a heat source for the bath through installing a heat exchanger in the way of the exhausting piping.

Problems and counter measures for them

1. The type of traditional traps dryer traps should be replaced by those of an intermittent type.

2. Counter measures for leaks from piston valves of the tyre curing machine: With regard to the amount of steam used for the curing machine, the difference in its amount is great, between its actual usage indicated by the flow meter and the discharged amount of drain measured by the tyre curing model machine. Results of the actually measured amount of steam and the indicated value by the flow meter are 42.8 kg/h and 71.8 kg/h, respectively, resulting in showing a loss of 29 kg/h. As the result of an investigation, it had been found that there is a leak at the high pressure piston valve. Thereupon, the valve is replaced. After that, the flow meter indicated 52.6 kg/h.

3. Equalisation of loads (installing of an accumulator): The said multi-and-parallel equipment operation got a heavy loading, even if each loading of them was light, resulting in a drop of the boiler pressure. Therefore, the reserved boiler had been usually operated for 15 min at the time of pre-heating in the morning and thus a two-boiler operation had been carried out. In order to solve this problem, an accumulator was installed. As its result, the loading have not exceeded 10 T/h, there have not been any pressure drops and the two-boiler operation have been disused.

4. Thorough heat insulation: Since the heat insulation had not been fully applied on those in factories such as the valves, flanges, pressure reducers and curing can covers, it had been executed on all of those.

5. Counter measures for leaks to the outside of valves: Since there had been marked leaks from the ground parts of valves for a high pressure and the parts of bonnets, some components are replaced and in case that any repair on the old ones was difficult, they are renewed. By taking the counter measures, all problems of leaks in valves and pipes can be solved.

24.1   Introduction

Rubber conveyor belting essentially consists of various fabrics like all cotton, cotton/nylon, rayon/nylon, all synthetic, etc., with a cover of various rubber grades like oil resistant, flame resistant, etc. The design of conveyor belt cover compound is most important in determining the service life of the belt in any particular application. Heat and flex-cracking resistance being the essential requirements of conveyor belts in service, different types of anti-oxidants are used for heavy, medium and light duty belts.

Conveyor belt is designed as per requirement, width of belt can be taken different size as well as thickness of belt can be determined as per basic need of belt design. Conveyor belt has more tolerated power to bear heavy load and it has long life, moreover it can transfer power from one pulley to another.

The demand for rubber conveyor beltings has been steadily increasing with the rapid industrialisation in various countries. These belts are commonly used for industrial purposes in thermal plants, fertiliser plants, railways, chemical plant, etc. Depending upon the properties of cover compound and the number of plies, these rubber conveyor belts are put to specific end uses in various industrial plants.

Raw material required for manufacture of conveyor belt: (i) raw rubber, (ii) synthetic rubber/nylon, (iii) whiting processing oil, (iv) zinc oxide, (v) stearic acid, (vi) rosin, (vii) anti-oxidant, (viii) accelerator, (ix) sulphur, (x) pine tar, (xi) carbon black, (xii) canvas/fabric, (xiii) colour, paints, pigments and (xiv) packing material.

24.2   Process of manufacture of conveyor belt

Rubber conveyor belts are generally manufactured by two methods. One method involves rubberising the cloth by calender machine and in the second method rubberising of the cloth is done by spreading machine. In the case of rubberising by calender machine the production is very high and there are no costs involved on account of solvent losses. After rubberising, the cloth in requisite number of layers is transferred to Hydraulic press for vulcanising to achieve the desired thickness. Different types of conveyor belting cover composition are available for heavy duty, medium duty, light duty, heat resistant type of belts.

Sometimes the products has to be manufactured according to specific customer demand.

As discussed above belt conveyor is constantly operating transporting equipment which is mainly used to convey mass bulk material like mineral, coal, sand, etc., in powder or block as well as packed freight in metallurgy, mining, building heavy industries and transportation industry. Belt conveyor is the most perfect conveying equipment for coal-mining, because it can work efficiently and continuously. Compared with other transporting equipments, belt conveyor not only has the merits of long conveying distance, big capacity, constant working operation, but also with the features of operational reliability, easy to have automated and concentrated control. Belt conveyor has become the key equipment especially for high-output and high-efficiency coal mine.

To start with, it is essential to understand what a belt conveyor is. As we all know, a conveyor can move material, like cardboard boxes, wood boxes and plastic boxes. This is called gravity conveyor (Fig. 24.1). When the conveyor can move boxes up against gravity, down or horizontal, it is a belt conveyor, whose moving belt is controlled by electric power.

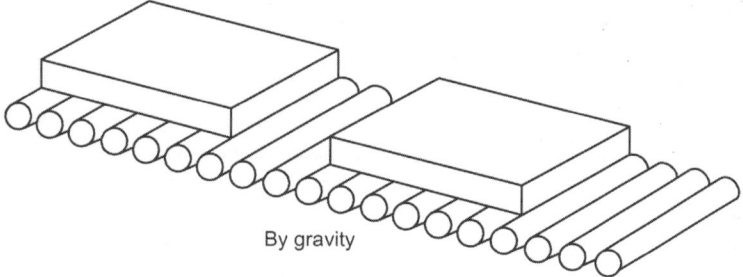

By gravity

Figure 24.1: Gravity conveyor.

A belt conveyor is a machine with a moving belt and the machine is made with these parts: a conveyor bed, which comes in variable sizes, lengths and widths; a pulley is like an iron pipe. Pulleys are put on each end of the bed and they are as wide as the conveyor bed. Each pulley has a steel shaft through it, the shaft turns on a bearing and the pulley turns with the shaft. Bearings are used to keep the pulley shaft and the conveyor from rubbing together; a conveyor driver is made up of a motor, a speed reducer and a drive pulley (Fig. 24.2). The drive pulley is driven by the motor. Two sprockets are put on both of the drive pulley shaft and the motor so that a chain can be put around the drive pulley sprocket and the motor sprocket. The chain moves when the motor is started and the drive pulley is turned by the chain. But, because a motor turns very fast (1750 rounds per minute), a speed reducer must also be used and it is put between the motor and the drive pulley. The motor is

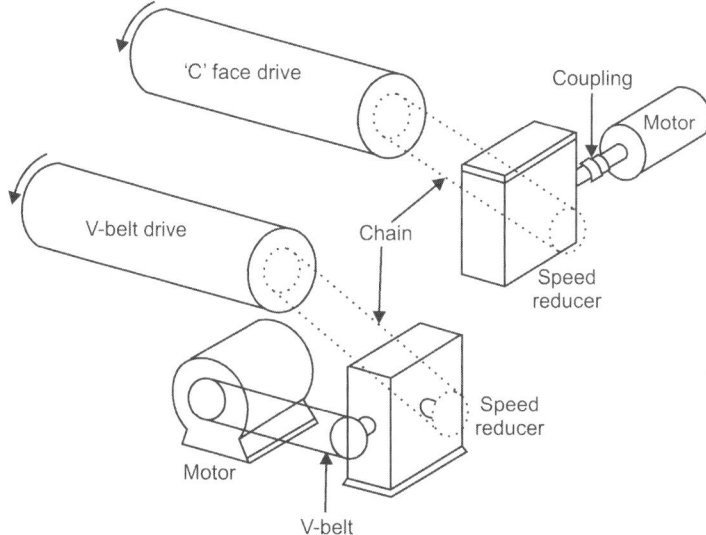

Figure 24.2: Conveyor drive.

connected to the reducer with a V-belt (like the fan belt in a car) or a 'C' Face coupling, the reducer is connected to the drive pulley with a chain, so the drive pulley turns more slowly.

Rubber mixer: Choice of rubber mixes will depend on the type of belt being manufactured, i.e., general purpose, heat resistant, oil resistant or food quality. The carcass compound for the general purpose belt is usually a high-quality natural rubber with low-modulus fillers and a high-scorch-time vulcanising system. For underground mining, conveyor belting is mainly based in PVC to provide flame resistance, which will prevent the danger of the spread of fire and has antistatic properties to minimise the risk of the explosion of mine gases due to static discharge.

Vulcanisation: Vulcanisation of conveyor belting is carried out in either a press or a continuous drum-curing machine.

24.2.1 Belt design

The overall design of a belt is determined by driving unit requirements and the requirements imposed by the nature and volume of material to be carried. Belts in general are produced as a multiply, single ply and solid woven. In all three types fabrics parameters are different. In realising an adequate design, there are a number of parameters which have to be considered and satisfied:

1. Adequate tensile strength and modulus to transmit the power and to carry the material.

2. Low elongation at working tension to give minimum take-up requirement.
3. Good load support and sufficient width to carry the type and volume of load.
4. Flexibility, both directions, to flex round the pulleys and transversely, for satisfactory troughing.
5. Dimensional stability to run straight and not to grow too much in service (low permanent elongation).
6. Good adhesion between all components to avoid delamination.
7. Good tear resistance to withstand damage.
8. Ability to be joined (mechanically or chemically) and be close in loop.

24.2.2 Multiply belt

It is most popular belt construction. It has 2–6 plies and can have breaking strength up to 3500 kN/m. Such belts are spliced (in loop) mechanically or chemically (cold glue or hot vulcanising).

24.2.3 Single ply belt

Not so popular like multiply belt construction, but more flexible, what is important for special applications. Breaking strength can be up to 1000 kN/m.

24.2.4 Solid woven belt

Construction companies commonly use in underground belt, especially as a impregnated by PVC for flame resistance applications.

24.3 Rubber textile composites

24.3.1 Actually used material

In the beginning cotton was the most popular material used for conveyor belts production. Cotton gives very good mechanical adhesion and good high temperature protection (no melting). Afterwards cotton was replaced by rayon. Actually mainly polyester, polyamide and aramides are used in this type of composite.

24.3.2 Polyamide (Nylon)

Compared with the cellulosic, the nylons are of much higher strength and also give much higher values of elongation at break. This latter property imparts to nylon fabrics a greatly improved impact resistance, higher work to rupture and much better tear resistance. It is largely on account of these properties that nylon has been adopted as the principal weft yarn for conveyor belting fabrics. Additionally, the lower modulus of nylon also contributes to very good troughing characteristics in the finished belt. One characteristic of nylon, not possessed by the cellulosics, is thermal shrinkage. Being a thermoplastic material, when heated nylon tends to shrink. On account of this, when processing nylon fabrics, either some restraint must be employed to control or limit this shrinkage or adjustment must be made in the design to allow for the subsequent changes in dimensions during processing. By choosing suitable conditions of processing, it is possible to modify the shrinkage characteristics to suit more precisely the final parameters to be satisfied.

Generally speaking, however, if nylon is allowed to shrink, the elongation at break will increase and modulus will fall, depending on the degree of shrinkage. Nylon is almost entirely used as continuous filament, but there is a small application for spun staple nylon, as with rayon, primarily for bulk and adhesion requirements.

24.3.3   Polyester

In principle, polyester combines the strength and elongation characteristics of the nylons with the modulus characteristics of the rayons. This combination of properties suits many applications, but there are two main areas where problems exist. The first concerns adhesion: being relatively inert chemically, it is somewhat more difficult to obtain adequate levels of adhesion with polyester than with rayon or nylon. However, methods of treatment have been developed to overcome this.

The second area relates to thermal shrinkage. Processes exist to modify the shrinkage characteristics of polyester and there are also various grades of fibres with differing shrinkage/modulus relationships, achieved by modification of the basic polymer.

24.3.4   Aramid

The properties of aramid are more akin to those of the inorganic materials than to the other textile fibres. The tensile strength, even assessed by the engineering method as strength per unit cross-section, is of the same order as those of steel and glass, so that when quoted as tenacity (where the advantage of low specific gravity is taken into account), the strength is exceptional. The modulus is also very high, but this is coupled with a very low value for elongation at break, which introduces some difficulties in certain applications. The major disadvantage of this low elongation occurs when aramid is used in several layers. When flat, each layer of textile is able to contribute its own share of strength, but on bending, the low elongation of the outermost layer prevents it from accommodating to the curve, which places the other layers in compression. This directly reduces the contribution of the inner plies to the total strength but also and more seriously, the performance of aramid in compression (especially its dynamic fatigue resistance) is not good. Under such conditions, premature failure of the inner plies is likely to occur. However, many applications have been developed which enable the excellent properties of aramid to be realised and much effort is being devoted to ways of overcoming the problems associated with this low elongation, so that other areas of use can be found for the unique combination of properties possessed by aramid.

24.3.5 Fabrics weaves

Majority of conveyor belts are made as a multiply construction. Therefore typical fabrics are commonly used. The simplest weave is the plain weave. In this construction, each thread, both warp and weft, passes alternately above and below the threads in the other direction and out of phase with the adjacent yarns on either side, running in the same direction.

The plain weave is the basic woven fabric structure and is very widely used. It forms the basis for the development of the more complicated interlacing patterns used in weaving. Most ply belting fabrics are plain weave as are most fabrics for coating applications.

The plain weave construction, using very low twist yarns, enables a very smooth and flat surface to be produced, the yarns flattening out at the inter-sections and spreading to give minimum gaps or interstices between adjacent yarns, thereby giving a very full cover fabric.

As the strength requirements for the fabrics increases, it becomes increasingly difficult to insert the necessary number of yarns into the space available, not-withstanding the use of folded yarns, so it becomes necessary to modify the interlacing to obtain the yarn density required.

For very high strength fabrics, particularly those used for single- or two-ply reinforcement of belting, alternative designs are required. There is a limit to the number of threads that can be woven into a given fabric design. Generally as the warp strength requirement increases, so also does the weft strength, thus greatly increasing the total bulk of yarn that must be woven into a given area. There are basically two routes whereby the yarn density, that is the total number of yarns, both warp and weft, that can be incorporated in a unit area of the fabric, can be increased – these are the 'straight' or 'stress' warp fabrics and the 'solid- woven' fabrics.

Considering the stress warp fabrics first, the main strength threads in the warp and weft lie in different planes and do not interlace with one another, thus allowing much greater density to be obtained, as it is the interweaving of these threads that limits the number that can be woven together. Effectively, the weft is divided into two layers, one above and one below the straight warp ends and much finer 'binder' warp ends, lying between the stress warp ends, interlace with the two layers of weft, so giving the fabric its integrity. In this type of construction, with only fine binder ends passing between the stress ends and interlacing with the weft, it is possible to use much heavier yarns in both the stress warp and in the weft than would be possible if it were necessary for these to interlace one with the other. There are many variations of this basic structure possible by altering the ratio of stress to binder ends. In the simplest form, there is one binder end to every stress end, but it is possible to reduce the number of binders, although this is likely to reduce the fastener retention strength of the final fabric.

24.4   Safety of using belt conveyor

The increase in production quality, work productivity, the capacity of the devices and efficiency in various regions of human activity must not conflict with safety. Good common sense is the key when working on any equipment and must be used while observing or servicing equipment. Some general safety guidelines of using a belt conveyor should be observed.

Lockout/tag out all energy sources to the belt conveyor, conveyor accessories and associated process equipment before beginning any work – whether it is construction, installation, maintenance, or inspection that is directly associated with the equipment you are involved in. The use of lockout device with one key for each piece of equipment should be used. The person actually doing the work should be the only person with the key to the lockout device. Operating and maintenance personnel should become familiar with the material being handled in the system along with the location and purpose of the safety devices before being allowed to operate or work on the equipment. A belt conveyor safety training session should be a portion of a comprehensive safety programme provided by the company to all employees that will be required to operate or maintain the equipment.

All safety devices should be in a good working condition, properly maintained and easily accessible. An emergency stop switch with safety pull cords should be mounted at a proper height. The equipment should be operated at its design capacity and speed.

Overloading belt conveyors result in spilled material and hazardous working conditions and premature failure of components. During and after maintenance of the equipment a safety 'walk around' is recommended as a precaution for leaving tools or work material prior to starting the equipment.

A formal maintenance and inspection schedule should be developed and followed for the equipment and associated safety devices. Required personnel safety equipment such as hard hats, safety glasses and steel toe shoes should be worn when in the area of the equipment to provide any type service or work.

Manual inspection, maintenance or repairs must be done at a time that can be taken out of service, properly lockout and tagged. In no case should belt conveyors or any operating equipment be serviced while in operation. Only visual inspection can be done during operation and care must be taken to be at a safe distance and not be wearing loose clothing. Inching drives provide an excellent method of visually inspecting the belting.

Rubber hose

25.1 Introduction

A hose is a flexible hollow tube designed to carry fluids from one location to another. Hoses are also sometimes called pipes (the word pipe usually refers to a rigid tube, whereas a hose is usually a flexible one), or more generally tubing. The shape of a hose is usually cylindrical (having a circular cross-section).

25.2 Manufacturing of hose

Hose is manufactured in the unvulcanised state by forming a cylindrical tube over which a reinforcement and cylindrical cover are applied. In its uncured form, a hose tube will often need support to maintain proper internal diametre (ID) and dimensional tolerances while being processed through the various stages of manufacture. Thus, the three basic methods of making hose have evolved: (i) non-mandrel, (ii) flexible mandrel and (iii) rigid mandrel.

In methods (ii) and (iii), the mandrels are used for support and as dimensional control devices for the hose tube during processing. Then after the hose building and if necessary, the vulcanisation are complete, the mandrels are removed, inspected and recycled.

25.2.1 Non-mandrel style of manufacturing hose

The non-mandrel method of manufacture is generally used for lower working pressure, smaller diametre, textile reinforced products not requiring stringent dimensional tolerances. Typical hose products in this category would include garden, washing machine inlet and multi-purpose air and water styles.

Essentially, the non-mandrel technique involves extruding the tube, applying the reinforcing and extruding the cover in the unsupported mode (without a mandrel). Frequently low pressure air is used inside the tube for minimal support, keeping the tube from flattening during the reinforcing process. In some cases, especially in larger sizes, the tube may be extruded with air injection along with an internal lubricant to prevent adherence to itself.

The non-mandrel tube extrusion process can be done continuously, if appropriate handling equipment is available, thus providing excellent length patterns for the finished product. In recent years with improvements in die design and cooling, dimensional control of non-mandrel rubber tube is approaching that of flexible mandrel style.

Most smooth bore thermoplastic hoses are extruded non-mandrel. The higher rigidity of most thermoplastics eliminates the need for mandrel support. In addition, with advanced cooling and dimensional sizing equipment, thermoplastic tube dimensions can be maintained quite accurately.

25.2.2 Flexible mandrel style of manufacturing hose

When moderate tube processing support is needed and more accurate dimensional tolerances are a concern, flexible mandrels may be utilised. These mandrels are rubber or thermoplastic extrusions, sometimes with a wire core to minimise distortion. Of the three flexible mandrel styles, solid rubber offers minimal support, while rubber with wire core and thermoplastic versions provides good dimensional control. In all cases, the flexible mandrel is removed from the hose with either hydrostatic pressure or mechanical push/pull after processing. The mandrel is then inspected for dimensional and cosmetic imperfections, rejoined into a continuous length and recycled into the hose making process. Although the flexible mandrel is continuous, limitations of expulsion from the finished hose rarely allow hose lengths above 1000 ft. (300 m). Either textile or wire reinforcements may be used. Examples of this style product are power steering, hydraulic, wire braided and air conditioning hoses.

25.2.3 Rigid mandrel style of manufacturing hose

In larger hose sizes, where flexible mandrels become quite cumbersome to handle, working pressures are high, or stringent dimensional control is required, the rigid mandrel process is the preferred technique.

The rigid mandrels are normally aluminium or steel. For specialty applications where cleanliness is a necessity, stainless steel mandrels are used. Because of weight considerations the mandrels are usually hollow. Mandrel lengths vary from 10 ft. to 400 ft. (3 m to 120 m) with 100 ft. to 200 ft. (30 m to 60 m) being the most common. The hose tube may be either extruded on the mandrel, pneumatically pulled onto the mandrel, or wrapped in sheets onto the mandrel. As with the flexible mandrel style, when the hose manufacturing process is complete, the mandrel is removed and prepared for recycling. Manufacturing with rigid mandrels offers two unique production opportunities. Rigid mandrels can be: (i) rotated on a stationary horizontal axis, similar to a lathe, so that material can be applied in bias style or (ii) fed horizontally through the tubing, reinforcing and covering operations as the various hose components are spirally fed onto the mandrel.

The former method is often referred to as hand built hose. The reference of wrapped ply hose can be associated with either method. Some hand built hoses, depending on the application, have special ends to accommodate its attachment to existing flanges in the field.

One traditional method of making wrapped ply hose is on a three roll builder. This machine consists of three long steel rolls, two of which are in a fixed parallel position in the same horizontal plane. The third or top roll is on pivotal mounts so that it can be raised or lowered. A mandrel supported hose tube is placed on the trough between the two bottom rolls. Then the top roll is rotated down with sufficient pressure to cause the mandrel and tube to rotate. This enables the reinforcement and cover to be bias wrapped over the tube in uniform fashion.

Yarns: Yarns are used in hose for reinforcement of the tube material to provide the strength to achieve the desired resistance to internal pressure or to provide resistance to collapse, or both. The basic yarn properties required for hose reinforcement are: adequate strength, acceptable heat resistance, dynamic fatigue resistance and satisfactory processability for the various methods of reinforcing hose. Other special properties such as stiffness, adhesion, conductivity, etc., may be developed depending upon the specific hose application. Yarn is available in two basic forms: staple (sometimes referred to as spun yarn) and filament.

Wire: Reinforcing wire is used in a wide variety of hydraulic and industrial hose, primarily where textiles alone do not satisfy the special engineering requirements or the service conditions for which the hose is designed.

25.3   Type of hoses

25.3.1   Molded hose

In the manufacture of molded hose, as the name implies, the hose is formed during the vulcanisation stage by being molded against a lead sheath. The lining (tube) is normally extruded in lenghts of upto 500 metre, depending on bore size. The extrusion may be carried out on conventional equipment, either cold or hot fed or a combination of the two and in either strip or pellet form. To meet the subsequent processing requirements, the compound has to be fairly hard and firm in the green uncured state.

After cooling and maturing, the linings are reinforced by either braiding or spiralling the textile around them. The braid is mainly of cotton for the general purpose hose (air, water, garden) or it is of rayon. Alternate left and right-hand laid textile reinforcement is also used.

A braider is a machine which applies the textile yarn to the hose lining in a manner often called maypole machines and in this instance are usually installed in a vertical position.

The outer-cover coat of rubber is next applied to the reinforced hose by means of a cross-head extruder. After covering, the hose is passed through either a lead press or a lead extruder. After lead covering and winding onto drums, the lining is filled with water (in some instances, air is used also), pressure is applied and the ends of the hose and lead are clamped. The drum and contents are placed in large pans for vulcanising.

The water inside the hose, in fact, expands and becomes super-heated and the hose is pressed against the lead, which acts as a mold (hence the name of the process). If the inside of the lead is fluted, then such a finish is imparted to the hose. If smooth dies are used to form the lead, then, of course, a smooth finish is produced.

After vulcanisation and cooling, the clamps are cut from the hose and then the lead is removed by slitting along its length in a stripping machine. The cured hose is coiled up, tested and inspected and the scrap lead is returned to the melting pot for reuse. Molded hose is used for comparatively low-pressure applications and is manufactured in lengths of upto 500 m, depending on size, of which the range is inclusive to 38 mm bore.

25.3.2   Hydraulic hose

Basically, hydraulic hoses, depending on pressure ratings, are reinforced with either textile (usually rayon) or steel wire, applied by either braiding or spiralling techniques, in this process, the braiders and spiralling machines are usually mounted horizontally.

The process consists basically of extruding the lining tube (which must be hard and firm uncured and must also have good compression-set characteristics when cured), blowing it onto a steel mandrel with compressed air and then braiding and/or spiralling with reinforcement. Covering, usually with chloroprene rubber (CR) and cloth wrapping is then carried out. After vulcanisation, the finished hose is stripped from the mandrel.

In this method, the production length of the hose is limited to 40 m by the steel mandrel length. The introduction of the flexible mandrel process has made lengths upto and even longer than, 300 m possible, depending on bore size. The limiting factor in this case becomes the handling weights and the wire capacity of the braiding machine carriers and bobbins.

Vulcanisation under lead is often used instead of cloth wrapping, particularly in the flexible-mandrel long-length process. In this process, the steel mandrels are replaced by long lengths of flexible cores of suitable polymeric material, on which the lining is applied. Braiding is carried out conventionally and then the cover is extruded into position. After the tubing has stood for some time for maturing purposes, the lead is applied through either a lead ram press or a lead extruder, the whole is vulcanised and then the lead and mandrel are removed from the hose. In modern hose plants, the curing may be carried out in steam pans, continuous cateneries, fluidised beds, or salt baths, or by high-frequency methods.

Flexible hydraulic hoses are used in almost every industry, the main ones being aircraft, automobiles, earth moving, materials handling and mining.

In the majority of hydraulic hoses, the linings consist of either NBR- or CR-based compounds, which are suitable for most of the hydraulic fluids in common everyday use. Because of the extremely low temperatures encountered, hoses made for outside use in some latitudes are specially compounded for temperatures below the normal specification limits of –35°C to –40°C. Special liquids, such as the phosphate ester type, require other materials, e.g., IIR or EPM, to be used.

Over the years, service requirements and conditions have increased in severity, particularly for burst pressures and for impulse performance. As a general guide, the universally recognised operating factor for burst pressure to working pressure is 3:1 for static conditions and 4:1 for pulsating conditions.

Considerable progress in design has been made over the past few years and wire-reinforced braided hoses capable of well over 5,00,000 impulse cycles are now commonplace, whereas such performance was only possible in the past with hose of all-spiral-wire construction. Under certain conditions, with attention to the operating and safety factors, cycles of well over 1,000,000 impulses can be achieved with braided constructions.

25.3.3 Machine-made wrapped hose

As the name implies, this type of hose is made on machines, consisting of three rollers whose centres are on the corners of an equilateral triangle, the base being formed by two of these rollers.

The process consists of extruding the lining tube in the straight position and after cooling, blowing it onto a circular-section mandrel treated to prevent sticking. The mandrel, the outside diametre of which is the nominal bore size of the hoses, may be made either of steel for the smaller-diametre hoses or of light alloy for the larger sizes to facilitate ease of handling and movement, its length is that of the available three-roll making machine, which can be as much as 40 m. The mandrel, together with its blown-on lining tube, sits nicely on and between the two bottom rollers, the third top roller applying pressure as the fabric reinforcement is applied lengthways in strip form into the revolving rollers. The woven fabric (or duck, as it is sometimes called), previously frictioned on a calender or spread-coated on a spreading machine, is supplied cut to the required width, at a bias angle of 45° to give the hose flexibility.

The cover rubber is now applied lengthways and is pulled into position and consolidated by the rollers, after which a wrapping cloth, usually of nylon, is applied by spiralling it through the rollers. After vulcanisation, the hose is stripped of the nylon cloth wrapping (which is used for future cures) and removed from the mandrel.

A new concept is the development of a machine which laps the lining, reinforcement and cover around the mandrel, the textile fabric is usually straight cut and may be applied at any angle as required. There is, therefore, greater flexibility in angle control than with the normal 45° bias cut and joined fabrics. Furthermore, weftless cords can be laid in position by this machine and if so desired, armouring wires can also be incorporated in one operation and on the run during manufacture. The hose is completed by applying a wrapping cloth in line with the other taping heads and thus all the hose manufacturing stages are completed, ready for cure, in simultaneous operations.

Hoses made by either of these methods are used, for example, for water, compressed air and steam delivery, for welding and shot blasting equipment and for conveying beer, wine and food.

25.3.4 Hand-made hose

Many special large-size hoses required in the oil industry are made by hand on large lathes or making tables which rotate as necessary, each layer of lining, carcass and cover being hand applied and rolled into position. Steelwire helices are incorporated by spiralling them into position when it is required to prevent collapse of the hoses under suction conditions. Oil suction and discharge (OS and D) hoses are currently produced with bore sizes up to a nominal 600 mm (24 inch) and development work is active on even larger sizes to keep pace with the requirements of the oil companies. Furthermore, with the advent of the super oil tankers and the need for loading and discharging on the open sea to avoid the cost of very expensive port and docking facilities, buoy mooring systems have been introduced.

In addition to conventional OS and D hoses used in the system, there is a need for some of the hoses (those between the buoy and the ship) to be capable of flotation. This is achieved by one of the two methods. The first is to place flotation beads around the hose. While this is very effective, it has the disadvantages that assembly is difficult and time consuming and the outfit is prone to damage. The other method is to build into the hose an integral flotation system by the use of expanded rubber, with a closed-cell structure, introduced into the carcass of the hose.

25.3.5 Circular woven hose

The carcass of this type of hose consists of long lengths of fabric woven on circular looms. The warp threads run in the longitudinal or lengthways direction and the weft threads are woven into the warps to produce a seamless circular weave. One type of kerbside petrol-pump hose is produced by weaving around an oil-resistant lining tube, with wire helices included in the weft. The largest volume user of this form of reinforcement is, however, fire hose and briefly, such hoses are made by weaving the jacket (which also serves as the cover), as a separate operation, on vertical looms.

The lining tube is extruded or hand-built from calendered material and either in an uncured or in a semi-cured state (backed with an uncured compound) inserted into the woven jacket. Consolidation and vulcanisation of the hose are completed by admitting live steam to each length, the ends of which have been suitably closed. The fabric is usually treated to prevent mildew, etc.

25.3.6   Radiator hoses

The function of radiator hose in a vehicle is to provide a flexible connection between the engine block and radiator. This is used for the efficient cooling of automobile engines. The hose must permit carrying of water at a high temperature and must be flexible in order to avoid transmission of any distorting loads to the radiator tank and not be too soft as to result in collapse and throttle of water supply. These hoses are shaped hoses and the size of the hose varies for different vehicles such as buses, lorries, trucks, cars, jeeps , tractors, etc.

Raw materials required for radiator hose: (i) nitrile rubber, (ii) zinc oxide, (iii) stearic acid, (iv) SRF black, (v) hard clay, (vi) calcium carbonate, (vii) process oil, (viii) pine tar, (ix) sulphur and (x) vulcanox SP, (xi) pilcure TMT, (xii) pilnox TDQ, (xiii) solvent (xiv) cotton fabric and (xv) miscellaneous chemicals like talc, etc.

Manufacturing process of radiator hose

The radiator hose consists of three basic components, viz., (i) the inner tube, (ii) the reinforcing fibre and (iii) the rubber cover.

The rubber lining is compounded to withstand the service temperature of hot water and cover compound to function effectively under the operating environment. The reinforcing fabric provides strength to withstand external and internal pressures.

Typical formulations are as follows:

The tube and cover compounds are first prepared in the mixing mill and allowed to mature for about 24 hr. The cover compound is sheeted out of the mill. The tube compound is pre-warmed in a mixing mill and fed into the extruder to produce tubing of required cross section.

A solution of the spread compound is prepared in a churning mill and is then applied to the fabric using a spreader. The spread fabric is dried, wound up on rolls and is cut to required size. The extruded tube is cut into specified lengths and blown onto the mandrel of appropriate shape, using compressed air. The required plies of fabric are then applied over the tube followed by the rubber cover. The whole assembly is finally wrapped with a wet cloth tape and placed over mandrels in an autoclave for vulcanisation. After curing, the hoses are removed from the mandrels and the cloth tape wrappings are removed. The finished hoses are inspected for defects and packed for storage and despatch.

25.4   Factors affecting hose service life

All hose has a limited life for a given application. This is true even if the proper hose has been selected for the application, it is used within rated pressures, temperatures and environmental conditions and it is properly inspected and maintained. This is because the elastomers and reinforcement used to construct the hose will break down over time and with use. In addition, there are a number of factors that can adversely affect the service life of a hose. The major ones are discussed below:

1. External abuse: Kinking, bending, high end pull, crushing, abrasion, exceeding the recommended minimum bend radius, exposure to chemicals and other abuse or damage will reduce the service life and performance of hose. This may be the case even though the hose may appear to be undamaged from exterior appearance. Hoses should not be stretched, run over by equipment, or used to hoist, carry or pull objects. Hoses should not be bent beyond recommended minimum bend radius. This could result in kinks which could increase pressure and hose damage that could reduce pressure resistance. Large diametre hoses may require additional support to reduce stretching, kinking and external abuse.

2. System pressures: Never use hose at pressures that exceed its ratings. A system (or device or application) can have varied pressures caused by source, operator action or mechanical components. It is the responsibility of the purchaser or user to accurately determine the maximum system pressure. Steady state pressure can be measured readily by gauges. Surge pressures are difficult to measure and may require the use of electronic pressure sensing devices. 'Hammer effects' refers to sudden blockage or stoppage of the system that causes pressure spikes. This can damage or even cause catastrophic failure of the hose or system.

3. High temperatures: Never use hose at temperatures that exceed its ratings. High temperatures can degrade a hose very quickly, resulting in shortened service life. These are for internal product temperatures and assume external or ambient temperatures do not exceed the recommended working temperature of the fluid. Fluid and environmental temperatures that are high, but within working temperature of hose, still shorten hose life.

4. Low temperatures: Never use hose at lower temperatures than recommended. Doing so could cause the hose to crack or break.

5. Internal abrasion: Applications involving abrasive fluids, particularly where the hose makes one or more bends, will reduce the service life of the hose.

6. Flexing and vibration: Flexing, twisting, vibration or other movement of the hose may shorten service life.

7. Modifications to the hose: Repairing the hose, improperly coupling or re-coupling of the hose, or use of inappropriate fittings and other modifications to the hose will shorten service life and possibly cause immediate failure.

8. Improper installation: Installing hose assemblies in a manner where the hose is subjected to a torqued condition (twisted layline), will reduce the life of the hose significantly.

25.5   Steam hose warning

Steam heat is hotter than boiling water (212°F or 100°C) and increases in temperature as pressure increases. The danger from steam in industrial applications is due to the great heat and pressures involved. Water changes to steam at higher temperatures when under pressure. If the steam escapes, massive quantities of heat are released. This, combined with high pressures, can prove to be dangerous for the operator. Use only steam hoses designed for these applications. A steam hose should never be used to carry pressures or temperatures higher than it is rated to handle, in spite of any safety factor.

When making a selection for this type of application, keep safety in mind. Be sure to select a hose identified as steam hose. There should be a permanent form of branding on the hose and not just on the package. The manufacturer's name, hose type and operating pressure should be readable. If not, don't use the hose. Also, be sure to identify the type of service the steam hose will be required to accomplish. What will the temperature of the steam be? Will the steam be superheated (dry) or saturated (wet)? What environment will this hose be used in? Be sure that you can recognise that spillage or accumulations of corrosive materials can have a detrimental effect on the hose cover.

Make sure the hose is installed properly by using hose couplings designed for steam service. Check the tightness with each use. Installing and using a shut-off valve between the steam source and the hose will maximise service life and operator safety.

Provide operators with adequate clothing which would include rubber boots, gloves, eye protection and full length protective clothing. Do not allow the hose to remain under pressure when not in service. Failure to depressurise and drain the hose when not in service can reduce the usable life of the hose. Continue to monitor hose to ensure it has not deteriorated to the point to where it can no longer provide safe service. Most, if not all steam hoses are date-coded by the manufacturer. It is recommended that assemblies be tagged with a date that it went into service. This information will be helpful in identifying those hoses that should be replaced due to age.

Couplings: Hose couplings are extremely important when steam is being handled. High temperatures and pressures inside steam hose act like a pressure cooker and cause the inside and outside diametres to shrink during use. Couplings must be specifically designed to combat this effect. Only couplings designed for steam hose should be used.

25.6   Chemical hose warning

Do not use chemical hose at pressures or temperatures above those recommended by HBD/Thermoid. All operators must be thoroughly trained in the care and use of these hoses and must, at all times, wear protective clothing and other appropriate safety equipment. A hose or system failure could cause the release of corrosive, flammable or poisonous material. Never allow chemicals to drip on the exterior of the hose or allow the hose to lie in a pool of chemicals since the hose cover may not have the same chemical resistance as the inner tube. If kinking or crushing occurs, immediately subject the assembly to the hydrostatic pressure test and examination. If the hydrostatic test is not an option, immediately replace the assembly. If the reduction of the internal diametre (ID) is greater than 20%, replace the assembly. Extreme care must be taken when flushing out a chemical hose with water or removing clogs. Some chemicals, such as concentrated acids may react with the water. Spattering may occur which could result in serious injury to the eyes or other areas of the body. When flushing the hose, care must be taken so that all chemicals or flushing fluids are disposed of according to EPA recommended guidelines.

25.7   Static electricity warning

Serious bodily injury, death, property damage or other loss, can result from the use of hose in hazardous or explosive atmospheres due to the buildup of static electricity from the movement of conveyed materials through the hose as well as movement or vibration of the hose against the other surfaces. Hose, as well as the entire system or application, used in such atmospheres must be properly grounded or bonded. For this reason, HBD/Thermoid recommends only hose with static wire be used.

Static electricity, as a source of ignition for flammable vapours, gases and dusts, is a hazard common to a wide variety of industries. A static spark can occur when an electrical charge accumulates on the surfaces of two materials that have been brought together and then separated (between two solids, between a solid and a liquid, or between two immiscible liquids, i.e., incapable of mixing). One surface becomes charged positively and the other surface becomes charged negatively. If the materials are not bonded or grounded, they will eventually accumulate a sufficient electrical charge capable of producing a static spark that could ignite flammable vapours, gases and dusts. Some common processes capable of producing a static ignition are as follows:

1. The flow of liquids (for example, petroleum or mixtures of petroleum and water as well as any flammable fluids) through hose, pipes or fine filters.
2. The settling of a solid or an immiscible liquid through a liquid (e.g., rust or water through petroleum).
3. The ejection of particles or droplets from a nozzle (e.g., water washing operations or the initial stages of filling a tank with oil).
4. The vigorous rubbing together and subsequent separation of certain synthetic polymers (e.g., the sliding of a polypropylene rope through PVC gloved hands).

Preventing and/or dissipating static electricity as an ignition source can be accomplished through bonding, grounding or possibly selecting a different non-static conducting material. Bonding is the process of connecting two or more conductive objects together by means of a conductor.

Grounding, or earthing, is the process of connecting one or more conductive objects to the ground.

Certain thermoid hose incorporates a static wire, which if properly coupled can be used to ground the hose assembly. Other parts of the application or equipment may have to be grounded as well. Hose that does not contain a ground wire will nevertheless have to be grounded if used in an explosive or hazardous atmosphere.

In all applications, it is the user's responsibility to ensure the hose assembly and equipment it is used on, is properly grounded to earth.

25.8   Care maintenance and storage of hose

Hose has a limited life and the user must be alert to signs of impending failure, particularly when the conditions of service include high working pressures and/or the conveyance or containment of hazardous materials. The periodic inspection and testing procedures described here provide a schedule of specific measures which constitute a minimum level of user action to detect signs indicating hose deterioration or loss of performance before conditions leading to malfunction or failure are reached.

General instructions are also described for the proper storage of hose to minimise deterioration from exposure to elements or environments which are known to be deleterious to rubber products. Proper storage conditions can enhance and extend substantially the ultimate life of hose products.

25.8.1 General test and inspection procedures for hose

Hose should not be subjected to any form of abuse in service. It should be handled with reasonable care. Hose should not be dragged over sharp or abrasive surfaces unless specifically designed for such service. Care should be taken to protect hose from severe end loads for which the hose or hose assembly were not designed. Hose should be used at or below its rated working pressure, any changes in pressure should be made gradually so as to not subject the hose to excessive surge pressures. Hose should not be kinked or be run over by equipment. In handling large size hose, dollies should be used whenever possible, slings or handling rigs, properly placed, should be used to support heavy hose used in oil suction and discharge service.

An inspection and hydrostatic test should be made at periodic intervals to determine if a hose is suitable for continued service. A visual inspection of the hose should be made for loose covers, kinks, bulges, or soft spots which might indicate broken or displaced reinforcement.

The couplings or fittings should be closely examined and if there is any sign of movement of the hose from the couplings, the hose should be removed from service.

The periodic inspection should include a hydrostatic test for one minute at 150% of the recommended working pressure of the hose. An exception to this would be woven jacketed fire hose.

During the hydrostatic test, the hose should be straight, not coiled or in a kinked position.

Water is the usual test medium and following the test, the hose may be flushed with alcohol to remove traces of moisture. A regular schedule for testing should be followed and inspection records maintained.

25.8.2 Safety warning for hose

Before conducting any pressure tests on hose, provision must be made to ensure the safety of the personnel performing the tests and to prevent any possible damage to property. Only trained personnel using proper tools and procedures should conduct any pressure tests.

1. Air or any other compressible gas must never be used as the test media because of the explosive action of the gas should a failure occur. Such a failure might result in possible damage to property and serious bodily injury.
2. Air should be removed from the hose by bleeding it through an outlet valve while the hose is being filled with the test medium.

3. Hose to be pressure tested must be restrained by placing steel rods or straps close to each end and at approximate 10 foot (3 m) intervals along its length to keep the hose from 'whipping' if failure occurs, the steel rods or straps are to be anchored firmly to the test structure but in such a manner that they do not contact the hose which must be free to move.

4. The outlet end of hose is to be bulwarked so that a blownout fitting will be stopped.

5. Provisions must be made to protect testing personnel from the forces of the pressure media if a failure occurs.

6. Testing personnel must never stand in front of or in back of the ends of a hose being pressure tested.

7. If liquids such as gasoline, oil, solvent, or other hazardous fluids are used as the test fluid, precautions must be taken to protect against fire or other damage should a hose assembly fail and the test liquid be sprayed over the surrounding area.

25.8.3 Storage of hose

Rubber hose products in storage can be affected adversely by temperature, humidity, ozone, sunlight, oils, solvents, corrosive liquids and fumes, insects, rodents and radioactive materials.

The appropriate method for storing hose depends to a great extent on its size (diametre and length), the quantity to be stored and the way in which it is packaged. Hose should not be piled or stacked to such an extent that the weight of the stack creates distortions on the lengths stored at the bottom.

Since hose products vary considerably in size, weight and length, it is not practical to establish definite recommendations on this point. Hose having a very light wall will not support as much load as could a hose having a heavier wall or hose having a wire reinforcement. Hose which is shipped in coils or bales should be stored so that the coils are in a horizontal plane.

Whenever feasible, rubber hose products should be stored in their original shipping containers, especially when such containers are wooden crates or cardboard cartons which provide some protection against the deteriorating effects of oils, solvents and corrosive liquids, shipping containers also afford some protection against ozone and sunlight.

Certain rodents and insects will damage rubber hose products and adequate protection from them should be provided.

Cotton jacketed hose should be protected against fungal growths if the hose is to be stored for prolonged periods in humidity conditions in excess of 70%.

The ideal temperature for the storage of rubber products ranges from 50° to 70°F (10–21°C) with a maximum limit of 100°F (38°C). If stored below

32°F (0°C), some rubber products become stiff and would require warming before being placed in service. Rubber products should not be stored near sources of heat, such as radiators, base heaters, etc., nor should they be stored under conditions of high or low humidity. To avoid the adverse effects of high ozone concentration, rubber hose products should not be stored near electrical equipment that may generate ozone or be stored for any lengthy period in geographical areas of known high ozone concentration.

Hose should not be stored in locations where the ozone level exceeds the National Institute of Occupational Safety and Health's upper limit of 0.10 ppm. Exposure to direct or reflected sunlight – even through windows – should also be avoided. Uncovered hose should not be stored under fluorescent or mercury lamps which generate light waves harmful to rubber. Storage areas should be relatively cool and dark and free of dampness and mildew. Items should be stored on a first-in, first-out basis, since even under the best of conditions, an unusually long shelf life could deteriorate certain rubber products.

25.9   Testing methods for hose

The Rubber Manufacturers Association (RMA) recognises, accepts and recommends the testing methods of the American Society for Testing and Materials (ASTM). Unless otherwise specified, all hose tests are to be conducted in accordance with ASTM Method No. D-380 (latest version). Where an ASTM D-380 test is not available, another test method should be selected and described in detail.

RMA participates with ASTM under the auspices of the American National Standards Institute (ANSI) in Technical Committee 45 (TC45) of The International Organisation for Standardisation (ISO) in developing both hose product and hose test method standards. Many of the hose test method standards published by ISO duplicate or closely parallel those shown in ASTM D-380. Many are unique and in those cases, the RMA may be able to provide the necessary test standard references which may be purchased from the American National Standards Institute (ANSI).

25.9.1   Hydrostatic pressure tests

Hydrostatic pressure tests are classified as follows:

1. Destructive type:
   (a) Burst test.
   (b) Hold test.
2. Non-destructive type:
   (a) Proof pressure test.
   (b) Change in length test (elongation or contraction).

(c) Change in outside diametre or circumference test.

(d) Warp test.

(e) Rise test.

(f) Twist test.

(g) Kink test.

(h) Volumetric expansion test.

Destructive tests

Destructive tests are conducted on short specimens of hose, normally 18 inches (460 mm) to 36 inches (915 mm) in length and as the name implies, the hose is destroyed in the performance of the test.

1. Burst pressure is recorded as the pressure at which actual rupture of a hose occurs.

2. A hold test, when required, is a means of determining whether weakness will develop under a given pressure for a specified period of time.

Non-destructive tests

Non-destructive tests are conducted on a full length of a hose or hose assembly. These tests are for the purpose of eliminating hose with defects which cannot be seen by visual examination or in order to determine certain characteristics of the hose while it is under internal pressure.

1. A proof pressure test is normally applied to hose for a specified period of time. On new hose, the proof pressure is usually 50% of the minimum specified burst except for woven jacket fire hose where the proof pressure is twice the service test pressure marked on the hose (67% of specified minimum burst). Hydrostatic tests performed on fire hose in service should be no higher than the service test pressure referred to above. The regulation of these pressures is extremely important so that no deteriorating stresses will be applied, thus weakening a normal hose.

2. With some type of hose, it is useful to know how a hose will act under pressure. All change in length tests, except when performed on wire braid or wire spiralled hose, are made with original length measurements taken under a pressure of 10 psi (0.069 MPa).

    The specified pressure, which is normally the proof pressure, is applied and immediate measurement of the characteristics desired are taken and recorded. Per cent length change (elongation or contraction) is the difference between the length at 10 psi (0.069 MPa) (except wire braided or wire spiralled) and that at the proof pressure times 100 divided by the length at 10 psi (0.069 MPa). Elongation occurs if the length of the hose under the proof pressure is greater than at a pressure of 10 psi (0.069 MPa). Contraction occurs if the length at the proof pressure is less than at 10 psi (0.069 MPa). In testing wire braided or spiralled hose, the proof pressure is applied and the length recorded. The pressure is then released and at the end of 30 seconds, the length is measured, the measurement obtained is termed the 'original length.'

3. Per cent change in outside diametre or circumference is the difference between the outside diametre or circumference at 10 psi (0.069 MPa) and that obtained under the proof pressure times 100 divided by the outside diametre or circumference at 10 psi (0.069 MPa). Expansion occurs if the measurement at the proof pressure is greater than at 10 psi (0.069 MPa). Contraction occurs if the measurement at the proof pressure is less than at 10 psi (0.069 MPa).

4. Warp is the deviation from a straight line drawn from fitting to fitting, the maximum deviation from this line is warp. First, a measurement is taken at 10 psi (0.069 MPa) and then again at the proof pressure. The difference between the two, in inches, is the warp. Normally, this is a feature measured on woven jacket fire hose only.

5. Rise is a measure of the height a hose rises from the surface of the test table while under pressure. The difference between the rise at 10 psi (0.069 MPa) and at the proof pressure is reported to the nearest 0.25 inch (6.4 mm). Normally, this is a feature measured on woven jacket fire hose only.

6. Twist is a rotation of the free end of the hose while under pressure. A first reading is taken at 10 psi (0.069 MPa) and a second reading at proof pressure. The difference, in degrees, between the 10 psi (0.069 MPa) base and that at the proof pressure is the twist. Twist is reported as right twist (to tighten couplings) or left twist. Standing at the pressure inlet and looking toward the free end of a hose, a clockwise turning is right twist and counter clockwise is left twist.

7. Kink test is a measure of the ability of woven jacket hose to withstand a momentary pressure while the hose is bent back sharply on itself at a point approximately 18 inches (457 mm) from one end. Test is made at pressures ranging from 62% of the proof pressure on sizes 3 inches (76 mm) and 3.5 inches (89 mm) to 87% on sizes under 3 inches (76 mm). This is a test applied to woven jacket fire hose only.

8. Volumetric expansion test is applicable only to specific types of hose, such as hydraulic or power steering hose and is a measure of its volumetric expansion under ranges of internal pressure.

Rubber footwear

26.1   Introduction

Footwear can be defined as garments that are worn on the feet. The main purpose of footwear is protecting one's feet. Of late, footwear has become an important component of fashion accessories. Although, their basic purpose remains that of protection, adornment or defining style statement has become their additional and a significant function. There are many types of footwear- shoes, boots, sandals, slippers, etc. They are further categorised into many more types.

26.2   Shoes and shoe making

Shoes are further divided into many categories such as athletic shoes also known as sneakers, galoshes, high heels, Stiletto heels, kitten heels, lace-up shoes, high-tops, loafers, Mary Janes, platform shoes, school shoes and many others. Shoe making can be considered a traditional handicraft profession. However, now it has been largely taken over by industrial manufacture of footwear. A variety of materials are used for making shoes-leather fabrics, plastic, rubber, fabrics, wood, jute fabrics and metal. However, with the development of modern machines, a pair of shoes can be made in very less time as each step in its manufacturing is generally performed by a separate footwear making machine.

26.2.1   Parts of a shoe

A shoe consists of sole, insole, outsole, midsole, heel and vamp (upper). They are the basic parts of a shoe that are mostly included in all types off shoes. Other parts of a shoe are lining, tongue, quarter, welt and backstay. These parts are included as per the design of the shoes.

Sole: The exterior bottom part of a shoe is the sole.

Insole: The interior bottom of a shoe, which sits directly beneath the foot, is its insole. They can be removable and replaceable too. In some of the shoes, extra insoles are often added for comfort, health or other reasons, such as to control the shape, moisture, or smell of the shoe.

Outsole: It is that layer of the shoe that is in direct contact with the ground. These can be made of various materials like leather, natural or synthetic rubber, etc. Often the heel of the sole is made from rubber for durability and traction and the front is made of leather for style. Special purpose shoes often have refined modifications, for example, athletic cleats have spikes embedded in the outsole to grip the ground, dance shoes have much softer or harder soles.

Midsole: The layer that lies between the outsole and the insole for shock absorption, is the midsole. Some special shoes, like running shoes have other materials for shock absorption, that usually lie beneath the heel where one puts the most pressure down. Materials used for midsoles depend on the shoe manufacturers. Some shoes can be made even without a midsole.

Heel: The rear part at the bottom of a shoe is the heel. It supports the heels of the feet. Heels of a shoe are often made from the same material as the sole of the shoe. It can be high for fashion purpose or for making a person look taller. They are also flat for comfort and practical use.

Vamp or upper: The upper part of a shoe that helps in holding the shoe onto the foot is the vamp or simply called the upper. This part is often embellished or given different styles to make shoes attractive.

26.3   Hand assembled shoe making process

A footwear company has mainly four departments in which a progressive route is followed for producing finished shoes.

These are: clicking or cutting department, closing or machining department, lasting and making department, finishing department and the shoe room.

26.3.1   Clicking or cutting department

In this department cutting of sole is done and upper part of shoe is made. The clicking operative is given skins of leather, mostly cow leather but not restricted to this type of leather. Using metal strip knives, the worker cuts out pieces of various shapes that will take the form of 'uppers'. This operation needs a high level of skill as the expensive leather has to be wasted at the minimum level possible. Leather may also have various defects on the surface such as barbed wire scratches which needs to be avoided, so that they are not used for the uppers.

26.3.2   Closing or machining department

Shoe closing or stiching: Here the component pieces are sewn together by highly skilled machinists so as to produce the completed upper. The work is divided in stages. In early stages, the pieces are sewn together on the flat machine. In the later stages, when the upper is no longer flat and has become three-dimensional, the machine called post machine is used. The sewing surface of the machine is elevated on a post to enable the operative to sew the three dimensional upper. Various edge treatments are also done onto the leather for giving an attractive look to the finished upper. At this stage only, the eyelets are also inserted in order to accommodate the laces in the finished shoes.

26.3.3 Lasting and making department

Shoe lasting: The completed uppers are molded into a shape of foot with the help of a 'Last'. Last is a plastic shape that simulates the foot shape. It is later removed from the finished shoe to be used further in making other shoes. Firstly, an insole to the bottom of the last is attached. It is only a temporary attachment. Sometimes, mostly when welted shoes are manufactured, the insole has a rib attached to its under edge.

The upper is stretched and molded over the last and attached to the insole rib. After the procedure completes, a 'lasted shoe' is obtained. Now, the welt- a strip of leather or plastic- is sewn onto the shoe through the rib. The upper and all the surplus material is trimmed off the seam. The sole is then attached to the welt and both are stitched together. The heel is then attached which completes the 'making' of the shoe.

That was the process for heeled shoes. When a flat shoe is in the making, there are considerably fewer operations. The insoles in this case is flat and when the uppers are 'lasted', they are glued down to the surface of the inner side of the insole.

The part of the upper, that is glued down, is then roughed with a wire brush to take off the smooth finish of the leather. This is done because rough surface absorbs glue to give a stronger bond. The soles are usually cut, finished and prepared as a separate component so that when they are glued to the lasted upper, the result is a complete and finished shoe. Soles can also be pre-molded as a separate component out of various synthetic materials and again glued to the lasted upper to complete the shoe.

26.3.4 Finishing department and the shoe room

Shoe finishing: The finishing of a shoe depends on the material used for making it. If made of leather, the sole edge and heel are trimmed and buffed to give a smooth finish. To give them an attractive finish and to ensure that the edge is waterproof, they are stained, polished and waxed.

The bottom of the sole is often lightly buffed, stained and polished and different types of patterns are marked on the surface to give it a craft finished look. A 'finished shoe' has now been made. For shoe room operation, an internal sock is fitted into shoe which can be of any length- full, half or quarter.

They usually have the manufacturers details or a brand name wherever applicable. Depending on the materials used for the uppers, they are then cleaned, polished and sprayed. Laces and any tags that might have to be attached to the shoes, such as shoe care instructions, are also attached. The shoes, at last, get packaged in boxes.

Compounds for hand-assembled products are normally based on natural rubber because good building tack is essential and excessive shrinkage can cause distortion of shaped component parts. Adhesives for hand-built products are also based on natural rubber. The grades of natural rubber selected depend on the end-use. For good-quality thin upper compounds, where flexing is a factor, a rubber with minimum dirt content is necessary, but lower grade rubber can be used for soles and heels. In molded products synthetic rubbers are used extensively, such as non-staining low-Mooney SBR, oil-extended SBR, IR and BR. Where heat- and oil-resistance properties are required, compounds are based on nitrile rubbers, blends of nitrile rubbers and PVC, or polychloroprene, the last of these being used mainly in hand-assembled products because of its superior tack. Whole tyre reclaim is used extensively in black solings and in some molded black upper compounds. Finely ground vulcanised crumb can be used for their cheapness and to help reduce porosity.

Carbon black is the reinforcing ingredient for black industrial boots, but, to meet the demand for non-black compounds, reinforcing siliceous fillers are used. A high-quality non-black soling will have fine particle size silica as a reinforcing agent and the medium grade solings either aluminium or calcium silicate. Good-quality clays or activated calcium carbonate will give reasonable reinforcement with natural rubber, but to obtain equal reinforcement with SBR and particularly with oil-extended SBR, the addition of some siliceous fillers is required. The main diluents are whiting and cheap clays and for whiteness it is titanium dioxide.

The accelerator system is normally a thiazole with a guanidine, thiuram, or dithiocarbamate as a secondary, depending on the rate of cure required. For white and light-coloured products, an antioxidant must be selected which is non-staining.

Cotton is still the most commonly used fabric in the rubber footwear industry. Blends of cotton and man-made fibres are sometimes used where additional resistance to abrasion is required. The leg and foot lining of an industrial boot is normally a square-woven fabric of 0.30–0.40 kg/m^2 in weight, rubberised by frictioning and topping and with sufficient elongation to enable it to be lasted onto the shape of the boot tree.

On molded boots, knitted fabrics (either plain knit or ribbed) are also used, giving the boot leg a more flexible feel.

The upper material of canvas-topped footwear is prepared from a square-woven face fabric combined to a twill backing fabric by means of a thick rubber solution or compounded latex. For top-quality sports shoes, the combined weight of the fabrics is approximately 0.50–0.60 kg/m^2. For slippers, which are very much a fashion article, a variety of fabrics are used such as cotton, wool, nylon, suede and lurex.

26.4 Types of rubber footwear

Rubber footwear can be: (i) plimsolls, (ii) built-up shoes, (iii) all-rubber shoes, (iv) direct vulcanisation process (DVP) shoes and (v) direct injection process (DIP) shoes.

26.4.1 Plimsolls

Plimsolls are the simplest type of hand-built shoes. The uppers are made of frictioned canvas uppers with roll (calendered) soles, edge foxings, toe caps, toe guards, etc. Such shoes are made on continuous conveyors or belt-type conveyors.

Components and their treatment

The components of plimsolls are as follows:

Canvas upper	Filler
Eyelet reinforcement	Sole
Toepuff	Eyelets
Counter	Lace
Binding	Toe cap
Tongue	Foxing
Insole	Toe guard

Preparation of uppers: Basically canvas uppers are made of two textiles. The upper canvas is of cotton, a cotton-synthetic blend, or a synthetic material weighing 200–250 g/m^2, but heavier.

Weights (e.g., 350–400 g) are used for specific purposes. The linings are in general of lighter weight between 120 and 200 g/cm^2 and can be of cotton or a non-cotton synthetic.

Insole: Insoles are calendered rubber compound pasted onto one side of the textile. The rubber compound can be solid (consisting of crushed textiles) or of cut-out sponge insoles from pre-blown sponge sheets.

Filler: Fillers are made of low cost rubber compounds consisting of crushed textile wastes, wood flour and so on. The compound is calendered.

26.4.2 Built-up shoes

The construction of built-up shoes is more or less the same as for canvas shoes, but the rubber sole is pre-molded. Uppers are of cotton canvas, synthetic canvas, cotton-synthetic blend to impart special properties (e.g., strength, low weight, scuff resistance). Shoes of this type are generally used for jogging and sports (e.g., basketball). For the bonding of sole, side foxing, toe cap, toe guard, etc., in built-up shoes manufacturers apply natural rubber latex as cement by dipping or brush of natural rubber base cement.

26.4.3 All-rubber shoes

Wellington boots and waterproof casual shoes are of common types of all-rubber shoes. For shoes made of rubber, special care must be taken in designing formulations that will have good flexibility and good cut resistance without posing problems in the building-up operation.

Rubber shoes are normally made on a conveyor belt. In preparing the upper components, stockinette-type materials are coated on one side with rubber friction dough on a spreading machine and cut to size. The cut pieces are pasted with pre-calendered black waterproof rubber compound 1.0–1.5 mm thick to form individual components of Wellington boots or waterproof casual shoes. To make the shoes, the different components are assembled manually. Rubber boot components are shown in Fig. 26.1.

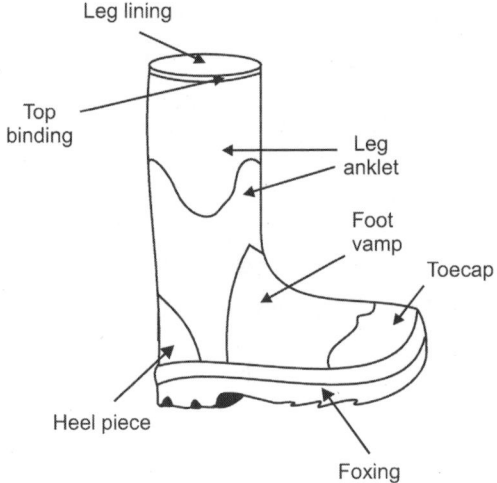

Figure 26.1: Rubber boot components.

26.4.4 DVP shoes

In DVP shoes, either canvas uppers are pre-vulcanised in an autoclave or rubber friction compounds of the self-vulcanising type are used.

The uppers are generally sock lasted, that is, the upper is stitched onto a sock, which is directly lasted on a two-part metal last (normally known as a broken last). A rubber sole is placed on the lasted sock and calendered edge foxing is built-up around the lasted margin.

The mold has three parts: A top piece, which is heated to 170–180°C and two side parts around the last, which are heated to 160°C. The molding operation takes about 2–3 min.

26.4.5 DIP shoes

As the name implies, the DIP shoe is made by direct injection of polyvinyl chloride (PVC) compound onto lasted upper material (Fig. 26.2). In the direct injection process, any conventional upper material can be used (e.g., as leather, PVC, leather cloth, textiles). When textile shoes are being molded with foxing strip, the upper must fit tightly over the last without creases. To avoid having the seam of the sock and the upper obstruct the flow of PVC, it is essential that the sock stitching be: (a) string lasted, (b) slip lasted or (c) reverse lasted (seam inside).

Figure 26.2: DIP shoes.

Before injection, the upper must be pre-heated to around 60°C to make sure that it is perfectly dry, since this is the surface at which the PVC will come into contact with the textile.

Compression molded industrial rubber boots

A complete boot is molded in one operation using vulcanising presses consisting basically of two lasts, two side molds, suitably profiled to give the required pattern and thicknesses and an engraved sole mold. Five components only are required by the molder, a pre-formed fabric lining shaped to the last, a sole blank, a rubber leg, a rubber vamp and a filler. A one-piece rubber leg and vamp is sometimes used. The lining fabric is first coated with rubber compound and then pre-cured to avoid any strike-through of the outer rubber leg during molding under high pressure. Correctly shaped legs, vamps and insole socks, die cut from the pre-cured roll, are machined together to form the boot lining. This lining must have sufficient stretch to enable the molder to fit it onto the last and yet, when on, it must be a reasonably tight fit to avoid pinching during the molding operation. Knitted fabric linings may also be used, suitably coated with rubber compound. The rubber leg and vamp compounds are calendered on a profiled roll and die cut to shape. The filler compound can be calendered or sheeted off a mill and die cut to shape. Correct distribution of the rubber in the mold is essential to give good mold definition. An economical method of preparing the sole blank is by the use of the Barwell precision performer, which can produce a pre-shaped sole and heel unit of accurate weight and dimensions.

The heel portion of the sole being the thickest in the boot, governs the vulcanisation time. Recently, a method has been incorporated whereby the sole and heel can be given double the vulcanising time set for the completed boot, reducing cure times by nearly half. While one boot is being vulcanised, a sole and heel blank is being molded and semi-vulcanised in a second sole mold for the next boot. Overcuring the sole must be avoided, otherwise the bond of sole to the upper could be impaired. Two lasts, normally collapsible or split to ease the fitting of the lining, are provided so that, whilst one last is in the press vulcanising a boot, the other is having the lining and rubber parts assembled onto it. The press is fitted with automatic temperature controls and timing devices.

To obtain efficient economical operation, compounds have to be rapid curing yet with delayed action to give good mold flow and perfect definition of pattern. With sole plate and mold temperatures ranging from 150–155°C, a complete boot can be obtained every 4 min, the sole and heel having had two 4 min cures. Compounds with Mooney viscosity and scorch values at 120°C of 45–50 and 3–4 min (for a 10 points rise), respectively, are satisfactory for such conditions.

The vulcanised leg compound must be very flexible to avoid any discomfort in wear, have good flex cracking resistance, good atmospheric ageing resistance and reasonable tear strength. The vulcanised sole and heel compounds must have adequate abrasion resistance and good cut growth resistance. Moreover, the specific gravity of the compound should be no more than 1.2 to 1.25 to avoid excessive weight in the boot.

Direct molded process for shoe bottoming

As the name implies, direct molded footwear is produced by vulcanising rubber compounds directly to lasted uppers. More shoes are made in this way than by any other method. Molding presses can be classified into two types, for cellular rubber-soled footwear, internal pressure machines which rely on pressure created by the blowing agent in the compound and external pressure machines, either pneumatically or hydraulically operated, for molding solid compounds.

Injection molded rubber sole and heel units

Injection molded rubber sole and heel units are produced on multi-purpose molding machines. These machines consists basically of a rubber injector which consists of a reciprocating screw working within a pre-plasticising cylinder, the screw and cylinder being fitted with independent temperature regulators. The molding stations vary depending upon the machine design, but four, six and ten-station machines are commonly used, the mold stations being carried on a rotary table. At the beginning of each cycle, the screw rotates and simultaneously moves to the back end of the cylinder and the uncured compound in strip or pellet form is fed into the injection unit (it is generally found that pellets are more satisfactory for harder stocks). The rotary action of the screw plasticises the compound and the movement backwards of the screw leaves hot plasticised material in front of the screw. The entire injection unit then advances to form a tight seal between the nozzle and the mold sprue. The screw, acting as a ram, moves forward without rotating, forcing the compound through the nozzle into the closed mold.

Pressure is maintained for several seconds after the mold is filled, to allow the compound to set up. The injection unit then retracts, the rotary table moves the next station in front of the unit and the cycle is repeated. At the same time, the station prior to the injection point opens, allowing the molded units to be removed from the mold cavities.

The number of mold cavities per station depends upon their volume and the maximum shot size of the machine. In addition, the total mold surface area must be such that the upward thrust is not in excess of the closing pressure of the presses themselves, which is in the region of 300–500 Mg, depending upon the machine.

Molds of varying capacity may be used in any one run, since the volume of compound injected into each mold is determined by the back pressure developed in the mold. Effective flash-free moldings are not produced from this type of machine, but by designing the molds for tear trimming, minimum spew waste can be achieved.

Whilst the process for bottoming footwear can be adapted as above, the manufacturing of full-length rubber boots presents difficulties in both engineering and rubber technology. Boots of this type are, however, being injection molded by using PVC, which lends itself readily to this method and has proved to be a very satisfactory material for lightweight wellington boots.

Thermoplastic rubbers can be satisfactorily injection molded and are now used for some casual footwear.

26.4.6 Special purpose footwear

Safety footwear

The need for added protection of the foot in the vulnerable areas has led to the manufacture of safety footwear, incorporating a steel toecap in the toe area and a spring steel midsole between the base of the foot and the outsole. The demand for this type of footwear in industry has increased recently. The inserts are fitted during the assembling of the product, which can be hand built, compression molded, or direct molded.

The steel toecap is shaped to fit over the toe area of the last and is fitted after the boot lining has been lasted. It is essential that the toecap is a tight fit so that no movement occurs during manufacture. The toecap must meet the impact test requirements as laid down in BS 1870. The steel midsole needs to cover as much of the sole area as possible and BS 1870 specifies that the margin between the protective midsole and the edge of the insole must not exceed 6.35 mm. To ensure that no rusting occurs during service, the steel parts are given a chemical anti-rust treatment and enveloped between two thin layers of rubber compound. Depending on the service requirements of the product, safety footwear is manufactured with steel toecaps only, protective midsoles only, or incorporating both toecaps and midsoles.

Conductive and antistatic footwear

Where it is necessary to eliminate the build-up of static charges in the body to avoid sparking, as in factories handling explosives and in hospital theatres, conductive or antistatic footwear is used. Conductive footwear should only be used where there is no danger from electrical equipment. Where potential electrical defects can occur, it is essential to have a lower limit on resistivity to give adequate protection against dangerous electrical shock. This type of footwear is referred to as 'antistatic'. In the manufacture of conductive and antistatic footwear, the upper parts are of standard materials but the base of the footwear is made of conductive materials. The insole, outsole and any fillings required are prepared from compounds possessing the required resistivity. For conductive footwear, BS 3825 specifies an upper resistance limit of $1.5 \times 10^5 \ \Omega$ and for antistatic footwear, BS 2506 specifies a lower limit of $5 \times 10^4 \ \Omega$ and an upper limit of $5 \times 10^7 \ \Omega$.

Cellular rubber

27.1   Introduction

Cellular rubbers or foam rubbers are porous rubber qualities with all-round closed cells. They are manufactured in accordance with the expansion process using gas-developing agents. The cellular structure distinguishes cellular rubber from sponge rubber skin, which contains partly opened cells and sponge rubber, consisting of completely opened cells. Thus cellular rubber in contrast to sponge rubber skin does not require an outer skin in order to use it as a gasket or lining. Cellular rubber is employed in the automotive industry as well as in the space industry, sanitary technique or ventilating and air conditioning fields. In all these fields it is used for sealing, isolating, confinement and shock absorbtion. Particularly, flat sealings can be fabricated at a low cost by splitting and die-cutting. Since cellular rubber is usually filled with soot, it appears black.

27.2   Manufacture of cellular rubber

Rates of polymerisation can range from many minutes to just a few seconds. Fast reacting polymers feature short cycle periods and require the use of machinery to thoroughly mix the reacting agents. Slow polymers may be mixed by hand, but require long periods on mixing, as a result industrial application tends to use machinery to mix products.

   Product processing can range from a variety of techniques including, but not limited to spraying, open pouring and molding.

   1. Material preparation: Liquid and solid material generally arrive on location via rail or truck, once unloaded liquid materials are stored in heated tanks. When producing slabstock typically two or more polymers streams are used.

   2. Mixing: Open pouring, better known as continuous dispensing is used primarily in the formation of rigid, low density foams. Specific amounts of chemicals are mixed into a mixing head, much like an industrial blender. The foam is poured onto a conveyor belt, where it then cures for cutting.

   3. Curing and cutting: After curing on the conveyor belt the foam is then forced through a horizontal band saw. This band saw cuts the pieces in a set size for the application.

4. Further processing: Once cut and cured the slabstock can either be sold or a lamination process can be applied. This process turns the slabstock into a rigid foam board known as boardstock. Boardstock is used for metal roof insulation, oven insulation and many other durable goods.

27.2.1 Physical properties

The main physical properties of foam rubber are generalised as being 'Lightweight, buoyant, cushioning performance, thermal and acoustic insulation, impact dampening and cost reduction'. Cross-linking technology is used in the formation of ethyl vinyl acetate (EVA) based foams, including LLDPE, LDPE, HDPE, PP and TPE. Cross-linking is the most important characteristic in the production of foam rubber to obtain the best possible foam expansion and physical properties.

Cross-linking is defined as chemical bonding between polymer chains and is used for foam rubber manufacturing to stabilise bubble expansion, enhanced resistance to thermal collapse and improve physical properties.

27.3 Sponge rubber

Sponge rubber is a cellular, softly elastic material made of caoutchouc. During vulcanisation the material inflates forming a partly open and partly closed cell structure as well as on its surface, a compact outer membrane.

The specific weight of sponge rubber varies according to its quality from 0.4 to 0.8 g/cm^3. Sponge rubber is produced among others from natural rubber (NR), chloroprene rubber (CR), nitrile-butadiene rubber (NBR) or ethylene-propylene-diene monomer (EPDM). The choice of an elastomer quality should be guided by its use and the requirements of the end product.

Sponge rubber products are manufactured in many different processes either by vulcanisation in molds or by extrusion. Molded parts or profiles of sponge rubber are light and resistant. Today, sponge rubber products or profiles combined with other rubber qualities or materials such as PVC, are used whenever sealing requirements are most important.

27.3.1 Sponge rubber (open cell)

Sponge rubber (open cell) is made by incorporating into the compound a gas-producing chemical such as sodium bicarbonate, which expands the mass during the vulcanisation process. Sponge rubber is manufactured in sheets, molded strips and special shapes. Sheets and parts cut from sheets will usually have a surface impression since sheets are usually molded against a fabric surface which allows air to be vented during the expansion of the sponge. Molded strips will have open cells exposed at the ends of the part unless otherwise specified.

Die-cut parts will have open cells on all cut edges. On parts where open cell surfaces cannot be tolerated this should be so specified.

Trapped air, which may affect the finish, is a universal problem of sponge manufacturing due to the fact that sponge molds are only partially filled with uncured rubber, allowing for expansion to fill the mold. For this reason long and/or complicated cross sections may require vents or multiple splices to effect low reject percentages. To minimise trapped air, it is common practice to use a considerable amount of a chemically inert dusting agent such as talc, mica, or starch, which is difficult to remove completely from the surface of the finished part, although molded closed cell parts prepared by transfer molding need not have this disadvantage. In addition to a normal mold skin surface, some parts are manufactured with an applied solid rubber skin or coating to give a more durable, water-resistant surface when exposed to weathering. This is usually applied by calendering a thin sheet of solid rubber compound (0.005 in.–0.040 in., 0.12–1.0 mm) and applying it to a sheet of sponge compound and placing this in a mold suitably parted to form skin on the exposed surfaces of the part.

Since the solid skin must stretch to cover the surface of the mold during the blowing of the sponge compound, there are practical limitations to designs which can be made by this process, as when skin stretches, the thickness decreases and may ultimately break through. In addition to the above method, an applied skin may be formed by dipping a molded and cured part in latex or cement and depositing a coating on the surface of the part, followed by suitable drying and curing. This coating may be built up to desired thickness by multiple dipping. Limitations on this method are those inherent in most dipping methods such as a tendency to bridge slots or holes, loss of detail of molding and uneven thickness of skin.

27.3.2 Sponge rubber expanded (closed cell)

Closed cell rubbers are made by incorporating gas forming ingredients in the rubber compound, or by subjecting the compound to high pressure gas such as nitrogen. Expanded rubbers are manufactured in sheet, strip, molded and special shapes by molding or extruding. Closed cell sheets are generally made rectangular and of sufficient thickness to be split into several layers for die-cutting.

Closer tolerances can usually be maintained on split sheets (no skin surfaces) than on sheets with a natural skin. Unless otherwise specified, the presence of the skin on the top or bottom surfaces of sheet and strip expanded rubber is optional. Die-cut parts will have exposed cells on all cut edges. On parts where exposed cell surfaces cannot be tolerated (appearance or abrasion, etc.) this should always be so specified.

Extruded closed cell rubber is made by extruding the raw compound through a die which determines the shape of the section. The extruded stock is carried from the die by a conveyor system in a continuous process through vulcanising chambers. As it moves through the vulcanising chambers the heat causes the blowing agent to decompose to produce an inert gas which expands the extrusion. The gas generation takes place in the middle section of the vulcanising process and the cure is completed as the extrusion completes its travel through the remaining chambers. On emerging from the vulcanising chamber the extrusion is cooled to create dimensional stability. Hole punching, coating, dribacking, buffing and cutting are additional operations which can be performed following the cooling. The extrusion can be placed onto reels in continuous lengths or cut to specific lengths depending on the needs of the customer.

Characteristics of extruded closed cell rubber are:

1. The surface of the extruded section has a natural skin formed during vulcanisation.
2. It is possible to produce the part in continuous lengths.
3. A great variety of complex and irregular shapes may be produced.
4. Air chambers or hollowed-out designs may be utilised, giving the advantage of reduction in weight of material.

The design engineer, by properly designing the cross-section with maximum air chamber space, can generally achieve considerable advantage in terms of performance and compression deflection. Molded closed cell parts are manufactured similarly to open cell molded sponge. They require venting of trapped air and possibly the use of inert dusting powder which is difficult to remove completely from the surface of the finished part.

Distinct advantages of closed cell products are their low water absorption characteristics, providing a tight seal and the ability to conform to curves, corners and varying cross-sections without bridging or creasing. This is attributable to the closed cells which do not collapse, losing air as in open cell sponge and yet deform sufficiently to conform tightly to irregular surfaces. Its thermal value is utilised in insulation applications. Design of extruded or molded shapes (uncored or cored) radically affects the compression of parts and leads to greater or less apparent compression set values.

27.3.3 Cellular silicone rubber

Cellular silicone rubber in sheet, molded or extruded forms can be made by processes similar to those for other cellular rubbers, or by foaming a liquid silicone polymer. A post-cure period in a hot air oven is usually used to ensure complete vulcanisation. Because dimensions can undergo some change during this postcure, wider dimensional tolerances must be allowed, particularly on molded items. Chemically blown cellular silicone is almost always produced with a closed cell structure. Cellular silicone rubber foamed from a liquid can be partially or completely open cell.

27.3.4   Sponge-dense sealing products

In recent years manufacturers of cellular sealing products have developed and are supplying a type of seal based on a co-extrusion of dense and sponge rubber. The majority of these types of seals are used in the automotive industry to seal doors, hoods and trunk lids. The major components of this type of seal are cellular compound, dense compound and reinforcing woven wire (or stamped steel) embedded in the dense portion. The continuous curing process usually requires two extruders with the utilisation of hot air, molten salt or fluidised bed curing mediums.

Sponge-dense sealing products are almost always closed cell. These products have many of the characteristics of expanded (closed cell) rubber mentioned in previous section. Due to the unique design and manufacturing methods, separate length tolerances have been developed for these products.

Compression set test

A compression set test has been in use for a long period of time on solid rubber and open cell sponge rubber products - 50% compression of sponge, for 22 hr at 70°C (158°F). The set test is used to determine the quality of those products and their applicability to certain types of usage. Because of this extensive use of the set test on other materials, it is frequently applied to closed cellular materials for the same purposes, namely to determine the quality and applicability of the closed cellular material for general usage or for specific jobs. However, due to the special characteristics of the closed cellular structure, the compression set test has an entirely different effect on closed cellular materials and requires an entirely different interpretation.

27.4   Standards for dimensional tolerances

This section discusses standard tolerances for basic dimensions of sponge and expanded rubber parts. Due to the complexity of design (coring, thick and thin cross-section in each part, etc.), it is recommended that tolerances be established for each part, between the manufacturer and customer, only after studying the clearances and the particular function desired in practical use. It should be noticed that tolerances are plus or minus and are related to the actual or theoretical center of the part. In extruded sections or molded strips, it is a good practice to use 10 times size shadow-graphs with tolerances emanating from a specific centerline. In all discussion of tolerances the high compressability of sponge and expanded rubber parts, as different from solid molded rubber parts, must be taken into consideration as well as the ease of stretching or crowding into sections where design has called for cellular sponge or expanded parts.

27.4.1 Factors affecting tolerances

Shrinkage

All sponge and expanded rubber have some amount of shrinkage after manufacture. The mold designer and rubber compounder must estimate the amount of shrinkage and incorporate this allowance into the mold cavity size, or extrusion die. However, the shrinkage is also a variable in itself and is affected by such things as rubber batch variance, state of vulcanisation, temperature, pressure and other factors. The shrinkage of various compounds varies widely. As a result, even though the mold or die is built to anticipate shrinkage, there remains an inherent variability which must be covered by adequate dimensional tolerance. Complex shapes may also cause irregular shrinkage.

Expanded (closed cell) materials are particularly affected by the gas under pressure (when first manufactured) in the individual cells.

Optimum conditions would be where the internal pressure is finally equal to atmosphere pressure. Manufacturers stabilise their products by prolonged room temperature ageing or by suitable oven conditioning before cutting to dimensions for shipment or fabricating. Since gas is trapped in each closed cell, due consideration should be given to possible changes in dimension resulting from atmospheric temperature and pressure variations.

Mold and die design

Molds and dies can be designed and built to varying degrees of precision. With any type of mold or die, the builder must have some tolerance and therefore each cavity will have some variance from the others. The dimensional tolerances on the part must include allowance for this fact. For molds or dies requiring high precision, the machining and design work must be done accordingly.

High pressure is not required for molded cellular pieces allowing the use of cast aluminium molds. For parts which require close register, greater precision can be obtained by other types of mold construction such as self-registering cavities. Tolerances and quality of finished article, are adversely affected by designs which have undercuts, abrupt changes in volume of cross-section, feather edges and sharp corners. A realistic consideration of tolerances required on the part will usually be more economical and will result in a more satisfactory production job. For extrusion dies the same general factors apply.

Trim and finish

Many different methods are used to remove flash and otherwise complete the finished part. This section is concerned with the effects of finishing methods on dimensions and tolerances.

The objectives of most trimming and finishing operations are to remove the flash, plugs or other rubber material which are not a part of the finished piece. Often this is possible without effecting the important dimensions, but in other instances some material is removed from the piece itself. It is therefore necessary to give consideration to trimming in setting dimensional tolerances.

In expanded products where hot splicing is necessary, there may be irregularity in finish and tolerances due to the temperature of splicing which causes expansion when heating and later contraction of the gas cells on cooling and also due to pressure which could cause some changes in dimensions.

Core dimensions

In molded products, core dimensions are determined by the cores in the mold which in turn form the interior of a hollow article. A core may be suspended individually in a cavity by bars, or pins, or attached to a core bar or other multiple unit. The nature of the part may prevent rigid suspension, causing the pressure of the stock to deflect the core, such as long tubing.

Parts may be deformed or stretched in removal from some types of cores. Realistic tolerances should be established between purchaser and supplier.

On thicker sections of expanded (closed cell) rubber, hollow extrusions should be considered for better control of compression and less material. Alternately, hollow cores of uniform cross-section can be obtained by extruding expanded, closed cell rubber. Floating of the I.D. (inside diameter) with subsequent variation in wall thickness may not adversely affect the overall dimensions and functions of the part.

Rubber insert dimensions

These are the dimensions from a rubber surface to an insert molded to or in the rubber. The accuracy with which these dimensions can be held depends upon the mold construction, method of locating the insert and the tolerances of the insert. Dimensional control is difficult when inserts are for an odd shape which causes difficulty in loading the mold.

The rubber supplier may wish to make slight revisions on an insert to allow use of locating pins, support pins or other devices to prevent inserts from drifting or 'floating'. Insert irregularities such as edges at formed radii, irregular edges from dies, or shearing often prevent good fit in the mold. The supplier's mold engineers can offer information and help on these details.

Other items

There are other items which affect dimensions. The ease of stretching or compressing cellular rubber parts can readily affect the measurements of length and cross-section which in turn affect the tolerances that may be set. In die-cutting closed cell parts over 12.5 mm (0.5 in.) thick, a dish effect occurs on the edges which may affect close tolerances on width and length.

27.4.2   Environmental storage conditions

Temperature

Rubber, like other materials, changes in dimension with changes in temperature. It has a coefficient of expansion which varies with different formulations. Compared to other materials, the coefficient of expansion of rubber is high. To have agreement in the measurement of products that are critical or precise in dimension, it is necessary to specify a temperature at which the parts are to be measured and the time required to stabilise the part at that temperature.

Humidity

Some rubber materials absorb moisture. Hence, the dimensions are affected by the amount of moisture in the product. For those products which have this property, additional tolerance must be provided in the dimensions. The effect may be minimised by stabilising the product in an area of controlled humidity and temperature for a period not less than 24 hr.

27.4.3   Recycling of cellular rubber

Due to the variety in polyurethane chemistries, it is difficult to recycle foam materials using a single method. Reusing slab stock foams for carpet backing is how the majority of recycling is done. This method involves shredding the scrap and bonding the small flakes together to form sheets. Other methods involve breaking the foam down into granules and dispersing them into a polyol blend to be molded into the same part as the original. The recycling process is still ever developing for foam rubber and the future will hopefully unveil new and easier ways for recycling.

Sports goods

28.1   Introduction

Sport goods provide the consumer with satisfaction, entertainment, sociability and achievement.

There is no formal definition of sports. It can be a game, a fitness activity or an organised competitive sport. Competitive sports are governed by a set of rules often designed by national or international sports organisations. Sports can also be a recreational activity. The same sport can be played as a competitive sport or a fitness activity. Sports can be an indoor activity or an outdoor activity. It can be a team activity or an individual activity (for example, squash). While most sports are associated with some form of physical activity, there are some sports, like chess, which do not need strenuous physical activity.

Globally, there are a large number of sports and each can be played in multiple formats. Some sports have received international recognition while others can be country specific, regional and even local, played by a limited number of people. The recognition of sports, its classification and formats may vary across countries.

All sports require some kind of infrastructure, but the requirements vary across different sports. Some sports like shooting and motor racing require sophisticated infrastructure and technology-oriented, expensive equipment while equipment for sports like cricket, badminton and tennis are available at all price ranges and are easier to purchase and store. Overtime, across all sports, equipment has become more sophisticated. In the past, many sports equipment/goods were unbranded, but with increased sophistication and research and development, branding and specialised equipment manufacturers have emerged. Proliferation of brands and specialisation in manufacturing has led to the growth of sports product retailing.

Peoples' passion for sports, their level of participation and their willingness to pay for sports products influences the sports retail market. In each country, certain sports dominate. These sports have huge fan followings and this drives the market for sports products and accessories.

Malaysia is being developed as a golfing destination. In the audiovisual sector, a major portion of revenue comes from the sports sector through dedicated sports channels and sports magazines that advertise and promote sports goods.

On every sports field, court and stadium, fans and athletes alike depend on durable sporting goods to play their game with pride. That's why it's crucial for sporting goods manufacturers, retailers, suppliers and distributors to strive for the highest standards in quality and safety. Winning brands not only comply with local and international regulations, but ensure their product performance stays on par with an ever changing competitive landscape. From design to game time, consumer confidence thrives on knowing which brands champion the methodologies needed to provide the tools of play they trust.

Thus, understanding market segmentation of the sports consumer market is important in order to sell products and services.

Some of the rubber sports goods such as: tennis balls, golf balls and squash balls are discussed in the chapter.

28.2   Tennis balls

A tennis ball consists of a hollow core covered with a cloth usually composed of wool and nylon. Tennis is a four-point game played on a court and which is divided in the middle by a net. For playing this game, a ball is hit by players (two or four) on both sides in such a way that it goes to others side without touching the net and falls in the specified marked area of the court. This ball is known as tennis ball (Fig. 28.1). These balls are made of vulcanised rubber and covered with melton cloth, which has high wool content or needle cloth, which is cheaper to produce and can have a greater content of synthetic fibres. It is available in white or optic yellow. Tennis is an internationally popular game. It is played in many countries of world. It is becoming in very popular in India also as amount of Award is huge in most of the events.

Figure 28.1: Tennis balls.

28.2.1   Process of manufacture of tennis balls

The manufacturing of tennis balls involves various processes. First of all a solution is prepared by thoroughly masticating (kneading) rubber with variety of powders to give the required properties, e.g., strength, colour and to enable it to cure and to make it softer to work and to ensure that subsequently the solution will flow correctly. The solution of this rubber compound is prepared with petroleum solvent.

Then a formulation of rubber compound is prepared which contains natural rubber, general purpose furnace' (GPF) black (a reinforcing filler), clay, zinc oxide, sulphur, diphenylguanidine (DPG) (an accelerator for the curative system), cyclohexyl benthiazyl sulphenamide (HBS) (also an accelerator). This compound is then extruded to produce pallets. The pellets are loaded into a hydraulic press which forms them into hemispherical 'half-shells' and partially cures them, typically for 2½ min at 150°C.

The half-shells are removed from the press, joined together in a sheet by the 'flash', which has spread out of the molds during the forming. The flash is removed by a hydraulic press fitted with cutting knives, which remove the half-shells from the sheet. The edge of the half shell is roughened (or buffed) with a grinding wheel to provide a key for the adhesive which is next applied. A vulcanising rubber solution is applied to the edge of the buffed half shell. For inflation of the ball, inflation chemicals like sodium nitrate and ammonium chloride are used which produce nitrogen during the molding process. These cores are then cured in molds.

The core is then buffed to provide a rough surface to act as a key for the solution which is applied next. The cores are then coated with uniform layer of rubber solution. These coated cores are then covered with Melton cloth or Needle cloth. The balls are then placed in a molding press and heated, curing together the rubber solution on the core and that on the back of the cloth. Finally, Tumbling the balls slowly through a steam laden atmosphere causes the cloth to fluff, giving a raised and softer surface and the ridge around the ball also disappears. The balls are then tested and graded brand name is marked.

Quality control and standard

The balls can be produced as per international standards or Indian Standards IS: 2216-19885.

However, following parameters are very important for a tennis ball:

Dimensions: The balls shall be of diameter 63.5 to 66.7 mm.

Mass: The mass of the balls shall be between 56.7 and 58.5 g.

Requirements: The ball shall have a uniform outer surface and shall be white or yellow in colour. If there are any seams, they shall be stitchless.

Bounce: The ball shall have a bounce of not less than 135 cm and not more than 147 cm when dropped from a height of 254 cm upon a 1:3:6 concrete base of 76 mm thickness.

Deformation: The ball shall have a forward deformation of more than 5.6 mm and less than 7.4 mm and return deformation of more than 8.9 mm and less than 10.8 mm at 8.165 kg load. The two deformation figures shall be the averages of three individual readings along three axis of the ball and no two individual readings shall differ by more than 0.8 mm in each case.

28.3   Golf balls

Golf, a game of Scottish origin, is one of the most popular sports in the world. In the United States alone more than 24 million people play golf, including over 8,000 professional players. Golf tournaments around the world are popular with spectators, as well as with players and since the 1960s, they have received wide television coverage. There is now even a cable channel devoted to golf, as well as numerous computer games. Figure 28.2 shows golf balls.

Figure 28.2: Golf balls.

The basic game involves using a variety of clubs to drive a small ball into a succession of either nine or 18 holes, over a course designed to present obstacles, in as few strokes as possible. A player is permitted to carry a selection of up to 14 clubs of varying shapes, sizes and lengths. The standard golf ball used in the United States is a minimum of 1.68 in (4.26 cm) in diameter; the British ball is slightly smaller.

28.3.1   Conventional wound golf balls

A 'wound' golf ball consists of four components:

1. The center.
2. The windings of rubber thread.
3. The cover.
4. The paint and markings.

Each component has an important contribution to make to the properties of the finished ball and these will be considered in turn.

Center

The center must be the nucleus for the winding operation. It must be large enough for this purpose, but not so large that the resilience of the core is adversely affected. It must contribute a significant amount of mass to the ball because the overall ball weight and rotational inertia are largely dictated by those of the center.

Windings

Rubber thread used in the winding of a golf ball must produce the highest possible resilience in the ball core. The thread, normally based on natural rubber or a blend of natural rubber with *cis*-polyisoprene, is made by dry rubber or latex techniques on the cut thread principle.

The percentage of *cis*-polyisoprene is limited by the requirement of good chafe resistance necessary for the winding operation.

Cover

The cover of a golf ball provides protection for the core and carries the surface pattern necessary to produce aerodynamic lift. Covers have been traditionally made from natural gutta percha and balata, until recently the only materials having suitable properties—in particular, moldability at 80–100°C, which is necessary to avoid degradation of the rubber core by heat. Synthetic gutta percha (*trans*-polyisoprene) is now also used for golf ball covers.

In conventional procedure, hemispherical shells are compression molded by putting heated blanks into a cold mold. The shells are then mounted on a core and the assembly is subjected to a compression molding process at 90°C, in which the shells are molded to the core and to each other and the surface pattern is molded in. The molds are cooled by refrigerated water before extraction takes place. The spew lines are ground by rotating frozen balls against a revolving emery disc.

Vulcanisation of the cover is then carried out by exposing the molded balls to carbon bisulphide vapour for several hours at a temperature of 32°C. The mechanism of the process is that of the Geer process, in which the carbon bisulphide reacts with a secondary amine and zinc oxide to form zinc diethyl dithiocarbamate.

Vulcanisation is completed by putting the balls in ovens for 48 hr at a temperature of 32°C.

With other plastic materials, special molding techniques must be used to avoid the problem of thread degradation.

Painting and marking

The balls are carefully graded for weight and compression, before being given a paint pre-treatment. White polyurethane paint is then applied by rotating the balls, held by metal prongs, in the path of a spray gun which oscillates through the arc of a circle and covers the whole ball surface. The name and numerals are stamped on and the ball is finished with a clear polyurethane coat, wrapped and packaged.

28.3.2 Raw materials for manufacturing golf balls

A golf ball is made up of mostly plastic and rubber materials. A two-piece ball consists of a solid rubber core with a durable thermoplastic (ionomer resin) cover. The rubber starts out as a hard block, which must be heated and pressed to form a sphere.

The three-piece ball consists of a smaller solid rubber or liquid-filled center with rubber thread wound around it under tension and an ionomer or balata rubber cover. During the 1970s the interior of the ball improved further, thanks to a material called polybutadiene, a petroleum-based polymer. Though this material produced more bounce it is also too soft. Research at spalding determined that zinc strengthened the material. This reinforced polybutadiene soon became widely used by the rest of the manufacturers.

28.3.3 Manufacturing process of golf balls

Three-piece golf balls are more difficult to make and can require more than 80 different manufacturing steps and 32 inspections, taking up to 30 days to make one ball. Two-piece balls require about half of these steps and can be produced in as little as one day.

Forming the center

The center of the two-piece ball is a molded core. It is a blend of several different ingredients, all of which are chemically reactive to give a rubber type compound. After heat and pressure is applied, a core of about 1.5 inches (3.75 cm) is formed.

Forming the cover and dimples

Injection molding or compression molding is used to form the cover and dimples on a two-piece ball using a two-piece mold. In injection molding, the core is centered within a mold cavity by pins and molten thermoplastic is injected into the dimpled cavity surrounding the core. Heat and pressure cause the cover material to flow to join with the center forming the dimpled shape and size of the finished ball. As the plastic cools and hardens, the pins are retracted and the finished balls are removed.

With compression molding, the cover is first injection molded into two hollow hemispheres. These are positioned around the core, heated and then pressed together, using a mold which fuses the cover to the core and also forms the dimples. Three-piece balls are all compression molded since the hot plastic flowing through would distort and probably cause breaks in the rubber threads.

Polishing, painting and final coating

'Flash' or rough spots and the seam on the molded cover are removed. Two coats of paint are applied to the ball. Each ball sits on two posts, which spins so that the paint is applied uniformly. Spray guns that are automatically controlled are used to apply the paint. Next, the ball is stamped with the logo. The final step is the application of a clear coat for high sheen and scuff resistance.

Drying and packaging

After the paint is applied, the balls are loaded into containers and placed in large dryers. After drying, the balls are ready for packaging in boxes and other containers.

28.3.4   Quality control

In addition to monitoring the manufacturing process using computers and monitors, three-piece balls are X-rayed to make sure the centers are perfectly round. Compression ratings are also used to measure compression-molded, wound golf balls. These ratings have no meaning when applied to two-piece balls, however. Instead, these balls are measured by a coefficiency rating, which is the ratio of initial speed to return speed after the ball has struck a metal plate. This procedure measures the coefficient of restitution.

Mechanical testing is also used to verify that the ball's performance meets the United States Golf Associations (USGA's) standards. Special equipment has been developed and some manufacturers even use wind tunnels to determine wind resistance and lift action. A machine called the True Temper Mechanical Golfer or Iron Byron, modelled after the swing of golf leg-end Byron Nelson, can be fitted for any club and can be set up at various swing speeds. For normal testing, the Iron Byron is configured using a driver, 5 iron and 9 iron.

Another machine called the Ball Launcher provides the capability to propel balls through the air at any velocity, spin rate and launch angle. This has the advantage of using launch conditions typical of a wide cross-section of golfers. Using both types of equipment, performance data associated with the flight of a golf ball can be measured and analysed. These include the apogee angle, carry distance, total distance, roll distance and statistical accuracy area.

The apogee angle indicates the height the trajectory of a ball reaches. It is measured using a camera with a telescopic lens pointing down range in conjunction with a gridded monitor. Carry distance is the distance a golf ball travels in the air and is measured using a grid system with markers in the landing zone. Total distance is the distance a golf ball travels in the air plus the roll distance. Roll distance is the total distance minus the carry distance. The statistical accuracy area (SAA) or dispersion area is used as a measure of a golf ball's accuracy. For a given ball, the SAA value is based on the deviations of the ball's performance in the directions of carry and left/right of the centerline. These deviations are used to calculate an equivalent elliptical landing area.

28.4   Squash balls

Squash ball (Fig. 28.3) is rubber ball used in playing squash. The ball is a round object that is hit and thrown and kicked in games.

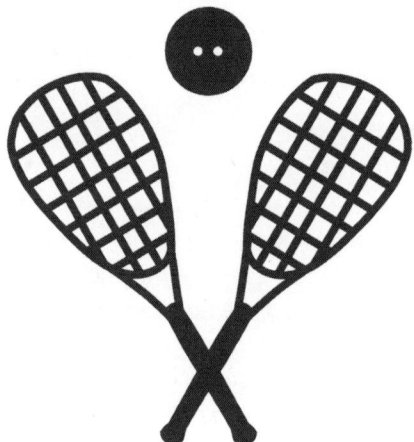

Figure 28.3: Squash balls.

Squash balls are unlike any other sports ball. The ball does not bounce until it is warmed up by hitting it hard 50–100 times. After that the ball reaches an equilibrium temperature and is quite bouncy. Unfortunately, you have to continue to hit the ball hard to keep it warm and lively.

Also make sure you wash your balls after every match, particularly if you notice it slides a lot. This is because it picks up fine dust which makes the surface shiny. A quick wash will restore the surface to its original grippy feel. Always play with the balls till they break.

28.4.1   Manufacturing process of squash balls

In manufacturing squash balls raw rubber is delivered to factory in 'bales' of about 25 kg – sufficient to make about 1200 balls. In its natural state rubber is very stiff and difficult to work, so it is first 'masticated' to a softer consistency. A variety of natural and synthetic materials and powders are then mixed with the rubber to give it the required combination of strength, resilience and colour as well as to enable it to cure (or 'vulcanise') later in the process.

The manufacturer's 'recipe' is, of course, a no less closely guarded secret than that of coca cola and different combinations of ingredients (as many as 15 ingredients are used, including polymers, fillers, vulcanising agents, processing aids and reinforcing materials) produce fast (blue dot), medium (red dot), slow (white dot) and super slow (yellow dot) balls.

The resulting compounds are warmed and loaded into an extruder, which forces them (rather like a mincing machine) through a 'die'. A rotating knife cuts the extruded compound into pellets, which are then cooled. The pellets, which now have a putty-like consistency, are dropped into a hydraulic press which subjects them to a pressure of 1100 lb per square inch and a temperature of 140–160°C for 12 min. The heat causes the material to cure and so retain its shape. Each pellet makes half a ball, known as a 'half shell'. 50% of these are 'plains' and 50% 'dots'. The mold for the dots has a pin in the bottom to create the tiny dimple which takes the different coloured paints that indicate the balls' speed. When the half shells are removed from the press, the excess compound (called 'flash') must be cut away before the dots can be glued to the plains to make complete balls.

First the edges of the half shells are roughened ('buffed') by a grinding wheel to provide a key for the adhesive. The buffed edges are then coated with rubber solution and a measured amount of adhesive is applied in three coats at thirty minute intervals. Both the adhesive and the dot paint are produced in a similar way to the rest of the ball; the adhesive, for example, is also made from raw rubber mixed with various powders before being ground, broken down into a fine web and 'wet mixed' for several hours with a solvent. At last the half shells can be stuck together – an operation called 'flapping'.

The flapped balls are then put through a second molding, heating and vulcanising process, this time subjecting them to 1000 lb per square inch for 15 min, to cure the adhesive. Further buffing, this time of the balls' exterior, smoothes the join and gives the balls their characteristic matt surface. After being washed and dried each ball is inspected. This is one of the few operations which is still carried out by hand, by a team of four persons, the only other manual operations being the loading and unloading of the presses, the final buffing and washing and most importantly, testing.

The balls are tested at every stage in the process and those that are unsatisfactory rejected. Those that pass are stamped with the logo, boxed in dozens and shipped all over the world, but a sample of them is given a final test to ensure that they conform to World Squash Federation (WSF) standards.

Eye protection: Eye-protection is neccesary to protect your eyes from the ball and from the racquet. Wearing eye-guards is mandatory in almost every pro tournament worldwide. Keep in mind that pros have spent their lives developing accurate shots with compact follow throughs. So if these players feel eye-guards are important than you can be certain you should be wearing them too.

Shoes: For beginners it is best to buy shoes with plenty of shock absorption. Thin-soled shoes or shoes with worn out soles should be avoided. Also do not wear running shoes as you are likely to sprain your ankle in them (no lateral stability and high sole increases twisting forces into your ankle should you sprain it).

28.4.2 Specifications of a standard yellow dot squash ball

The following specification is the standard for a yellow dot ball to be used under the rules of squash.

Diameter	(Millimeters)	40.0 + or − 0.5
Weight	(Grams)	24.0 + or − 1.0
Stiffness	(N/mm) @ 23°C	3.2 + or − 0.4
Seam strength	(N/mm)	6.0 minimum
Rebound resilience	From 100 in/254 cm @ 23°C	12% minimum
	@ 45°C	26–33%

Note: The full procedure for testing balls to the above specification is available from the WSF. The readers are requested to obtained these specification from WSF.

Specifications for faster or slower speeds: No specifications are set for faster or slower speeds of ball, which may be used by players of greater or lesser ability or in court conditions which are hotter or colder than those used to determine the yellow dot specification. Where faster speeds of ball are produced they may vary from the diameter and weight in the above specification of a standard yellow dot squash ball. It is recommended that balls bear a permanent colour code or marking to indicate their speed or category of usage.

It is also recommended that balls for beginners and improvers conform generally to the rebound resilience given below:

Beginner	Rebound resilience @ 23°C	Not less than 17%
	Rebound resilience @ 45°C	36% to 38%
Improver	Rebound resilience @ 23°C	Not less than 15%
	Rebound resilience @ 45°C	33% to 36%

Specifications for balls currently fulfilling these requirements can be obtained from the WSF on request.

The speed of balls may also be indicated as follows:

Super slow   –   Yellow dot

Slow   –   White dot or green dot

Medium   –   Red dot

Fast   –   Blue dot

Yellow dot balls which are used at world championships or at similar standards of play must meet the above specifications with players of the identified standard to determine the suitability of the nominated ball for championship usage. The slowest speed of balls intended for elite players and championship usage may if required be identified by a double yellow dot. Such balls will be deemed for the purposes of this specification to be yellow dot squash balls.

Anti vibration mounts

29.1   Introduction

Vibration is a magnitude (force, displacement, or acceleration) that oscillates about a mean reference. The maximum displacement from the mean is called the 'amplitude' of the vibration. The interval of time within which the motion sequence repeats itself is called a cycle or period and the number of cycles executed in a unit of time (e.g., second or minute) is called 'frequency,' Any kind of vibration is commonly characterised by 'frequency,' in cycles per second or hertz (Hz) and 'amplitude.'

Shock is defined as a motion in which there is a sharp, sudden change in velocity. Shock usually entails a single impulse of energy of short duration and large acceleration.

Vibrations can be isolated but shocks need damping; hence the two are treated separately. Many applications involve both vibration and shock in varying proportions, however and here a judicious blend of the two principles becomes necessary.

Damping is the difference between deforming work and elastic recovery and is represented by the hysteresis loop. In many a vibration reducing system it is necessary to tackle isolation and damping at the same time and to find a real manifestation of the viscoelastic property of rubber.

Rubber, in dynamic condition, can be conceived of as a hybrid of steel spring and liquid-filled dashpot—the former represents the fully elastic three-dimensional network of vulcanisate having a modulus, dependent on amount but not rate of strain, that is retained for long periods without relaxation and without heat buildup and the latter represents raw rubber having a modulus dependent on rate but not on amount of strain, which relaxes with time and loses heat. In vulcanised rubber both these characteristics are present in varying proportions, depending on the amount of cross-linking and network formation.

Rubber anti vibration mounts are crucial components in most types of heavy machinery and equipment which help dampen noise levels and vibration frequency while safeguarding fragile components from external vibrations. Large motors and industrial machines generate powerful vibrations and excessive noise. Those vibrations cause damage to equipment and create unsafe working conditions. Eventually this adversely affects product quality and can even bring production to a stop.

Vibration isolation mounts protect the machinery by reducing the amplitude and frequency of the vibrational waves. Acting like a shock absorber, vibration mounts allow motors and machines to operate smoothly and efficiently. Vibration damping mounts are commonly made with an elastomer such as natural rubber which is then bonded to metal.

The task of a rubber technologist is to design a suitable compound and maintain modulus at a level that will satisfy the load–deflection requirement of a given design. If no damping is involved, the best elastomer will be natural rubber, which can be compounded to ensure that optimal elastic properties will be maintained over a long period of service.

There is an increasing demand to reduce noise emitted by plant and machinery. The use of vibration isolators is an important part of reducing noise and vibration from industrial and marine plant and from mechanical equipment located in offices, residential apartments, studios, theatres, auditoriums, schools and universities, hospitals and research laboratories and other critical low noise spaces.

Machinery vibration from rigidly mounted equipment can be transferred to the supporting structure and travel large distances to be emitted as noise elsewhere in a building or structure.

The aim of vibration isolation is to reduce the transfer of vibration to the supporting structure and a correctly designed vibration isolation system can reduce vibrations by more than 95%.

Equipment which is resiliently mounted will have at least one natural frequency – the frequency or frequencies at which it will naturally oscillate. When the disturbing frequency and the natural frequency coincide, vibrations are amplified. The best vibration isolation, therefore, is achieved when the disturbing frequency is significantly higher than the natural frequency.

Most machinery can be effectively isolated provided the vibration isolation system is selected in accordance with the simple design procedures.

29.2   Role of rubber in anti vibration mounts

Rubber is available in a wide range of different types, each with specific properties. The vibration absorbers are normally manufactured in various types of rubbers.

As rubber cannot be compressed, a spring effect can only be achieved if the rubber under pressure is able to expand. The rubber's degree of deflection is determined by its size, shape and resilience. The resilience indicates the rubbers' resistance to changes in its form when under pressure. Rubber can tolerate a great pressure load with a deflection capacity of around 20% of its unloaded height.

Sideways pressure or 'shear' can normally take between 10 and 15% of the pressure load. One should ensure, however, that the rubber is not exposed to tensile load as the rubber's lifetime will then be significantly reduced.

The following description of the design procedure is a useful and detailed tool, but should not be viewed as complete. It should, however, be sufficient for the vast majority of design challenges when:

1. The vibration mounts are significantly softer than the sub-floor and the machine's feet.
2. The force of the vibration goes approximately through the centre of gravity.
3. The distance between the vibration mount is greater than the vertical distance between the level of the vibration mounting and the centre of gravity.
4. The design does not directly aim for a degree of isolation that is lower than 70%.

Very low rotating speeds require a very high degree of deflection to achieve a high degree of isolation. However, as a high deflection can lead to positional instability, a common solution is to choose a vibration mounting that is so hard that the resonancy frequency becomes higher than the disturbing frequency.

This gives zero isolation in terms of the fundamental frequency but shock and high frequency vibrations are significantly absorbed – and these are often the most disconcerting.

Supporting medium or base frame: Insofar as the machinery is made up of several units, they should all be mounted on the same supporting medium or base frame. The vibration mountings are then mounted between the frame and the sub-floor. If the various individual units employ different rpm's, the isolation must be selected to correspond to the lowest rpm.

A heavy supporting medium or base, which markedly increases the overall weight of the machine, appears to be stabilising when positioned. If, however, the force of the vibration is great and/or the rpm is low, then it is recommended that an extra mass is added in the following order of size.

Sub-floor: An apparently correct vibration isolation can entirely fail, if the sub-floor lacks sufficient mass or stiffness. If the layer rests directly on the earth, it is permissible to double the mass.

Flexible connections: When a machine is mounted on vibration mountings, it has to be ensured that the vibrations are not transferred via stiff, inflexible connections. Pipe joints need to be fitted with flexible connecting links, (such as rubber hoses or compensators). Cables also need to be flexible and axle connectors should be fitted with flexible couplings.

Shock disturbances: When selecting a vibration isolator for mobile installations, the design also needs to take into account shock disturbances, that is to say, the vibration mountings, in addition to the mass of the machine and the force of the vibrations, need to be able to deal with any shock disturbances. It is recommended to reckon with an additional 50 to 100% to deal with shock.

Placement of the vibration mounts: Ideally the vibration mounts should be placed symmetrically around the centre of gravity and following the same layout. If the load placed on the vibration mountings is unequal, isolators should be selected so that the deflection remains the same at all points. In a purely practical sense, it can be difficult to position the vibration mountings on the same horizontal level as the centre of gravity. If that is the case, the distance between the vibration mountings needs to be greater than the horizontal distance between the vibration mountings and the centre of gravity.

29.3   Manufacturing process of anti vibration mounts

The manufacture of vibration isolators and shock mounts is much simpler than their design and formulation. Mainly it involves molding by compression, transfer, or injecting, depending on the design and the volume of production. A vast majority of these devices are metal-to-rubber bonded products. Some are partly bonded and partly assembled and some are simple assemblies.

In the case of metal bonding, surface preparation of the substrate is of vital importance. Ideally this is done by blasting of chilled angular iron grits, normally preceded by degreasing with chlorinated solvents and followed by application of chemical bonding agents, usually proprietary formulations of polymeric materials, different resins, reinforcing fillers and so on, in a solvent base.

Metallic inserts so prepared must be handled carefully to avoid contamination before taking up for molding alongwith the rubber compound. Take-up should be completed within 24 hr of metal preparation.

Normally such products are manufactured by a molding process, by compression or transfer molding (injection molding in a compression molding press), or injection molding. Injection molding is normally used only in the case of mass-produced articles like shaft seals used in automotive transmissions. Some shaft seals are still molded by compression and some are molded by transfer methods, but the trend has been toward injection molding, particularly when very large quantities are involved.

29.4   Types of mounts of anti vibration mounts

Rubber mounts are not normally effective for very low frequency applications because of the impractically high frequency ratio. In such cases the natural frequency of the isolator must be lowered, by changing either the mass of the isolator (since it is inversely proportional to natural frequency) or its stiffness. In many cases it is impractical to change the mass of the isolator. Thus one changes the other directly proportional quantity (i.e., the stiffness) by, for example, adding fillers or cork powder (which incidentally also improves damping by hysteresis), or by using reinforcing wires or inserts. Such cases always require a special design.

While an elastic composition is recommended for vibration isolation, a certain amount of damping is wished for in some applications, to depress resonance rise or to cause rapid die-out in case of shock. In such cases the same compounding principles used for low frequency applications will apply for particular elastomer. But elastomers themselves are characterised by differing degrees of transmissibility at different ratios of disturbing frequency to natural frequency.

Cylindrical rubber mounts: This vibration mount is used to dampen noise and vibrations from stationary machinery such as ventilators, pumps, electric motors, converters and compressors with frequencies above 1200 rpm. It is constructed to be used in compression, but can handle minor shear forces.

Cylindrical vibration mounts are manufactured from natural rubber, vulcanised to galvanised steel plates with central male or female threads. Available in different rubber compounds for environments with oil or ozone, acid or for the food industry. They are also available in different hardness's where the standard hardness is 45 Sh. The same mount in a harder compound can carry more load.

Machine isolators: These types are manufactured by neoprene rubber, where the rubber is vulcanised onto galvanised steel discs with a central thread.

Machine feet can be used for the absorption of noise and vibrations from any installed machinery. The shape of the feet gives the machine feet a high static deflection and therefore excellent vibration insulation.

Formula for machine mount is given in Table 29.1.

Table 29.1: Formula for machine mount.

Ingredients	*phr*
Natural rubber	100
Zinc oxide	5
Stearic acid	1.5
PBNA	0.8
IPPD	0.8
Black FEF	25.0
Oil	5.0

(Cont'd...)

Ingredients	phr
CBS	1.5
Sulphur	1.5
Physical properties	
Shore hardness	52°
Tensile strength	21.19 MPa
Modulus at 300% E	12.04 MPa
% Elongation	550

Rubber buffers: Rubber buffers or blocks vulcanised to a galvanised steel plate with a central thread. The rubber buffers are manufactured in natural rubber of different densities. These are used to dampen all types of drop and impact shocks from cranes, rail wagons and stopping mechanisms. These buffers also absorb the kinetic energy, minimising shock from the impact. These rubber buffers can also be used in machinery as a limiting element against any extreme oscillations for instance with the shifting of resonance frequency or as part of a flexible guide rail.

These rubber buffers can be fitted as required in either fixed or mobile sections of machinery. If the buffer does not supply sufficient damping, then two buffers can be fitted in series opposite each other so that the necessary damping is obtained.

Polyetherurethane (PUR) buffers:

The polyetherurethane buffers are used in following operations:

1. As end stops for traverse cranes and elevators.
2. Can also be used as machine feet.
3. Because the buffer can be compressed to such an extreme extent it offers really good shock absorption.
4. Is less suited for isolation of dynamic forces from vibrating machines.
5. The maximum compression force should not be exceeded.

Rubber turret mounts: This vibration mount is manufactured by natural rubber. These mounts are also available as hollow versions with lower maximum loadings than the standard solid version.

This vibration mount is used to insulate against noise and vibrations of smaller machinery and installations, such as smaller fans, pumps, electric motors, compressors, convertors and vibration tables. These vibration mounts offer good deflection and therefore good vibration isolation. These range can also be used for passive insulation against vibrations that come from the outside. For example, with radio transmitters, measuring equipment, navigational instruments as well as insulating against vibrations during transport.

Round cone mounts: These shock absorbers offer high elasticity in both radial and axial directions. The horizontal elasticity of the product is very similar to its vertical elasticity, allowing very fine isolation of equipment with rotational vibrations and low speeds.

These mounts are ideal for indoor stationary equipment and can be used for a variety of machines running at low speed such as: (i) compressors, (ii) generators, (iii) air-conditioners, (iv) fans and (v) shakers.

Sandwich mounts: The sandwich mount is made of 3 steel plates that are zinc plated and vulcanised to layers of natural rubber. The vibration mount is made in 3 different hardnesses where the colour of the rubber indicates its hardness. The sandwich vibration mount is used for reducing noise and vibration from large stationary machinery such as fans, pumps and diesel engines. The mount is designed to be used in compression but can also be used in shear.

Captive rubber mounts: Captive mounts are designed with an integral overload stop, which controls the movement of equipment during transit and provides a fully safe arrangement suitable for both mobile and static applications. These are used in diesel engines generator sets, pumps, compressors, vehicles, marine engines, construction plant, marine plant, mobile plant and fans.

Marine mounts: Marine mounts were originally designed for use with marine propulsion engines and are also safe to accommodate thrust.

Marine mounts are used in generator sets, diesel engines, vehicles, pumps and compressors. Also for other very corrosive environments such as offshore agricultural and coastal applications.

Spring mounts: When installing a diesel engine generator set, or other rotating equipment, it is important to reduce vibration and structure borne noise to the building in which the engine is located. When designing an isolation system for a generator set some of the factors to consider include:

1. Mass of equipment.
2. Mass of floor.
3. Stiffness of floor.
4. Vibratory force produced by equipment.
5. Frequency of rotating equipment.
6. Natural frequency of isolators.
7. Natural frequency of floor.

Restrained spring mounts: The range of spring vibration isolators consist of a spring, steel plate and rubber mats. The two side connecting guide threads join the upper cover and base support together into a whole restraining vertical movement. The threads also provide control of horizontal movement.

The rubber mat on the base plate aids in isolating against higher frequencies and also has anti-skid properties. These are used in pumps, cooling towers, chillers and HVAC plant equipment.

Suspending spring mount: The suspending spring mounts consist of a large outer diameter spring and other materials which can isolate vibration and reduce noise. These suspension mounts have a simple structure with an obvious vibration isolation performance making them an ideal component to reducing noise. These are used in small air conditioner units, gas/water pipes and fans.

Rubber rollers

30.1   Introduction

In the broad field of rubber technology, the application of rubber to the surface of metal cores to make rubber-covered rollers, generally referred to as rubber rollers, is a highly specialised area comprised of unique methods, practices and standards using a well balanced combination of the rigidity of the metal core and the elasticity of rubber.

Although rubber rollers may vary in importance, their versatility ensures that they are used almost in all sectors of industry for a wide variety of purposes and on almost all kinds of machines – from tiny rollers for making envelopes and movie film to giant rollers used in paper making. The functions of rubber rollers in various applications are diverse, demanding different properties from the rubber that is used. This, in turn, means that rubber rollers must be made for specific applications, which makes standardisation difficult.

One of the most extensive uses of rubber rollers is in the control of liquids by pressing or squeezing, as exemplified in the manufacture of paper and in the processing of metals, textiles, plastics and leather. Such typical devices as the press roller, the touch roller (paper making), the padding roller (textile) and the transferring roller (printing) all take advantage of the elasticity of rubber.

Rubber rollers are used for many conveying purposes, such as feed rollers and guide rollers of various kinds, as used in the paper (table rollers), steel (sink rollers, deflector rollers, tension rollers), textile (guide rollers, tension rollers, expander rollers) and plastics industries.

The properties of rubber that are important for such applications include elasticity, friction, abrasion resistance and chemical resistance. In addition, resistance to solvents and plasticisers as well as resistance to heat are desired in applications such as laminating, embossing and coating (textile, plastics, paper and printing).

The design of rubber rollers involves consideration of general factors such as the work place, the working conditions, the desired performance, the required durability. The core is designed depending on factors such as the load, the speed and the required precision. The cores may be built of seamless, welded, or cast pipe and may be made from mild steel, cast iron, stainless steel, bronze, aluminium, etc. However, the use of drawn steel pipes and mild steel as the core material is desirable because of the ease with which these materials can be bonded with rubber. For rollers that are to operate under heavy loading, high speeds, or both, it is customary to design the core so that water cooling can be used to carry away the heat of hysteresis that is a generated when the rubber goes through the nip.

In addition to the conventional type of roller body, there are types that have unique designs for specific industrial applications. For example, the suction press roller used in paper making is generally made of bronze or stainless steel and has holes drilled through the surface of the metal core and later, through the rubber covering. A stationary suction box is provided inside the roller to aid in the removal of water by drawing it into the rubber cover.

In another example of a specially designed roller body, internal oil pressure is used to ensure uniform nip pressure across the face of the roller, which permits crown adjustments while the roller is running. These rollers consist of a smooth bore in a cylindrical shell, a stationary centre shaft and accurately machined hydraulic components. In such applications the roller covering manufacturer generally bolts 'dummy heads' onto the shell, thus providing support so that the roller can be covered.

After the core has been made ready, the rubber covering is applied on its surface. It is important to consider what is expected of the rubber covering. The following areas are significant: (i) chemical resistance, (ii) heat resistance, (iii) physical properties, (iv) hardness, (v) thickness of the rubber and (vi) economics.

30.2 Process of rubber roller manufacturing

The process of rubber roller manufacturing consists of the following major steps:

1. Preparation of core and rubber-to-metal bonding.
2. Formation of rubber covering.
3. Curing.
4. Grinding and finishing.

30.2.1 Preparation of core and rubber-to-metal bonding

Excellent adhesion between rubber and core is decisively important to the satisfactory performance of most rubber rollers. Irrespective of the method of bonding, the core must be suitably prepared. The first step of core preparation is to expose the cores to steam at about 150°C for several hours. This is required to remove the anticorrosive materials applied on new cores, which are normally rough machined.

The prepared core is applied with bonding agent as soon as possible, to avoid reoxidation of the surfaces. In the manufacture of rollers, good rubber-to-metal bonding is ensured by two methods: chemical bonding agents and use of an ebonite base layer.

Chemical bonding agents offer short curing times (unlike ebonite) and give high adhesion values even at elevated temperatures (contrary to ebonite bonding, which begins to soften at a relatively low temperature: 70–80°C). Chemical bonding agents are, however, somewhat sensitive to mechanical influences. If rollers, particularly those having a relatively thick rubber cover, are subjected to dynamic stresses, the increasing shear forces may cause the cover to break loose in the course of time.

30.2.2   Forming of rubber covering

The following methods are normally used for covering the metal cores that have been prepared and treated with bonding agent:

1. Plying of calendered sheeting.
2. Plying of extruded sheeting.
3. Covering with extruded profile.
4. Covering with extruded hose.

Rubber cover compound 1.0–1.5 mm thick taken from a calender is passed through cooling drums, then wound with a liner (normally polyethylene sheet). Such thin sheets are preferred because they can be removed from the calender without forming air bubbles. The cover sheet is next applied over the core, mostly using hand rollers, until the built-up diameter is roughly 4–6 mm greater than the desired finished diameter of the rubber roller. During application of the sheet, some solvent wiping is carried out in case the compound does not have sufficient building tack. Sufficient care is taken to ensure that solvent is fully evaporated before plying up. Alternatively, such sheets can be warmed by passing over hot platens or by heating with an infrared bulb for reviving the tack of the compound.

To ensure that the sheeting is rolled on satisfactorily, the pressure roller usually has a somewhat softer covering, in most cases, the pressure is adjusted hydraulically. Finally, the material protruding at the ends of the roller is cut off to measure and if necessary, the side surfaces of the rollers are also covered. To prevent the compound from flowing away at the ends of the roller, blocking disks are normally fixed at the ends of the roller.

30.2.3   Curing

Roller covers are usually open-cured in horizontal autoclaves, the curing media being steam, hot air, a combination of steam and hot air, or water. Except for small rollers that must be made in large numbers, curing in the mold generally is not practical for three reasons: (i) it is uneconomical to prepare molds for the production of small quantities, (ii) the inset of the metal core, which has large heat capacity, makes it very difficult to cure the rubber in a short time and (iii) since the rubber layer is thick, slow curing is desirable.

For the most part, saturated steam is used in directly heatable curing autoclaves because of simplicity of operation. The heat transfer of saturated steam, compared with that of other gaseous media, is the most favourable for the curing process. Steam curing offers the desired uniformity of the temperature with considerably shorter curing time.

Simple hot air curing, however, has the disadvantage of unfavourable heat transfer associated with oxidation of the top layers of the rubber cover.

Water curing has limited applications when a combination of low curing temperatures and high pressure is desired, as in the production of thick ebonite linings. In water curing the wrapping process can be eliminated and there is hardly any premature ageing induced by oxygen. The main disadvantage of this method is that it makes heavy demands on space and time because the low curing temperatures necessitate very long curing times.

The following factors determine vulcanising time and temperature: (i) type of rubber, (ii) thickness of cover, (iii) bonding system employed and (iv) size and condition of the core, particularly whether it is solid or hollow.

The curing time for rubber rollers can be divided into four stages or steps, namely rising time, pre-heating time, curing time and evacuation time.

The thickness of rubber covering is important in general because of the poor heat transfer properties of elastomeric materials, it is particularly important with large undrilled or solid core rollers. The amount of heat that reaches the cover from the inside of the roller is determined by: (i) the amount of heat circulated from the core, (ii) the condition of the interior surface and (iii) the thickness of the shell itself.

The largest influence in curing is exerted by the core. If the rubber cover has been applied on a hollow or tubular core, the roller can be heated both from the cover and from the core and under such conditions it is generally unnecessary to extend the pre-heating time unless the rubber cover is very thick.

30.2.4 Grinding and finishing

The diverse uses of rubber rollers in different industries call for finishes suitable to the production methods and the service conditions in those industries.

Customarily the roller surface is ground in a single operation, using a grinding wheel or power-driven abrasive belt and one size of grit. This finish usually shows grinding and chatter marks.

The surface finish and tolerance of a roller covering must satisfy very strict requirements. A precision grinding machine, as encountered in the machine tool and engineering industries, is suitable, although special cylindrical belt grinding machines have also been developed which grind the entire roller length at once. The factors that affect the grinding operation are the abrasion resistance and hardness of the rubber, alongwith its tear strength and modulus. The size and peripheral speed of the grindstone, its grit size, the grinding depth, the traversing speed and the rpm of the roller are contributing factors, as well.

Surface treatment: As a practical matter, there are three methods by which the surface of a roller can be treated: heat treatment, irradiation (e.g., UV) and chemical treatment (e.g., chlorination).

Surface treatments usually are intended to reduce the adhesion of material that comes into contact with the roller. For example, the UV irradiation of spinning rollers is intended to reduced the yarn's tendency to wind up and the chlorination of printing rollers reduces the absorption of printing ink.

30.2.5 Roller balance

Vibration from unbalance shortens roller covering life, damages bearings and other parts of the machine and has seriously adverse effects on the product. Because of high rpm, all rollers must be balanced to reduce vibration. If the weight of the roller is doubled, the centrifugal force is double, if the velocity is doubled, however, the centrifugal force is quadrupled, thus showing the importance of balancing at high speeds.

Roller balancing is the manipulation of the centre of gravity so that it falls along the axis of revolution.

Static balance

If a roller supported without friction remains at rest wherever the rotation is stopped, all the local irregularies are self-compensating and static balance is said to exist. If a roller is not in static balance, sufficient weight must be added to the 'light' side to bring it into balance. Usually this is accomplished by drilling hole and adding lead before the rubber covering. Static balance is used chiefly for rollers operated at a surface speed of less than 180 m/mm. However, rollers that are balanced for one speed are not necessarily in balance for other speeds.

Dynamic balance

Even if a roller is statically balanced, one end of it may have an eccentric centre of gravity, with the result that correction has been applied to the other end. This change cannot be detected by simple static balance and will become noticeable only when the roller is speeded up. The resulting oscillation, known as dynamic imbalance or couple or moment imbalance, requires investigation to determine whether one end or both requires counter-weighting.

In such a situation each end might require its own particular counter balance, with the sum of the two again producing static balance. The rubber cover of a press roller that is out of balance may develop a flat spot across the face or may separate from the core. Separation may also occur if the top roller is out of balance.

Kinematic balance

When a roller is made from a pipe, the walls vary in thickness along the face of the roller and also around the circumference. When such a roller is statically and dynamically balanced and then speeded up, local conflicts in balance along the face appear, with the heavy spots in the shell being slung farther from the centre. When the shape of the roller is deformed by its own weight irregularities, the resulting condition is called kinematic imbalance. To avoid this problem, the use of a machined core (both inside and outside), balanced at operating speed before covering, is recommended.

30.3　Industrial rollers

Some of the rollers used in various industires are discussed below.

30.3.1　Paper mill rollers

Rubber rollers used in the paper industry must satisfy a great variety of requirements, the specific requirements to be fulfilled depending on the function of the individual roller. The basic principles of operation are essentially the same for both the types of machine used for paper making, namely, the Fourdinier machine and the cylinder machine. In each case a sheet is formed on a travelling wire screen or a cylinder, dewatered under rollers, dried by steam-heated rollers and finished by calender rollers.

30.3.2　Steel mill rollers

Rubber rollers used in steel mills, in continuous electrolyte tin plating machines and in galvanising plants are not crowned but are ground to a uniform diameter throughout. Rubber compounds used for covering rollers that handle metal sheets and strips must be designed to resist the cutting and tearing action of rough, sharp edges. Most steel mill rollers are covered with polychloroprene rubber and some with NR in the hardness range of 45–80 Shore A. In the rugged service of sheet steel rollering mills, oiling rollers made from nitrile rubber are used for lubrication of finished sheet metal.

In a strip steel mill, rubber rollers are used as conveyors and hold-down rollers in acid pickling tanks and as wringer rollers at tank ends, for removing excess pickling solution from the strip. In stainless steel manufacture, rubber roller covering prevents the stainless alloy from touching ordinary steel or iron and eliminates scratching and other damage. To prevent marking, light coloured rubber stock is used for handling stainless steel.

Continuous electrolytic machines use rubber rollers of three types: (i) tension rollers, to pull the sheet through the line, (ii) wringer rollers, to remove water and chemical solution from the plated surface and (iii) guide rollers which act as deflectors in changing the direction of travel of the sheet. The rubber stock must not mark or discolour the tinned surface. Such rollers are made from SBR and CSPE.

In galvanising plants, rubber rollers remove excess water from the surfaces of the iron sheets, after they have cleaned, to hasten drying with hot air.

30.3.3 Printing rollers

The function of printing rollers is to apply ink to a selected and prepared surface at a controlled rate, the quality of printing depending very much on the quality of the roller cover. The specific requirements for the rubber coverings of printing rollers depend on the printing process (letterpress, lithography, offset, photogravure printing, duplication process, etc.). Offset printing uses an inking roller, which transfers ink to the plate and a watering roller, which supplies water to the nonprinting area of the plate, to repel ink.

The factors to be considered for such rollers are: strength, resistance to oils and solvents, resistance to glaze, hardness and dimensional accuracy.

30.3.4 Tanning industry rollers

All tanning industry rollers require nonstaining cover compounds. Degraining, buffing and splitting machine rollers are mostly covered with natural rubber, whereas in fleshing and dehairing machines, rollers with nitrile rubber covering are used. The hardness of splitting and fleshing machine roller covers usually is 45–50 Shore A, while those of the buffing and degraining machines are in the range of 60–70 Shore A. The roller covers used on fleshing machines to separate the flesh from leather are required to satisfy rather exact demands with respect to tensile strength, compression set, tear propagation strength and resistance to swelling.

30.3.5 Rollers for plastic industry

Polyethylene extrusion and lamination

In the making of polyethylene film with a ring die, known as the inflation process the extrudate is always inflated by air that caused to flow into the cylinder. This process is very effective in the production of extremely thin films as well as in the manufacture of bags. Rubber rollers work as tight rollers to retain the air fed from the centre of the ring die and at the same time to pinch the folded tube-shaped film and to feed it onto the winding machine. When the film touches the roller, it is already cooled. The functional requirement of such pinch rollers is excellent elasticity, so that air can be sufficiently retained. Rubber rollers with hardness of roughly 60 Shore A made of natural, chloroprene and nitrile rubbers are frequently used. In the case of thick films, nitrile rubber with a superior abrasion and heat resistance is employed.

In applications calling for embossing or lamination of the extruded sheet, the T-die process is used. Here, fused polyethylene flowing from the T-die is received first by the cooling roller and is embossed or laminated with the web in a half-fused condition. Hence the rubber rollers should have tear resistance and non-tackiness.

Since normally polyethylene film is made wider than paper in the paper lamination process and the excessive thick polyethylene film is cut off at the edges, the fused polyethylene is likely to stick to the surface of the rubber rollers. Under such conditions, silicone rubber rollers are preferred. When the working width of the rubber roller is restricted to slightly smaller than the width of the paper, the polyethylene of both edges will not stick to the roller surface and the rubber rollers made of chloroprene or nitrile rubber may be employed.

Polyethylene printing

The polyethylene film surface is oxidised by corona discharge to overcome defects due to nonpolarity and the adhesive property and printing ability of the material are thus improved. In the corona discharge process, rubber rollers work as the negative electrode and CSPE covering (2.5–5.0 mm thick) is used because it has superior resistance to high voltage and to ozone. Careful attention must be given to the selection and mixing of the compounding ingredients, since poor dispersion causes a short discharge, resulting in damage to the cover.

PVC processing

The rubber rollers employed of the calendering of polyvinyl chloride (PVC) consist of touch rollers (lamination rollers), embossing rollers, sand rollers and so on, where properties such as heat resistance, plasticiser resistance and abrasion resistance are required. Butyl and EPDM rubbers are commonly employed as the covering material for touch rollers with hardness to 65–80 Shore A and embossing rollers with hardness of 55–85 Shore A.

Sand rollers are embossing rollers used for matting the backside of the film and sheet, to prevent them from blocking. Sand rollers are prepared by compounding a large amount of emery sand in SBR or EPDM, with special attention to dispersion and the prevention of splitting off of the emery particles.

30.3.6 Textile mill rollers

Rubber rollers are used in the textile industry for different purposes to withstand widely different conditions involved in desizing, scouring, dyeing, mercerising and finishing. The media used in the textile processing include not only water and acidic and alkaline solutions but also oxidising agents, softeners, resins and such organic solvents as gasoline, chlorinated hydrocarbons and aromatics.

Coverings with a natural rubber base are widely used in the removal of water-soluble sizes and for rollers needed for some bleaching processes (using cold hypochlorite, chlorite and peroxide liquors), mercerising, dyeing (with any dyestuffs that do not attack NR and finishing). Roller covers based on nitrile rubber are particularly suitable for use in desizing with fet dissolving soaps, mercerising (e.g., in neutralisation compartments), special dyeing processes, e.g., pigment padding, finishing and some types of special finishing (e.g., flame proofing). The rubber covering is made in the range of 65–80 Shore A. Guide rollers of jiggers are usually covered with ebonite, while the pair of squeeze rollers pressing on the dyestuff (rinsing agents, etc.), must be covered with soft rubber.

30.3.7 Rollers for food industry

Rollers covered with blends of natural and polybutadiene rubbers, nitrile rubber and carboxylated nitrile rubber are used in machines for the dehusking of rice. The covers of these rollers must be quite hard (~ 90 Shore A), yet show sufficient elasticity and low compression set, as well as high abrasion resistance.

Dairies used heated dying cylinders with nitrile or silicone rubber covers for the evaporation of water in the production of powdered milk. Rollers covered with nitrile and silicone rubber are also used in bread, cake and cookie factories – for example, in the application of fat-containing cookie or cracker, fillings to the products and in pressing the baked top layer onto the filling. In this operation, the flavour of the products cannot be influenced by the rollers.

30.3.8 Other applications

In the coating industry rubber rollers are used either in spreaders using a doctor knife or in roller coaters, solvent resistance and non-stickiness are important requirements of such rollers. Another interesting application, employed in various textile processing industries, is as expander rollers for controllering the tension across the width of textiles. In the plywood industry rubber rollers play an important function in wide belt sanders.

The end users of the rubber rollers are located in an extremely wide range of industries and information from all raw materials suppliers, including the manufacturers of synthetic rubber and compounding ingredients, together with interaction with the end users of rubber rollers, are valuable to the rubber roller manufacturer for the necessary future development of this important product.

30.4 Maintenance of rubber rollers

Rollers regardless of their function must be properly cared for to assure optimum performance of the entire roller assembly. Optimum performance yields high quality finished product and better utilisation of equipment which translates to reduced downtime and maintenance costs along with higher product output.

Careless handling and haphazard maintenance can be very expensive to a roller user in time and profits. A scheduled maintenance programme is an efficient and cost effective way to eliminate potential problems before they become actual issues. Breakdowns are always costly and disruptive to scheduling.

Storage: Rollers should be stored in a cool, dark area where they won't be exposed to direct sunlight or large temperature or humidity variations from normal conditions. With rubber rollers, sunlight causes oxidation which makes the covering age prematurely. It hardens, shrinks in size and begins to crack. When rollers remain out of service for long periods of time, they should be refurbished by grinding off 1 mm (1/32") from the cover thickness, if dimensionally possible. Reducing the roller diameter by 2 mm (1/16"), will usually remove the oxidised layer.

Whenever a roller is not in use, it should be supported on its shafts. This rule applies whether the roller is cradled vertically or positioned in a stand horizontally. Particular efforts should be made to protect the bearing surface or other specially machined areas (i.e., threads, keyways, internal bearing fits or seal surfaces). For example, a burr or displacement of metal on the shaft will not allow the precision bearing to be mounted on the roller.

Another acceptable way to support a roller in a vertical position is to set the end of the roller on a rubber mat and lean the roller, at an angle, against a wire screen or cyclone fence type material. The roller should be positioned so the rubber and specially machined portion of the shaft will not be damaged.

The storage area should be other than a work or high traffic area to avoid accidental damage. Where possible, rollers should be kept in their shipping boxes. For additional protection, the roller covering should be kept in the wrapping supplied by the manufacturer.

This covering should be a waxed polyethylene coated paper or a similar type of material. Areas of ozone concentration, such as areas with motor generators or other electrical arc producing machinery should not be used for storage. Ozone is a gas that attacks rubber, causing deterioration of the surface.

With urethane coverings, care should be taken not to store rollers in high temperature or highly humid surroundings. Under these conditions, urethane could revert. Reversion is a chemical reaction which causes the elastomer to go back to a liquid state. The process varies in degree depending on the urethane formulation and the ambience. The damage will be the same in both cases. Rollers should not be laid flat. If so, the elastomeric surface will develop a flat spot or, more technically, the covering will take a 'set.'

Surface protection: Elastomeric coverings may deteriorate if oil, grease, kerosene, solvents, or other chemicals are allowed to remain on the surface. Similarly, the bond between the core and covering may be weakened or broken by contact with these materials.

Cleaning solutions should not be used indiscriminately in washing rollers. Elastomeric rollers should always be washed with an accepted commercial roller wash-up solution or solvent recommended by the roller manufacturer. After cleaning, rollers should be placed in a roller rack in such a way that they are exposed to good circulation

Cleaning by hand with abrasive paper is not desirable. Usually the rubber or urethane material is removed unevenly from the surface with resulting flat spots and in turn, operating difficulties.

Precautions should also be taken in handling rollers to avoid damage by bumping or knocking the covering or critically machined surfaces. Many times the critically machined area of the shaft is damaged when we least expect it.

Coating rollers: Choosing a roller compound compatible with the coating solvents is critical to minimise swell in operations that require a uniform laydown of the coating material. Cleanup solvent should also be compatible. The wrong cleaning agent may remove the coating but affect the roller covering, causing extraction of plasticisers, premature roller failure and an increase in durometer or cracking.

If organic solvents are absorbed into the covering, the roller should be rotated periodically and allowed to air dry (preferably 24 hr) before grinding. This will prevent surface irregularities due to uneven or incomplete solvent evaporation.

Maintenance of rubber rollers: Roller coverings should be used only for the service and operating conditions for which they were supplied. If these conditions are to change, the roller manufacturer should be consulted for changes in the application.

Rollers should be removed from service as soon as damage or deterioration is observed. Continuing to use the roller will accelerate the loss of performance and lead to premature failure.

30.5 Production of rubber sheets from roller lines

Roller head lines have been used for many years for the production of sheets for technical rubber articles. Typical products include sheet goods, tank linings, conveyor belts, V-belts, printing sheets, wear protection panels, preliminary pressing products, shoe soles, pneumatic springs and many more.

In contrast to multiple-roller calender lines, in the roller head principle the calender gap is fed with compound from an extruder with a preform head. Due to the preliminary forming of the cross-section in the preform head, rubber sheets up to a thickness range of 20 mm can be produced without bubbles.

Modern roller head lines offer a high level of automation and rationalisation possibilities in comparison to conventionally fed calenders. It should be emphasised in particular that roller head lines can be used for almost every type of compound. With the wide range of compounds available, the selection of the individual line components requires both experience and extensive knowledge of the handling of the relevant compound.

30.5.1 Extruder technology

The cold-feed extruder, sometimes equipped with a vacuum zone for venting of the compound, has established itself as the preferred system for the production of technical rubber articles. Hot-feed extruders in combination with mills are used for compounds which are difficult to plastify, such as compounds based on natural rubber or those with a high material feedback proportion.

The state-of-the-art technology for roller head lines today is the cold-feed pin-type extruder. Pins in the extrusion barrel serve to distribute the flow and thus ensure good distribution and uniform material temperature.

The performance optimum between homogeneity and output can be adjusted by the screw geometry, the pin configuration or the set temperatures of the screw and the barrel. The selection of the screw geometry is based on the type of material to be processed and can be adapted to the customer's requirements.

Cold-feed extruders are also used in medical technology as vacuum extruders in order to remove volatile components from the compound.

When compounds have to be cleaned of impurities or inadequately mixed constituents of the compound prior to forming, this is achieved with the aid of a strainer process, in which the compound is forced through a fine-mesh screen.

The straining can take place at various process stages:

1. Directly after the mixing process with an extruder-gear pump combination.
2. After storage and before extrusion by means of an independent gear extruder.
3. During extrusion and before the preform head.

A cold-feed gear extruder before the extruder enables the greatest process window in terms of output, melt homogeneity and maximum permissible processing temperature. The rubber compound is strained through the volumetric transport in the gear pump under minimum shear load.

30.5.2   Preform head

The optimum feeding of the roller gap is achieved if the thickness and width of the compound fed can be adjusted to the sheet to be calendered. This requirement is fulfilled by the preform head. The distribution of the compound is carried out by the so-called fishtail contour in the preform head. Easily replaceable extrusion dies are mounted at the end of the preform head. The thickness is adjusted with the aid of the different outlet heights of the extrusion dies. The different flow characteristics of the various compounds can be compensated for by profiling of the extrusion dies. Different product widths are achieved by means of easily replaceable insert parts. The cleaning of the preform head is particularly easy, since both halves of the head are swung open when changing compound. By this means, the compound is released from the flow channel when opening the head halves.

The roller head process is characterised particularly by its ability to achieve very accurate thickness tolerances. The influence of the calender gap on the achievable tolerances is particularly great at low sheet thicknesses. In contrast, the influence of the calender gap is reduced at greater thicknesses and the thickness tolerance of the material sheet is determined largely by the material distribution after the extrusion preform head. The production of elastic NR compounds with a thickness of over 10 mm places particular demands on the roller head system. On the one hand, the roller gap has no great influence, while on the other, reset forces in the transverse direction adversely affect the sheet tolerances. In this case, the design of the flow channels is particularly important. The channels in the preform head must be designed so that the reset forces are minimised, although no impermissible stress is placed on the compound. This requires accurate material distribution over the complete width at the head outlet. In order to achieve this, flow channels today are designed with the aid of FEM (finite element method) flow calculations. The particular challenge in this case and the necessary experience, lies in optimising the flow kinetics of the peripheral area. For the optimum design of the flow channel therefore, the rheological properties of the compounds must be known in advance.

30.5.3   Calender technology

The calibration of the material sheet takes place in the calender gap. The number of calender rollers is determined according to the process task. Two or three-roller calenders can be used.

In order to be able to produce a plane parallel sheet with accurate thickness tolerances, the roller head calender is equipped with one or more of the following devices for compensation of the roller deflection:

1. Crowning on one or more calender rollers.
2. Crossing of one calender roller.
3. Roller bending of the calender rollers.

The crowning of the calender rollers and the crossing device form part of the standard equipment of a roller head calender. The roller bending device of the calender rollers is used for high-viscosity rubber compounds, which are formed into thin sheets.

Modern roller head calenders are also equipped with a hydraulic roller adjustment device, which enables accurate positioning of the calender rollers. This also enables the monitoring of the maximum gap force, which ensures reliable protection of the rollers against breakage resulting from overloads. Roller head calenders comply with the latest safety regulations and in addition to various emergency-off circuits, are also equipped with programme controllered monitoring of the brake device for the drive of the calender rollers.

30.5.4   Downstream following equipment

After the forming of the sheet, it must be cooled down to the winding temperature. Cooling drum assemblies, each with two drums, are frequently used in roller head lines. Cooling drums are very effective in the case of low sheet thicknesses. In the case of higher sheet thicknesses, the cooling characteristics are determined largely by the heat conduction in the rubber compound. In this case, the dwell time in the cooling section must be increased. This can be achieved for example by a multilayer air cooling section. This also enables self-threading, which is not possible with drum cooling.

Within the roller head line, it is particularly important that the material is transported carefully. The transfer points between the individual conveyor belts are designed with the aid of dancers and head rollers so that the material sheet is not subjected to tensile stress and thereby deformed. Winding can be carried out using contact winders or centre winders. In combination with a cross-cutter, the cross-cutting, the residual winding, the replacement of the winder and the winding take place fully automatically. The centre winders are equipped with an edge control device for the material and the liner. The liner is also unwound under tensile force control.

30.5.5   Line control and regulation

An essential component of a modern roller head line is the line control with fully automatic process regulation. The optimum design of the user interfaces, both on the PLC screen and on the PC screen, provides ergonomic system operation for the operator. The display language can also be switched to the relevant national language. Due to the selected hardware of the system controls, very high system availability is ensured.

During the project planning and commissioning of the system, the controls are adapted and optimised to the relevant customer requirements, in order to be able to operate the system with short product change times and optimum start-up technology. The graphic process visualisation and the disposition and recipe management represent further essential elements for efficient production. For the evaluation of past production, the production planner and process engineer are provided with online and historical trend displays of the different process parameters, production reporting and evaluation and statistical process control. These process data are available not only on the system itself, but can also be made available to other computers of the customer by networking via internet, enabling either centralised or decentralised data storage.

With the aid of remote maintenance, Troester engineers can monitor every system worldwide, provide assistance and if necessary carry out programme modifications. This ensures quick reaction times following system stoppages and rapid resumption of production.

Maintenance personnel are provided with an online maintenance management system, which draws attention at the relevant time to what maintenance work must be carried out. The system also provides information on which auxiliary materials are needed for this purpose and displays instructions for the performance of the work.

Thus, the standard systems can rarely be used for the production of technical rubber goods, this must be carried out instead with roller head lines specially designed and tailor-made to the relevant customer requirements. All the requirements can be fulfilled thanks to the wide range of extruder designs and sizes available, with screw diameters of up to 250 mm and the corresponding calender equipment with roller diameters from 80–700 mm.

Rubber to metal bonding

31.1   Introduction

Rubber to metal bonding is a generic phrase that covers several interdependent processes used to manufacture a wide variety of automotive and industrial products used to isolate noise and vibration. Rubber to metal bonding has three essential elements—the rubber, the bonding agents and the substrate. The polymer base and the associated compounds that are chosen depends on the product being produced. Assuming the rubber can flow into the mold with less than 2% cross-linking, a bond can be formed using any rubber compound.

While there are few restrictions on the use of any individual compounding ingredients, it is generally a good idea to avoid substances that could bleed rapidly to the surface of the uncured stock. A rubber compounder, like delta rubber, is most often used to match the physical requirements of the cured rubber to the molding process.

Eventually, bonded rubber to metal (R to M) composites came to be employed in the mechanical world to not only minimise the damaging effects of vibration and to protect against shock, but also to seal against leakage of oil and other fluids and to simplify assembly by providing a tolerance for misalignment.

The motor vehicle industry is by far the biggest user of bonded elastomer components. Typical applications such as engine mounts, suspension bushings (bushes), transmission and axle seals, couplings and body mounts are engineered largely to each vehicle's requirements. Manufacturing usually involves molding of the elastomer to shape, vulcanisation and bonding, all in a single-stage press operation.

In the aerospace sector, R to M mounts and bushings fulfill similar and at least as critical functions. Well-engineered assemblies are vital for reliable sealing of the fuel systems in rocketry and missiles. Sophisticated mounts are necessary for aircraft engine isolation. Less esoteric but no less important to aircraft operation are a wide variety of bonded avionics mounts. For helicopters, advanced R to M technology has resulted in such highly specialised devices as low maintenance rotor bearings, embodying alternating layers of rubber and metal in special geometric configurations.

R to M assemblies are further necessary in their aforementioned functions for the nonaeronautical vehicles of war (tanks, submarines, etc.), and for rail and rapid transit rolling stock, as well as for trackbeds and rail crossings.

Bonded parts are essential for materials handling machinery, shipping containers, construction and agricultural equipment, leisure devices, air systems and business machines of various types.

On a more prosaic plane, rubber roll of all sizes with bonded metal cores are used widely for processing rice and other food and in the production of paper and continuous metal sheet stock. Finally a typical but significant specialty area of bonding is that of tank and chute lining (with relatively thin sheets). Rubber-lined vessels or conduits are required for mining and chemical processing operations and also for scrubbers in the utilities industry.

31.2   Bonding layer

For many years the bonding agents have consisted of proprietary polymer/solvent solutions, with a primer coat based on phenolic-style resins and a polymer topcoat. Bond formation is produced via the development of a very high modulus layer in the rubber immediately adjacent to the surface of the substrate.

31.2.1   Selection of bonding agents

The selection of bonding agents depends on the type of rubber to be bonded, the modulus of the rubber and the component design. The selection process is critical to the robustness of the bonding process. With the advent of greater concern for environmental safety, water-based versions of bonding agents have been introduced, which after much development are now effective replacements for the solvent-based bonding agents. Bonds tend to be up to 10% lower but components show good resistance to the usual environmental conditions found in automotive applications.

The application methods for water-based bonding agents are similar to those for the solvent systems, but inserts do need to be pre-heated to 60–80°C before spraying with the primer and reheated before applying the topcoat. Drying times are quick and are no barrier to high volume production.

Choice of substrate

The choice of substrate depends on the strength and durability requirements for the component being produced. The most common substrate has often been steel, but the use of aluminium alloys and polyamides to save weight is increasing dramatically. Almost any material can be bonded to rubber, provided that it can withstand the heat and pressures of the rubber molding process. For practical purposes this eliminates polyolefin plastics. Polyacetal inserts can be bonded but require careful etching and rubber molding temperatures below 150°C. Polytetrafluoroethylene (PTFE) provides a useful low friction material for use in anti-roll bar bushes. It can be bonded successfully to the rubber by chemically etching the surface of the plastic prior to application of the bonding agents. However, its use in such applications has been largely superceded by woven PTFE/Terylenefibre material, which offers a mechanical bond and more recently, by slip agents that bond freely to the surface.

31.3   Bonding process

The process begins by loading a specific volume of inserts into a perforated, motorised stainless steel (SS) drum. The volume is determined by the size of the inserts. The drum is equipped with a motor and timer which tumbles the inserts for a specified time in each subsequent operation of the treatment process. The first operation is to pre-clean in a 400 gallon, heated tank of alkali. This removes all traces of oil, grease and solid lubricants to promote good bonds.

At the end of the pre-clean cycle, the drum automatically rises out of the tank and is then manually moved through two water rinse cycles. The drum of inserts then goes through an acid dip followed by two water rinses to etch the metal, preparing the surface for good attachment when immersed in a heated zinc phosphate solution. The zinc phosphate coats the metal surface providing a good foundation of the application of the adhesive.

After the phosphate dip there are two water rinses and an air dry prior to immersion in either a solvent or water based adhesive. The combination of the type of metal and of rubber determines the selection of adhesive. The coating of the zinc and adhesive not only bonds the rubber to the metal but also imparts rust resistance. Following the adhesive dip the inserts are air dried prior to use. The temperature and concentration of each of the chemical tanks are measured and plotted daily. The per cent solids of the adhesives are also checked daily and recorded.

During molding, a bond pull test using pliers is made three times each shift by the press operator to be sure the bond is such that the rubber fails before the bond is broken. Additionally, samples of the rubber bonded inserts are put through a salt spray chamber according to ASTM B117 daily to check for corrosion resistance.

31.3.1   Rubber molding

The rubber molding operation is the most critical point in the process. If there is a problem with even a single element in the production cycle for the inserts or bonding agent, it is likely that the product will fail. For automotive production injection molding is the most common method employed.

This method provides a significant amount of control over the process by tailoring the condition of the rubber as it enters the cavity to produce consistent product quality.

Molding

The three main types of press continue to be used for bonded assemblies–compression, accounting, transfer molding. Injection molding of course enables faster throughputs, with vulcanisation cycles of 1 or 2 min at $200°$ C^+ being feasible for many synthetic elastomer assemblies. But with NR's continuing dominance for mounts and bushings, milder cure cycles of 4–6 min at $175°C$.

For seal production there continues to be considerable compression molding of NBR assemblies. Injection molding is increasingly being used for ultrahigh performance fluoroelastomer seals.

It should be noted that the mechanics and conditions of molding place special demand on adhesive systems. As molten rubber enters mold cavities in advance of vulcanisation, it may move laterally across adhesive-coated metal surfaces. The tendency of the adhesive to become thermoplastic under these conditions can cause it to be wiped or swept along by the rubber–to the detriment of the final adhesive bond. Some commercially available adhesives are much more resistant to this mold 'sweeping' phenomenon than others.

Also important is the ability of the adhesive to withstand exposure to near vulcanisation temperatures before the coated metal is in actual contact with the vulcanising rubber. Since most adhesives are reactive chemical systems, 'prebake tolerance' is an important practical requirement.

Another mold-related problem that sometimes arises also is attributable to the reactive constituents of adhesive systems. Certain ingredients may volatilise or sublimate from the heated metal surface prior to rubber contact, thus causing fumes at the press and occasionally mold fouling. Postvulcanisation bonding requires manufacturing setups different from normal assembly fabrication. The best quality bonds, for example of vulcanised natural rubber, are achieved by using the same metal primers employed for normal vulcanisation bonding. Appropriate adhesive covercoats are applied either to the primed metal or to the surface of the vulcanised rubber article. The assembly is then jigged under compression and heated in an oven at a temperature high enough to establish strong forces of adhesion–primer to metal, covercoat to primer and covercoat to vulcanised rubber. But for less demanding applications there is also some use of instant-setting cyanoacrylate adhesive and of two-part urethane or epoxy cements that cure at room temperature.

Postvulcanisation bonding is nevertheless a practical means of simplifying the assembly of struts or mounts with two or more bonded interfaces (i.e., as designs become more complex).

Mold design

Molds do need to be designed to ensure exact balance between cavities and the elimination of trapped gases. The presence of gases causes a high incidence of bond failures through the 'diesel effect', whereby elements of the bonding agent film burn under the combined effects of heat and high-pressure gas. Lack of balance between cavities will result in some components that are imperfectly formed and give rise to bonds that may fail.

Problems such as these are readily avoided by study of the rubber flow through the mold using computer flow simulation packages. These model have an effect of mold design on pressure gradients and the cross-linking behaviour of the rubber. Computer analysis allows the optimisation of runner sizes and molding conditions before the mold is manufactured, so that components are produced to specification.

Plated inserts

The use of plated inserts is popular with some designers. Good bonds can be obtained with plated inserts, but some problems can arise in service if the bond edge is subjected to frequent exposure to electrolytes. Salt solution from roads will produce 'battery' cells between the plated metal and the carbon black in the rubber. Up to 0.8 V has been measured under salt spray test conditions. The production of nascent hydrogen at the electrode surface causes a localised de-bonding process, known as cathodic disbondment, that will cause eventual failure.

31.4   Rubber to metal assemblies

Rubber-metal composites for shock and vibration isolation are sometimes referred to as bonded rubber springs. The basic designs of such R to M mounts have not change greatly over the years, particularly for the less demanding applications such as isolation or damping of industrial machinery. There has been substantial progress but more in an evolutionary rather then revolutionary vein. In some instances, thermal degradation of the rubber after years of service (and as aggravated by the softness of the compounds) resulted in catastrophic failure of the mountings such that the engine could break loose from the chassis. Many engine mount designs were then changed to embody a fail-safe cage-type assembly that caused the engine to be retained even in the event of total rubber loss.

To facilitate the manufacture of cage mounts, two-step bonding came to be developed. By this method union was established to a first metal member during normal vulcanisation and to the second metal part by post-vulcanisation bonding. This procedure placed new demands on adhesive systems.

To maintain or upgrade car handling and ride quality, tighter specifications became necessary for such critical properties of the rubber as change in dynamic modulus with temperature. These also began a clear trend to increased numbers of mounts and bushings for vehicle drive trains and suspensions.

The most advanced automobile engine mounts widely used are the so called hydraulic or fluid mounts. These assemblies not only have the traditional bonded rubber springs but also embody the fluid dashpot principle. Confined liquids such as mixtures of ethylene glycol and water are able to move back and forth through an orifice or through baffles. A secondary mechanism of damping thereby imposed results in a more effective control of vibration. Such dual action mounts are being designed not only for motor vehicle engines but also as isolators (e.g., for truck tractor cabs) and now for suspension bushings, as well. Even more sophisticated mounts have been or are being developed that combine electronic feedback with dual-action control and damping. 'Hydromounts' now in production with electromagnetic switches are being further engineered to make use of the electrorheological properties of confined liquids.

In effect, the viscosity of this new type of fluid can be changed by application of an electrical field so that its viscoelastic damping is even more closely attuned to changes in the engine's operation occurring in a fraction of a second. For the aerospace and defence sectors, there have been advancements not only in mechanical design but also in materials. Demands of extremely high operating temperatures have resulted in mounts and isolators with elastomers other than natural rubber (NR) or the early synthetic diene types.

After the broad categories of assemblies for control of vibration or shock, seals probably comprise the next most important class of elastomer-metal composites. Seals are of course used in engines pumps, compressors and axles to prevent leakage of fluids or gases. The terms 'radial seals,' 'shaft seals,' and 'bore seals' describe the main types, which are generally circular with open centers to accommodate rotating shafts. The sealing function necessitates the use of elastomers which maintain resistance to permanent set in service.

Advancements in seals for transmissions or differential housings have been less striking in regard to mechanical design than in regard to rubber usage and adhesive application. The availability of ultrahigh performance specialty elastomers has also made possible sealing elements of new types (e.g., for fuel injection systems and for engine valve stems).

31.4.1   Elastomers of rubber metal bonding

Two unusual classes of synthetic elastomer, both embodying polyether linkages in their main chains, have been tested extensively in mounts and bushings because of attractive combinations of performance characteristics. Copolymers of propylene oxide and allyl glycidyl ether submit good dynamic properties over a wide range of operating temperature along with good resistance to thermal degradation and ozone attack. Other elastomers that have found use in dynamic R to M applications include, more or less respectively, polychloroprene rubber

(CR), styrene–butadiene rubber (SBR) and nitrile rubber (NBR). In many cases the synthetic is blended with NR or less often with another synthetic, for compromises in performance and/or economics. Silicones are used to some extent in vibration control isolators, particularly for aerospace/defence applications. These provide exceptionally constant performance – resonant frequency and transmissibility – over an extremely broad temperature range.

Butyl rubber (IIR) finds use in mounts requiring high hysteresis loss such as certain dampers and isolators (e.g., cradle mounts for front-wheel drive automobiles). It is also used in truck body mounts, in bonded overload pads and in elastic damper hang-straps.

EPDM, like butyl rubber, is important in applications calling for excellent ageing characteristics such as bonded weatherstrips, window channelling, boots, bellows and body mounts–that is, where dynamic properties are less critical. Its capacity to be heavily oil extended with no great sacrifice in elastomeric qualities is advantageous in selected bonded applications.

Other elastomers used in R to M parts, especially where shock control is targeted and ease of fabrication is a consideration, include polyurethanes (particularly the reactive 'castable' sort) and a few of the newer elastomeric thermoplastics (or 'melt-processible' elastomers).

The thermoplastics have been among the more significant developments in specialty rubbers. Important commercially available elastomeric thermoplastics include various formulations of elasticised polyvinyl chloride, urethanes, styrenics, polyesters and a broad range of thermoplastic polyolefin (TPO) types. The advantages of thermoplastics for manufacturing operations are obvious: reuse of scrap or trim waste and absence of requirements for capital–intensive compounding, milling and curing facilities (i.e., the ability to be fabricated into products using plastics processing machinery.

Thermoplastic elastomers are obviously not practical for bonded R to M assemblies that will do service at very high operating temperatures. But they are beginning to receive attention in applications having reactively modest performance demands and adhesives are available to bond most types to metal. Ultrahigh performance vulcanisable elastomers for seals have been successful in meeting the more and more stringent demands of the aerospace (and automotive) engineers. The polyacrylates found application, as did NBR, in automatic transmission seals and in seals for crankshafts, pinions and oil pans. The NBR elastomers, like NR for mounts and bushings, have not been superseded for many motor vehicle seal applications. This remarkable steadfastness is due to the superb oil resistance imparted by the nitrile group. Carboxylated NBRs are not uncommon for seal applications. Offering even greater durability under seal service conditions are the more recently developed hydrogenated nitrile rubbers (HNBR). Ethylene-acrylate ester elastomer is also used in bonded seals because of good temperature and oil resistance, along with reasonable cost.

During past the most demanding seal requirements were being met by new synthetic polymers of very high fluorine content. Break throughs in new methods of curing enabled fluoroelastomers to be offered with very low permanent set for extended service under the worst conditions of high temperature and contact with corrosive gases, fuels, oils and hydraulic fluids.

31.4.2 Metals and other rigid substrates

The rigid members for bonded rubber springs continue to be predominantly steel, particularly for the high volume assemblies of the motor vehicle industry. Castings are used as well as stampings, including a few castings fabricated via the powder metallurgy route. For seals, low carbon cold-rolled steel is often the framework of choice. More and more of these various steel parts are today chemically treated (i.e., phosphatised) to minimise corrosion of the entire metal piece.

Lead is specified occasionally for bonded parts requiring a substantially concentrated center of mass and is sometimes a minor constituent of steel alloys to facilitate forming.

Very little aluminium has been used over the years for automobile R to M assemblies. But there are reports recently of aluminium beginning to be specified for mounts or bushings in Japan because of improvements over steel in dynamic performance.

Treated aluminium alloys are certainly important in aerospace applications (e.g., for avionics mounts), as are stainless steel and even titanium for highly sophisticated applications. There has been considerable interest in recent years in reducing the weight of bonded rubber springs by replacing metals with rigid engineering plastics. For example, glass fibre reinforced thermoplastic polyamides are beginning to be used in a small way for automotive bushings. Manufacture in many cases is by regular bonding in the press, but some firms are using postvulcanisation bonding.

One other type of bonded elastomer assembly that has recently assumed importance in the automotive sector makes use of glass rather than metal as the rigid member. Polyurethane elastomers, thermoplastic olefinic rubbers and elasticised polyvinyl chloride are employed as the gasketing for prefabricated automobile windows, the so called modular window assemblies.

31.5 Manufacturing methods

As noted in the introduction, most R to M assemblies are produced by single-stage press operations wherein shaping, vulcanisation and bonding all occur more or less simultaneously. The sequence of manufacturing operations is straight forward. The major changes of the past 30–40 years have been in the areas of metal preparation and adhesive application and in the development of advanced injection molding techniques.

31.5.1   Metal preparation

It is axiomatic for any type of industrial bonding operation that substrates should be carefully prepared and cleaned. Even autogenous metal surface contaminants such as heavily hydrated iron or aluminium oxides can interfere with establishment of the necessary adhesive bond, one that not only will be strong initially but will withstand various adverse agencies in its end-use service environments.

In most cases, metal preparation involves at least scouring or solvent degreasing, often in combination with grit blasting or some other type of abrasive roughening.

The better two-coat adhesive systems require only that steel parts be sufficiently clean for good primary (initial) bond formation and for robust resistance to service environments. But chemical pre-treatment can lessen not only the corrosion tendency of exposed metal surfaces of the final assembly but also the demands on the adhesive system (i.e., in ensuring the durability of the bond for the lifetime of the part).

A typical metal pre-treatment sequence involves mechanical descaling (grit blasting), solvent degreasing or alkaline scouring, rinsing, phosphate treatment, rinsing, chromate sealing and drying. The chromate sealing step is not entirely desirable in light of today's environmental concerns – sewage effluent contamination or disposal problems – but it does provide the highest order of service protection of the final R to M part.

Phosphatisation prior to bonding must be carefully controlled. Very thick layers of inorganic phosphates-iron phosphate structures or, in the case of the applied zinc type, mixed iron-zinc composites – can have much less cohesive integrity than the R to M adhesive. In other words, improperly applied inorganic phosphates can act as weak boundary layers. In such cases, the strength of the bonded assembly is limited to the force necessary to start delamination within the phosphate conversion coating layer.

31.5.2   Adhesive application

One or two-coat adhesive systems are applied to metal parts by spraying, dipping, bushing and even by roll-coating. Spraying predominates, particularly for high volume motor vehicle assemblies. But whatever the method, film thickness control is essential for satisfactory adhesion and service performance.

Advances in spray application have been notable over the past 10–20 years. Electrostatic spraying, the first advance held great promise for considerably improved economy of use of adhesive systems. However, the electrostatic method places unusual demands on solvent selection and manipulation, since solvent polarity must be within defined limits. And the spray tips easily become clogged. Certain adhesives therefore lend themselves better than others to electrostatic application.

The availability of chain-on-edge machines with automated spraying stations has made possible the most significant advance in this method of adhesive application. The chain on edge machine is very well suited for small and not particularly complex metal inserts that can be hand-arrayed, continuously conveyed to stations where electronic trip mechanisms result in spray coating of just the right amount of primer or adhesive and dried before removal to bins. Roller-coating, also by hand, is the most practical technique for tank lining, whereas adhesive application to coil metal (open width) is typically done by reverse roll coating.

The frames or inserts for seals generally involve methods of adhesive application different from those used for the metal parts for mounts or bushing. In one technique the parts to be immersed are contained in a revolving basket at the end of a metal treatment line. Another more complicated method spirals the metals through a reservoir of adhesive and from there to a conveyor line drying station.

Much if not most bonding is still being done with sulphur-vulcanising natural rubber, which adhered spontaneously to metal surfaces that had been suitably plated with copper alloys.

Polyisocyantes, particularly triphenylmethane triioscyanate, had been effective in press-bonding steel to several of the synthetic diene rubbers. It was hypothesised that the isocyanate group combined during vulcanisation with hydrated metal oxides to form polyureas and triazine ring structures that were strongly adsorbed on the metal surface. The vulcanising synthetic rubbers were assumed to offer various active hydrogen sites that were also able to react chemically with this type of adhesive.

Aromatic isocyanates, probably because they do form true chemical bonds across the interface, can provide R to M assemblies with fairly good resistance to underbond corrosion and attack by water or organic solvents. In a practical sense they are difficult to use under less then ideal conditions because of their great reactivity with adventitious water or even moisture in the air. If metal parts are coated with triphenylmethane triisocyanate and laid over prior to bonding under even moderately humid conditions, much of the adhesive activity is lost.

Special adhesives had to be developed for the newer non-sulphur-vulcanising specialty elastomers. More significantly, broadly versatile two-coat adhesives emerged that were able to bond all the important high volume diene rubbers over a wide range of curing conditions. They were not sensitive to normal variations in humidity. Nor did these newer adhesive systems restrict the part designer's choice of metals. Steel, phosphatised steel, galvanised steel, stainless steel, treated aluminium, brass, titanium – all were able to be successfully bonded as long as the parts were clean.

31.5.3   Broad-purpose adhesives

Nitrile rubbers (NBR) and polychloroprene (CR) were observed to be less demanding of the adhesive composition than SBR, NR, or IIR – elastomers that are significantly less polar. Also very important in determining how easy an elastomer may be to bond is its total stock recipe: vulcanisation system (including sulphur level and accelerators), types and amounts of fillers and antidegradants.

This variation in 'bondability' defined the challenge that was met so successfully by adhesive formulators. Because they do bond a wide variety of elastomer types as well as different stock recipes, broad-purpose R to M adhesives tend to be complex and highly proprietary in their makeup. Unlike many industrial structural adhesives, they cannot be categorised generically as 'epoxy,' 'phenolic,' 'urethane,' or 'acrylic.' The versatile R to M formulations can contain as many as six to eight components in addition to the solvents. Certain of these ingredients are able to become chemically active at elevated curing temperatures. Thus versatility and ability to perform under a variety of vulcanisation conditions appear to be due to more than one mechanism of bonding and in particular, to one or more processes of chemical reaction across the rubber-adhesive interface. This type of directed chemical reaction can be described as cross-bridging. Thus, depending on the elastomer in question and its particular vulcanisation system, one or more of the adhesive's multiple bonding mechanisms will generally be operative to ensure the formation of strong chemical bonds.

The widespread use of various halogen-containing polymers in primers and covercoats for vulcanisation bonding is gaining importance. Highly chlorinated resins are known to wet metals efficiently and also to provide good barriers to chemicals that may attack and undermine the adhesive-to-metal bond. Thermo-setting resins or reactants bolster performance, possibly by formation of interpenetrating polymer networks. For interaction with vulcanising rubbers, there has been speculation about the ability of chlorinated resins to act as Friedel-Crafts reagents and to thus bond to aromatic sites across the interface. It has further been conjectured that dehydrohalogenation may play a mechanistic role, particularly if sites are thereby created in the adhesive phase for attachment by polysulphidic chains extending across from the vulcanising rubber side.

Polyfunctional chemicals (nonisocyanato) that are not in themselves adhesive film formers but do appear to function as cross-bridging agents are also disclosed to R to M adhesive patents. By analogy with what is known of diene rubber cross-linking, it is likely that the polyfunctional compounds react across the interfaces in the fashion of many nitrogenous 'ene' reagents.

Most R to M bonding for motor vehicle assemblies is done with robust two-coat systems (i.e., using both a primer and a covercoat). The primer is designed to wet and chemisorb strongly on clean or chemically treated metal surfaces. Its chemistry, again, is such as to achieve high cross-link density during the vulcanisation cycle and to become strongly knitted to the covercoat that is applied over it.

Also available are single-coat adhesives that enable high quality bonded parts to be manufactured without a separate application of a primer. They are often used on metals that have been both phosphatised and chromate-sealed.

Some of the newest developments in general-purpose adhesives are in formulations that offer superior resistance to very specific service conditions or to environments that are not encountered by most R to M assemblies. For fluid engine mounts (i.e., those embodying contained fluids as well as rubber springs), two coat adhesives are now available that better withstand interfacial attack by the contained liquids at high operating temperatures. Very hot mixtures of ethylene glycol and water, for example, can slowly disrupt the R to M bond provided by some of the older commercial adhesives.

Also developed in recent years are adhesives for assemblies that will do service under conditions of cathodic disbonding: for example, rubber bonded to metals for under water service, where the metals are protected from corrosion by the use of sacrificial anodes.

Probably the most important adhesive development efforts of recent years have been directed to water-based products for the high volume diene elastomers. The organic solvents that are used for virtually all bonding agents today are becoming less and less acceptable in the industrial world.

The same forces militating against solvent borne coatings – air pollution, groundwater contamination, adverse health exposure for workers, flammability hazards and even high costs – eventually will dictate use of solvent-free bonding methods.

31.5.4   Specialty elastomer adhesive

High performance specialty rubbers such as fluoroelastomers, silicones, polyphosphazene types and castable or RIM polyurethanes, generally require adhesives that are tailored to each elastomer's specific chemistry and mode of cure. Special adhesives are even offered for the diene rubbers such as NBR and HNBR, which find extensive use in seals.

Adhesives for the specialty elastomers are more often than not able to function satisfactorily as single coats, that is, they do not require separate application of a primer. Functionally, one or more of the adhesive ingredients is able to chemisorb strongly and irreversibly on clean metals as driven by the heat of vulcanisation. For knitting to the elastomer, there is almost invariably a reactive moiety that can cross-bridge chemically with active sites in the elastomer or through its active vulcanising agency (e.g., with the free radicals that are generated during peroxide curing of silicone rubbers). For suitable durability of the bonded specialty elastomers in service, it is typical that the adhesive film also undergo internal condensation or cross-linking.

Adhesives for the high performance specialty elastomers are often based on organofunctional silanes or reactive phenolic resins. As in the case of adhesives for the high volume diene rubbers, a few of the newer bonding agent prototypes for specialty elastomer make use of water as the carrier rather than organic solvents.

31.5.5   Thermoplastic elastomer adhesives

As already discussed, thermoplastic elastomers of various types have become increasingly important in recent years. Adhesives are available to bond most of the thermoplastic rubbers, but they generally function rather differently from vulcanisation bonding agents. Physical chemistry must be relied on through processes of diffusion and adsorption – as set apart from cross-bridging or interfacial chemical reaction – to achieve bonds strong enough for durable practical composites. For the manufacture of metal assemblies with thermoplastic elastomers, very limited heat is available to drive the physical or chemical processes that are believed to be important for good bond formation. Even so, some of the thermoplastic elastomer adhesives are formulated as two-part systems, for cross-linking internal to the adhesive. This enables bonds having better resistance to service environments (e.g., moisture) than would otherwise be the case.

The more nonpolar thermoplastic elastomers generally require two-coat adhesive systems. Others polyurethanes and elasticised polyvinyl chloride, for example – are in some cases effectively bonded to metal or to glass by single-coat formulations.

31.6   Testing of rubber to metal bond

31.6.1   Primary bond

The primary R to M bond can be arbitrarily defined as that which can be characterised or measured on completion of the assembly operation. It can sometimes be distinguished from the bond that is (or is not) retained after long-term service or after environmental exposure. Test protocols for primary bonds have been recognised by the American Society for Testing and Materials for many years. As delineated under Adhesion to Rigid Substrates (D42-81), there are five standard methods:

1. Rubber part assembled between two parallel metal plates.
2. 90° stripping test – rubber part assembled to one metal plate.
3. Measuring adhesion of rubber-to-metal with a conical specimen.
4. Adhesion test – postvulcanisation bonding of rubber to metal.
5. 90° stripping test – for rubber tank lining, assembled to one metal plate.

Test method 1 is often referred to as the butt joint test, the force of rupture is straightforward tension.

Method 2 involves peeling forces, whereas the conical test (method 3) is most discriminating of bond quality in concentrating shear forces within a very narrow zone of the bonded surfaces.

The standard method 4 for postvulcanisation bonding also involves preparation of butt joint test pieces. For tank lining, bonded sheets are arranged vertically and peeled in the same direction.

These tests are valuable to adhesive suppliers for purposes of new product development and technical service (i.e., for comparative measurements).

As important as static tests to the fabricators of end products such as engine mounts are special dynamic fatigue tests, which often are tailored to very particular applications and service conditions. The most advanced dynamic testing of highly engineered R to M assemblies involves servo-controlled electro-hydraulic test trains. These devices make use of advanced electronic controls and are of course highly computerised. Dynamic properties can now be logged over a wide range of frequencies and service conditions in a fraction of the time required for a single frequency–response curve with the test equipment.

It is interesting to note that design, production and quality engineers often view R to M bond testing as a yes/no proposition – the assembly either fails within the rubber (good) or interfacially (bad). Yet changes in the forces of rupture together with changes in the nature of the failure can convey information that is of great value in the course of systemic analyses or experimentation– for example, in development or improvement of manufacturing processes.

An excellent primary bond is of little or no value if it does not hold up in service. Accordingly several accelerated tests have evolved over the years that focus on the resistance of the bonded assembly to various service environments. Most of these tests involve exposure to heat, to salt spray (resistance to underbond corrosion), to boiling water, or to organic fluids. The latter category subdivides into lubricating oils, hydraulic fluids (several types), fuels and antifreeze mixtures.

Method 5 90° stripping test is useful for screening purposes is a modification of the ASTM 90° stripping test. For exposure to aggressive environments, the rubber strip is tensioned back on itself. The highly stressed rubber is tightly tied down (e.g., with a stout section of wire) and 'scored' or cut directly at the exposed interface with razor blade. It is then subjected to the service condition (immersion in boiling water, e.g., or salt spray, cycling condensing humidity, hot oils, or hot mixture of ethylene glycol and water). After an appropriate period of exposure – which may be as trying as 250 hr immersion in organic fluid at 150°C the stressed part is removed and peeled mechanically to failure. If the assembly still doggedly ruptures within the elastomer section, its bond is rated as having satisfactorily robust resistance to the environment in question.

Another accelerated environmental test involves jigging of butt joint assemblies under variable and quantifiable tension. Jigs have been designed to accommodate as many as a dozen test pieces simultaneously. By immersing the jig in an aggressive environment such as dry heat or boiling water, resistance can be measured quantitatively in many instances, simply as a function of time to failure. (With appropriate levels of tension, the parts generally fail catastrophically – as a result of attack on the bond or not at all.

31.7   Applications of rubber to metal bonding

While there are many industrial applications for rubber to metal bonding the largest market for the technology is motor vehicles. Motor vehicles use a large number of rubber components, many of which are bonded including:

1. The engine and gearbox units are mounted on a rubber bonded unit that may incorporate hydraulic damping systems to damp out engine noise and vibration more perfectly over a wide spectrum of frequencies.

2. The strut units mount the wheels to the vehicle chassis via a shear style rubber bonded bush.

3. Sub-frames need to be coupled by stiff mounts that allow some flexibility.

4. The steering wheel is joined to the steering rack by bonded components and the various link arms in the suspension and steering all incorporate bonded bushes. They act together to provide the comfort and road handling characteristics that are demanded in the modern motor vehicle.

5. s found in the chassis of the lighter, less rigid bodies. Using a rubber bonded mass, called a harmonic damper, that is connected to a vibration node allows the damping of these vibrations.

Manufacture of miscellaneous rubber products

32.1   Introduction

This chapter discusses the large number and variety of other products consuming relatively little rubber because of their small size, their small quantity usage, or their low content of new polymer. Many of them are commercially very remunerative to the manufacturer, others because of their technical importance for the engineering applications in for example the aircraft, electrical and electronic industries, receive a great deal of effort in government and co-operative research laboratories, as well as in individual manufacturer's laboratories. Because of their variety, it is difficult to group or classify them, furthermore, it is not the intention to attempt to discuss or to list every product.

Some of the miscellaneous products discussed are engine mounting, auto tubes and flabs, rubber cables, rubber gaskets, rubber matting, latex gloves, microcellular rubber sheets and products based on spread fabrics.

32.2   Engine mountings

Engine mountings play an important role in the efficient functioning of automobile systems and vehicles. These mountings are not only used in original equipment but also required for subsequent replacements as well. It is a heartening fact that the automobile sector is expanding very rapidly not only in the existing capacities but also in the creation of new capacities.

The demand for various types of mountings for automobiles has been increasing steadily. These products are directly linked with the industrialisation of the country. The demand-rise in the various types of vehicles is fast increasing with the living standards of the people. The engine mounting essentially consists of two parts. One is metallic and another is rubber part. Subsequently, they are jointed together to form the end product.

32.3   Process of manufacture of engine mountings

As mentioned above the engine mounting consists of two parts namely metallic and rubber. First of all, metallic part is made from the suitable iron plates, according to size and thickness desired as per standard specifications of the end product. For making the rubber compound, first of all various chemicals are checked for their respective percentage purity. Subsequently, they are mixed with rubber on a standard mixing mill as per standard weights and formulations adopted. Some chemicals are added to soften the rubber or sometimes rubber is warmed to lighten the load on the mixing mills before mastication process.

Mixing mills generally consist of two steel rollers or chilled cast iron rolls which rotate towards each other at different speeds so that any material passing between the rollers is subject to tearing action. After the pre-mixing, usually the raw compound sheets are stored for one day. For final mixing, raw-compounded sheet is again mixed for 10 to 15 min. Subsequently, these sheets are pasted on metallic part with rubber solutions as per size and weight required. These are then placed in suitable mold in a press and cured. Finally, these mountings are inspected for any defects, tested for physical parameters and then packed.

Raw materials used are: (i) natural rubber, (ii) synthetic rubber, (iii) carbon black, (iv) rubber chemicals like rosin, stearic, acid, calcium, carbonate, etc., (v) curing agent like sulphur, vulcanising agent, etc., (vi) synthetic adhesive, (vii) paints, lacquers, etc., and (viii) CRCA (metallic parts).

32.4   Auto tubes and flaps

32.4.1   Process of manufacture

Auto tubes are manufactured by molding method. First of all rubber along with other materials is properly mixed in the two roller mill or a Banbury mixer. This compounded rubber is fed into the extruder and the rubber compound takes the shape of long tube, then proper length of this green tube is cut and tube valve is fitted in this green tube and the end of the tubes are join by means of butt joining machine. This green tube is vulcanised in the mold having air pressure inside. After proper vulcanising, it is tested by filling specific amount of air inside for leakage, if any.

Auto flaps are manufactured by the pressure molding technique in the mold, after making the rubber compound on a two roll mixing mill.

Raw materials used are: (i) smoked natural rubber, (ii) synthetic rubber, (iii) china clay, (iv) carbon black, (v) stearic acid, (vi) zinc oxide, (vii) sulphur, (viii) valve fitting, (ix) processing oil, (x) chemicals like accelerator antioxidant, etc., and (xi) packing material.

32.5   Rubber cables

The cables are manufactured with EPDM, silicone, fluroelastomer, CSP, PCP, NBR-PVC, EVA and other types of rubbers. Processing of a good quality cable needs right technology starting from manufacturing of conductor, compounding, insulation, laying up, sheathing and vulcanisation. The process used contributes significantly to build inherent strength of the cable for long life and performance. The cables are used for critical applications in the areas like mines, railways, defence, steel plant, windmill, earth moving equipments, etc.

32.5.1   Machinery and equipments

Latest machinery and equipments should be used in all stages of processing starting from manufacturing of conductor, rubber compound, laying, extrusion and vulcanisation to the final testing.

Wire drawing

Copper wires are drawn from electrolytic grade CC rod of 99.96% purity level on machines equipped with in-line annealer. This process enables to produce wire with uniform annealing.

Electro tinning

Tinning of fine copper wires is carried out by electro tinning process. This process ensures smooth, uniform and homogenous tinning with bright finish to the wire.

Bunching

Long lengths of bunches are produced on high capacity bunching machines which result in joint less final conductors.

Floating carriage stranding

All flexible conductors are rope stranded on floating carriage type machines. The conductors produced on floating carriage machines should be highly flexible and are free from back twist and kinks.

Compounding

The rubber compounding is carried out in an internal mixer, widely known as intermix, This is unique rubber compounding machine where mixing is accomplished inside a closed chamber by rotating rotors of special design to ensure complete sheering, thorough mixing and uniform dispersion of various compounding ingredients into the rubber matrix. Batch to batch variation of properties, loss of ingredients during mixing are minimum in compounds produced in intermix.

Extrusion and continuous vulcanisation

The pressurised liquid salt continuous vulcanisation (PLC) process: In this process three layers can be extruded simultaneously by using dual tandem extrusion and continuously vulcanised by passing the cable through eutectic molten salt mixture bath maintained at 200°C under nitrogen pressure.

Multi layer co-extrusion and co-vulcanisation: Three layers can be extruded simultaneously. This offers interfaces free from contamination and voids. The process also offers physical as well as chemical adhesion between the layers.

Cable produced in PLCV process offer excellent flexibility and smooth surface finish.

Batch curing: In the batch or discontinuous method, the insulation or sheathing is cured in an autoclave in steam at a pressure corresponding to 130–170°C, curing cycles vary between 15 and 90 min. The cable or core in its uncured state is spirally overlapped with rubberised cotton tape or polyester film for better consolidation, then coiled in trays or reeled on drums. Where maximum consolidation and improved surface finish are required, a metal sheath is applied. Commonly, a lead tube is extruded onto the cable by either a ram or a screw-type press. The lead is stripped off after the cure and reused.

Continuous curing: Conventionally, a continuous vulcanisation (CV) plant consists of a curing tube attached to the die face of the extruder head. There are three types of continuous vulcanisation plant: horizontal continuous vulcanisation (HCV), catenary continuous vulcanisation (CCV) and vertical continuous vulcanisation (VCV). The HCV plant has an obvious disadvantage, namely for reasonable tube diameter and with practical tension, the cable must touch the bottom of the tube.

Infrared curing: Another important cross-linking method used for the cable manufacture is infrared curing. By this technique, the elastomers are continuously cross-linked by the infrared energy source after extrusion at ambient pressure and at relatively low temperature. Silicone rubber cables, especially, are made by applying this technique.

Curing under lead sheath

Multi core cables where high compactness is required with reinforced sheath for stacker and reclaimer, LHD machines, earth moving equipments, etc., are vulcanised under lead. In this process the rubber compound fills all the interstices and gap in the cable under high pressure of lead during vulcanisation. This process provides smooth surface finish and compact cable. Summary of various manufacturing process is given in Table 32.1.

Table 32.1: Summary of manufacturing process.

Stage	*Normal*	*Unistar*	*Added feature*
Wire drawing	Batch annealing	In-line annealing	Consistent and uniform annealing resulting in high flexibility
Tinning of wire	Hot dip tinning	Electro tinning	Uniform, smooth and bright surface finish
Bunching	Low loading capacity	High loading capacity	Longer lengths of finished cables

(Cont'd...)

Stage	Normal	Unistar	Added feature
Stranding	Fixed carriage standing	Floating carriage stranding	Highly flexible conductors without back twists and kinks, etc., resulting in better performance and longer life during reeling-unreeling operation.
Compounding	Open roll-mill or kneader type mixer	Inter mix	Complete sheering and uniform dispersion of ingredients into the rubber matrix, resulting in superior and consistent thermo-mechanical and electrical properties.
Extrusion and vulcanisation	Uni-layer extrusion and co-vulcanisation in CV or batch vulcanisation by autoclave by high pressure steam	Uni-layer as well as multi layer extrusion and co-vulcanisation in PLCV line	Dry cure, high degree of cross-linking void and contamination free inter-face good adhesion between layers, superior flexibility and smooth surface finish
Curing of reinforced sheaths	Under tape in autoclave	Under lead	Highly compact cable with smooth surface finish giving longer life in reeling unreeling application

Applications of some specialty cables are given in Table 32.2.

Table 32.2: Applications of specialty cables.

Type of cable	Application
Flexibly cables	General wiring, trailing, control and power distribution
Mining cables	Land line, LHD and SDL machines, drills, coal cutters, stacker-reclaimers and earth moving equipments
High temperature silicon cables	High temperature location furnaces and steel plants
Wind mill cables	Wind energy generators
Ship wiring cables	Power and control cables for ships and defence application
OH-rigs cables	Power and control circuits for oil rigs
200°C fluoroelastomeric cables	Traction motors
Fluoroelastomeric lead wires	Brush gears connector cable in traction motor
120°C thin wall cables	Tap changer electric locomotives/EMU's /AC coaches
Cadmium-copper catenary cables	25 KV ac electric traction
Radiation resistant cables	Nuclear power plants
Fire survival cables	Circuit integrity during fire hazards for critical application

32.6 Rubber gaskets

A rubber gasket is a type of mechanical seal that is used to prevent leaks between distinct substrate sections.

Although there are several types of processes in manufacturing gaskets depending on the type of sealing gasket to be manufactured, the manufacturing process of cylinder head gaskets includes all stages concerning the rest of gaskets (light gaskets, metal gaskets).

A brief description of the same is given below:

1. Steel punching: As a basic component of the cylinder head gasket, the steel in coils is punched by means of a stamp.

2. Sandwich assembly: The punched steel coil and two coils of aramide fibre are assembled sandwich like, so that the punched plate is between two fibre plates.

3. Calendering: The above sandwich goes through a calendering line, in order to get a definite thickness depending on the gasket to be manufactured. The outcome of this process can be a coil of material or formats of several sizes.

4. Cutting/stamping: The coil or the formats are cut according to the geometrical profile of the part with the corresponding tool: stamp or die.

5. Gasket pressing: The gaskets are pressed to get a definite thickness.

6. Assembly and riveting of reinforcements: As an auxiliary process, some plates of different types of steel are cut and drawn to shape the metal reinforcements, which are assembled and riveted in the cylinder area.

7. Pressing of reinforcements: Like the gaskets, the reinforcements are pressed in order to get a definite over-thickness.

8. Assembly and riveting of rings: Other auxiliary pieces that may be included in the cylinder head gasket are the rings, which are cut and drawn like the reinforcements. The rings are assembled and riveted around the holes for lubricating fluids (oils).

9. Impregnation: The mission of a gasket is to ensure the correct sealing of the part. This is achieved by sinking the material into a silicone bath that fills the pores of the fibre material.

10. Curing in static oven: The gaskets are cured at high temperatures with the resulting emission of hydrocarbons to ensure the correct application of the impregnating compound.

11. Silk-screen printing: Once the gaskets are cured, they receive a surface silicone beading to get the best sealing. The cleaning of the used tools should be considered as an auxiliary process to this stage.

12. Curing in dynamic oven: Lastly and as the final stage of the manufacturing process, the cylinder head gaskets are cured again in dynamic ovens.

13. Labelling and packing: Once the final product has been manufactured, this goes through the packing lines. The package includes a cardboard support and a polystyrene or PVC cover.

In regard to the manufacturing process of thermal shields, this is as follows:

1. Stamping: The steel and the fibre are stamped with the appropriate profile according to the part to be manufactured.

2. Assembly and riveting of the parts: The steel plates and the fibre plate are riveted sandwich-like.

3. Sandwich shaping: Lastly, the three plates are shaped by bending and the necessary rings are added.

There is a third type of manufacturing process for the production of oil seals and molded rubber gaskets. In this process, a band of synthetic rubber goes through a rubber press where it is ground. Then it is subjected to high temperatures and becomes a semi-fluid. Further it is injected into a mold in which the part is shaped for a definite time and at a definite temperature, in order to be cured. Then the parts are taken out from the mold and trimmed. Lastly, they are post-cured in a static oven.

32.6.1   Environmental issues related to rubber gaskets

Some of the environmental issues are briefly discussed below.

Air emissions

Emission points are those pipes and outlets to the outside that carry and throw out to the air the gases and vapours generated in the plant. The generated air pollution is mainly due to the combustion gases (CO_2, NO_x) from the combustion boilers (heating and sanitary hot water) and to volatile organic compounds (VOC's), specially hydrocarbons generated in the curing ovens and during the cleaning of silk screens. Both combustion gases and VOC's contribute to the greenhouse effect. Furthermore, particles from an evacuation point for gases resulting from the laser cut procedure of metallic materials are thrown out to the air.

Sewage

There are four points of treatment of sewage generated by the following activities:

1. Sewage resulting from the cleaning of rolls and emulsion of silk screens: It is purified through a physical-chemical treatment based on the precipitation of paint and emulsion remains by means of a flocculation agent. The water drained to the sewage system is free of noxious substances.

2. Sewage coming from compressor purges: A mixture of water and lubricating oil generated in the compressors that supply compressed air to the plant machinery and that circulates through a water/oil separator.

3. Sewage coming from the purges of the pump of the water jet cutting machine: A mixture of water and oil generated from leaks in the equipment, that goes through a water/oil separator.

4. Sewage from the rinsing of molds: It comes from the cleaning of molds for rubber injection, which includes three tubs: degreasing (closed), rinsing (recirculation) and passivate (closed). The rinsing water is recirculated after going through a series of filters that keep the solids and fats that the water may contain. The chemical oxygen demand (COD) and the alkalinity are corrected through active carbon filters and the controlled injection of CO_2 respectively. The excess water generated is drained to the sewage system after going through a second filter of active carbon, to prevent eventual overflowing in the rinsing tub. In all cases the sewage is drained to the internal system of the plant after its treatment and from here it goes to the sewage system of the industrial estate.

Noise

The main noise generating points are the machines in the production and packing departments and in the auxiliary services (air conditioning, etc.). Considering the features of the production processes the level of internal noise in dB(A) in the plant is not homogeneous. Therefore, annual measurements of max level of internal noise are carried out by an external service of prevention.

Waste

The waste generated should be treated in compliance with the applicable regulations in force and under consideration of their possibilities of minimisation and recycling, in order to reduce the environmental impacts associated to its removal and to the space in the dump.

Another type of waste generated in the daily activity of the plant allows its external recycling. Therefore, it is separated in different containers according to its type (metals, paper and cardboard, wood) and it is regularly collected by authorised organisations. Inert and urban-like waste is removed by the authorised agent and carried to the local dump.

32.7   Rubber matting

The insulating rubber mats are especially designed to be used in front of switch boards and high voltage equipment. These long life mats are designed to protect operators from electrical shocks thus acting as personal protection equipment. Safety from electrical shock is required for workmen whether they are involved in electricity generation, transmission, distribution or its use. Safety mats are highly recommended for total safety of workmen from electrical shock when working in or around environment like high voltage panels, substation, power transformer rooms, LT and HT labs, near bus bars, near control panels, anti-skid/flooring, etc.

These long lasting, easy to install, clean and maintain mats are resistant to oils, acids, alkalis and low temperature and are flame retardant as per the provisions of IS 15652:2006.

Conductive rubber sheets: These sheets come in standard sizes of 1000 × 2000 mm but can further be customised to any size as per customers requirements in a variety of thickness ranging from 2 to 25 mm. These sheets normally have a plain finish and they come in a variety of colours depending on your requirement.

These products are extensively used in the transformer stations, power plants and electricity distributing stations. EPDM, neoprene and natural rubbers are considered to be good polymers for this use.

Floor mats: Rubber mats and rubber rolls for home and commercial installations designed for quality and value. Rubber flooring is a good fit for durability factory of rubber and to provide cushion on floors for home and commercial applications. Much of black rubber sheets and rubber rolls are made with recycled rubber content, which is great for the environment, while full colour rubber is manufactured from virgin or synthetic material.

Stud mats: These mats are primarily used in stables to ensure the horses or cows live in a totally clean, sterile and smell free environment made from high quality natural rubber, these mats are designed to be soft, dry, smell and dust free. All the materials are tested for tensile strength, elongation at break and tear resistance.

Kids play area mats: These mat protect children from injuries by installing one of rubberised floorings. Assuring durability and high skid resistance, these floorings are ideal for children play areas and parks. Clients can lay down these rubber floorings on different kinds of surfaces, including sand, soil, wood and tarmac. To meet the variegated requirements of the clients, these floorings are available in various colours, patterns and designs.

Gym mat: Gym mats add comfort, safety and appeal to exercise areas. These gym mats include single rubber mats for isolated areas such as weight lifting mats and exercise machine mats, or rubber mat rolls covering larger areas, or even interlocking rubber mats going wall-to-wall.

32.8 Latex gloves

Natural rubber latex glove production is an interesting process that starts with nature and ends with comprehensive barrier protection. Each step along the way ensures the gloves are of the utmost quality when they arrive to distributors and end users.

32.8.1 Harvesting phase

The process begins with the *Hevea brasiliensis tree*, which mostly grows in Southeast Asian countries like Vietnam, Thailand and Indonesia. Farmers extract the trees' milky white latex sap from mature trees through a process called tapping. This occurs in the early morning, as the sap coagulates faster when temperatures rise later in the day. Farmers start by stripping bark from the tree at a downward curve. This directs the sap to a spile, which then allows the latex to drip into a cup affixed to the tree. Then, farmers boil the milky white latex to make it more concentrated, which gives the sap a consistency similar to syrup. Rubber trees are suitable for tapping for five years. 'Farmers remove latex from trees through a process called tapping.'

32.8.2 Production phase

Once farmers collect the sap, it goes to a factory for production. This phase includes several steps:

1. Preparing the latex: While latex gloves come from natural rubber latex, they are not 100% pure. This is because manufacturers combine the latex concentrate with a number of compounding chemicals during the initial step of the production process. This step enhances the latex's properties, such as the elasticity, as well as stabilises the material and its shelf life.

2. Cleaning the formers: To mold the latex into the shape of a glove, manufacturers use hand-shaped ceramic formers. The first task is to wash these formers by dipping them in water and then bleach. This ensures no residues are left from the previous batch. Afterward, formers dip into a chemical solution of calcium carbonate and calcium nitrate to help the latex stick.

3. Dipping in latex: Once the formers are ready, manufacturers dip them into a tank full of latex, with the length of time the former is immersed in the tank varying based on the desired glove thickness.

4. Vulcanising the rubber: To ensure the rubber does not crack while drying, the formers enter an oven to dry and solidify. The development of the vulcanisation process is an integral to the creation of the latex rubber.

5. Leaching the gloves: This process involves dipping the gloves in water tanks and removing excess latex proteins to lower the risk of wearers having an allergic reaction and enhance the feel.

6. Beading the cuffs: Once the gloves are done with leaching, the manufacturers roll the cuffs to make the gloves easier to remove. The gloves may undergo leaching again after beading.

7. Applying powder: If the gloves are powdered, they enter a wet food-grade cornstarch powder slurry. Afterward, manufacturers dry the gloves again.

8. Chlorinating or polymer coating the gloves: If the gloves are powder free, they undergo alternative processes to facilitate easier donning. The first is chlorination, which makes the latex less tacky. The second involves coating the gloves with a polymer, which makes the surface smoother.

9. Stripping the gloves: Once the gloves are finished, workers remove them from the formers by hand.

'Medical-grade gloves are subject to more rigorous testing.'

Quality control phase

To ensure the gloves are of the highest quality, manufacturers test them. Workers test gloves using methods from the American Society for Testing and Materials (ATSM) and the U.S. Food and Drug Administration (FDA) regulates these standards. The pinhole leak test is one of these methods.

Workers first fill the gloves with one litre of water. Then, they close and hang the gloves to check for leaks.

The tests adhere to guidelines regarding acceptable quality limits (AQLs). These standards designate a percentage to evaluate a batch of gloves. If a batch's failed gloves exceed this percentage of the total batch, all the gloves in that batch fail.

The results of these tests determine whether the gloves will be industrial- or medical-grade. The latter are subject to more rigorous testing.

Packaging phase: Once the gloves are done with production, workers and pack them for shipping.

32.9 Microcellular rubber sheets

The use of microcellular soles is becoming very popular because of its wear and tear resistance properties. The units manufacturing these can be ancillary to some large scale footwear manufacturing unit. Though the large scale units do manufacture microcellular sheet, their production usually falls short of their requirements thereby necessitating purchases from outside sources.

Raw materials used: (i) natural rubber, (ii) SBR, (iii) microcumb, (iv) zinc oxide, (v) china clay, (vi) stearic acid, (vii) titanium dioxide, (viii) accelerator, (ix) antioxidant, (x) paraffin wax, (xi) ethylene glycol, (xii) CI resin, (xiii) calcium silicate, (xiv) sulphur, (xv) process oil, (xvi) DPT, (xvii) miscellaneous chemicals like talc, etc., and (xviii) packing materials.

32.9.1 Manufacturing process

All the rubber chemicals are compounded along with rubber (both natural and synthetic, masticated previously) and measured quantities of the compound is molded in suitable molds in a hydraulic press, which is generally of multi daylight type. After first curing, the sheets may be cured again second time (if required) in vulcaniser under steam pressure. The sheets are taken out and kept under load to avoid deforming while cooling.

32.10 Products based on spread fabrics

Coating of fabrics by calendering is usually done, but in addition to that method, fabrics can even be coated with rubber by a spreading technique.

In the spreading process, the rubber compound is dissolved in a suitable solvent to achieve the form of dough.

When calendering the total thickness of the coating is normally made in one pass in the machine but by spreading it is common to build up the total thickness by several passes and drying, evaporation of the solvent between each pass. Spreading is one of the first techniques used in rubber product manufacturing. Spreading is still an important technique mostly used for high quality products as protective clothing for chemical industry, fire brigades and military purpose, coated fabrics for production of inflatable products and sheets for offset printing.

Typical for fabrics used for these products is coating with many thin layers for absolute tightness (free of pinholes) as well as different polymers used in the different layers.

In the calendering process, fabric and rubber material is passed through a series of rollers to flatten, smooth and commingle the two or more materials. Calendered sheets can have multiple layers of both the elastomeric and polymer 'sandwiched' together.

Sheeting produced by the calender process is typically divided into two classes: either fabric inserted, or unsupported. Unsupported calender goods contain only layers of rubber or plastic that have been joined without cord, cloth or textile being inserted for strength or tear resistance. Tyre cord can be used in special cases: rayon, cotton, nylon, or polyester cords are arranged in parallel and bound together by rubber on a calender.

Depending upon intended use and required look or feel, the sheets of rubberised fabric are then smoothed, glazed, polished, or given a more or embossed surface. The resulting surface characteristics depend on the pressure exerted by the rollers, on their temperature, composition and surface designs and on the type of coating or glaze previously applied to the material to be calendered. Following the calendering process the material is then wound into rolls and placed into storage.

These rubber sheets are also manufactured from over 15 polymers including: nitrile/buna-N (NBR), fluorocarbon (viton®, fluorocarbon (FKM), silicone (VMQ), fluorosilicone (FVMQ), polyurethane (AU/EU), natural rubber (NR), ethylene propylene (EPDM), hydrogenated nitrile rubber (HNBR), styrene butadiene rubber (SBR), chloroprene (neoprene®), etc.

Section V

Pollution control and energy conservation in rubber industry

Pollution control in rubber industry

33.1 Introduction

Natural rubber may be obtained from the latex of about 2000 species of plants containing rubber as its constituent. However, only a few species (*Heveabrasiliensis*, *Parthenium argentatum*, *castilla elastica*, *Ficus elastica*, etc.), are identified for commercial production of rubber. Natural rubber is obtained in two forms, namely field latex and field coagulm (or scrap rubber). Field latex and field coagulum are highly susceptible to degradation due to oxidation and contamination and hence are processed to obtain the following forms (products) that allow easy storage and marketing for utilisation by the rubber product manufacturing units.

Rubber manufacturing generally comprises the following operations: Raw materials handling, weighing and mixing, milling, extruding and calendering, component assembly and building, 'curing' or vulcanising, inspection and finishing, storage and dispatch (Fig. 33.1).

33.2 Natural rubber processing industry

33.2.1 Production process and pollution generation

Sieving: Latex collected from the field is first sieved through a mesh to remove impurities like bark shaving, leaves sand particles, etc.

Bulking: Bulking is the process of collecting latex from different fields in a common tank in order to have uniform properties.

Standardisation: The bulked latex is diluted to a standard dry rubber content (DRC) of 12.5%. The dilution helps in settling denser and finer impurities at a faster rate and in obtaining a softer coagulums.

Preservation: Before coagulation, chemicals such as sodium bisulphite and paranitrophenol are added to the latex as preservative. Preservation is also accomplished by adding ammonia.

Latex concentration: The process of concentration involves removal of substantial quantity of serum from the latex and increase its rubber content to about 60%. This is accomplished either by creaming method or by centrifuging. In creaming process, a creaming agent (tamorind seed powder) is added and allowed to remain undisturbed till the desired level of creaming (concentrated

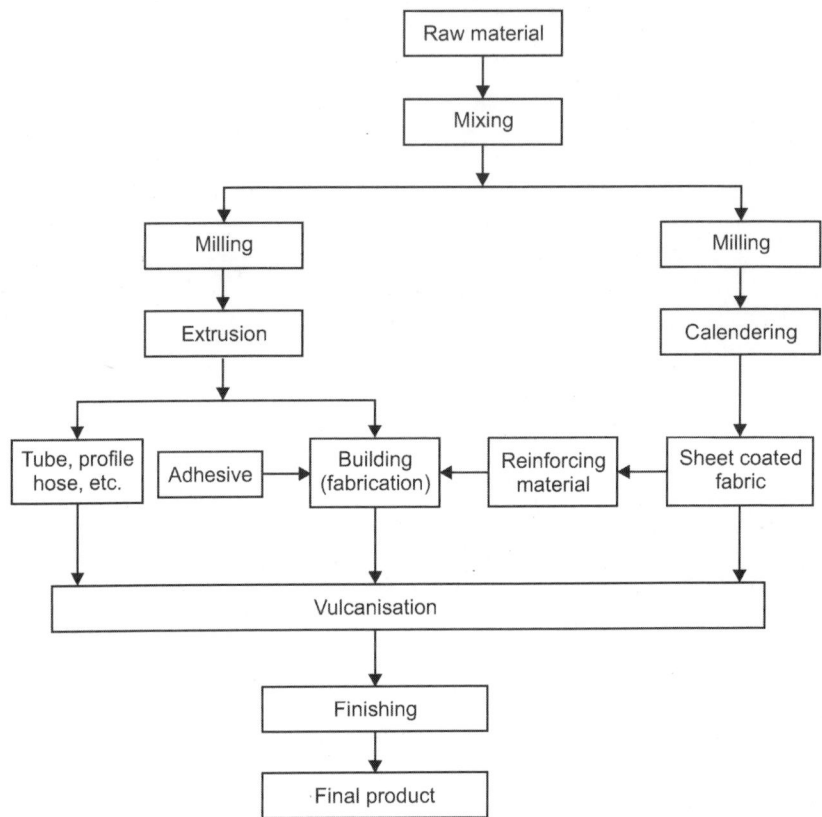

Figure 33.1: General process flow diagram for manufacturing of rubber products.

rubber) is obtained. In centrifuging, the latex is subjected to centrifugal force in a bowl rotating at speed around 6000 rpm whereby individual rubber particles tend to separate into a layer surrounding the axis of rotation.

Coaguluation: Coaguluation is the process of destabilisation by some means with a view to recover rubber from it. Acetic acid and formic acid are generally used for coagulation.

The above unit operations are most or less common for manufacturing of various products.

33.2.2 Pollution abatement

Pollution abatement involves: (i) in-plant control of waste and (ii) end-process treatment of wastewater. Few in-plant control measures are suggested to enable reduction in consumption of water, generation of pollutants and to increase the efficiency of the end-of-process wastewater treatment.

In-plant control measures

In the crepe and crumb rubber units, in which field coagulum is processed, higher than the required water quantity is generally used for soaking and also the soaking time allowed is not adequate. If the raw scrap rubber is properly soaked and primary, dirt removal is done by scrap-washer, the quantity of water consumed in milling can be reduced.

In the crumb units wastewater from final milling can be collected separately from the effluent of the other milling section and can be used either for soaking the scrap rubber or for the first milling process. This is compratively clean and amounts to reduction of about 25% of the total water consumption.

In centrifuge machine bowl washing is done at the interval of 3–4 hr to remove the sludge. About 0.5% rubber is lost during this washing. This washing can be done at two stages. The first washing which is more concentrated may be segregated and collected in a separate tank and coagulated for recovery of the rubber lost during washing. This will result in reduction of pollution load in the effluent. The possibility of diverting this waste stream into the skim coagulation tank can also be considered.

The quantity of acid used for coagulation of the latex specially skim latex kit after centrifuging operation is generally found to be higher than the actual requirement. The incomplete coagulation results in escape of rubber particles into the effluent alongwith skim serum. The excess acid not only causes acidic effluent but also redissolves the rubber protein and causes delay in coagulation. Hence, it is suggested that proper acid dosage and sufficient coagulation time should be provided so as to obtain more or less clear liquid after complete coagulation. The skim latex if demmoniated before coagulation, acid requirement can be reduced and the ammonia concentration in effluent may also be reduced. In the latex process units the segregated first washing of the coagulum may be diverted to the skim coagulum tank where after skim coagulum recovery, the effluent may join the other wastewater streams.

Extruding and calendering: The extruders force the rubber compound through a die into various forms, which are then cut to appropriate lengths. Strips of softened rubber compound are fed into multiple-roll milling machines (calenders) to form rubber sheeting, or to apply the rubber directly onto woven textile fabric, which can then be wound off onto a roll. During such manufacturing operations, fumes are often generated.

Component assembly and building: At this stage, solvents are frequently used, with the possibility of inhalation of solvent vapours or of direct effects of the solvent on the skin of the operator.

Curing or vulcanising: Heat is applied to the product, usually by use of steam, in a curing mold, press, or autoclave. Operators working in the area are exposed both to heat from the presses and to fumes from the heated rubber products. Chemical reactions take place throughout the manufacturing process and may give rise to new, more volatile chemicals.

Inspection and finishing: This involves the handling of cured rubber products, often while still hot. It usually involves direct and extensive skin-contact with the surface of the finished article (during inspection) and may also involve exposure to vulcanising fumes. Grinding, trimming, repair, painting and cleaning may also entail exposure to rubber dust, fumes and solvents.

Storage and dispatch: Large quantities of stored rubber goods may release considerable amounts of toxic substances, either as vapours or as constituents of the 'bloom' on the surface of finished goods.

Chemicals used in the rubber production process

A wide variety of natural or synthetic elastomers, fillers (e.g., carbon black, precipitated silica or silicates) and additives are used in compounding to create the necessary properties of the final rubber product. The actual chemicals used in this process have changed over time and vary extensively depending on the manufacturing sector (e.g., tyres, general rubber goods, retreading) and on the specific plant.

Compounding ingredients are classified as vulcanising agents (e.g., elemental sulphur, sulphur donors such as organic disulphides and higher sulphides, peroxides, urethane cross-linking agents), vulcanisation accelerators (e.g., sulphenamides, thiazoles, guanidines, thiurams, dithiocarbamates, dithiophosphates and miscellaneous accelerators such as zinc isopropyl xanthate and ethylene thiourea), vulcanisation activators (e.g., zinc oxide, magnesium oxide, lead oxide), retarders and inhibitors of vulcanisation (e.g., benzoic acid, salicylic acid, phthalic anhydride, *N*-nitrosodiphenylamine (NDPA), *N-(cyclohexylthio) phthalimide*), antidegradants, antioxidants (e.g., phenolics, phosphites, thioesters, amines, bound antioxidants such as quinone-diamines, miscellaneous antioxidants such as zinc and nickel salts of dithiocarbamates), antiozonants (e.g., *para*-phenylenediamines, triazine derivatives, waxes), anti-reversion agents (e.g., zinc carboxylates, thiophosphoryl derivatives, silane coupling agents, sulphenimide accelerator, hexamethylene-1,6-*bis*-thiosulphate disodium dehydrate and 1,3-*bis* (citranimidomethyl) benzene), plasticisers and softeners (e.g., petroleum products such as petroleum waxes and mineral oils, coal-tar products such as coumarone resin, pine products, synthetic softeners and other products such as vegetable oils and fats) and miscellaneous ingredients (such as peptising agents, blowing agents, bonding agents and pigments).

End of process treatment

The rubber trap used for arresting suspended matters should have holding capacity of at least 12 hr with proper baffles to induce continuous up and down flow pattern. If designed properly, this can reduce suspended solids by 40 to 60%. The equalisation tank should have at least one day detention time. It is preferred to have two eqalisation tanks, each of one day detention time. For a latex processing unit, effluent from the equalisation tank to be sent for neutralisation and chemical treatment by alum and iron salt (about 200 mg/1). Combined wastewater of latex process units also need neutralisation by user of lime and settling of suspended solids by user of coagulants. The settlers/ clarifier should have adequate detention time for removal of suspended solids. The sludge may be taken to sludge drying beds for dewatering.

After this, the effluent should be subjected to the biological treatment. If sufficient land area is not available then the effluent after primary settling may be subjected to an extended aeration activated sludge type biological treatment process.

Before going for biological treatment, it must be ensured that: (i) all the in-plant control measures are adopted and (ii) physicochemical/primary treatment, e.g., rubber trap equalisation neutralisation and clarification steps are incorporated. The above measures will reduce substantial quantity of pollutants particularly BOD and suspended solids. The primary treated affluent can be treated in a secondary/biological treatment unit. It is envisaged to render secondary treatment by adoption of extended aeration activated sludge process. The biological treated effluents should be settled in a secondary settling tank. If there is no constraint of land, the biological treatment could be anaerobic followed by aerobic pond system with the proper dimensions, holding capacity and adequate detention time (10 to 15 days) for anaerobic pond followed by 5 to 10 days for aerobic ponding system. The type of soil and proximity to the wastewater and ground water table condition should be taken into consideration before going for these treatment systems. Protective lining is recommended to eliminate any risk. In place of the anaerobic-aerobic system, an oxidation ditch of detention time of 2–3 days can also be considered as an alternative for treating the effluents of the crumb rubber unit.

33.2.3   Pollution control in rubber manufacturing

Basic polymer and ingredients used for making rubber products, either individually or after combinations make some carcinogenic gases and fumes during mixing, vulcanisation process and even during storage. Also, chemical dust comes in contact with the air, some gets contaminated in water and thus, directly or indirectly, affects living things.

However, waste disposal management is the big issue in the rubber manufacturing industry. The international standardisation forum is continuously reviewing the effect of the use of rubber chemicals in the industry. There is need for use of biodegradable materials in compounding, avoiding human beings direct contact with ingredients, developing techniques that will reduce contamination of particles, gases and solvents released during the different processes of rubber product manufacturing and hence avoiding them from polluting the environment.

Rubber products are manufactured in different stages. Basically, natural or synthetic rubber is used as the basic rubber for the manufacturing process.

The typical manufacturing processes in rubber industry are as follows:

1. Raw material storage, handling and weighing.
2. Milling of raw rubber, mixing rubber and rubber chemicals.
3. Intermediate processes, such as sheeting, extruding and calendering.
4. Component assembly, building and pre-shaping.
5. Vulcanisation or curing of rubber compound.
6. Product finishing, inspection and testing.
7. Packing, storage and dispatch.

In each of the above mentioned operational stages, there is the possibility of environmental pollution through dust, vapours, gases and fumes. They came in contact with air, water, soil and ultimately the environment. Many rubber product manufacturing units discharge their wastewater on land, rivulet and tunnel. Few examples, this pollution includes (but not limited to):oil mixed water, water from mixing mills, water with chemicals from testing labs, product wash water and water with different ingredients from latex plants. This pollutes drinking water resources, land for irrigation, air and ultimately the whole environment. However, no information is available regarding the long term effect on soil, ground water and also on crops. A detailed study needs to be carried out in this regard. There is a need to upgrade production processes so as to decrease the amount of pollution from rubber industry. In the latex industry in order to increase dry rubber content (DRC) of latex, it is essential to remove water and other impurities. Also, in washing the product water is needed. Although maximum care is taken, recycling of water is done, many times, this contaminated water is released into the open ground or tunnel.

This may contain ammonia solution, Tetramethyl Thiuram Disulphide (TMTD), zinc oxide and diamonium phosphate. Appropriate and low-cost methods need to be explored in dealing with wastewater from latex industry as well as in molded, extruded and other fabricated products, tyre and tube industry, reclaim rubber industry. Wastewater generation sources and the major water pollutants are given in Table 33.1.

Table 33.1: Wastewater generation sources and the major water pollutants.

Type of industry	Wastewater sources	Contaminant
Tyre and tube industry	Cooling water bleed and boiler blow down, floor wash down of different units (batch wise), leakages and spills contaminating cooling water, accidental overflow from antitack water tank, sanitary and other miscellaneous water	Oil and grease, rubber fines, antitack agents, etc.
Molded, extruded and fabricated products	Cooling water bleed and boiler blow down (batch wise), rinse and product wash water (batch wise), steam condensate with organic leach (batch wise), sanitary and other miscellaneous water	Oil and grease, rubber fines, antitack agents, powder as fillers, etc.
Latex based products	Cooling water bleed and boiler blow down, product wash water, wash down from latex storage, compounding and transfer areas, from wash and rinse waters, accidental overflow water, sanitary and other miscellaneous water, coagulation/coagulating agent tank wash/over flow	Latex spills, surfactants, coagulants, etc. Ball milling and washing containing flow chemicals including zinc, acetic/formic acid, etc.
Reclaimed rubber	Cooling water bleed and boiler blow down, steam condensate from auto-clave, area wash down of all processing areas, sanitary and other, miscellaneous water	Oil and greases, soluble and insoluble inorganics, rubber particles

During the manufacturing and storage of rubber products, various gases, vapours, fumes and aerosols are emitted due to the leaching out of chemicals and also during vulcanisation that occurs at high temperatures. These emissions consist of volatile components from the original compounding ingredients such as– plasticisers, antioxidants and residual monomers or polymer oligomers, as well as primary and secondary reaction products from the cross-linking systems. The formulae and vulcanisation conditions of individual substances as well as their origin have been identified in numerous studies. These substances are amines, amides, aliphatic/aromatic hydrocarbons, highly volatile sulphur compounds, aldehydes and ketones, phenols and esters. Fumes of solvents may also be present, depending on the manufacturing process, as these solvents are used for assembly and cleaning. Monitoring of the air quality at the workplace requires the implementation of a measurement method that is able to cope with the extremely complex situation in the rubber industry. From material inward stage and storage process, pollution starts.

Generally it is due to the shortage of storage area, humid climate, storage temperature, improper ventilation, storage in contact with other chemicals, inadequate knowledge of product and its maintenance, loading and unloading of material and the use of hooks and thus damaging the bags, etc., are some reasons for contamination as well as spillage of these ingredients. Sometimes they are very casually handled by operators, e.g., without packing, they are openly taken towards the weighing machine and mixing machine. Some of them get in contact with air and may spread unpleasant odour and dust. During the mixing process, fillers are dumped on polymer in the mixing mill. During that process, a lot of chemical dust gets spread into the air and the working area is affected, due to these dust particles.

During the vulcanisation process, fumes from heated or previously heated rubber create unpleasant odour which is a complex mixture of components and the associated hazard unknown. Epidemiological studies suggest that exposure to rubber fume may be a significant factor in the increased incidence of certain type of cancers and the cause of some other diseases. Airborne contamination from rubber fumes is either visible or non-visible. The invisible pollutants are gases and vapours of low molecular weight organic compounds (carbon disulphide and amines) or inorganic (hydrogen sulphide) compounds. The visible pollutants arise from aerosols. These aerosols evolved during the mixing, milling, blending of elastomers or combination of them with chemicals involved in the process. The aerosol fraction of rubber fumes is complex and with unknown composition. The concentration of airborne aerosols is very considerable at the workplace in the rubber industry and it should be monitored by approved trapping and detection methods. The major component of fumes is due to accelerators and their reactions with other ingredients of rubber compounds. For example, the major volatile component of tetramethyl thiuram disulphide (TMTD) and zinc dimethyl dithiocarbonate (ZDMC) accelerated cure, will include carbon disulphide, dimethyl amine and sometimes hydrogen sulphide. In case of the latex industry ammonia solution added to latex causes a strong smell. This can have adverse effect on workers health, especially the respiratory system.

33.2.4   Compounding ingredients and their effect

Generally, rubber compound is made from basic polymer, activators, accelerators, fillers, plasticisers, antidegradants, curatives, etc. The followings are some few examples of the precursors for pollution:

Polymers

Raw or base rubber is the major part of rubber product as far as the weight of complete rubber product is concerned. Natural rubber and synthetic rubbers are used in the rubber industry. Within each type, based on the chemical structure, method of production and functional group present in the structure, there are many grades. Generally, each rubber has its own characteristic smell. We can experience the same in the storage area, during the rubber mastication process and vulcanisation process. It releases complex fumes and smell, which may either be individual or reactant products.

Rubbers are highly flammable. After coming into contact with heat and flame, it ignites spontaneously and generates considerable quantities of smoke while burning. Basic raw materials for synthetic polymer making are petro based products. It should be stored away from energy-generating sources such as boiler and oil storage room.

Fillers

In the rubber industry different types of fillers are used in order to improve the properties of rubber compound as well as for cost reduction. Few examples of such fillers are: carbon black, white fillers such as: silica, whiting, barytes, hard clay, soft clay, etc. These contribute (in weight) the main part of rubber compounds.

Carbon black: Approximately 65% of the carbon black is used in tyre production. The International Agency for Research on Cancer (IARC) conducted a comprehensive review of carbon black and concluded that there is limited evidence for carcinogenicity from carbon black, based on studies in humans. They concluded and classified carbon black as IARC classification 2B-possibly carcinogenic to humans and definitely carcinogenic to animals. Non-cancer respiratory effects in workers who work with carbon black have been reported include, cough, sputum production, bronchitis, chest radiographic opacities (e.g., Pneumoconiosis) and decrements in lung function. Short-term exposure to high concentrations of carbon black dust may produce discomfort to the upper respiratory tract through mechanical irritation. As carbon is the precursor for CO_2 emission and thus the warming of environment, there is the need to partially or completely replace carbon black with ecofriendly and biodegradable material or its use should be kept as minimum as possible.

Silica: In case of crystalline silica dust, dust particle can cause fibrosis (scar tissue formation in the lungs). Silicosis is the respiratory disease of the lungs that caused due to the inhalation of airborne crystalline silica dust. Individuals with silicosis are at increased risk of developing pulmonary tuberculosis, if exposed to person with tuberculosis.

Titanium dioxide (TiO_2): There is inadequate evidence of carcinogenicity of titanium dioxide in humans, but there is sufficient evidence through experimental work on animals, for the carcinogenicity of titanium dioxide. Titanium dioxide is possibly carcinogenic to humans (Group 2B).

Talc: In the case of talc, it is generally used as filler as well as dusting on rubber sheet in order to avoid sticking through adhesion of rubber sheet during storage. During dusting process, dust particles come in contact with the environment at working area.

Plasticisers

Plasticisers represent a special group of ingredients. They are used in the rubber industry as processing additives. They help in improving the processing of rubber as well as modify the physical and chemical properties of the rubber compound. The emission of carbon dioxide due to the consumption of fossil fuels is considerable in the rubber industry. In oils, based on aromatic content the important types are: aromatic oil, naphthenic oil and paraffinic oil. Also, there is one important group, i.e., phthalates that are mostly used in polyvinyl chloride (PVC) and nitrile rubber. These are polar materials. They are released into the environment, either by spillage, in gaseous form, as odours, form complex structure with other ingredients and evaporate during the vulcanisation stage at high temperature. The followings are some of the examples.

Aromatic process oils: They contain polycyclic aromatic hydrocarbons (PAHs) or polycyclic aromatic (PCA) which are carcinogen and can cause mutation. Regulations (European Directive 2005/69/EC) imposed ban of process oils containing ≥ 10 mg/kg (ppm) of PAH since 2011. Aromatic process oils are labelled as carcinogenic under OSHA regulation.

Phthalates: Within the phthalates there are different types. Some of the commonly used are dioctyl phthalate (DOP), di-butyl phthalate (DBP), etc. It has been well-documented that endocrine disruptors, such as phthalates can be additive, therefore even very small amounts can interact with other chemicals to have cumulative, adverse 'cocktail' effects. Phthalates in pure form are usually clear liquids, some with faint sweet odours and some with faint yellow colour. With respect to health effects, phthalates are often classified as endocrine disruptors or hormonally-active agents (HAAs) because of their ability to interfere with the endocrine system in the body. In many countries there is ban on phthalate plasticisers.

Activators

In the rubber industry zinc oxide and stearic acid are used as activators. Generally, zinc oxide is used as a 5 phr (parts hundred rubbers). Recently the tendency has been to use between 2–4 phr and 3 is adequate in most purposes. The reduction is due to its high cost and high density. There is a more recent reason for the reduction of the zinc oxide quantity in rubber compound formulations. The U.S. Environmental Protection Energy (EPA) has classified zinc oxide as toxic chemical. The earliest symptoms of this metal oxides fume is metallic taste in mouth, accompanied by dryness and irritation of the throat in addition toflu-like symptoms, chills, fever, profuse sweating, headache and weakness. These symptoms usually occur within several hours after exposure to zinc oxide fumes and can persist for between 24–48 hr.

Accelerators

These types of ingredients are used to accelerate the rate of vulcanisation during the curing of rubber products. There are many types of accelerators, such as: amines, dithiocarbamates, guanidines, sulphenamides, thizoles, thioureas and thiurams, etc. It is known that rubber fume can contain N-nitrosamines. Many N-nitrosamines are carcinogenic. They are formed incidentally, when certain precursor compounds coexist under favourable reaction conditions. Most commonly, they are formed in the rubber industry by the reaction of an amine with a nitrosating agent. A number of accelerators are based on secondary amines. Few examples: thiurams, dithiocarbamate, sulphanamides-based accelerators, are likely to generate nitrosamines during curing. Nitrosating agents can include atmospheric oxides of nitrogen (NO_x), which are common pollutants from internal combustion engines, particularly diesel engines. The volatile N-nitrosamines are of most concern and they are the focus of the regulation. Measuring of N-nitrosamines emitted from elastomeric compounds is still an important issue that rubber industry has to deal with.

The chemical pathways of nitrosamine formation and degradation in the atmosphere (particularly in the aqueous phase) remain uncertain. Exposure to nitrosamine has been shown to form tumours in laboratory animals and have been linked in epidemiological studies to human cancers, including pancreatic cancer, childhood brain tumours. Nitrosamines are classified by a number of international organisations and regulatory authorities in accordance to their carcinogenicity. Under the International Agency for Research on Cancer (IARC), which is part of the World Health Organisation (WHO), N-nitrosodimethylamine (NDMA) and N-nitrosodiethylamine (NDEA) are classified as Group 2A substances (probably carcinogenic to humans). Earlier studies have shown that N-nitrosamines are evolved from heated elastomers. More recent studies by Ireland and others showed that N-nitrosamines can be extracted in water from elastomers.

Curing agent

Generally, rubber curing is done by sulphur and peroxide. There are also other chemicals used such as: metal oxides, resin curing agent, sulphur donors as a curing agent. They generate fumes and smell during storage as well as vulcanisation process. During sulphur vulcanisation with guanidine type accelerators (e.g., Diphenyl Guanidine-DPG, N,N' Di-O-Tolyguaniidine-

DOTG) and antioxidants based on phenylene diamine, cancerogenic primary aromatic amines (PAA) and isothiocyanates (ITC) are formed inevitable. Respiratory morbidity has been associated with exposure in the rubber industry. Chronic and acute bronchitis, decreased lung function and airway symptoms have been observed. Smith and others also surveyed other symptoms (e.g., headache and nausea) among workers doing vulcanisation.

However, environmental conditions have improved during the last few decades in rubber industry. During peroxide curing, it generates fumes and characteristic smell. In case of dust inhalation, continuous exposure may cause light irritation to the eyes and skin, irritating to nasal mucous membranes and upper respiratory tract. Generally peroxides are heat sensitive. These are to be protected from heat, as they easily catch fire and burn rapidly. In peroxide cured rubber products, sometimes, there is need for post vulcanisation. During this process, a lot of fumes come in contact with air.

Antidegradants

This group of ingredients is used to protect the rubber article from degradation due to oxidation, ozone. These chemicals improve the dynamic properties of rubber products. According to their types, they are phenolic, phosphates, thioesters, amines and other multi functional types. Generally, they are volatile under severe temperature conditions, migrate towards surface of products and may give rise to staining and blooming. Antiozonant, e.g., *N*-phenyl-*N'*-(1,3-dimethylbutyl)-*p*-phenylenediamine (6PPD) partly decomposes and undergoes further reactions after migration in aqueous food simulants forming Primary aromatic amines (PAA), which is a critical hygiene concern.

These are the only few sources of pollution spreading. Besides this, there are many other sources such as: from scraps, cuttings, flash of rubber product, dust from rubber products grinding, different types of solvents that are used for cleaning and adhesive preparation, spillage from bags and oil drums, different types of dusting powders, surface finishing agents, wastewater and chemicals from testing labs, oil leakages from machineries, cuttings, rejected and scrapped products, e.g., tyre.

Suggestions and basic precautions: There is lot of scope for improvement. But some basic attention can minimise the issue at following levels:

1. Processing and storage: At the work area, during operation, masks, goggles, gloves and other safety wares should be used. The work area to be kept neat and clean. To prevent spreading of spilled chemicals, it is to be cleaned immediately. For collecting the material spread during storage and manufacturing process, dry vacuuming is the recommended method. Arrangement should be made for collecting fumes and dust. Fire extinguishers should be kept at working area at all times. Wastewater

treatment related care should be taken. The recycling of wastewater must be done.

2. Selection of material and process: Rubber compound ingredients that are degradable, environment friendly, creates minimum pollution are to be selected. While designing compound formulation, the pollution and health hazard nature of the individual ingredient, should be taken into account. Material safety data sheet should be referred to, for precaution. Product making process should be properly designed where human contact with process-generating pollutants, will be minimum, e.g., the use of machines having closed chambers from which contaminated air, fumes can be collected by suitable collecting arrangement should be employed.

3. Scrap: By using rubber as an auxiliary fuel in cement plants and heating plants, the fuel value of the waste can be utilised. However, large quantities of waste rubber still end-up in landfill. Recycling methods that are applied, to some extent, include: the reuse of scrap rubber in the form of finely ground powder in various rubber materials and the mixing of rubber waste into asphalt. Chemical recovery methods are also used to a limited extent. Examples of these methods include pyrolysis, devulcanisation and the production of reclaimed rubber. The problem, however, is that vulcanisation is an irreversible process.

33.3   Emissions from rubber industry

33.3.1   Air emissions

The manufacture of rubber products exposes workers to a large variety of chemicals as discussed above.

Volatile organic carbon (VOC) emissions are also an environmental concern in the rubber product manufacturing process. VOCs are added to rubber compounds in order to aid in mixing, promote elasticity, produce tack or stickiness and extend or replace a portion of the raw material. VOCs are also used as solvents to degrease equipment and tools and as a type of adhesive during building and fabrication. Typically releases of solvents occur either when the spent solvent solutions are disposed or when degreasing solvents are allowed to volatilise. In some facilities, mold release compounds, sprayed in the cavities of compression molds, produce significant fugitive emission. However, solvents are becoming less of an issue as water, silicon and non-solvent based release compounds are now common.

Emission sources

All possible sources of air emissions from rubber product manufacturing processes are shown in Fig. 33.2.

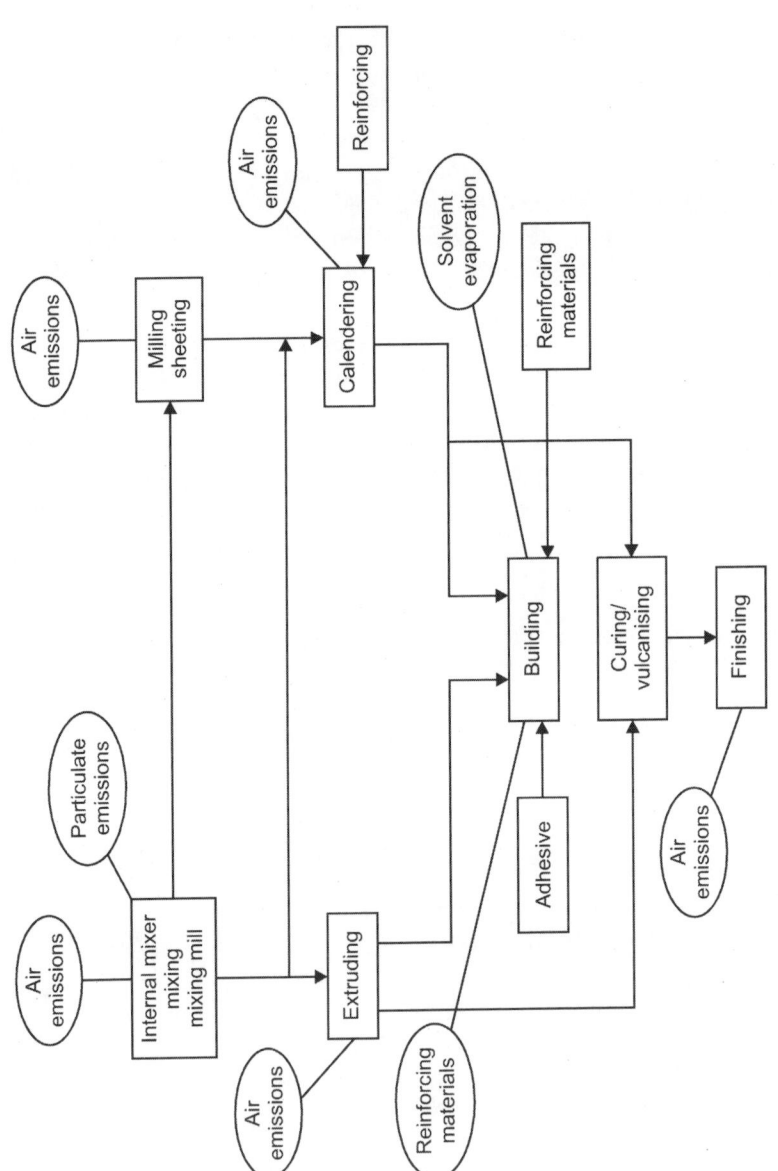

Figure 33.2: Rubber products manufacturing processes showing air emission sources.

33.3.2   Noise pollution

Process operations involving machineries are a source of noise in the in-plant environment. The sources of noise in the rubber products manufacturing industry are attributed to:

1. Machine operation.
2. Generator operation.
3. Boiler operation.
4. Mechanical workshop.
5. Cutting, grinding and finishing operations, etc.

Industrial clusters may also contribute to ambient noise levels.

33.4   Control of air emission and noise mitigation

33.4.1   Air pollution control

Prevention of emission is source controls. Transmission or path control interferes with transfer. Worker or operated oriented controls reduce exposure and uptake by the receiver. Source control is generally accepted as being most effective. Putting control in the transmission or exposure/uptake stages, without controlling the source, means that someone else could be unexpectedly exposed from the same source. Thus the classification according to pollution based process provides a classification of the solutions in terms of where the intrusion takes place.

The following approach should be adopted to control occupational hazards from emissions can take place at any of three locations:

1. At the source of the hazard (source emission control).
2. Along the path from the source to the worker (transmission control).
3. At the worker (reception exposure control).

33.4.2   Source control

Source control can be achieved by eliminating the use of materials causing air emissions by reducing the amount of material used (if possible) by substituting them with less polluting ones, or by changing their form. Some of measures at source control are described below:

Elimination

Elimination generally means avoidance of hazardous substances so that workers are no longer exposed to contaminated air. Elimination can be encouraged by local and national legislation.

Substitution

If elimination is impossible, substitution of less hazardous materials by nature and form is potentially the best way to reduce pollution risk. Elimination or substitution of harmful materials is perhaps the most effective means of engineering control of hazards in rubber manufacturing.

For example, beta-naphthylamine was removed from use because it is known to be carcinogen. Another known carcinogen, benzene, was once widely used as a solvent throughout the rubber industry, it is being steadily replaced by toluene.

Reducing toxic chemical usage by reformulation

Reformulation of compound master batch through trail may help to reduce use of toxic chemical without compromising the quality of product.

Form

It may be possible to effectively eliminate or decrease hazardous exposure by changing the form in which a substance is used, such as:

1. Some dusty materials (fine powders) can be palletised or used in liquid suspension.
2. This control method is particularly useful in the compounding and mixing area of a rubber plant.
3. Powdered raw materials can be pre-packed according to batch intake.
4. Powdered materials and chemicals in rubber industry can be incorporated in a rubber pre-mix for addition to the process, which can minimise the possibility of exposure.
5. Organic solvent can be substituted by water based solvents.

Process control

Significant reductions in fume levels may be achieved by avoiding compound temperatures in excess of process requirements. Fuming products should be cooled as soon as they have finished processing using water, air or by passing cured items over chilled surfaces.

33.4.3   Solid waste generations and disposal

Rubber product industries have most of waste with salvage value and face no problem for its disposal. Cured and off-specification rubber wastes serve as raw materials for inferior quality products in industry. Solid waste like discarded materials other than rubber such as sludge from wastewater treatment plant is disposed as land filling.

33.4.4   Noise control

The noise levels can be mitigated by adopting: (i) engineering measures and (ii) administrative measures.

Engineering control measures involve

1. Structural and mechanical modifications.
2. Proper maintenance of the machines.
3. Provide silencers on compressed air exhausts.
4. Replace old gear drive motors with new belt drive ones.
5. Plug air leaks.
6. Reducing the transmission path of noise from the source to the receiver.
7. Providing proper acoustic enclosures, vibration insulator paddings for the noise generating machines, generators, etc.

Administrative control measures involve

1. Ear mufflers/plugs for workers working in the noisy area.
2. Regulating the length of noise exposures by proper shift scheduling, job rotation and/or by restricting the operation of the noise source.
3. Education of workers.

33.5   Wastewater generation in tyre industry

33.5.1   Water consumption

Tyre and tube industry is basically dry process and water is mainly required for cooling of machinery used for rubber processing at different stages of final product. There is considerable heat generated by compounding, extruding, calendering and molding processes of solid raw materials involved in the manufacturing of tyre and tube and it must be dissipated and controlled to ensure the quality of final product. The other main area for water consumption is boiler for steam generation used in curing/vulcanisation process. The water uses in all possible areas of tyre and tube industry are listed as below:

1. Steam generation (boiler).
2. Non-contact cooling water of machinery.
3. Contact cooling of tread tyre.
4. Make-up water for anti tack solutions and water based spray of green tyres.
5. Washing of floor and machinery.
6. Domestic use in toilets, canteen and gardening, etc.

The industrial water consumption varies widely mainly due to different cooling water systems adopted by tyre and tube industry. The consumption variations are mainly due to the use of once through cooling water in certain plants as compared to recirculation cooling in others.

33.5.2 Wastewater generation sources

The wastewater from the process areas includes water and steam leakages, overflows, runoff from oil-storage areas soapstone solution spillages and wash down and runoff from process or storage areas.

Water leakages occur at various water-cooled machinery units including mills, Banburies, extruders and tread cooling tanks. In addition, water can escape from the hydraulic water system used in Banbury and Press areas. Water and steam leakages occur in the process area due to broken seals, failing bladder bags and overflows from the collection pumps. Oil and solid matter, which have collected on the floor area, scavenged by these various water streams and are carried, untreated to the drainage system. Oil on the floor spaces is lubricating oil, which is dripped or leaked from oil seals of mills, pumps and like equipment, from open gears, from gearboxes and from the hydraulic water system.

In mixing and compounding area, soapstone solution is used for coating rubber sheets to prevent them from sticking together during storage. Spills in the soapstone area are common and may create a wastewater problem. Wastewater problem in the tread processing area arise from the spillage of the solvent base cements, from oil and water leakages from the various mills and from accidental overflows from the cooling water system. The cooling water overflow would not normally be a problem since the rubber tread is relatively inert and therefore does not contaminate the water. However, it does serve as a wash-down agent for an area contaminated with the cements and oils.

Wastewater problems in tyre cords and belts arise due to the latex dipping operation in addition to problems with oil and water leaks and spillages, which are similar to those of the tread process.

In the final operation of grinding and painting involved in tyre manufacturing process, relatively small grinding particles and runoff due to over spraying of the paint may find their way to local drainage system creating wastewater problem. In general wastewater problems arising from compounding, extrusion, molding and curing operations in tube manufacturing are very similar to that of the tyre manufacturing.

Another component of wastewater arises in tyre and tube industry from utility services such as:

1. Once through cooling waters.
2. Boiler blow down.

3. Cooling tower/pond blow down in case of recirculating system.

4. wastewater treatment.

The third component of wastewater is generated from toilet and canteen as domestic wastewater stream.

33.6   Control and treatment technology

1. The first approach adopted by some plants is to combine domestic and industrial wastewater and to treat the entire plant effluent. Generally, the reasons supporting this approach are as follows.

   (a) In-plant drains for domestic and industrial wastewaters are usually combined, thus making combined treatment more attractive.

   (b) Industrial flows in case of recirculated cooling water management particularly for low production line of industries is highly intermittent and of low volume as compared to continuous domestic wastewater generation.

2. The second approach generally adopted by the plants of lower production lines is to use recirculating cooling water pond as a sink and diluting pond of intermittent and low volume industrial wastewater generated from process areas and utility services. Such industries claim to have zero industrial discharge and no treatment to industrial wastewater is given.

3. Once through cooling water as used by some plants is mixed with low volume and intermittent wastewater generated from other sources. Pollution load is diluted to such an extent requiring no further treatment as claimed by some of the industries.

4. The fourth approach adopted by very few plants particularly of high production lines is to control and treatment of a segregated and undiluted wastewater. This approach has been followed in plants having partially or wholly segregated industrial drains. This would, of course, include any plant using re-circulated cooling water. Treatment units adopted by such plant generally include oil segregation followed by settling tank.

5. The fifth in-plant control technology covers segregation and measures for handling, reuse, modification of processing and disposal of various types of wastewaters including spills and leakage, wash downs, control of runoff and house keeping practices. End-of-pipe treatment technology covers the primary treatment given at end of pipe after combining wastewater generated from different sources.

33.7   Reclaimed rubber

Reclaimed rubber is the product resulting from the treatment of ground scrap tyres, tubes and miscellaneous waste rubber articles with heat and chemical agents where the rubber compound is converted to its plastic state by devulcanisation (depolymerisatioin) process. It is then compounded and vulcanised like regular natural and synthetic rubber and in most cases is blended with new rubber to achieve balance in processing and physical properties. Reclaimed rubber manufacturing process involves the following main steps, which are shown in Fig. 33.3.

1. Debeading: Beads in scrap tyres are removed manually for its further processing. In most of the existing plants, debeaded tyres are being received.

2. Size reduction: Debeaded scrap tyre is reduced by mechanical chopping or cracking on a very heavy cracker mill to a suitable size for the devulcanisation step being used. The cracked ground stock is conveyed to a vibrating screen of a given mesh size. The oversized material is returned to the crackers for further grinding. The stock, which passes through the screen, is conveyed for the storage to use in the depolymerisation process. Some reclaiming plants use a series of screens, air separators and sizing equipment to remove/ reduce fibre content from ground rubber scrap.

3. Depolymerisation (devulcanisation): Rubber scrap separation and size reduction is followed by appropriate depolymerisation process. Three basic techniques are used to produce reclaimed rubber are used to produce reclaimed rubber as discussed below.

33.7.1   Wet digester process

The digester process consists of placing the ground scrap, water and reclaimed agents into a steam-jacketed agitator-equipped autoclave (digester). The batch is then cooked for 5–24 hr at 370–405°F. Reclaiming agents used include petroleum and coal tar-based oils and resins as well as various chemical softeners such as phenol alkyl sulphides and disulphides, thiols (mercaptans) and amino compounds.

The reclaiming agents generally function by catalysing the oxidative breakdown of the polymer chain and oxidative disruption of sulphur cross-links. Sometimes, defibreing agents such as caustic soda or chlorides of zinc and calcium and plasticising oils are added to the digester to complete the charge. At the end of the digestion period, the contents of the digester are screened, frequently washed, dewatered and dried in a hot-oven prior to further processing.

Chemical defibreing and the subsequent washing process create an effluent problem. Hence, this process is not opted by reclaim industries these days.

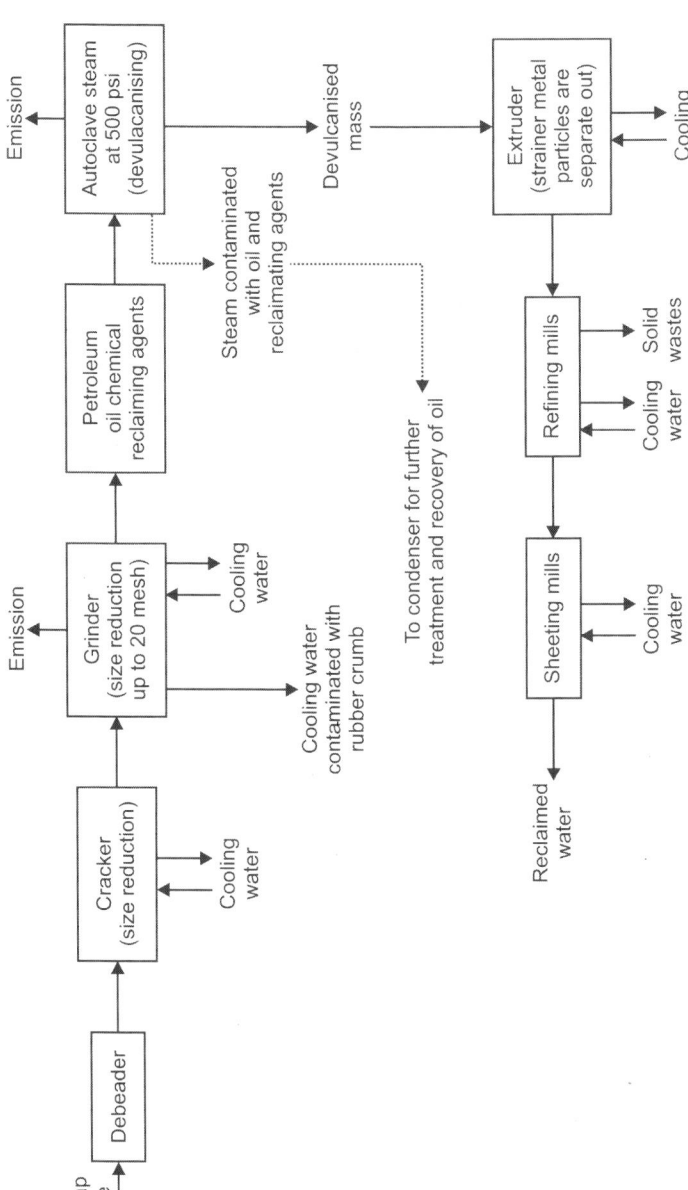

Figure 33.3: Process flow diagram for the production of reclaimed rubber.

33.7.2 Pan (heater) process

The finely ground scrap which is usually free from fibre is blended with the correct amount of reclaiming agents and placed in a single shell pressure vessel (autoclave) into which live steam is passed. Depolymerisation is carried out at about 365° F for 2–18 hr. After this treatment, the heater is vented, the pressure vessel discharged and the crumb of rubber sent on further processing. Since the condensate from this operation is highly contaminated with oils, resins, etc., it cannot be returned to the boilers and therefore, must be treated as a waste. This process is relatively inexpensive because the equipment is simple and the washing and drying steps are eliminated. This process is mainly used in India with some variation, known as dry digester process.

33.7.3 Mechanical reclaiming process

The mechanical reclaiming process, unlike the other two preceding processes, is continuous. The fine ground, fabric free rubber scrap is fed continuously into a high temperature, high shear machine. The discharged reclaimed rubber needs no drying and is ready for further processing. This process is not common in India.

Final processing

The final stage of the reclaiming operations involves straining to remove foreign matter before going to refining and sheeting mill. The strainer is an extruder, which contains a wire mesh screen held between two strong perforated steel plates in the head of the machine. The reclaim is then given a preliminary refining on a short two-roll mill having a high-fraction ratio between the roll surfaces. The reclaim is then sheeted for the final thickness on a sheeting mill. Sheeted reclaim is cooled, dusted with talc to prevent sticking with each other, tested, packed and finally dispatched to the customer.

To sum up, in rubber industry for spreading pollution, different chemicals contribute. Also at different stages it occurs. The described data gives few examples of the chemicals from particular group, e.g., polymer, fillers, plasticisers, activators, accelerator, curing agent and antidegradants. During different stages of manufacturing, there is possibility of spreading pollution and same can be avoided and minimised by proper precautions. To avoid and minimise the pollution, there is need of proper storage, processing techniques, proper knowledge of chemicals, manpower training, proper selection of material and developing the new techniques and methods for scrap disposal and reuse of scrap.

Energy conservation in rubber industry

34.1 Introduction

Rubber is a yellowish, amorphous, elastic material, composed almost entirely of an isoprene polymer, obtained from the milky sap or latex of various tropical plants, especially the rubber tree.

There are many different kinds of rubber, but they all fall into two broad types: natural rubber (latex—grown from plants) and synthetic rubber (made artificially in a chemical plant or laboratory). Commercially, the most important synthetic rubbers are styrene butadiene (SBR), polyacrylics and polyvinyl acetate (PVA), other kinds include polychloroprene (better known as neoprene) and various types of polyurethane. Although natural rubber and synthetic rubbers are similar in some ways, they're made by entirely different processes and chemically quite different.

Synthetic rubbers: Synthetic rubbers are made in chemical plants using petrochemicals as their starting point. One of the first (and still one of the best known) is neoprene (the brand name for polychloroprene), made by reacting together acetylene and hydrochloric acid. Emulsion styrene-butadiene rubber (E-SBR), another synthetic rubber, is widely used for making vehicle tyres. It takes several quite distinct steps to make a product out of natural rubber. First, you have to gather your latex from the rubber trees using a traditional process called rubber tapping. That involves making a wide, V-shaped cut in the tree's bark. As the latex drips out, it's collected in a cup. The latex from many trees is then filtered, washed and reacted with acid to make the particles of rubber coagulate (stick together). The rubber made this way is pressed into slabs or sheets and then dried, ready for the next stages of production. By itself, unprocessed rubber is not all that useful. It tends to be brittle when cold and smelly and sticky when it warms up.

Further processes are used to turn it into a much more versatile material. The first one is known as mastication (a word we typically use to describe how animals chew food). Masticating machines 'chew up' raw rubber using mechanical rollers and presses to make it softer, easier to work and more sticky. After the rubber has been masticated, extra chemical ingredients are mixed in to improve its properties (for example, to make it more hard wearing). Next, the rubber is squashed into shape by rollers (a process called calendering) or squeezed through specially shaped holes to make hollow tubes (a process known as extrusion). Finally, the rubber is vulcanised (cooked), sulphur is added and the rubber is heated to about 140°C (280°F).

Vulcanisation of rubber: Vulcanisation is a chemical process by which the physical properties of natural or synthetic rubber are improved, finished rubber has higher tensile strength and resistance to swelling and abrasion and is elastic over a greater range of temperatures. In its simplest form, vulcanisation is brought about by heating rubber with sulphur. In modern practice, temperatures of about 140–180°C are employed and in addition to sulfur and accelerators, carbon black or zinc oxide is usually added, not merely as an extender, but to improve further the qualities of the rubber. Antioxidants are also commonly included to retard deterioration caused by oxygen and ozone. Certain synthetic rubbers are not vulcanised by sulfur but give satisfactory products upon similar treatment with metal oxides or organic peroxides.

34.2   Energy saving in rubber industry

Energy saving and reducing plays an important role in rubber industry. Energy generally represents up to 10% of direct costs in the rubber processing industry. By introducing best practice, companies can save between 10 and 20% of their overall energy bill. Better control of energy use is achieved by gaining a greater understanding of processes through improved monitoring, management and training.

This closer monitoring of overall processes can have additional positive side effects, such as: (i) better specification of product and (ii) fewer rejects.

Reducing energy use does more than just save money. It also helps your company to reduce its effects on the environment by consuming fewer finite resources and reducing the generation of harmful greenhouse gases. Together, steam, high pressure hot water (HPHW), cooling systems, chilled water, hydraulics and compressed air are directly responsible for some 60% of the energy bill of a rubber processing company. Reducing the energy costs of these utilities is the main focus of this chapter. There are three main areas of a utility system where energy is a consideration–conversion, distribution and use.

34.2.1   Maintenance matters

Some key maintenance matters that affect energy efficiency are:
1. Component wear and mechanical alignment.
2. Bearing condition and lubrication.
3. Water, steam and air leaks.
4. Electrical insulation integrity.
5. Condition of motor commutators, slip rings and brushes.

6. General cleanliness.

7. Optimisation of transmission arrangements.

Rubber processing machinery generally has crucial moving components, such as– extruder screws and injection rams. Wear leads to energy wastage from friction and lower product quality. Misalignment can lead to major failures. Routine visual inspections can identify any problems with wear or alignment. Vibration sensors can be fitted that will identify problems early on. Bearings on all large motors and equipment, including fans and pumps, should be checked regularly. Elevated bearing temperatures can indicate problems which could cause significant damage and vibration techniques can also indicate deterioration. For very large equipment it may be worth installing permanent on-line condition monitoring sensors. Many companies choose to use external contractors rather than maintain in-house expertise and equipment.

Lubricate bearings and gearboxes to a set schedule as recommended by the manufacturer, noting that frequency, quantity and type of lubricant are important. 'Energy-saving' lubricants are available, although additives that reduce friction may degrade some materials in the long-term, for example oils containing chlorine will damage bronze. If you decide to change your lubricant, satisfy yourself that the supplier has test results showing long-term effects (or lack of them). Run some initial trials on equipment that is not critical for production purposes, inspecting the condition after a reasonable interval.

One of the main causes of electric motor burnout is electrical insulation breakdown due to excessive temperature or temperature cycling. Periodic off-line checks of insulation integrity with a meter, although this can be difficult to fit into a tight production schedule, or infrared thermography are useful non-invasive techniques for detecting the first symptoms, or even the causes, before deterioration occurs.

Poor or intermittent electrical contact in electric motors wastes energy. Draw up a schedule for inspecting and cleaning/replacing commutators, slip rings and brushes. Cleanliness of machinery does not generally affect efficiency, but it allows problems to be spotted much faster (e.g., developing oil leaks). Many factories have 'clean room' conditions in selected areas and so their processing machinery is kept spotless. Consider applying this way of thinking to air compressors and boilers. Most rubber processing equipment is driven indirectly (hydraulically), or by electric motors either directly or through a gearbox. Direct drives are inherently more efficient as there is no intermediate link to increase friction losses. However, for ancillary (product handling) or utility equipment, drive is sometimes transmitted through chains or belts.

1. Regularly check chain and belt conditions and adjust tensions correctly.

2. Grease chains at correct intervals or as required.

3. Regularly check conditions of drive pulleys or sprockets.
4. Ensure correct chain or belt is fitted and make sure multiple sets are matched.
5. Check pulleys are still aligned correctly.
6. Some companies fit perspex guards to drive systems, so that visual checks of alignment and tension can be carried out with the equipment on-line. Make sure that any alterations to drive guards comply with all the safety regulations.

34.2.2　Achieving continuous energy saving results

Every employee can save energy and greater commitment by all employees gives higher cost reductions. In today's competitive markets continuous cost reductions are vital and it makes sense to commit to good business management of energy, which is probably the third highest cost to the company. For energy efficiency to be effective on an ongoing basis, it must become part of a company's culture and processes, rather than simply a one-off success as part of a rigorous, but short-term energy efficiency campaign.

The energy management message becomes more effective when reinforced on a regular basis and when all employees become involved, rather than responsibility being left to one enthusiast or energy 'champion'.

34.3　Importance of steam in mixers and mills

Steam is the most common form of heat used in mixers and mills (start-up), in extruders and molds and for space heating and domestic hot water. In the rubber processing industry steam generation accounts for 67% of energy usage. The common elements of a steam heat supply system are:

1. Boiler plant.
2. Steam distribution.
3. Condensate return.

The overall delivery efficiency of a conventional steam heat supply system is at best about 70%, based on the heat ultimately delivered to the user compared with the energy input to the three elements listed above. The key issues for energy efficiency relate to the optimisation of boiler efficiency and the control of heat delivery systems.

34.3.1　Boiler plant

Most boiler plants in the rubber processing industry are gas or oil-fired, with some larger plant using coal. Electrical and compressed air systems support the fuel conversion process. At all stages improved efficiency will bring energy efficiency results. The main stages in operating a boiler plant are outlined below.

Fuel preparation

Fuel preparation is particularly important for coal and oil, less so for gas. For coal, choice of fuel is critical since grates are often designed for quite a narrow selection. Characteristics such as size range and ash content are important. For oil, viscosity is the key characteristic to be controlled. Check on the correct temperature needed to optimise oil atomisation with oil supply companies and monitor the burner oil temperature on a regular basis. Clean oil filters in each shift.

Fuel combustion

Fuel combustion is dependent on both fuel and air provision and contributes significantly to system efficiency. Keeping an even spread of coal on the grate, ash removal, under and over-bed combustion, air pressure and coal height are all critical for optimising coal systems. For heavier oils, cleanliness of the rotary cup, pressure jets, etc., is essential. The pressure of fuel and air is critical, particularly at low turndowns when mixing is poor. Poorly maintained burner systems require higher excess air levels giving lower efficiencies. For fluid systems, oxygen trim systems are now finding wide application, although maintenance is an important cost factor.

Heat exchanger

Heat exchanger performance requires a check on flue gas exit temperature. For steam or hot water boilers a good figure is one which is 30–55°C above the highest steam/water temperature with the boiler at full load. As the thermal load reduces, this temperature difference should decrease since the heat exchange surface area remains unchanged. If this temperature is seen to increase, this indicates that air flow rates are too high and the heat exchanger is operating too far outside its design parameters.

Both fireside and waterside heat transfer surfaces must be kept clean. Fireside with gas firing does not usually present a problem, but oil and coal-fired boilers may need their tubes cleaned on a 12-week and 6-week schedule respectively. In particular, tube cleaning should be carried out when the flue gas working fluid temperature difference increases to 750°C due to fouling. For oil-fired boilers, water injection may help to keep the fireside tubes significantly cleaner.

Gas-fired boilers tend to scale on the waterside, but the timing of cleaning will still be governed by the degree of efficiency loss and its value compared with the cost of descaling.

Boiler feedwater

Boiler feedwater must comply with the relevant standards (e.g., BS 2486 for shell boilers below 25 bar pressure) and the relevant water treatment company's standards. Good control over boiler water conditions based on TDS (total dissolved solids), phosphate level, alkalinity (pH 4–11), etc., will minimise scaling. This benefits not only heat transfer but also the action of float-type level control systems. Packaged boilers generally need to be operated with a TDS of 3000 ppm. Regular checks are required on boiler water to allow feedwater and blow down conditions to be adjusted.

High pressure hot water (HPHW) boilers: For HPHW boilers, testing of boiler water conditions is required less frequently. Both steam and hot water boilers should be scheduled to operate at maximum efficiency, using the minimum number of boilers to meet the demand. This way fuel is burned more efficiently, excess air levels are reduced, air purging before and after burner switching on/off either in high/low/off or modulating burner systems is avoided and ventilation losses are eliminated.

Steam generation

For steam generation, steam and hot water lines should be insulated and steam leakage should be minimised. Lost condensate and flash steam are responsible for significant energy losses. Isolate heat supplies that are no longer needed with automatic valves and also sections where significant losses occur when there is no demand. Check steam trap operation and replace units as necessary. Steam should be generated at the pressure necessary to meet the maximum required by the equipment in the system. Control process and space heating systems to optimum operating times and temperatures. For hot water generation systems, minimise supply temperatures and match pump operation to demand.

Distribute steam at a pressure higher than that required by any user to ensure that the final required pressure is provided. Losses are due to heat emission from pipework and leakage. Heat loss from the boiler casing represents between 2–5% of boiler steam output in well-designed and operated systems. In the rubber processing industry, where steam use has been reduced through introduction of modern processes, many steam systems are poorly loaded so that boiler surface losses are higher than expected and can easily be between 5–10% of steam heat content. Overall efficiency of heat up to the delivery point is therefore 65–70%.

Condensate return

Use condensate return wherever possible. Condensed steam under pressure flashes when reduced to atmospheric pressure conditions in collection tanks. This flash is usually lost and accounts for 43% of the heat in the condensate.

As the condensate at 7 bar pressure carries 26% of the heat in the delivered steam, then the flash represents 11.2% of this figure. If the condensate (now at atmospheric pressure) is lost to drain, this is another 14.8% of input steam heat content lost. Therefore, in sites where all flash steam is lost and only 50% of condensate returned, the overall loss is 18.6%. Heat is also lost from condensate lines. In tyre plants, returns are lower due to the loss of steam from the molds which is vented to atmosphere. Also, condensate that has been in contact with the tyre is considered to be contaminated and is rejected to drain.

If condensate is sent to drain, the overall efficiency of heat supply is therefore 45–50%. Sites with long distribution lines, insulation in poor condition and steam leak problems could have efficiencies significantly lower, making heat supplied as steam at the point of use very expensive.

Leaks: Check the cost of leaks by measuring the length of the steam plume.

Monitoring and targeting

Three measurements are essential for steam systems:

1. Fuel - to measure the major energy input to the boiler plant.
2. Feedwater - to provide data to allow the calculation of a minor energy input to the boiler plant.
3. Steam - to provide data on the major energy output from the boiler plant.

If the values of these inputs are known, the overall plant efficiency can be calculated:

$$= \frac{\text{Heat in steam} - \text{Heat in Feedwater}}{\text{Heat in fuel}} \quad \text{(Aim to achieve 75\% or better)}$$

Blow down will be dictated by the quality of boiler feedwater and will influence the efficiency figure. If, however, there is heat recovery from blow down the change in plant efficiency will be insignificant. Consider electricity use as part of overall energy efficiency (i.e., include pumping and ancillaries for steam generation), although it is common practice to use just fuel. Suggested methods for measuring electricity use are: (i) kWh/litre fuel, (ii) kWh/T steam and (iii) kWh/m³ feedwater.

All these ratios highlight a change in performance. Figures can be calculated based on manufacturers' design data for pumps and fans, not just for boilers but also for auxiliary water treatment, fuel transfer, etc. Target figures need to be assessed for individual boiler plants and take into account plant type and age. Large sites may justify steam metering to specific business areas or even to specific processes, e.g., compounding, molding and curing. A feedwater meter is particularly important, providing a check on water mass flow through the boiler.

Since: Feedwater (kg) = Steam output (kg) + Blow down (kg)

As blow down quantity is usually small and a constant proportion of feedwater input, this relationship can be used to cross-check the main steam meter.

34.3.2 Saving ppportunities in rubber industry

Fuel preparation savings

Heavy fuel oil tank and line insulation: These systems are heated constantly and insulation will usually have a payback period of about two years. Some users experience the formation of an oil layer or skin on the inside of a tank which provides some insulation effect, so it is recommended that tank surface temperatures are checked and calculations made of the value of the heat loss.

Heavy fuel oil temperature control: With some older oil circulation systems, excess oil is returned to the storage tank at a higher temperature than normal oil storage temperatures. Oil at the upper levels of the tank can then be held at a higher temperature than necessary, thus causing higher tank losses. Excess oil should not be returned to the tank but should if possible be returned to the suction of the local circulation pump.

Oil heating by steam: Oil heating systems often use steam (or hot water) with electricity as a standby. Lock settings on thermostats in position to ensure electricity is used only as a stand by.

Gas oil standby to gas: Storing gas oil as a stand by to gas as boiler fuel can save energy and costs, as the costs of heating and pumping heavy fuel oil are avoided.

Fuel combustion savings

Fuel to air ratio control: Excess air is needed to ensure complete combustion. Many systems for controlling excess air levels depend on mechanical linkages between fuel and air flow control equipment. These systems are frequently set up to use more excess air than necessary to avoid poor combustion as linkage settings alter with wear and vibration. Oxygen trim systems avoid these problems, showing savings of 1% when compared with an optimally set burner. In reality, few manually-trimmed burners can be maintained in optimum condition without regular attention between normal service intervals and automatic oxygen trim could add savings of between 1 and 3% of fuel usage.

Water injection: These systems are believed to make savings but the benefit is difficult to quantify, as measurements prior to installation were probably made on the burner in a poorly adjusted state. Various types of system are available using water atomisation and injection into combustion air to improve combustion and claim reduced excess air levels and cleaner fireside heat transfer surfaces as well as reduced NO_x emissions.

Combustion air pre-heat: The conventional method of pre-heating combustion air is heat exchanging with the hot exhaust gases. An alternative way to improve boiler efficiency is by ducting combustion inlet air from high level in the boiler house using the temperature gradient to allow temperatures of 10–20°C above ambient to be obtained. The estimated saving is 1% for a 20°C temperature lift.

Heat exchange savings

Boiler scheduling: Operating a multi-boiler installation so that boilers are always well loaded (above 50%), avoids inefficiencies due to standing losses and boiler purging. It is essential to be able to isolate boilers that are not in use to avoid ventilation losses. Some large plants have management systems that predict heat demand, but the same result can be achieved manually. Boilers should be shut down if the total load, including peaks and troughs, can be handled by fewer boilers. Savings will depend upon the load change but could be in the range 2–5%.

Heat exchanger: Build-up of deposits on the heat transfer surfaces, particularly of coal-fired boilers, results in efficiency losses because exhaust temperatures are higher than with clean surfaces. Vibration at the infra and ultra frequency levels reduces this type of fouling, saving around 1% of fuel. This technique is likely to be viable only on larger coal-fired plants.

Isolation dampers: Fit dampers into boiler exhausts, or build them into combustion air/burner systems to avoid ventilation losses. Plants that have economisers may already have flue gas dampers that could be automated.

Note: Damper operation must be built into the firing control system for safety reasons.

Heat recovery from flue gas: There are three methods of heat recovery from flue gas:

1. Recirculation of exhaust gas (gas and oil).
2. Indirect recovery using a heat exchanger (gas).
3. Direct recovery using sprayed water into the exhaust from a gas-fired boiler.

Recirculation: Gas recirculation is currently used mainly to reduce NO_x emissions in fluid-fired boilers, with higher boiler efficiency being a secondary, but valuable, additional benefit. Since furnace tube sizes need to be larger and extra convection heat exchange surface is needed to cool the recirculating gas, this option can only be considered in new plant.

The cost premium for the new design may be partially off-set by the improved energy efficiency of the system. Careful analysis of manufacturers' data and usage requirements is recommended.

Indirect heat recovery: Indirect heat recovery is perhaps the most common system and is usually applied to packaged gas-fired boilers. Designs are available for oil-fired boilers, but because of the corrosive conditions produced when sulphur dioxide from oil exhaust gases is cooled/condensed, construction materials are expensive and equipment capital cost is high. It is possible that large, well-loaded oil-fired plant could generate sufficient savings to justify heat recovery. Usually where a plant is dual-fuelled, recovery units are installed with a by-pass arrangement for oilfiring where these units cannot withstand the corrosive conditions. In the case of packaged gas-fired boilers, run-around coils, shell and finned tube-type heat exchangers are generally used to recover heat from the exhaust. Savings of up to 4% of fuel fired are available. Heat is usually recovered into boiler feedwater on the pressurised side of the boiler feed pump.

Direct heat recovery: Heat recovery through direct contact between exhaust gases and water may be viable in some large tyre plants. Spray recuperators can produce quantities of warm/hot water in a closed loop system in the temperature range 30–50°C, the end-point temperature being dictated by the energy balance for the system. A mixture of direct and indirect heat exchange systems may be justified on large plants where higher end-point temperatures of about 120°C could be possible, again dependent on the energy balance.

Consider this type of opportunity for heat recovery in large tyre plants at the same time as heat recovery from exhausts on tyre molding and curing plant. In the latter case, heat is recovered essentially from steam/vapour and equipment could therefore be more compact and possibly less expensive per unit of heat recovered.

Feedwater savings

Blow down heat recovery: Water at temperatures equal to that in the boiler is purged from the boiler to control TDS levels. Flashing the water at low pressure (0.2–0.5 bar g) allows clean flash steam to be recovered directly into the feedwater storage tank, while residual hot water is heat exchanged with cold incoming make-up boiler feedwater.

Molding plant heat recovery: A significant quantity of steam is used in 'bag-o-matic' presses in the tyre industry and for autoclaving in rubber processing plants. For heat recovery from the exhaust of these molds a regular flow of waste heat is needed. Flow rates vary significantly and viability is dependent on the heat available.

Distribution savings

Line rationalisation: Optimising steam supply systems will reduce standing losses. Review demands in conjunction with line diagrams of the steam distribution and condensate return system for this updating process.

Line utilisation control: Optimising line usage through the use of slow-opening/closing automatic valves is viable for the larger rubber processing sites or tyre sites where operation is intermittent. These valves need to be linked to central or local control units depending on the size of the site, i.e., consider whether a building management system is needed.

Line pressure control: Sites with a variety of operations working in different shift patterns will cause pressure at the extremity of the system to vary significantly due to pressure loss changes. In this situation, varing the boiler pressure in response to the pressure at the most sensitive part of the system will increase boiler and steam distribution system efficiencies.

Trap maintenance: In general, resources limit the attention which can be paid to steam trap operation to an annual overhaul. However, severe problems need to be identified by checking for steam plumes from condensate receivers that are 'forced' rather than 'lazy' and should be attended to as soon as possible. More sophisticated systems are available to check steam trap conditions, using central monitoring. Alternatively portable equipment can be used for spot checks on individual traps.

Condensate/flash steam recovery: By recovering as much condensate as possible, savings are made in energy, water, water treatment and effluent costs. Apart from condensate that is available at relatively great distances from the boiler house (say 500 m as an upper limit), most condensate is generally worth returning, although each case must be considered separately. Flash steam is more difficult to recover economically and usually depends on considerations at the design stage of a project. Heating boiler feedwater and combustion air will improve overall boiler efficiency. Recovered heat may also be used for space heating in low temperature hot water or low pressure steam systems.

HPHW flow and pressure control: Two factors influence the demand conditions (flow rate and pressure) for hot water in space heating systems:

1. Space heater design.
2. Method of space temperature control.

A heater has an in-built pressure drop that is set by the equipment designer. Since most units are off-the-shelf, the system designer only has the option to oversize the system to decrease pressure drop. However, because there is a square law relationship between pressure drop and flow rate and because most systems are over-designed, spare capacity does exist and if used, can significantly affect pressure drop.

Space temperature can be controlled by varying convective heater fan speeds (rare) or by varying the flow of hot water to any type of heater. Two-way modulating valves are preferred to allow variations in flow rate, or on/off control, but not three-way valves that maintain constant flow rates. When flow rates vary, variable speed drives (VSDs) can be applied to booster pumps in the system to maintain the required system pressure.

34.4 High pressure hot water in tyre manufacturing

High pressure hot water (HPHW) systems are used in the tyre manufacturing process for molding and curing.

34.4.1 Plant and equipment

The systems are pressurised by constantly operating pumps and the water is heated by steam in large expansion reservoirs. Boilers can also be used to generate HPHW and these are similar to those for steam boiler plants. The HPHW operation and maintenance tips are similar to those for steam.

Monitoring and targeting

Heat meters can provide output information where generators are used or where steam/hot water heat exchangers are used to provide HPHW. For hot water boilers, efficiency calculations are more simple than those for steam systems.

However, an overall boiler plant efficiency target of 75% should be used for both. HPHW supply requires more electricity (for pumping) than do steam heating systems and it is necessary to check usage in terms of total energy, i.e., kWh (electricity) + kWh (heat) and total energy (kWh/hour of system operation).

Total electricity usage should cover fans and pumps (boiler or heat exchange circulation and any booster pumps in the circuit). The savings opportunities are the same as those for steam.

34.5 Cooling systems in mixing milling

Cooling is required to remove process heat from rubber and represents almost 2% of all electricity usage in the rubber industry. Heat is generated during compound mixing, milling and extruding. During the mixing/milling process, cooling is applied to the chamber walls on the inside of both the rotors in the mixer and the drums in mills. The extrusion process requires cooling of the extrusion tool and subsequent cooling of the extrudate.

34.5.1 Plant and equipment

Cooling water is usually provided by a recirculating water system. Warm water returns to a cooling system, has heat removed and is then returned to the process. Most rubber processing facilities use wet cooling tower systems where warm water returning from the process is cooled by direct contact with the air. Evaporation is fundamental to the cooling performance of this type of tower and can only occur at the surface of the water.

The warm water is distributed by spray nozzles over a packing material inside the tower. A good distribution of water over the cooling tower and the evaporation of a relatively small amount of water (approximately 3 kg/kW of heat rejected) contribute to 80% of the cooling load. The remaining 20% is achieved by sensible heat transfer to the ambient air.

Water can be only be cooled to 4–6°C higher than the wet bulb temperature of the entering air. The temperature difference between water entering and leaving ranges from 3.5–11°C depending on the application.

34.5.2   Operation and maintenance

Water distribution in the cooling tower: It is important that re-circulated water is evenly distributed over the entire cooling tower packing and that the packing is correctly stacked and undamaged.

Air distribution: Air inlet louvres should remain free of obstructions to allow air to pass through the packing. The tower-fill should also be free of obstructions and must be thoroughly cleaned periodically to ensure correct air and water distribution.

Water temperature control: During periods of reduced process load or when ambient temperatures are low, the cooling tower circulation fan can be cycled on/off according to the cooling water temperature required. Check the thermostat sensing the pond temperature is working correctly and its set point is correct. Avoid short cycling of the fan and long off periods. Short cycling causes excessive 'wear and tear' on the fan drive system and is less energy efficient, while excessively long cycles allow water temperatures to rise above the maximum process requirement.

Water treatment: As the recirculating water comes into contact with a heated surface, a proportion of the water evaporates. Depending on the water hardness, solids are deposited on the heated surfaces.

This build-up reduces the effectiveness of heat transfer resulting in increased energy consumption. Treat water to prevent build-up of solids and microbial growths. Trace heating can be used to prevent freezing of exposed pipes. Trace heating and sump immersion heaters should be temperature controlled so that they do not operate unless dictated by weather conditions.

Control is necessary to ensure the correct temperature for the process, poor control will affect product quality and lead to high energy consumption. Cooling operation should coincide with hours of production as unnecessary operation adds to high energy usage. When cooling water is fed to a refrigeration plant providing chilled water, temperatures should be as low as possible. Chilled water is increasingly being applied to mixer cooling circuits.

34.5.3 Monitoring and targeting

Measuring key parameters allows simple comparisons to be made of typical values for cooling tower performance. Monitor energy usage against set targets, but remember to take into account the relevant variables.

Electrical consumption (kWh) gives the total energy consumed by the cooling tower system over a period of time. Multiply this by the average electricity unit cost to find the utility running costs. These costs will vary according to production levels and weather conditions.

Pressure (cooling water): Pressure gauges at the inlet and the outlet of the cooling water circulating pumps allow the total pressure developed by the pump or pumps to be determined. This pressure can then be checked against the pump curves to give the flow rate of the pump. The pump efficiency can be checked by closing the pump discharge valve and relating the total developed pressure to zero flow on the pump curve. If the developed pressure is less than that shown on the pump curves, an overhaul may be required.

34.5.4 Saving opportunities

Fans

Temperature control: Cooling towers and dry coolers are designed/selected to achieve a specified amount of cooling at a maximum design dry bulb/wet bulb temperature. In reality, these conditions only exist for a small part of the year so for most of the year fans can be cycled on/off according to water temperature. This creates an excellent opportunity for saving.

Pumps

Pumps are selected to deliver a flow rate to meet the maximum cooling requirement of the process overcoming the distribution pipework resistance. Design safety factors are often applied to both the cooling requirement and the pipework resistance, resulting in an oversized pump and motor. This in effect has the motor running at low load and operating at typically 50% efficiency instead of 70%. Pumped volumes can be matched to demand by:

1. Pump sequencing.
2. 2-speed single pump.
3. VSD to single pump.

Variable speed drives (VSDs) should be considered for motors over 10 kW. Combinations of the three options above can be used, e.g., for a set of pumps, the lead pump could have a VSD and the others could be sequenced to maintain a system pressure.

Distribution

Few energy-saving opportunities exist in the distribution of cooling water.

1. Deliver cooling water at the highest possible temperature for the process. Examine separate 'chilled' supplies for small users requiring lower temperatures.

2. Insulate pipes running through high temperature areas to prevent heat gains, e.g., above autoclaves or where they run parallel with steam pipes.

Process

Opportunities for energy saving at user level are mainly related to varying production levels and production times. Cooling water to processes must be controlled if savings are to be made.

34.6 Chilled water in rubber processing

Chilling accounts for just 1% of overall utility energy consumption in the rubber processing industry, but this figure is increasing.

Chilled water can be maintained at a constant temperature all year round, whereas cooling water from cooling towers or dry coolers can vary between 10–30°C during the year. For this reason, chilled water is beneficial for some processes such as mixing.

Typical chilled water temperatures suitable for most rubber processing are 6°C flow/10°C return.

Chilling requires a refrigeration system and most, if not all, is carried out by vapour compression systems.

The plant items can vary in type but operating principles remain the same, incorporating a closed system with the refrigerant fluid contained at all times. In the rubber industry, the chilling requirement is usually quite small and the duties are mostly served by factory-assembled packaged units.

34.6.1 Operation and maintenance

Monitor system refrigerant pressures as this could highlight refrigerant leakage, among other conditions that affect operational efficiency. Loss of refrigerant results in a total loss of cooling but only a slight reduction in power consumed by the compressor. The condenser heat exchange surface must be kept clean. A dirty air-cooled condenser affects plant efficiency in two ways. Firstly, the compressor absorbs more power at higher condensing temperatures and secondly, condenser fans consume more electricity because they operate for longer periods. Similarly water-cooled condensers must be kept clean and free of fouling.

Water temperature controls: It is important that the control thermostat is operating correctly, that the sensing bulb is in the correct location, away from extraneous heat sources. If the set point is too low it will cause the refrigeration system to run for longer and less efficiently.

34.6.2 Saving opportunities

Minimise demand for chilling: Use cooling water where possible. If the process requires cooling over a wide temperature range to a final temperature below that of cooling water temperature, then cooling water can be used for the upper part of the range and chilling for the lower part to minimise the chilling load.

Refrigeration system: Low condensing pressure coupled with high evaporating pressure offers the most efficient conditions for a refrigeration system. Evaporative condensers give lower condensing pressures than air-cooled condensers and should be used where possible. Condensing pressures should be allowed to fall as the ambient temperature falls, to a level that can still operate the expansion device. Electronic expansion valves allow 'floating' head pressure on a system which gives the biggest savings in compressor power (in the region of 20–30%).

Distribution: The chiller must have a constant flow rate through the heat exchanger part to prevent water freezing. Pump control is therefore not straight forward.

34.7 Hydraulic systems in rubber processing

Hydraulic systems account for 2% of the total electrical energy used in the rubber processing industry, forming part of the main process equipment used in the industry. Small individual machines and larger centralised units are both used. Hydraulic systems are used mainly to operate hydraulic presses for rubber curing. Often two hydraulic systems are used: 30–40 bar hydraulic fluid for press closure and about 200 bar on the press piston. The resulting pressure in the mold is usually 35 to 100 bar, which is necessary to exclude air bubbles and achieve accurate molding.

34.7.1 Plant and equipment

Pumps create a flow of fluid, although not pressure (other than that needed to overcome the resistance to flow in the circuit). There are two groups of pumps non-positive displacement and positive displacement.

1. The use of non-positive displacement pumps (typically centrifugal) in power hydraulic circuits boosts the supply to the main positive displacement pump.

2. Positive displacement pumps are the main pumps in hydraulic circuits. The two major categories are rotary and reciprocating.

Hydraulic accumulators are devices for storing energy in the form of hydraulic fluid under pressure. They act as buffer vessels and provide a high flow rate of fluid over a short period of time. A pump with a low delivery rate can be used to charge the accumulator over a long period of time. Accumulators also dampen delivery pulsations and pressure surges in the circuit.

34.7.2   Operation and maintenance

Hydraulic pumps: Lack of maintenance accounts for most efficiency problems with reciprocating pumps. The efficiency of rotary and screw pumps does not deteriorate quite so rapidly.

System cleanliness: About 80% of all breakdowns in hydraulic systems can be directly or indirectly attributed to fluid contamination. Efficiency can be severely affected by increased pressure drop through filters. Carry out regular checks at sampling points built into the system in lines where there is relatively constant flow. Valves with very large pressure drops allow sampling to atmosphere with low fluid velocity. Analyse contamination by particle count or chemical analysis. Clean oil to reduce breakdowns using an in-circuit cleaning filter, or by hiring or buying a portable filter pump unit.

Hydraulic circuit leaks: Hydraulic fluid leakage occurs both internally and externally. Excessive internal leaks reduce system efficiency and generate heat, causing deterioration of fluids. Some internal leakage is designed for component lubrication although wear leads to excessive internal leaks which reduce system efficiency. Checks should be made during regular maintenance. External leaks are not only messy but are often hazardous and costly because of lost fluid replacement. Vibration, mechanical knocks and hydraulic shocks cause leaks as a result of loosened fittings and components, fatigue of pipes and wear between touching parts. Leaks occur through seal deterioration, due to excessive temperatures and/or contaminated fluid.

34.7.3   Monitoring and targeting

The instrumentation requirements and monitoring for a hydraulic system are similar to those for a compressed air system.

Saving opportunities: Good maintenance is responsible for the major energy savings for this utility.

34.8   Compressed air in rubber processing

Compressed air systems account for almost 4% of the total electrical energy used by rubber processors. One of the main uses for compressed air is to power the rams on the mixers, other users are fluidised beds for powder transfer, automatic weighing stations and for control air. Compressed air is used at pressures between 3 bar g and 15 bar g, but is always generated at the highest pressure required in the plant. The most popular compressors are reciprocating and screw compressors with auxiliary equipment being used to provide dry and oil-free qualities. Generally, a ring main is the most effective method of supplying compressed air to a point of use.

34.8.1 Operation and maintenance

The following is a summary of maintenance matters that can affect the efficient operation of compressed air systems.

Compressors: Lack of maintenance significantly affects the efficiency of reciprocating piston compressors, particularly the oil-free type. Typical deterioration is between 10 and 12.5%.

Air treatment: Excessive pressure drop through the dryers/filters wastes energy. Check pressure drops across such components regularly and replace filters as required.

Distribution system: Carry out routine surveys to find any sources of leaks.

Good places to look for leaks are:

1. Condensate traps: Check they are functioning correctly. Automatic drain traps are not always reliable. Modern, electronically operated condensate traps are very reliable.
2. Pipework: Ageing pipework is a prime source of leaks. Replacement of corroded pipework will not only improve the system but also improve system safety.
3. Fittings and flanges: Leaks here are frequently caused by pipe strain due to inadequate supports, inadequate joints or twisting.
4. Flexible hoses: Hoses are used to make connections between the rigid pipe network and points of use. Leaks can be caused by abrasion, deterioration (e.g., heat sources) or sudden mechanical impact.

Dryers

Ambient air typically contains 12.5 g of water per 1 m^3 of free saturated air at 15°C. The rise in air temperature in the compressor prevents condensation, but when air passes through an after cooler a large amount of water condenses. Some water remains as vapour, but if the air temperature falls below its starting temperature (i.e., 15°C) there will be further condensation.

There are many different types of dryer available to remove moisture:

1. Desiccant dryers.
2. Sorption dryers.

3. Deliquescent.

4. Refrigeration dryers.

Cool, clean, dry intake air will lead to more efficient compression. Outside air should be used wherever possible. For every 4°C drop in intake temperature there is a 1% increase in efficiency.

34.8.2 Saving opportunities

Pressure setting: The compressed air system may be running at higher pressure levels than actually required at the point of use. Air should be generated at the lowest possible pressures, allowing for line losses at peak demand. The result is lower line losses and 7% generation energy saved for every bar reduction. Some applications may require air at a lower pressure than the main system and if a separate system cannot be installed, use a pressure regulator.

Avoid no-load operation: A compressor that loads and unloads is using power all the time. When off load, the power can be as much as 60% of the full load power. For compressors up to 1000 litres/sec no-load operation can be minimised when the off-load cycle is relatively long by turning off the compressor. Automatic stop/start control stops the compressor after a period of no-load running, usually 10–15 min and then automatically restarts the machine on demand for air. Examine the on and off-load running times to ensure that the number of stops and starts are within the recommended criteria for the motor.

Piston machines with unloading give the best part-load efficiencies. Screw machines are available with both two-step unloading and modulating control with a changeover switch. Modulation is inefficient at low loads and should only be used if the load is over 75% below this, two-step unloading is more efficient.

Variable speed control of drive motors: Using variable speed drives (VSD) on piston and screw compressors offers many control and efficiency advantages. Consideration should also be given to new integrated VSD compressors.

Multiple compressor plant: In the case of multiple compressor installations, base compressor sequencing on as narrow a pressure band as possible to achieve the minimum generation pressure at all times.

Computer based controls: For multiple compressor control, various micro-processor based systems are available. These have much more accurate pressure control than pressure switches and avoid large pressure differentials and energy waste. Energy is also saved by reducing the period of time that machines in a multiple installation are running off-load using predictive switching which shuts down a machine immediately it goes off-load. When demand increases, the next machine in a rotational sequence will start, enabling the first to stay off, thus preventing excessive starting and eliminating off-load operation.

Heat recovery: Over 90% of the energy consumed by a compressor is turned into heat. This heat is low grade and is usually wasted, but in many cases it can be recovered. Possibilities include pre-heating of domestic water and air heating for space heating. Where the heat recovered can be fully utilised, simple payback periods of under two years are frequently achieved.

Air quality

Drying: Air treatment costs in terms of energy depend on the quality of air required. A desiccant dryer consumes up to 15% of the compressor power for heated regeneration and causes pressure drops on the system of up to 1.5 bar.

Lower quality air can be produced by sorption and refrigeration dryers, which also use energy and create pressure drops. Deliquescent (absorption) dryers do not consume energy and create pressure drops of only 0.1 to 0.4 bar, but they are not regenerative and incur both material and labour costs. These are the least expensive and are very energy efficient, but only produce dewpoints about 60°C below the inlet temperature.

Use the minimum quality air needed. The higher the quality, the higher the energy cost.

Filtration: Filters cause pressure drops and to save energy it is recommended that only the minimum filtration requirement is met. Make sure filters are adequately sized for the duty.

Distribution

Leakage: Leakage is not only a direct source of waste, it is also an indirect contributor to operating costs. As leaks increase, the system pressure drops. Often the solution is to increase the pressure to compensate for the losses, which also increases generating costs. The first step in tackling leaks is to recognise the costs involved and make a commitment to a plant-wide awareness programme to stop them. Regular, continuous attention to the compressed air system, coupled with proper maintenance will lead to effective progress in minimising leaks. Distribution pipework should follow the shortest and most direct route possible to the process.

34.9   Ventilation systems in rubber processing

Ventilation (including local exhaust ventilation systems) account for 4% of the electrical energy used in the rubber processing industry. Ventilation systems are required for compliance with health and safety legislation for many stages of rubber processing, e.g., the removal of airborne powders during mixing, cooling of compound in coolers, extraction of solvents from proofing equipment, extraction of fumes from salt bath curing and cooling of extrudates.

34.9.1 Plant and equipment

Two types of ventilation are used:

1. Dilution ventilation.
2. Local exhaust ventilation.

Ventilation systems can be used for purposes other than the control of airborne contaminants, e.g., removal of heat from extrudate in the coolers.

Dilution ventilation provides a flow of air into and out of, a work area, but does not give any control at the source of the contaminant. The background concentration is reduced by the addition of fresh air but it has little effect on the concentration levels at the process.

Local exhaust ventilation (LEV): Reduces the concentration levels of airborne contamination by removing process dusts, mists and fumes as they are generated. Materials of concern to the rubber processing industry come mainly in powder or pellet form and are handled both manually or in bulk systems. In manual systems, bags are opened and their contents emptied into hoppers above weigh stations. Bulk systems use blowers to convey materials from road tankers to silos for transfer to high-level storage bins. In both handling methods, supply to weigh stations is by gravity. The generation of dust is inevitable whatever the transfer method.

Dust collectors are used in LEV systems to separate large quantities of dust from an air-stream. This is usually achieved with fabric filters in the rubber processing industry. The contaminated air passes through the fabric which captures the particles of dust on its surface. Most of the collected dust can be released by automatically shaking or by blowing high-pressure air (from the compressed air system) back through the filter. Airborne dust falls within the size range 0.1–75 μm, the main dust size likely to be handled in the rubber industry.

34.9.2 Saving opportunities

For electrical energy, fans are the only area to consider.

Usage: Fan power is proportional to the cube of the volume flow rate, consequently large energy savings can be made for a relatively small reduction in volume flow rate, e.g., by modifying inlet terminals to create a high velocity for particulate capture. More efficient dust capture minimises the air flow requirement and a pulley change can slow down the fan accordingly.

Minimise air volumes: Reduce air volumes required to achieve effective dust/vapour capture. Fan power is also proportional to volume flow system pressure. Close down any inlet terminals not in use.

Distribution: Excessive duct leakage will result in a lower air volume flow rate at the inlet terminals and an increase in fan speeds, which unnecessarily uses extra power for effective dust capture at the inlet terminals.

Heat recovery: Large amounts of heated air are extracted and discharged to atmosphere. This is wasteful, as an equal supply of heated outdoor air is required to replace it. Consider using an air-to-air plate heat exchanger or run around coil to recover up to 60% of the heat before it is discharged and reuse it to pre-heat the supply air. The choice of heat recovery device will depend on the ductwork layout, plate heat exchangers are preferred. For every 1 m³/s at 18°C discharged to outside at 0°C, an equal supply of air at 18°C will require 22 kW of heating.

34.10   Role of insulation in rubber processing

Heat plays a major part in rubber processing. Thermal insulation works by providing a barrier that slows down the rate of heat transfer. There are many reasons for insulating process plant and steam and hot water service pipework, the most important of which is cost. Thermal insulation can minimise heat losses and hence reduce the energy input needed to maintain the temperature of an item being vulcanised or 'cured'. The thermal conductivity of insulating materials varies considerably according to the type of material, its density and operating temperature. As most of the savings are achieved by the insulation nearest to the heat source, only 25 mm insulation has been considered. Thicker insulation will save more energy but the return on investment is much less.

34.10.1 Operation and maintenance

Insulation should always be maintained in good condition and needs continual supervision by the maintenance engineer. Take care to ensure insulation (including that of valves) is replaced after maintenance, especially break-downs. Make sure new pipework installations and modifications are insulated immediately.

34.10.2 Saving opportunities

The whole purpose of insulation is to save money and with the help of the preceding charts, savings can be easily quantified.

34.11   Motors and drives in compounding and extrusion equipments

Most rubber factories have large electric motors, particularly within compounding, milling and extrusion equipment. Because of this, demand profile, control and efficiency should be considered.

34.11.1 Demand profile

To assess the demand profile of a motor, its operation needs to be monitored. Portable monitoring systems are available, although for motors above 20 kW permanent metering should be considered. Process demand can be reduced to the minimum through timely maintenance and good settings. Once a motor has been optimised, the demand profile can be compared with subsequent efficiency curves. Control switch off motors during idle periods, either by training operators to do so manually, or installing electronic controls. These can be set to react to production schedules and switch off machine and ancillary drive motors, or to machine drives and switch off or isolate services such as compressed air, cooling water and air extract fans.

Automatic isolation valves are a very good way of eliminating leaks and pressure drops in utility supplies to machines that are not running and so to reduce demands on pump motors, etc. Thermostatic control valves on cooling water systems should also be considered.

Demand for utilities usually varies over a range of values and is seldom on/off. However, utilities are often supplied by several motors, e.g., several air compressors or several water pumps feeding a main.

Match supply more closely to demand by staged switching off of motors, triggered by flow rate reductions or rises in pressure. Look at soft-start systems where motors are likely to be switched in and out regularly.

34.11.2 Motor efficiency

The efficiency with which a motor converts electricity to motive power depends on four main factors:

1. Design and manufacture.
2. Load factor.
3. Operating temperature.
4. Wear and tear.

Design and manufacture

Typical 'standard' modern induction motor efficiencies range from 80–90%. 'High efficiency' motors are a few per cent better at up to 93%. Always verify the actual quoted efficiency (some 'high efficiency' claims can be misleading) and compare manufacturers. The increased cost of a genuinely more efficient motor can be more than offset by the reduced electricity consumption.

34.12 Lighting in rubber processing industry

In the rubber processing industry, electric lighting accounts for 9% of the total electrical energy used. This section considers industrial lighting in general terms and then discusses the application of lighting specifically in the rubber processing industry. Illuminance is the amount of light per unit area. The unit of measurement is lux. There have been substantial developments in lamp technology across a range of lamp types used in industrial and domestic environments: improved lamp energy efficiency coupled with improved colour rendering being the main areas. General lighting is the simplest layout that provides an approximately uniform illuminance over the whole working area. This has the advantage of flexibility but the disadvantage that energy can be wasted in lighting non-critical areas.

Localised lighting is designed to provide the required illuminance in specific work areas or for specific tasks, with a lower illuminance for surrounding areas. This system is not as flexible but would normally consume less energy. Local (or task) lighting provides illumination for a small task area and its immediate surroundings.

General background lighting is provided for circulation and non-critical tasks. This system can be an efficient method of providing adequate task illumination, particularly where high illuminance is required.

34.12.1 Controls

Methods of control fall into three broad categories: manual, automatic control and processor control (intelligent). Manual methods rely on good housekeeping. Automatic controls include time-based, photoelectric control and occupancy-based controls. These turn lighting on and off as appropriate. Processor control, or computer-based lighting management systems, can address every luminaire to programme the appropriate lighting in individual areas.

This intelligent control is usually a function of a building energy management system (BEMS) which would also control lifts, fire alarms, air conditioning systems, etc.

The typical approach to lighting for rubber processing is to provide uniform illuminance over the whole plant area and supplement with local lighting as required. Lighting considerations mainly concern the condition of the environment being lit, such as high ambient temperatures (e.g., positioning near autoclaves) and areas with large quantities of airborne dust, or combinations of both. Use lamps, luminaires and controls appropriate to each situation.

Note: The light output of fluorescent lamps reduces at high ambient temperatures.

Replace discharge lamps as they fail, as energy will still be consumed by the control gear, even with no light output. When lights are not protected by a luminaire, the lamps themselves will need cleaning. Good housekeeping is essential to ensure that areas are not over-illuminated and lights are switched off when not required.

34.12.2 Saving opportunities

Lights should not be switched on longer than necessary, the hours switched on can be reduced in some areas where natural daylight is sufficient. In many cases light output exceeds the actual requirement for a task. Daylight from windows may contribute to lighting, causing lighting levels to be too high. Arrange lighting circuits to allow lights near windows to be switched off when there is sufficient daylight.

Distribution: Consider local lighting to reduce general illuminance levels. Make lighting systems more flexible so that large out-of-use areas do not have to remain lit to provide light for small, in-use areas.

Fittings: Luminaire design for fluorescent lamps has resulted in improvements in efficiency compared with older luminaires. Refurbishment of older luminaires using modern equipment can often result in substantial energy savings in addition to improved visual conditions.

34.13   Heating in curing and mixing rubber

Within the rubber processing industry there is a large variation of heating requirements as some processes, molding, curing, extruding, mixing, etc., give out process heat. However, in the industry generally it can still account for up to 12% of the total fuel energy used.

Heating requirements vary from site to site but are usually associated with the storage temperature requirements for natural and synthetic rubbers and their associated adhesives, chemicals, etc. Heating is also provided for comfort. Direct-fired heaters have never been popular in the rubber processing industry due to possible fire risks. This should always be a serious consideration. The industry prefers steam or hot water heaters.

34.13.1 Operation and maintenance

Maintain burners in the direct-fired units twice-yearly to ensure correct combustion and continued reliability. Units mounted at high level may be difficult to access. Nevertheless, do not neglect them.

Steam and hot water mains may have very high standing losses due to site layout, poor or badly maintained insulation, extended running hours and excessive circulation when demand is low. Maintain doors and draught seals in good condition. Repair water steam leaks and steam traps.

34.13.2 Saving opportunities

This Section is used to discuss cost-saving opportunities in heating systems. There are three main areas for consideration: the building fabric, the boiler plant and space heating systems.

Space heating

1. Avoid overheating.
2. Check thermostats are set correctly. Turning the air temperature down by just one degree can save 5% of heating costs.
3. Turn off space heating in unused rooms.
4. Use frost protection thermostats outside occupation periods.
5. Install zone controls for areas with differing times of use or temperature requirements.
6. Install weather compensating controls. Install accurate thermostats and thermostatic radiator valves (tamper proof type).
7. Reduce temperature gradients.
8. Install de-stratification fans in high buildings.
9. Reduce distribution losses.
10. Install weather compensation controls.
11. Isolate space-heating circuits in summer months.
12. Use local heaters for hot water.

Section VI

Carbon footprints and nanotechnology in rubber industry

Carbon footprint in rubber industry

35.1   Introduction

The term 'carbon footprint' has become a topic of hot discussion all over the world. Carbon foot print can be described as the extent of damage caused to the environment due to some actions. It is the measure of severity of our activities on the environment, especially on the climate change. Many of the activities in our everyday life produce emissions, through the burning of fossil fuels for electricity, heating, etc. These activities have carbon footprint, producing large amount of greenhouse gases, causing a disastrous effect on the environment.

Greenhouse gases and global warming: Greenhouse gases are produced by human activities, which result in global warming. Carbon dioxide is a major gas that accounts for almost 80% of the emissions. Burning of fossil fuels, oil, natural gas and petrol releases carbon dioxide, methane, nitrous oxide, sulphur hexafluoride, perfluorocarbons, etc., are a few other greenhouse gases originating from industrial processes. These gases accumulate and absorb infrared radiation from the atmosphere, affecting the balance between energy received from the sun and the energy that escapes.

The green house gas emission is caused by the production and consumption of fuels, manufactured goods, materials, wood, roads and services. For simplicity of reporting, it is often expressed in terms of the amount of carbon dioxide, or its equivalent of other GHGs emitted. Just as walking on the sand leaves a footprint, burning fuel leaves carbon dioxide in the air, which is called a 'carbon footprint'. Thus the carbon footprint basically relates to the amount of carbon released into the air based on the fuel consumption.

35.2   Carbon footprint in rubber industry

As already discussed above, the carbon footprint is a measure of the amount of greenhouse gases (GHG) produced by our activities in relation to carbon dioxide (CO_2) or carbon. All activities caused by mankind from building our homes, using our cars to flying on holiday can be the subject of carbon foot-printing.

The carbon footprint of rubber is an estimate of all the emissions caused by the production (e.g., farming), manufacture and delivery to the consumer and the disposal of packaging. For example a 1.4 L petrol car emits about 160 g CO_2 per kilometer. So a carbon footprint of 80 g CO_2 for a standard packet of crisps is about the same as driving a typical petrol powered car half of a kilometer. Thus, the carbon footprint is a measure of the exclusive total amount of carbon dioxide emissions that is directly and indirectly caused by an activity or is accumulated over the life stages of a product.

This includes activities of individuals, populations, governments, companies, organisations, processes, industry sectors, etc. Products include goods and services. In any case, all direct (on-site, internal) and indirect emissions (off-site, external, embodied, upstream, downstream) need to be taken into account.

35.3   Adverse effects of carbon footprint

Due to these emissions, there is a rise in the temperature. During the past 100 years, the earth's temperature has risen considerably. It is estimated that if the current scenario continues, by 2100 global temperature may rise in the range of 1.4–5.8°C. This will result in floods in low coastal areas, unpredictable and extreme weather changes with storms, drought and sudden wild fires. The ecosystem will be disturbed and may put some species to extinction. Vital diseases may spread across the globe.

The carbon footprint is assessed in 2 layers:

1. Primary footprint: Primary footprint monitors carbon emission directly through energy consumption - burning fossil fuels for electricity, heating and transportation, etc.

2. Secondary footprint: Secondary footprint relates to indirect carbon emissions (life cycle of products and sustainability).

As a course of concern apropos survival of mankind in our planet on the product carbon footprint (PCF) and life cycle assessment (LCA) is proliferating worldwide through society as a topmost priority concerned in long term survival. The impression that is appertains to carbon footprint (CFP) offer and clues about where we have approached from and where we are heading in the course of an organisation's activities. It is interesting to analyse that organisational environmental behaviour needs an aim within the current economic system.

The manufacturing processers are concerned on sustainable development more than ever before and it that appears to aware of people, planet and profits. Environmental factors concerning the organisation focus more attention on natural resources to keep a balance between corporate biodiversity and process performances. A development necessarily means and that these are social, environmental and economic aspects in the three different perspectives in the organisation.

No expressive development destroys nature. In the current business scenario while businesses aim at result and in a healthy environment growth this is a necessary optimisation of the available resources through an integrated development plan. The ultimate objective of such a plan is to provide of right or propitious opportunities for the environmental well-being.

How much greenhouse gases are associated with a product along its entire life cycle? This question has increasingly become more and more important over the past few years. PCF can help manufacturers to decide which products, processes and organisational innovations they should focus on to reduce greenhouse gas emissions during the supply chain. A carbon footprint measures the total greenhouse gas emissions caused directly and indirectly by an organisation, product, event or person during their life cycle. The starting of carbon management in any activity is the commencement of calculation of carbon footprint. Establishing a carbon footprint (CF) in an organisation means insistence on the organisation to reduce carbon emission and improve efficiency. A CF is measured in tons of carbon dioxide equivalent (tCO2e). CO2e is calculated by multiplying the emissions of each of the six greenhouse gases by its 100 year global warming potential (GWP).

A carbon footprint considers all the six of the Kyoto protocol greenhouse gases: Carbon dioxide (CO_2), methane (CH_4), nitrous oxide (N_2O), hydro fluoro-carbons (HFCs), per fluorocarbons (PFCs) and sulphur hexafluoride (SF6).

There are different varieties of carbon footprints, mainly organisations use are discussed below:

1. Organisational carbon footprint: It considers emissions from all the human activities including buildings, energy usage and industrial process and company vehicles within the organisation boundaries.

2. Value chain carbon footprint: It represents emissions from both the suppliers and the consumers, including all who use and end the life of emissions.

3. Product carbon footprint: In this footprint emissions over the whole life of a product or service, from the extraction of raw materials and manufacturing right through to its use and final reuse, recycling or disposal. Land filling solid waste, industrial waste water sludge, heavy metal included electronic items and hazardous chemicals are adversely affect soil pollutants. Climate change is the main baffling components in environmental cause for concern – its impacts globally are quite menace us.

There has been an increasing national and international concern over the accumulation of green house gasses (GHG) particularly CO_2 and it effect on global warming. The rapid loss of forest cover in world had been a major cause of concern in terms of the environmental impacts. Natural rubber fertiliser input is very low and the surrounding soil appears to be enriched by the abundant leaf fall, biodiversity due to monoculture, excellent agronomic technique and vide variety of crops during immature period, further enhance and environmental justice of credential.

Rubber tree crops as in the case of forest trees are known to function as natural sponges for absorbing carbon dioxide from the atmosphere. Carbon sequestration is achieved through the uptake of carbon dioxide from the atmosphere and its conversion by plants into cellulose and organic matter. The rubber tree botanically known as *Hevea brasiliensis* was first planted at Henarath goda garden in Gampha such as exotic plant from Brazils many years ago from the wilderness of the amazon basin. Later rubber tree established itself as crop for plantation agriculture. Hence, one can expect *Hevea brasiliensis* to behave as a typical tropical rain forest tree that would at least function as efficient as forest trees in carbon sequestration slowdown in soil carbon oxidation and increase carbon fixation and storage. Previous studies indicated that, a rubber tree can fix about 1 MT during its 30 year cycle. Also rubber trees add about 23 MT/hectare of CO_2 to the soil through annual leaf fall, but part of which decomposes and is recycled to the atmosphere. About 23 MT of carbon (84 MT of CO_2) are removed by the trees as latex yield in 30 years. Rubber falls under 'Cash crop-forest cover' category contributing directly to reducing of CO_2 and organically derived natural latex is a unique gift from mother nature in the form of liquid material it is tapped and collected as an environmental friendly raw material.

Carbon footprint from synthetic rubber: On the other hand synthetic rubber requires petrochemicals as a feedstock for its manufacture, using roughly 3.5 times more oil than what is required for a rubber tree plantation. The most significance in natural rubber is low energy in raw material processing and amazing effect of sequestration carbon in their life. There are many significant environmental credits of NR such as ability to lock carbon both in biomass and rubber and functioning as self-sustaining eco-system, with annual leaf fall, branches, fruit, twigs, root hairs. Its main potential lies in its significant capacity to sequester CO_2 in soils and in its synergies between mitigation and adaptation. This potential is best utilised employing sustainable agricultural practices. 'Green credentials' for natural rubber over its synthetic rivals. Some may be confused it has some hidden energy in transportation and other processing stages. When comparing raw material energy consumption, Gigajoule (GJ)/T, with natural rubber it is 16 (very low compared to synthetic rubber processing), polychloroprene 120, styrene butadiene SBR 130, polybutadiene 108, poly-urethane 174 and butyl rubber 174, polypropylene 110 GJ/T respectively.

In developing nations like China, India and Brazil, per capita consumption of raw rubber shows an increasing trend highlighting an increased global demand for all kinds of natural rubber (NR) goods.

Rubber being such an important product that it had paved the way for industry providing employment to millions, one must also question its position as to how much of environmental damage it causes in relevance to other industry? However, the rubber plantations that give us the renewable raw material for an indispensable industry unconsciously help to purify the air we breathe by removing harmful carbon dioxide (CO_2) and indeed a major contributor in reducing global warming. If we can market the green image of natural rubber highlighting the true eco-friendly credentials and carbon sequestration potential of natural rubber plantations, tangible financial gains resulting in rich economies could be achieved by countries blessed with this 'golden gift' of a tree. Rubber manufacturing processes are based on steam obtained through burning fossil fuel and using electricity to generate power in the manufacturing process.

Rubber processing beginning from rubber cultivation and centrifuging is the next critical phase before processing the products. Organic plantation highly encourages farmers to protect environment through low emission and soil protection. Compared to conventional farming organic rubber plantation reduces their emission by 50% due to prohibition of chemical fertilisers.

35.4 Carbon footprint calculation

Once a company has determined its organisational boundaries, it is important that categorising them as direct and indirect emissions and finally choosing the scope of accounting by considering indirect emissions. Direct GHG emissions can be categorised as emissions from sources that are owned or controlled and indirect GHG emissions are emissions which could own or controlled by another company.

There are three defined scopes for GHG accounting and reporting purposes. Emissions from chemical production in owned or are controlled process equipment, emissions from combustion in owned or controlled boilers, furnaces and vehicles.

Companies can reduce their application by different technologies and energy conservation projects. Energy audits are also a very effective technique for controlling energy in organisations. Mini hydro-plants, solar techniques, wind mills and sea wave energy are used by organisations to mitigate this problem to a certain extent. There are three different types of calculation methods for CF, norm these are discussed below:

Bottom-up: The bottom-up approach is based on LCA, a method that estimates the environmental impact of products by 'cradle to grave' analysis. This method is mainly used for estimation of the CF of products and small entities. top-down and hybrid approaches.

Top-down: The top-down approach is used for calculating the CF of large entities such as sectors, countries and regions. Input-output analysis (EE-IOA) is the main method for top-down.

Hybrid approach: The hybrid approach to CF accounting combines the specificity of process analysis (using LCA) with the system completeness of EE-IOA. This approach retains the details and the accuracy of the bottom-up approach (which is especially relevant in carbon-intensive sectors). In the hybrid approach, the first and the second-order process data are collected for the product or service and higher order requirements are covered by input-output analysis.

The methodologies for carbon footprint calculations solve as important tools for greenhouse gas mitigation. The concept of product carbon footprint is commercialised by industries and several other authorities, but there is confusion in deferent stagers and still there is no clear-cut definition. In the process of calculation there are different methods, different methodologies and there is no proper stranded or much disagreement in greenhouse gas selection in calculation.

35.5 Cradle-to-grave

The system boundary for the assessment is cradle-to-grave and the simplified process map (Fig. 35.1) illustrates the key processes and activities in the life cycle of the products assessed. For each life cycle stage the following generic emission sources and sinks are considered and where relevant, the associated emissions are quantified, with reference to embodied emissions of raw material inputs, electricity use, stationary fuel combustion, mobile fuel combustion, land use change, emissions from waste disposal, other fugitive emissions (e.g., refrigerator leakages) and emissions due to use of fertiliser and pesticides.

Some leaders in rubber industry, strongly believed there is no proper standard or policy for carbon mitigations in industry level. But every organisation now concern about pollutions and environment more than ever before due to protect them through given targets. Few leaders suggested, carbon foot print is a window dressing in organisation within a given frame work to achieve marketing agendas in business, no genuine commitment to society.

Organisation which are leading in field or rubber base manufacturing those are still not much aware about greenhouse gas emission of products which they produced, because in the point of export it's not a big damage for them or no big competition when comparing to textile or apparels products. Some of leaders clearly suggested that carbon footprint is not a compulsory for organisations to control under environmental norms, therefore if an organisation would voluntary engaged with this kind of activities and then it will add values to them in the sense of customer satisfaction or marketing. Some of them

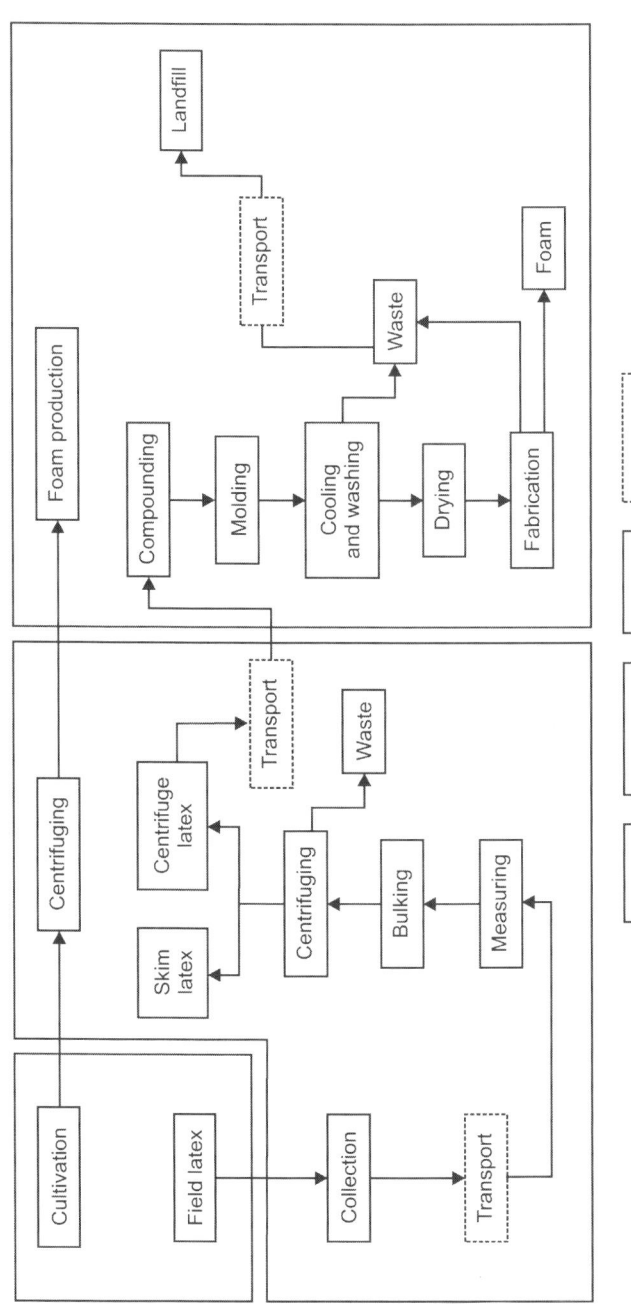

Figure 35.1: Process flow map for foam manufacturing.

commented that industries not ignoring carbon emission due to supply chain, but there is not properly set standard to reduce emission given by government or authorities. Driving and restraining forces related to reduction of greenhouse gas emission in product life cycle are given in Table 35.1.

Table 35.1: Driving and restraining forces related to reduction of greenhouse gas emission in product life cycle.

Driving forces (+)	Restraining forces (−)
Conduct and energy audits and implement electricity controls	No proper established policy
Introduce biomass boilers for steam generation	Huge investment for changers
Reduce emissions from raw material stage	Reluctant to change
Minimise other gasses emissions	No top management support
Management awareness programmes and training of employees	Organisation not allow for money for green activities
Increase sales	No financial benefits
Attitude of owners	Roll of the government

35.5.1 Cultivation, centrifuging and foam manufacturing

The only notable and emission in the cultivation phase of rubber is the inorganic fertiliser applications. Generally nitrogen: phosphorus: pottassium (with ratios of 12:14:14, 18:06:24) is used in all rubber fields with only additional Mg inputs in the form of Keiserite. Emissions related to fertiliser are resulting of both embodied emissions of the fertiliser and also the use phase in the rubber plantations.

The quantity of application differs, since the centrifuging plants have different centrifuged latex contribution to the foam factory, the emission as per allocation from each centrifuge factory is calculated. Each centrifuging facility has different process emissions depending on the production quantity, material uses and the use of electricity (power generators) as indicated in the comparison of process emissions by each centrifuging facility.

Table 35.2 summarises the product carbon footprint for the functional unit of 1 kg of 100% natural foam considering 100% input material supply from each centrifuge plant. Total product footprint was found as 6.67 kg CO2e per kg of foam produced. Total carbon emissions in producing 1 kg of foam was 5.07%. Emissions accounted at the cultivation stage are the emissions due to the application of fertilisers. The value 6.67 kg CO2e means during the process of cultivation to manufacturing of 1 kg of 100% natural latex foam it will have 6.67 times impact than 1 kg of CO_2 would course during a period of 100 years after manufacturing. The value 3.336 kg CO2e means during the process of cultivation to manufacturing of 1 kg of organic latex foam it will have

3.336 times impact than 1 kg of CO_2 would cause during a period of 100 years after manufacturing.

Table 35.2: Total emission in cradle to grave of 1 kg of foam-non organic.

Process	Emission type	Emission kg CO2e/kg of foam
Cultivation	Fertiliser-upstream (cultivation)	1.589
	Fertiliser-yse (cultivation)	1.5231
Centrifuging	Field latex transportation	0.286
	Grid electricity consumption	0.0357
	Use of power generator	0.0009
	Transport and use of machine oil	0.0087
	Use of chemicals (ammonia)	0.1538
	Centrifuged latex transportation	0.0708
Foam production	Emission at compounding	1.0593
	Emissions at molding	0.0426
	Emissions at cooling and washing	1.5509
	Emissions at drying	0.0123
	Emissions at fabrication and finishing	0.0371
	Emissions by other electricity uses	0.2048
	Emissions by waste disposal	0.0067
	Emissions by waste water treatment	0.0025
	Emissions by use of fork lift	No
	Emissions by use of power generator	0.0244
	Emissions by use of packing material	0.0047
	Emissions by use of machine oil	6.6734
Total emissions in producing 1 kg of foam		6.6734

That means non organic foam emissions double due to fertiliser application in rubber plantation. Figure 35.1 also clarifies that both centrifuging and foam manufacturing process do not show differences in emission levels.

Thus, the application of fertilisers during the cultivation would double the emissions in the final product. It further revealed that carbon emission in 1 kg of foam contribute and to global warming of 47% due to cultivation, 45% due to foam production and 8% due to centrifuging, which are the three stages involved in manufacturing of rubber latex foam.

When it comes into foam manufacturing process, GHG emissions due to drying is 52%, molding 35% and 8% due to waste disposal and the rest due to transportations, waste water treatment and other activities and with reference to centrifuging of latex, the emission generated increased to 50.49% in due to chemicals used in the process and the rest used for transportation and power generation. During the product carbon footprint assessment, it was identified that cultivation process is the main contributing factor to company's total carbon emission. However, foam manufacturing process is under direct control of the company, emission reduction opportunities which involves in foam manufacturing can be achieved easily and effectively. Force field analysis revealed that restraining forces are high compared to driving forces for product carbon foot print implementation.

35.6 Greenhouse gas emissions from rubber tree plantations

According to latest findings, 1 kilogram of fresh rubber latex triggers the emission of greenhouse gas equal to 321 gCO2e, 190 gCO2e and 235 gCO2e respectively. Excessive use of fertilisers tends to increase the carbon dioxide emission whereas the size of the planting area has no effect on the release of carbon dioxide. Global warming is presently a major problem spreading throughout the world, causing all nations to take various measures to action in order to reduce its intensity under the KYOTO protocol. The life cycle assessment (LCA) methodology has proved to be a valuable tool in the analysis of environmental considerations. The main purpose of an LCA is to provide a quantitative assessment of the environmental impact. The information on greenhouse gas emissions during rubber trees cultivation is the important and also useful data for the life cycle assessment.

A study of greenhouse gas emission of rubber tree plantations by using life cycle assessment is divided into the following steps:

Goal: To obtain the environmental information on the process of acquiring fresh rubber latex to analyse the environmental impact. The results of input-output based on life cycle inventory (LCI) can potentially be used as indicators.

Scope: The aim of this study is to calculate the amount of greenhouse gas emission during the fresh rubber latex acquiring process, which is measured in units of carbon dioxide equivalent and to study on the factors affecting greenhouse gas emission from the plantation. The greenhouse gas emissions from rubber tree plantations are given in Fig. 35.2.

Quantity of resources and raw materials: The term resources and materials refers to the resources and raw materials needed for each subprocess.

Energy consumption and utilities: Energy and utilities in the process of acquiring fresh rubber latex refer to the energy and utilities requirements for each subprocess, which might be in form of electricity, steam and fuel and the amount of energy is reported in terms of mass or volume.

Air pollution: The air pollution from the acquisition of fresh rubber latex is caused by the supportive systems such as the energy production system and

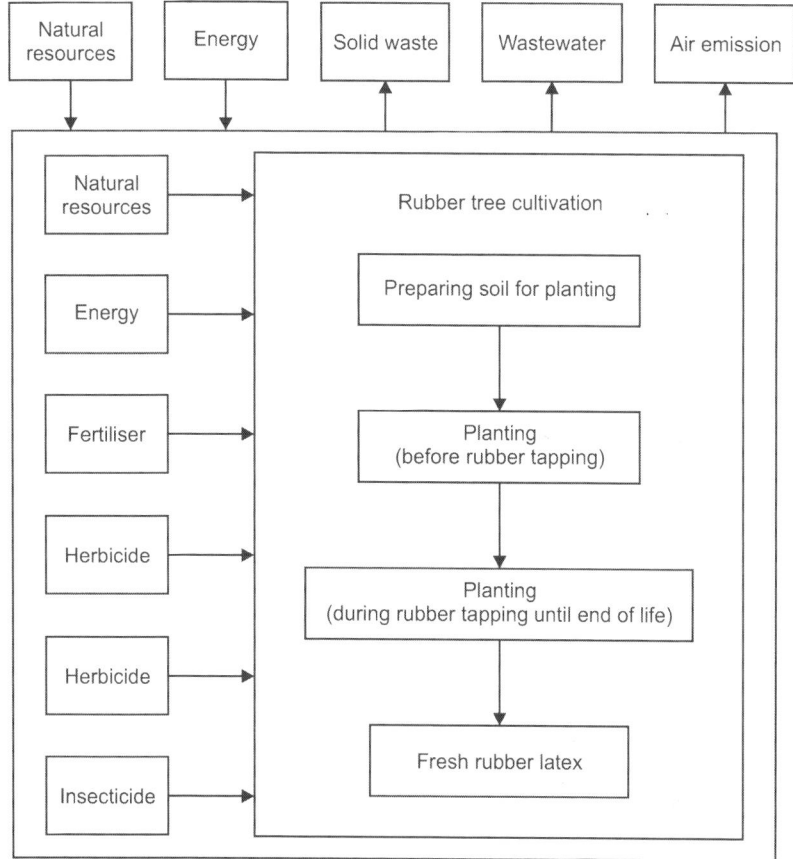

Figure 35.2: Greenhouse gas emissions from rubber tree plantations.

the process of waste disposal. The amount of pollutants emitted to the air is specified in terms of weight per unit of the products.

Life cycle inventory (LCI) analysis: A life cycle inventory is a process of quantifying energy and raw material requirements, atmospheric emissions, water-borne emissions, solid wastes and other releases for the entire life cycle of a product.

35.6.1 Impact assessment

Assessment of environmental impacts will consider raw materials used and energy consumption in the process from inventory analysis.

This study focuses on the evaluation of global warming potential in units of kilograms of carbon dioxide equivalent.

$$\text{Impact score} = \text{Emission factor} \times \text{Inventory value}$$

The carbon dioxide emission of small-scale rubber plantation reveals that the emission from planting before rubber tapping causes the highest environmental impacts. The components contributing to the impact of carbon dioxide emission come from fertiliser consumption more than 98% and almost 2% from pesticide and insecticide usage. While the highest environmental impacts of medium and large-scale plantations are from planting during rubber tapping until end of its life. The environmental impacts of medium scale rubber plantations are 99% a result from fertiliser consumption while the large-scale rubber plantations are 97% caused by fertiliser consumption and almost 3% by pesticide and insecticide usage. Therefore, it is rather explicit that the main environmental impacts of rubber plantation in three areas come from fertiliser consumption. The average fertiliser consumption: nitrogen, phosphorus and potassium in each area shows that the areas with higher fertiliser consumption prone to the increased carbon dioxide emission.

This can be explained by the fact that the nitrous oxide can be produced and emitted to the atmosphere because of nitrogen fertiliser usage. Although there is a little nitrous oxide emissions, when compared to the global warming potential in unit of carbon dioxide equivalent, it is higher than ever. Because nitrous oxide is 298 times the global warming potential of carbon dioxide. Use of fertilisers more than the amount that rubber trees need is even more harmful to the climate.

Thus, it can be concluded the analysis of greenhouse gas emission from the rubber tree plantations by using life cycle assessment (LCA) with the considered reference unit at 1 kilogram of fresh rubber latex and three groups of samples, categorised by the size of areas: small, medium and large. It was found that the emission of carbon dioxide equals to321 gCO2e, 190 gCO2e and 235 gCO2e respectively.

In addition, the results also unveil that the excessive fertiliser consumption over the rubber tree needs causes even higher greenhouse gas emission. Another important factor that affects carbon dioxide emission is the yield of fresh rubber latex. It also suggests that even though the fertiliser consumption is reduced, the quality of product is still sustained.

35.7   Reducing of carbon footprint

The methods of reducing carbon footprint are summared below:

1. The most effective way to decrease a carbon footprint is to either decrease the amount of energy needed for production or to decrease the dependence on carbon emitting fuels.

2.. By improving their environmental and social performances, brands can improve their reputation.

3. Linking business to social and environmental projects enables companies to build a strong connection with consumers by involving them in sustainability initiatives.

4. Technological innovation in production processes, along the supply chain which contribute to improve the environmental footprint of processes and which may save costs, enabling the use of more recycled materials.

5. Developing production processes using lower amounts of water, pesticides, insecticides, hazardous chemicals or lower releases of GHG, etc.

6. Low carbon foot print processes cut costs by reducing waste of raw materials and energy. Water and energy usage reductions by rubber processing and finishing can help reduce global carbon dioxide emissions.

7. By saving energy and water, the rubber industry cannot only save a lot of money, but also help to slow down climate change.

8. By reuseing effluents of chemically treated rubbers.

9. By creating a carbon free environment. This can be achived by planting trees. Trees are the gift of nature to filter our air. They absorb carbon-dioxide and release oxygen. Apart from filtering the air, sustainably managed forests aid multiple environmental and socio-economic functions which will be crucial at the global level in creating a sustainable development. They provide recreational, aesthetic and spiritual benefits.

Role of nanotechnology in rubber industry

36.1   Introduction

In the recent past, the term 'nanotechnology' has become a hot topic among scientists, engineers and technologists and it is widely considered as an enabling technology to produce new materials and products with promising/unique properties in all disciplines of science and technology. Nanomaterials are the core of the nanotechnology and unique/unusual chemical and physical properties of the nanomaterial compared to that of their bulk materials have created a new pathway of developing new products and systems with unique and functional properties. By definition, at least one dimension of the nano material should be in the nanometer range, 1–100 nm. However, even before the emergence of this novel technology, rubber industry had recognised the advantage of nanomaterials in reinforcement as primary particles of carbon black and silica falls in this category.

36.2   Nanomaterials applications in rubber

In recent past, synthesis of different nanomaterials with unique properties compared to that of their bulk counterpart and their applications with polymers have become promising in Research and Development activities in the field of material science and technology. With the understanding of unique properties and functionality associated with the nanomaterials, rubber scientists and technologists have realised their potential as an alternative to conventional materials commonly used in the rubber industry. Rubber nanocomposites based on different nanomaterials have attracted great interest in academia and industry because they exhibit improved mechanical and functional properties with a lower volume fraction of nanomaterials.

Rubber nanocomposite is a new class of composite material in which at least one dimension of the dispersed materials/particles, known as the effective particle, is in the nanometer range (1–100 nm).

Extraordinary higher surface area to volume ratio of nanomaterials and strong interaction between nanomaterials and the rubber matrix would result in superior/unique properties, especially reinforcement and different functionality such as thermal, electrical, barrier resistance with a lower volume fraction of nanomaterials, etc. In general, rubber nanocomposites have following major advantageous over conventional highly filled rubber composites.

1. Lower density and as a result lighter weight products.
2. Improved reinforcement whilst maintaining inherent elastic properties.
3. New functional properties without sacrificing strength and elastic characteristics.
4. Improved flow properties, viscous and elastic properties and as a result easy processing.

Over the past few years, widely researched nanomaterial with rubber is layered silicates which belong to the structural family of 2:1 layered silicates (smectite group). Rubber nanocomposites based on layered silicates, especially montmorillonite clay, have been developed to accomplish a higher reinforcement, to improve barrier resistance, to impart flame retardancy, etc.

However, in addition to layered silicates, other potential nanomaterials such as carbon based nanomaterials like graphene and graphite derivative, carbon nanotubes and carbon fibres have been investigated to make a value addition to rubber, especially to make functional elastomeric material whilst maintaining inherited elasticity and strength characteristics of the rubber.

Commercialisation of nanomaterial enabled rubber products shows a slow progress due to many reasons such as inability to achieve uniform dispersion of nanomaterials within the rubber matrix and weaker interface created between nanomaterial and rubber molecules. Most of the nanomaterials are not compatible with rubber, hence aggregation and phase separation is inevitable, resulting in imperfect property enhancement.

Novel and efficient methodologies are required to improve the efficiency of nanomaterial dispersion in the rubber matrix and to create a stronger interface between rubber and nanomaterials. Relatively little understanding of the behaviour of the interface and nano-reinforcement are the other barriers to develop rubber products with nano-enhancement.

Novel research and innovations: Recently, rubber nanocomposites reinforced with a lower volume fraction of layered silicates have shown a tremendous potential as an alternative to conventional NR composites. However, present layered silicates filled nanocomposites have exhibited limited reinforcement, especially tear strength and abrasion, due to incomplete dispersion/exfoliation of layered silicates and limited interaction between silicate layers and rubber which are prerequisite to achieve desired enhancement in reinforcement.

Nanocomposite group at Rubber Research Institute Shri Lanka RRISL has developed a novel methodology to prepare natural rubber nanocomposite based on montmorillonite clay in which rubber molecules and layered silicates are modified simultaneously during processing. The layered silicate filled natural rubber nanocomposite synthesised with the novel technique has shown extraordinary exfoliation of layered silicates and much stronger interface as a result of good adhesion/interaction between NR molecules and nanostructures of layered silicates. The reinforcement achieved with nanostructures of layered silicates created by the novel methodology is superior to the conventional NR nanocomposites reinforced with layered silicates.

For example, all reinforcement parameters such as tensile strength, modulus, tear strength and abrasion resistance are significantly improved with a lower volume fraction of layered silicates. The nanocomposite developed is targeted to replace conventional highly filled rubber compounds, especially for making environment-friendly greener tyres.

Nanotechnology enabled raw natural rubber, similar to natural rubber (NR) latex crepe rubber, has shown an excellent green strength and improved thermo oxidative stability in comparison to the conventional crepe rubber.

NR sole crepe, a value added form of crepe rubber, used in un-vulcanised form for shoe soles, can further enhance its value addition with nanomaterial applications. Studies are in progress to develop NR sole crepe with improved abrasion resistance, better hardness and improved resistance to thermal oxidation which are important quality control parameters of a shoe sole.

In addition, yellow fraction (off grade), a by-product generated during the production of fractionated bleached crepe rubber, is a significant percentage in crepe rubber industry. Nanomaterial is being researched to accelerate the fractionation process and also to improve the raw rubber properties. Nanomaterial applications to yellow fraction would help to add value to this low grade of raw rubber and thereby introduce new applications for the off grade rubber.

36.3   Benefits of rubber industry from nanotechnology

As already stated above nano-science and nanotechnology is to discover novel behaviours and properties of materials with dimensions at the nano scale which ranges from 1–100 nanometers (nm) and to control and manipulate the nano scale material into useful products/structures.

Materials that are in nano meter scale are extremely smaller than anything we see under conventional microscopes. A nano meter is one-billionth of a meter (i.e., 10^{-9} m) and approximately 10 times larger than the size of hydrogen atom. The size of nano meter scale is very small in comparison to micro scale materials. To elaborate further, it is like the size ratio between nano particle and a football, which is same as the size ratio between football and the earth planet. In the recent past, in addition to thermoplastic materials, rubbers have also demonstrated as very promising engineering materials for various industrial applications.

Although, rubber nanocomposites have showed slow progress in comparison to thermoplastic nanocomposites, a number of recent studies revealed that rubber nanocomposites have a good commercial potential for various applications.

Therefore, it is crucial that we explore the superiority of this novel technology for use in the rubber industry by developing rubber nanocomposite materials with unique properties.

Three major advantageous of rubber nanocomposites over conventional rubber composites are given below:

1. Lighter weight due to low loading level of nano particles.
2. Improved material properties (i.e., mechanical, thermal, electrical) and new functionalities (antimicrobial, barrier, flame retardant).
3. Easy processing in comparison to conventional composites.

Applications: Presently thermoplastic cups are used in estates for field latex collection. Development of new latex collecting cup with nano structured surface which provides a zero adhesion of latex to the latex collection cup and with antibacterial properties would be an advantage in improving the quality of field latex.

Preservation of field latex is another area where nanotecnology can be applied successfully. Application of preservatives, e.g., ZnO and TMTD at nanoscale to field latex would enhance the effectiveness of the preservative system.

Nanotechnology has potential applications in latex dipping. A point to remember when working with nanomaterials is the vast increase in surface area to volume which results as the particle size of the material is reduced to nanoscale. In latex technology, several physical phenomena are dependent on surface area, for example the extent of reinforcement and surface absorption of materials. One of the consequences of the large surface area of any nanofiller is it's effect on latex stability.

Another practical application of this technology has been the blending of an aqueous dispersion of vermiculite with butyl rubber latex. The composite latex has an exceedingly low permeability and is used in the manufacture of tennis balls. Besides these, we would expect many other benefits and openings for the rubber products sector if this technology is developed further.

Challenges: Nano particles and nanocomposites are the trend in polymer technology today. They show surprisingly enhanced material properties and novel functionalities which cannot be seen in conventional fillers and composites. However, commercial applications of rubber nanocomposites are yet to gain real momentum, at present, since there are a few unresolved issues that need to be resolved in order for successful conversion of these novel materials into commercial artifacts.

Most of the rubbers are not compatible with nano particles and as a result, dispersion of these particles into nano scale within the rubber matrix is a difficult task. Therefore, development of new and effective methods to achieve homogeneous dispersion of nano particles such as layered silicates and carbon nanotubes in rubber matrix is a challenge, since the degree of dispersion is one of the key factors to enhance the material performances.

Although rubber nanocomposites exhibit remarkable improved material properties, cost/performance ratio of some rubber nanocomposites is commercially unattractive since some nano particles (e.g., carbon nanotubes, metallic nano powders) are relatively more expensive. This is one of the major obstacles to commercialise these rubber nanocomposites. However, it is expected that the cost of nano particles will come down with the development of new process technologies. In addition, there are few other specific problems in applying this novel technology for certain rubber products, for example tyres. Replacing carbon black with nano scale layered silicates, which impart remarkable reinforcement at low loading levels, is not economically attractive for the tyre industry, at the present stage of development. The rubber nanocomposites based on layered silicates require large volume of rubber in order to keep the total volume of the final product unchanged. All these need to be looked into.

Thus, the rubber industry might have been the first one to use nanotechnology on a large scale, without even knowing it, because the term nanotechnology did not exist then. The use of carbon black to reinforce rubber in fact is an application of nanotechnology. It has been known for a long time that smaller particle sizes of carbon blacks lead to higher reinforcement of the rubber .This has to do with the active surface of the material which increases rapidly with decreasing size of the particles.

Now-a-days, all over the world nanotechnology is a hot item. In almost every field people expect much of it. Apart from the use of carbon blacks, in the rubber industry the influence of nanotechnology is still limited. Generally spoken, reduction of particle sizes to nano dimensions will make the added materials more reactive. This applies not only to carbon blacks but also to activators, accelerators and other compounding ingredients.

Increasing the reactivity means that the same effect can be achieved by the use of less material. So the quantities of chemicals in a rubber product can be reduced which is a good thing for the environment. Whether such a reduction of the quantities of chemicals will also lead to a reduction in costs is doubtful because the prices of nano chemicals are much higher than those of the common ones. In plastics, the use of nano particles produces an increase in strength, ageing resistance and permeability properties. The use of nanoparticles in relation to controlled release of antioxidants and UV stabilisers is presently being investigated. Experiments with nano clay in natural rubber did show already that the use of such materials indeed leads to an increase in stiffness and some improvement of the permeability properties. Furthermore it is claimed that nanotechnology can play an important role in fire resistance of plastics and rubbers. All of this will, of course, require further research. Therefore, considering the present economical situation, the heavy competition in the rubber and plastics industry and the relative conservative nature of the rubber industry it is not soon to be expected to have results in this area. However, we are convinced that in the long run some positive effects of nanotechnology will become visible also in applications with natural rubber.

36.4   Nanotechnology applications in automotive industry

Nanotechnology deals with manipulation of matter at near atomic level to produce new structures, materials, systems and devices that exhibit properties that are unique at these scales. It also involves the production and application of physical, chemical and biological systems at atomic or molecular scale to submicron dimensions and also the integration of the resulting nanostructures into larger systems. Therefore, nanotechnology deals with the large set of materials and products which rely on a change in their physical properties as their sizes are so small. The potential benefits of miniaturisation and the unique properties of nanoparticles open the possibilities of innovation in various diverse fields from health to security and from automotives to airplanes.

This section focuses on the advantages of using nano-sized materials in automobiles to increase their durability and efficiency. It briefly explains diverse venues of application of this new technology in the automotive sector.

36.4.1   Advantages of nano dimensional materials

Mechanical properties

In nanostructured materials there is significant increase in mechanical properties like higher hardness, increased breaking strength at low temperature. This behaviour is attributed to the decrease in grain size when nano-materials are used, which is below the dimension where the deformation does not occur in the grain itself. Hence, traditional polymers can be reinforced by nanoparticles leading to novel materials to be used as lightweight replacements for metals. Such enhanced materials will enable a weight reduction together with an increase in durability and enhanced functionality.

Geometric properties

At nanometer scales, the surface properties start becoming more dominant than the bulk material properties, generating unique material attributes and chemical reactions. Interactions in different media therefore require special physical and chemical properties of the surface of the particles, fibres, pores and the products. With regard to protection function, these demands include resistance against oxidation, corrosion, mechanical abrasions and high temperature.

By increasing the surface area the number of surface atoms increases dramatically, making surface phenomena play a vital role in materials performance influencing the chemical activity of the materials. This is because a greater amount of a substance comes in contact with surrounding material.

Optical properties

Fundamentally, the electronic structure of materials becomes size-dependent as the dimensions enter the nanoscale. Delocalised electronic states as in a metal or a semiconductor are altered by the finite dimensions. Hence, the optical properties, including light absorption and emission behaviour, will be altered. The size of the nanoparticles is comparatively smaller than the wavelength of visible photons also impacts the than the wavelength of visible photons also impacts light scattering, enabling the design of nanocrystalline ceramics that are as transparent as glass.

36.4.2   Nanotechnology in the automotive industry

The automotive industry is an important global driver of growth, income, employment and innovation. Nanotechnology contributes significantly to necessary developments and to the production of innovative materials and processes in automotive sector. The crucial advantages generated by the application of nanotechnology in the automobile sector are: lighter but stronger materials (for better fuel consumption and increased safety), improved engine efficiency and fuel consumption for gasoline-powered cars (catalysts, fuel additives, lubricants), reduced environmental impact from hydrogen and fuel cell-powered cars, improved and miniaturised electronic systems and better economy (longer service life, lower component failure rate, smart materials for self-repair). Nanotechnology if incorporated in the manufacture of automotive components can result in enhanced efficiency levels in different aspects of the automobile.

Nanotechnology for an improved chassis

The safety of the car and the passengers is one the major objective of development of nanostructured materials and structures. Apart from this more flexible, light weight yet high strength nanostructures can be used to increase the performance ultimately reducing the fuel consumption and hence the economy of the operation.

High strength steels for car bodies

Steel is one of the most important materials in the construction of the body of the automotive. Several companies have attempted using high strength steels in car bodies. However recasting the high strength steels in cold state is a difficult task as the size accuracy changes and undesirable spring-back effects may occur. Recasting in a hot state (at 1000°C) helps us to avoid such disadvantages during recasting of high-strength steel parts.

However, the scaling of this kind of steel is difficult at high temperatures. Nanotechnology can be used to solve this situation, a multifunctional protective coating can be produced based on nanotechnological approach with the principles of conventional paint technologies. This multifunctional coating is produced using bonded and connected nano sized vitreous and plastic like materials together with aluminium particles.

Nanotechnology for the shell of the car

Nanotechnology offers an effective solution for scratch resistant, dirt repellent and self-healing car paints. Nano coatings consist of very small particles which facilitates flexibility, quick adhesion and resistance to corrosion and microbial growth.

Nano coats have the following advantages in cars:

1. The better grip of the substrate on the surface does allow another substance to penetrate the surface.
2. Since the size of the nano particles is less than 100 nm it fills glossy sized pits in glossy surfaces and improves the gloss.
3. Nano-particle UV inhibitors are perfectly clear yet they absorb UV light energy.

Another major problem faced by automobiles is driving during heavy rains can be very dangerous due the poor visibility due to the raindrops on the windshield and water sprayed by wheels of other vehicles. This issue has been addressed by the invention of permanent hydrophobic nano-coating for glass surfaces by German scientists.

Nano-coatings with anti-corrosion properties

Conversion coatings are used to prevent corrosion and protect metal body against corrosive materials. Conversion coatings are coatings for metals where the part surface is converted into the coating with a chemical or electro-chemical process. The most significant of these coatings are Cr(VI) and phosphate conversion coatings together with electro-deposition coating. The high anticorrosion performance of Cr (VI) coating is due to its high self-healing behaviour in corrosive environment.

However, Cr (VI) compounds have found to be extremely toxic and their usage has been banned. Phosphate coatings are also toxic. Also, the bath containing these materials leaves a huge amount of sludge. Cr (III) is another option which is less toxic compared to Cr (VI) but it doesn't ensure a long term protection. Nanotechnology has been applied to decimate these disadvantages. A three layer system consisting of zinc layer, Cr (III) enriched layer and a layer of nano-SiO_2 particles is used for this purpose. Each layer has a specific role in the protection of steel. Zinc has a higher negative potential than iron. When it is exposed to a corrosive electrolyte, it produces electrons needed for cathodic reaction and prevents iron from oxidation. As a result, Zn cations produce positive charge at the surface. Whereas, SiO_2 nanoparticles have negative charges. These hence migrate to the corroded area and deposit on it. This phenomenon is called self-healing by nano-passivation.

36.5   Nanotechnology for tyres

Need of mobility all across the world is increasing exponentially. This is also an important prerequisite for the progress of modern society. In the past, automobile has played a crucial role and shall continue to play a dominant role in the progress of society. The demand of automobiles is increasing rapidly across all countries of the world. In order to achieve safety, comfort and environment friendliness, automobile companies are investing heavily in research and development. In this context, nanotechnologies are likely to play an important role. Nanotechnology is opening new doors for innovative products and imaginative applications in automobile sector. This section focuses and critically analyses the improvement in the tyre quality and to increase the life of the tyre by using nano materials in every layer of tyres.

Recently, the innovation trend is moving down the supply chain to the material suppliers, with new additives and nonmaterial's making their appearance, promising to expand further the 'magic triangle' of tyres. Green tyres have now-a-days a market share of about 30% and the demand for tyres of lower rolling resistance, lower weight and superior performance is likely to grow with the market uptake of electric cars.

Tyre performance depends on its cover composition as it is in continuous contact with the road. The composition of rubber plays an important role in its properties. Different properties like abrasion resistance, grip and resistance against tear propagation are important. Incorporation of 30% filler can improve such properties. Type and loading of filler as well as chemical and physical interactions between the filler and rubber are influential parameters.

While the tyre resistance against grip should be high, its rolling resistance has to be low. Also the resistance should not be so low that it allows for car slide. However, reducing friction can negatively influence car safety.

Carbon black, silica and organosilane are the important examples of materials used to enhance rubber properties. Adding such materials to rubber composition at nanometric dimensions can significantly improve tyre properties. The size and surface modification of the particles can affect their chemical and physical interactions with the rubber matrix. This varies the particles cross-linking with natural rubber molecules, affecting its properties. Nano sized soot particles can significantly enhance tyre durability and fuel efficiency. These particles have coarser surface due to their higher surface energy, hence produce stronger interactions with rubber matrix. As a result, inner friction can be reduced which results in better rolling properties.

It is known that strain vibration will occur within tyre material at high car speed. Nanoparticles can reduce this strain vibration and results in superior traction, especially on wet roads. The surface modification of the particles is important which will affect their interaction with rubber matrix and its final properties. It has been found that carbon nanotube (CNT) can improve mechanical properties such as tensile strength (600%), tear strength (250%) and hardness (70%) of styrene-butadiene rubbers. Tyres with higher stiffness and better thermoplastic stability can be produced using lamellar nano-sized organoclays like montmorillonite. The other nanoparticles used to enhance car tyre properties are nanoalumina, carbon nano fibres (CNF) and graphene.

The rolling resistance of tyres can be significantly improved using silane-treated silica compared with traditional carbon black based tyres. As a result, the stopping distance of car on wet roads can be reduced by 15–20%. Also there is a reduction of 5% in fuel consumption.

To sum up, automobile industry is set to be influenced by the development taking place in the field of nanotechnology. Due to the small size of nano-materials, their physical and chemical properties (e.g., stability, hardness, conductivity, reactivity, optical sensitivity, melting point, etc.), can be manipulated to improve the overall properties of conventional material. Nanotechnology is science and engineering and it is all about practical applications of physics, chemistry and material properties. Nanotechnology will influence the auto industry initially on a very small scale, but will certainly be developed to deliver features, products and processes that are almost unimaginable today.

36.6   Nano particles in automobile tyres

36.6.1   Advancement or level of usage of nano particles in present tyres

Different nanoprene grades can be used to satisfy different requirement profiles for tyres and various tyre components (tread, side wall, carcass, etc.), in line with their glass transition temperature. A Japanese company Lanxess recently began commercial production of the material. Its first customer, toyo tyre and rubber, will use it in winter tyres.

1. Nanobase is a nano-molecular structure at the bottom of the strong cap of the tyre, improving grip and steering properties, while also reducing heat emission and therefore rolling resistance, used in the Nokian WR A3 tyre.

2. Nanopro-tech (nanostructure-oriented properties control technology), a nano coating for the tyre tread, which reduces heat generation, used in the new Ecopia tyre range of Bridgestone.

3. Tyres enhanced with CNT (carbon nanotubes) appear to have improved mechanical properties, such as tensile strength, tear strength and hardness of the composites, by almost 600%, 250% and 70% respectively, comparing with those of the pure SBR composites (styrene-butadiene rubber).

4. A nanoclay containing BIMSM (brominated isobutylene- co-para-methylstyrene elastomer), developed and commercialised by ExxonMobil, shows increased air retention properties that exceed those of halobutyl rubbers by about 50%.

5. Lamellar nanomaterial organoclays, e.g., Montmorillonite Clay (MMT) developed by Pirelli give the tyre an isotropic behaviour (equal performance in longitudinal and lateral directions) and a better trade-off between handling and comfort, while also exhibiting higher stiffness, better thermoplastic stability and reduced decay.

6. Polyhedral oligomeric silsesquioxanes (POSS).

7. Nano oxides (silica, alumina).

8. Carbon nano fibres (CNF).

9. Graphene (delaminated Graphite).

10. Poly(alkylbenzene)-poly(diene) (PAB-PDM) nanoparticles (polymer nano-strings).

Scheme 1

1. The tread slab is placed on top of the belt system in the manufacturing process.

2. The tread usually contains two rubber compounds:

   (a) The tread base compound adheres to the belt system when the tyre is cured, is cooler running improving durability and helps stabilise the under tread area of the tyre.

   (b) The tread cap is typically made with an abrasion resistant, higher grip rubber compound, which works with the tread base and tread design to provide traction and mileage.

Here to increase the wear resistance and grip nano silica is added to the rubber. Polymer nano composites of natural rubber or styrene butadiene rubber (SBR) with 10% silicon dioxide (nano silica) and 3% multiwall carbon nanotube (Fig. 36.1).

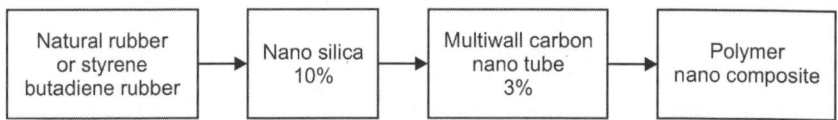

Figure 36.1: Polymer nano composite by adding nano silica and multiwall carbon nano tube in NR (SBR).

To increase the wear resistance and grip, 10% silicon dioxide is added with SBR or natural rubber. To increase the tensile strength and hardness of the tyre 3% mutiwalled nanotube is added to SBR or natural rubber .with these 10% silicon dioxide and 3% carbon nanotube we are going to get a new tyre with improved properties. The improvement by 600% in tensile strength , 250% in tear strength and 70% in hardness is expected.

Scheme 2:

In this scheme 10% silicon dioxide and 3% montmorillonite clay is added to the natural rubber or (SBR) addition of this will give an isentropic behaviour, i.e., equal performance in longitudinal and lateral directions and better trade off between handling and comfort, also improving the stiffness, thermoplastic stability and reduced decay. The nanosilica will improve the wear resistance and grip (Fig. 36.2).

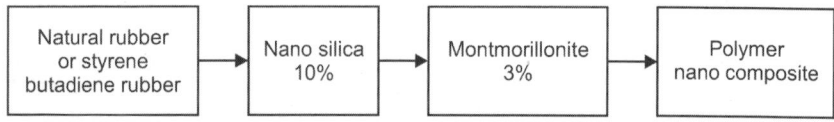

Figure 36.2: Polymer nano composite by adding nano silica, montmorillonite and NR (SBR).

Process for scheme 1 and scheme 2 is shown in Fig. 36.3

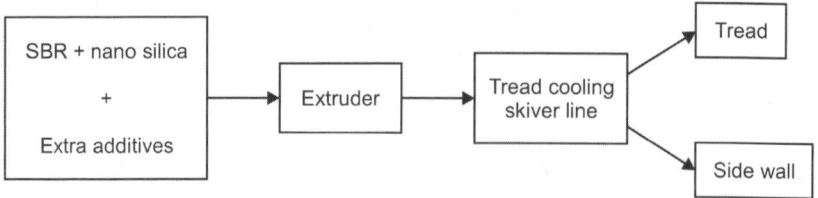

Figure 36.3: Process for scheme 1 and scheme 2.

Scheme 3:

The system, which relies on slurry of tiny particles of magnetite, a form of iron oxide. Here the magnetite nanofluid flowed (Fig. 36.4) through tubes and was manipulated by magnets placed on the outside of the tubes. The magnets attract the particles closer to the heated surface of the tube, greatly enhancing the transfer of heat from the fluid, through the walls of the tube and into the outside air. Without the magnets in place, the fluid behaves just like water, with no change in its cooling properties. But with the magnets, the heat transfer coefficient is higher. Now with these improvements, it is going to be a better tyre in automobile industry.

Figure 36.4: Magnet attracting particles closer to heated surface of the tube.

36.7 Reducing zinc oxide in rubber industry by mixed metal oxide nano particles

Zinc oxide is a widely used compound in rubber industry due to the excellent properties that shows as activator for sulphur vulcanisation. The tyre industry remains the largest single market for ZnO. Traditionally, ZnO is used in rubber formulations in concentrations of 3–8 parts per hundred rubber (phr).

Despite its superior characteristics, there is an increased concern about the environmental effects that zinc oxide causes and over the years lower levels of zinc have been tried in order to decrease its impact and to minimise the production costs. Different approaches have been considered for reducing zinc levels. Between all the alternatives proposed, the use of nanosized ZnO particles with high surface area seems to be promising. However, it was found that the use of more active forms of zinc oxide did not substantially reduce further the minimum zinc content that can be achieved with conventional zinc oxide although the dispersion of high surface area ZnO during mixing was found to be significantly better, which could enable low levels of this zinc oxide to be used in industry with more confidence.

There have been a number of studies comparing different metal oxides as vulcanisation activators in order to find substitutes for zinc oxide. Several metal oxides have been used, CaO, MgO, CdO, CuO, PbO and NiO. Among them, MgO is the most promising candidate since it is a non heavy metal oxide that provokes the breakdown of the accelerator to be faster than when ZnO is used and it is able to form active sulphurating agents. However the cross-link level achieved is lower than that obtained with zinc oxide, which has limited its industrial application.

In this section, a new approach to overcome the problems between ZnO and MgO is presented. It consists in the development of a new activator based in the mixture of both mixed oxides at nanoscale to take advantage of the behaviour of both zinc and magnesium oxides as nanoparticles. The new activator is nanometer – sized mixed metal oxide particles of zinc and magnesium ($Zn_{1-x}Mg_xO$) with very precise stoichiometry prepared employing a polymer–based method. In this accelerator, magnesium is incorporated into the ZnO structure and this inclusion and its size are expected to show a better performance taking advantage of the behaviour of both ZnO and MgO in sulphur vulcanisation.

Basically, the method consists on the preparation of a polymer/metal salt complex that is water-soluble, its purification by precipitation/redissolution cycles and finally the calcination of the dried purified complex to give nanosized crystals. The polymer used to form the polymer/metal salt complex is poly [acrylic acid]. Magnesium nitrate hexahydrate and zinc nitrate hexahydrate are the starting materials.

Dynamic light Scattering was employed to measure the particle size of the $Zn_{1-x}Mg_xO$ particles, which was found to be is in the range of 100 to 175 nm with a narrow distribution. No apparent dependence of the particle size with the magnesium content was found.

X–Ray diffraction can be employed to characterise the crystal structure of the mixed metal oxide particles. The X-ray diffraction patterns of the pure ZnO are indexed according to the known hexagonal phase (zincite) and that of MgO is indexed according to its cubic phase (periclase).

The model compound vulcanisation (MCV) approach with squalene as a model molecule for natural rubber and N-Cyclohexylbenzothiazole-2-sulphenamide (CBS) as accelerator has been used to study the role of the mixed metal oxide along the reaction. The results obtained with $Zn_{1-x}Mg_xO$ nanoparticles as activator for sulphenamide accelerated sulphur vulcanisation have shown when $Zn_{1-x}Mg_xO$ nanoparticles are used it is possible to take advantage of the behaviour of both ZnO and MgO in sulphur vulcanisation. It has been seen that the reactions that take place during the scorch time, the breakdown of the accelerator and the formation of MBT occur faster, which could be due to the presence of magnesium into the zinc oxide structure.

Nevertheless, the cross-link degree achieved is higher than those obtained with zinc oxide nanoparticles. It is worth noting that mixed metal oxide nanoparticles of zinc and magnesium lead to around a 30% higher cross-link degree than the one obtained with standard zinc oxide. This effect can be partly attributed to the small particle size of the $Zn_{1-x}Mg_xO$ since Bhowmick and others found that ZnO nanoparticles (30–70 nm) increased the cross-link degree by 15 % compared with standard ZnO.

On the other hand, the fact that, even with bigger sizes, higher amounts of cross-linked products are formed suggests that $Zn_{1-x}Mg_xO$ nanoparticles are more active and more effective transporting sulphur into the hydrocarbon chain than ZnO nanoparticles. Therefore, $Zn_{1-x}Mg_xO$ nanoparticles not only overcome the disadvantages of the use of a mixture of ZnO and MgO reported, which shows a cross-link degree similar to the one obtained with magnesium oxide, but a better performance is achieved.

36.8   Nanotechnology in sports equipments

Nanotechnology has even found applications in the wide field of sports. Within the niche of sports, nanotechnology has proven to very useful and has the potential to improve a broad range of aspects of the sports world.

Scientists are always looking for new and innovative ways to improve existing products and sports equipment is no exception. Already, scientists have found numerous applications of nanotechnology to improve current sports technology. These improvements range from creating stronger, yet lighter, golf clubs to taking away the odour normally associated with dirty sports clothing after it has been used.

The game of tennis is a prime example of how nanotechnology is having an interesting impact on sports equipment. According to present research, equipment producer Wilson has been able to create tennis racquets that are twice as stable conventional racquets and up to 22% more powerful. This increase in the racquet's performance capabilities can lead to big speed increases in what is already an extremely fast paced game. In addition to racquet research, Wilson is also conducting research to find innovative ways to improve the tennis balls used today. This research has allowed them to design balls that hold their bounce much longer than is seen in balls currently available on the market. This is accomplished by bonding microscopic balls of butyl rubber with clay particles. This mixture is then applied to the inner layer of the ball, creating an airtight but still flexible boundary that keeps the gas inside the tennis much longer. A visual explanation of the results of this new airtight boundary is shown in Fig. 36.5.

The impact of nanotechnology on the level of competitiveness of sports in present times became more evident than ever during the 2012 olympics.

Nanotechnology is allowing scientists to create new, ultra-lightweight swimwear that allows the swimmers to practically glide through the water. In fact, testing has shown that the developers have been able to reduce the water

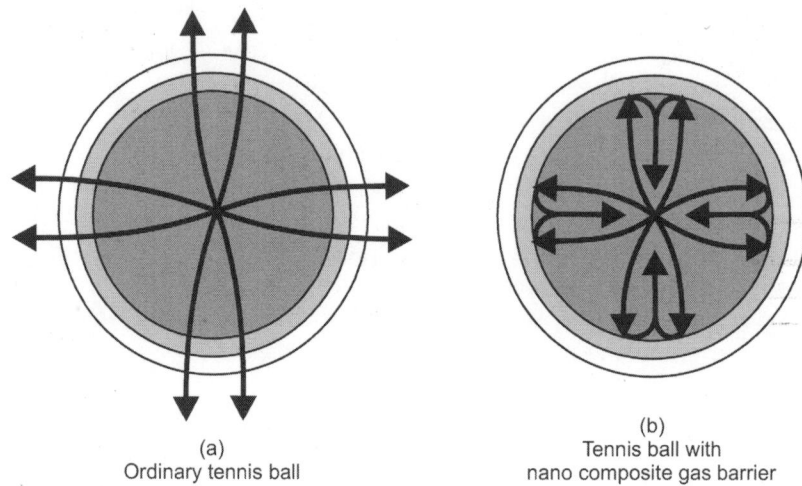

(a)
Ordinary tennis ball

(b)
Tennis ball with
nano composite gas barrier

Figure 36.5: (a) ordinary tennis ball and (b) new Wilson tennis ball coated with gas barrier.

absorbed by the new fabrics to only 2% of fabric weight. This is an amazing breakthrough compared to the 50% absorption of previous materials made specifically for low drag swimwear. These new materials also have huge potential in other sports and applications. Because the material rejects moisture, these fabrics can be used to manufacture clothing that repels sweat, leaving the athletes dryer. It can also be used for bicycle riders who wish to cut down on the drag caused by their clothing.

Like swimming, where friction reduction is an important concern, nano-technology has made it way into the world of racing. In particular, it has made land speed record setting cars even faster. To prove this, Nanotec-USA applied their Nano-Bionics treatment to the Nish Motor Sports' speed car prior to the speed week 2012 at the Bonneville Sand Flats in Utah. Despite using a 50% less than usual mixture of Nitro Methane fuel mixture, they were able to obtain a speed that was 37 miles per hour faster than their previous top speed. This state-of-the-art treatment makes any surface that it is applied to ultra smooth. 'The self-assembling nano-particles dock directly to the molecules of paint and metal, within the pores of the material. The NanoBionics Smart Particles/ molecules self-align themselves, forming new structures.' This product also has the beneficial side-effect of making any surface it is applied to much easier to keep clean. Simple soap and water methods of cleaning will keep the surface looking very good. This technology also has great potential in markets such as airliners, automobiles, high speed trains and watercraft by increasing fuel efficiency due to largely decreased drag. When tested on a commercial aircraft, it was found to decrease fuel consumption by 3%. This will reportedly reduce airliner operational costs by over $5,000,000.00 a year on fuel alone.

Athletic shoes are an interesting product to design because of the fact that they have to be soft in order to absorb the impact of usage and yet they still have to be hard enough to still maintain their shape and not 'deaden out' as they are used. This has been a difficult problem for designers to overcome, but nanotechnology has proven to be useful in solving this problem. The solution to this problem is the shoes are made of a mix of both hard and soft particles. Nanotechnology comes in to play because the particles used are the size of molecules and it enables the designers to control the mixture. Like the high-tech swimwear, these new nanotechnology shoes have already seen their application in Olympic level competition during the past year. Jeremy Wariner, using a new shoe designed specifically for him using nanotechnology by Adidas, was able to win the silver medal in the men's 400 m race.

This new shoe 'is believed to be one of the most technologically advanced and lightest running shoes to ever hit the track.' The shoe 'provides Wariner more stability, comfort, better torsion, safety and increased flexibility while minimising the energy loss.'

One of the biggest and most universal problems in the sports world is that of smelly gym clothes and sports equipment. Sports equipment is traditionally one of the best breading grounds for bacteria and fungi that 'cause infection, odour, itchiness sores and smelly feet.' If these bacteria and fungi could be kept under control, so could the smell of the clothes and the spread of infections. 'Silver has been used for the treatment of medical ailments for over 100 years due to its natural antibacterial and anti fungal properties.' However, incorporating the medical properties of silver into the equipment has been a challenge. This is where nanotechnology comes in. 'NanoHorizons of State College, Pa., this fall said it developed silver nano-particles that can mesh with the cotton, plastic or nylon material in shoes, pads, jerseys, helmets, socks or other pieces of sports equipment.' This is accomplished by making nano-silver particles that typically measure around 25 nm.

This gives the particles a small volume, but a very large surface area at the same time. The large surface area enables the particles to interact with more bacteria and fungi, which greatly improves its effectiveness in killing them. 'The nano-silver suppresses respiration, basal metabolism of electron transfer system and transport of substrate in the microbial cell membrane.' This process inhibits the multiplication and growth of the bacteria and fungi, therefore leading to much cleaner and better smelling equipment. The sport of golf is also impacted by nanotechnology is a similar fashion as tennis. Breakthroughs in the production of lighter, yet strong, composite materials has greatly impacted the design of golf clubs. For example, Wilson is using a nano-composite to replace the titanium crown used on its current golf clubs.

The result of this is to lower the weight and center of gravity of the club, thus increasing the power and accuracy potential of the club. In addition, the stronger materials used to produce the shafts used on golf clubs will render them more reliable and increase their service life. The golf balls have also reaped benefits of advances in nanotechnology.

The ability to control the production of materials on the molecular level has enabled companies to develop a golf ball that does not suffer from having an uneven spin. This allows for a ball that flies along a much straighter path.

Section VII

Applications, safety and testing of rubber products

Engineering and other applications of rubber products

37.1   Introduction

Rubbers are polymeric materials characterised by their ability of reversible deformation due to external deforming forces. Their deformation rate depends on the structure and molar mass of the deformed rubber and on external conditions of the deformation. By compounding and vulcanisation of rubber various products with specific properties of flexibility, elasticity, toughness or hardness, stretch, rebound and abrasion resistance, etc., can be made.

Some of the most significant uses of man-made rubber include – tyres, fuel missiles, scientific and medical field, adhesive tapes and liquids, protective clothing and tubes and hoses, etc.

Man-made rubber is used to manufacture tyres of various types. Today synthetic rubber is used to manufacture tyres as it is known to be tensile and durable. It is thus considered to be one of the best options for automobile tyres. Rubber based fuels were used for launching missiles during the Second World War. Man-made rubber can be formed into various forms so it is widely used for space travels. Man-made rubber is also used in the field of science and medicine. It is used for manufacturing rubber gloves, orthopedic braces and various other medical items. Due to its adhesive qualities, it is used to manufacture adhesive tapes and liquids. Synthetic rubber is also used to manufacture various types of protective clothes, boots, sleep sacks, diving suits and so forth. Man-made rubber is also used to manufacture garden hoses, gaskets, mechanical seals, belts and hoses. It is also used to manufacture inner tubes, garden hoses and laboratory tubes.

Apart from the above mentioned uses, man-made rubber can also be used to manufacture inflatable boats, diving suits, all types of protective clothing and so forth. Thus, rubber has multiple application in various industries.

37.2   Applications of rubber in aerospace

Rubbers find wide application in aircraft of all kinds.

1. Tyres.
2. Cabin and window seals.
3. Seals for fuel, oil and hydraulic systems.
4. Flexible hose.

5. Flexible fuel tanks.

6. Integral fuel tank sealants.

7. Sealants for air-ducting.

8. Diaphragms in fuel pumps and pneumatic devices.

9. Electrical insulation.

10. Erosion protection.

Apart from the usage in tyres, the requirements are largely for 'speciality' rubbers and in many of the applications, such as seals, sealants and hose, crucial properties are resistance to a wide range of aircraft fluids and retention of highly elastic properties over a wide temperature range. Speciality rubbers finding most significant usage are the medium and low nitriles, polysulphides, neoprenes, methyl silicones, fluorosilicones, ethylene propylenes and fluoro-hydrocarbons, with the nitriles still being the principal workhorse because of their inherent fluid resistance and the ability to formulate to meet an appreciable part of the wide temperature range requirement.

Of the newer speciality rubbers the fluorohydrocarbons, as exemplified by the commercially available Vitons and Fluorels, have probably attracted the most attention, their particular virtues from the standpoint of aerospace application being an exceptional resistance to numerous chemicals and to ageing and their outstanding high temperature stability. These properties singled them out for use in aircraft fuel and hydraulic systems (excepting those operating with Skydrol fluids) where they would find application as sealants and static and dynamic seals.

37.2.1   Nuclear and aerospace applications for rubber seals

Custom engineered rubber seals have a wide range of uses, from robotic systems to simple barriers. While seals are primarily for the automotive industry, the nuclear and aerospace industries serve as prime examples of how rubber seals can have a multiple applications.

There are few areas where custom engineered rubber seals are more valuable than the nuclear industry. Seals for airlock doors, equipment hatches, pool gates and other access areas are vital in helping maintain a safe work environment. The use of these seals helps prevent leaks, contamination and other issues.

Rubber seals are also valuable in the aerospace industry. For example, inflatable seals around the door frame of an aircraft help reduce wind noise and decrease the intensity of vibrations caused by wind force. Other seals guarantee the proper functioning of vital equipment such as landing gear and engines. These seals are used in commercial, private and military aircraft.

Custom engineered rubber seals such as fluorosilicone are especially valuable for aerospace applications, as they can resist the corrosive effects of fuel-based lubricant at temperatures exceeding 400°F.

37.3   Applications of rubber in railway engineering

The long range elasticity and resiliency of rubber and rubber-like materials, offer advantages in eliminating undesirable shocks, vibrations and noise associated with vehicles moving at high speed. Shocks and vibrations in rail vehicles are much more severe due to rail-wheel intersections, track irregularities, crossings and rail joints, firing impulses of engines, etc.

Railway engineers too have taken advantages of rubbery state of matters in designing and construction of all types of rolling stock and track. Railways utilises a large number of functionally different rubber components of which one of the oldest use of rubber which immensely interested the railway men is buffer spring against the longitudinal shocks.

Function of the rubber pad is critical, providing entire elasticity to the track and minimising undesirable vertical and lateral vibrations.

The ability of rubber to change shape under load, to resist this change in shape and subsequently to recover from the change in shape left no competitors for the purpose of elimination of shocks.

37.3.1   Rubber metal spring

Spring is an elastic element that deflects under the action of load and returns to its original shape when the load is removed. Rubber-metal springs are assemblies of rubber blocks and metal plates joined in series. The rubber and metal plates may joined by vulcanisation or they may have only the free contacts. Rubber-metal springs are good shock absorber as they possess elastic and damping characteristics simultaneously.

Springs are very suitable mechanical components to absorb shock and vibrations. Conditions of high vibration and shock in automobiles, railway vehicles, heavy machinery, pipe suspension systems at power plants and steel plants are the common phenomenon in engineering. Springs are thus used to cope up with those problems. Rubber being an elastic medium is the most universal material used for vibration damping. Rubber elements also absorb considerable amount of overloads for a short time without suffering any damage. During dynamic loading rubber converts the absorbed energy into heat by internal molecular friction. This phenomenon is known as damping and is continuous. This property is particularly helpful when shocks have to be reduced quickly. In rubber-metal spring, both rubber and metal plates are to be joined systematically. Such assembly of rubber and metal can be accomplished in various ways. It is also possible to use disc springs or Belleville springs in place of metal plates. Disc springs give wide variety of load deflection curves not readily obtainable with conventional forms of springs.

Rubber-metal springs are conveniently used in road and railway transport vehicles. Over metal springs they have the advantage like reduced weight, reduced cost, improved absorbing and damping capacity of shocks and overloads. In railway vehicles rubber-metal springs are used as primary and secondary suspensions, elastic supports of aggregates, buffers and draw gear applications.

Rubber-metal springs at railway vehicles are used for primary and secondary suspensions and elastic supports of aggregates. A specific application of rubber-metal springs in railway vehicle is to maintain wagons connection in a train. This is known as the buffing and draws gear.

The railway vehicle suspension represents an elastic connection between vehicle parts and it provides the stability and comfort during the ride as well as suppresses vibration and noise.

Buffing and draw gear of railway vehicle

The buffer gear maintains mutual distances between railway vehicles in a train. Their properties are of the crucial influence on transferring and reducing impact loads by which stability and safeness of rail roading and maneuver are affected. The draw gear of railway vehicle transfers the traction force from a locomotive to wagon in a train. The draw gears are subjected to change in very large axial force. Thus, the elasticity of the draw gear as well as its shock absorbing capacity is the important characteristics to be considered.

Rubber hysteresis

During the compression and subsequent load relaxation of any viscoelastic material an interesting phenomenon, hysteresis, is observed. Figure 37.1 shows a typical load-elongation curve for rubber. The clockwise loop, formed in between loading and unloading path, indicates the hysteresis. Hysteresis is the energy dissipation capacity of rubber. By virtue of this characteristic rubber can absorb shock and damp vibration which enables rubber to be used beneficially in rubber-metal springs and other engineering applications where shock loading and vibration are prominent.

An important consequence of the unique load-deflection properties of rubber-metal is its ability to store large amount of energy and to release most of the stored energy on retraction. However, the loading path and unloading path never coincides. Hence, there is a loss of energy (hysteresis) which appears as heat and can lead to a damaging rise of temperature in unfavourable circumstances.

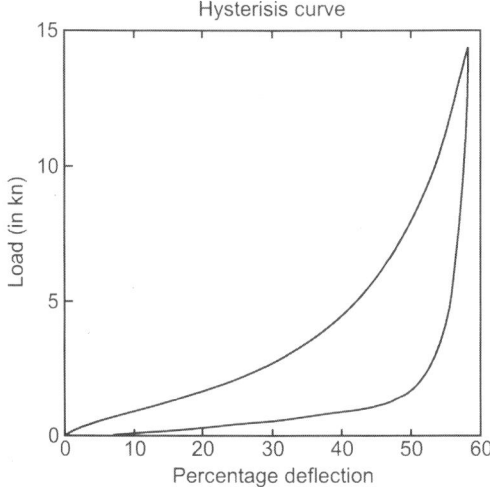

Figure 37.1: Typical hysteresis curve of rubber.

37.3.2 Rubber pads for ballastless track

In ballastless track assembly, rubber pad provides entire elasticity to the track. The rubber pad is placed between the bearing plat and the concrete bed. From assembly design consideration, thickness of the pad has been kept 12 mm nominal. Semi circular grooves in the transverse direction have been provided on the pad to obtain 'shape factor' and serviceability. With this typical design, the fatigue is expected to be minimum.

37.4 Anti-static and conductive uses of rubber

We all are aware of the insulating properties of rubber—most of our domestic and industrial electrical cables are made using rubber as insulant. Rubbers can be modified to make them conduct electric charges or currents, with the usage of certain additives and modifiers.

The major use of conductive or anti-static rubber is the dissipation of electrostatic charges, usually known as static charges. This is the reason why rubbers for this type of application are termed 'anti-static'. These conductive rubbers, in some cases, exhibit a thermo-emf and also Hall effect. Hall effect is the generation of an emf in a body when this is subjected both to a magnetic field and to a perpendicular current flow.

37.5 Applications of rubber sheets

Rubber sheets are utilised in a wide assortment of applications across various industrial sectors. Rubber sheets are majorly used due to the properties of rubber as an insulator and the ease with which it can be folded, molded, cut and shaped according to various requirements. Rubber sheet is ideally suited for applications where the sealing or protection is necessary. The use of rubbers is on a rise owing to the increased costs involved in making use of other alternatives. Hence, the combination of the efficiency and cost-effectiveness of rubber makes it widely used in industrial sectors.

There are particular forms of rubber products that have gained prominence over others. Rubber sheets, for instance, are products that are used in many areas and can be used flexibly for different purposes. Rubber sheets are available in the market today, in various designs and specifications. These variations are made in order to make them suitable for specific applications.

The major differences in their production divide them into the following types:

Natural rubber sheets: These are characterised by excellent dynamic, rebound and mechanical properties.

Neoprene rubber sheets: Neoprene rubber sheets offer excellent resistance to petroleum products, oil and flame, ageing, acid, heat, alkalis and cold mineral oils.

EPDM rubber sheets: These are ideal for outdoor and high zone environmental applications and are resistant to acids, alkali and ketones.

Silicon rubber sheets: These are resistant to high temperature and have a low permeability. They are widely used in pharmaceutical industry.

Nitrile rubber sheets: These sheets are specially designed for high resistance to oil and solvent products.

37.5.1   Popular uses of rubber sheets

One of the most popular uses of rubber sheets is in flooring applications. Rubber provides good amount of resistance to substances like oil and petroleum, ultraviolet lights, oxidising elements, etc. Also, the ability to maintain its form even in cold temperatures is a reason why rubber is widely used.

Rubber sheets are used in a wide range of applications – automotive, defence, marine, etc. The rubber sheets are also used for making orthopedic footwear and sand blasting curtains. They can be laid outside washrooms to prevent tripping, or can be laid under showerheads for the same reason.

37.6   Applications of rubber bearings for bridges

Bearings for bridges must be so designed that they permit expansion and contraction of the bridge with changing temperature in summer and winter, they must be able to rotate slightly from deflection under load and they must transfer thrusts from heavy traffic smoothly to the substructure of the bridge.

Modern vehicles travel at high speeds and have improved braking and acceleration. These factors, together with increasingly heavy loads, are increasing the problems to be overcome. In the past bridge bearings have been mainly in the form of mechanical devices designed to roll or slide, in order to permit movements of the ends of bridge. Such bearings are difficult and expensive to maintain and need frequent lubrication and attention to keep them free from corrosion.

They suffer from the very small movements which they are required to perform, which result in a high rate of wear over small areas. Concrete bearings have been used also, but these take up a large amount of space and are expensive. Bearing pads consisting of layers of canvas and red lead have been used as well.

These are able to compensate to some extent for minor irregularities in alignment and to prevent moisture, grit and dirt from getting between the bearing surfaces and causing abrasion and rusting.

They have, however, little or no cushioning effect and as they are built up on the spot, considerable time and labour is involved. They are also susceptible to weather conditions during installation and should only be built up during good weather to obtain the best results.

Advantages of rubber bearings: The comparatively recent introduction of rubber bridge bearings has enabled engineers to obtain the necessary flexibility in bearings without the attendant disadvantages of frequent attention, and/or high cost, cited above.

37.6.1   Types of bridge bearing

1. Steel-reinforced neoprene.
2. Steel-reinforced natural rubber.
3. Butyl rubber bearing pads.
4. Fabric-laminated pads.
5. Butyl bearings.
6. Neoprene bearings.

37.7   Applications of rubber in piers and fenders

For energy absorption of either impact or vibration rubber is pre-eminent. It provides a variable spring rate which, whilst yielding to a light blow, stiffens up as the shock pressure increases. It is largely self damping, its reaction to a blow dying out quickly and it needs no lubrication, painting or other attention. Rubber correctly compounded for the job has stood up for years, even decades, to tropical sunshine or arctic cold, to salt water, oil scum, Toredo worm and other agents to be found in enclosed waters, with the result that there is now a diversity of fender designs.

37.8   Applications of rubber in buildings

The term 'acoustics' in building practice refers to the study of listening conditions in auditoria. It is concerned with the perfection of desired sounds and the elimination of echo. Rubber possesses properties which commend its use for such purposes in wall and ceiling linings, upholstery and floors.

Sound insulation: Sound insulation deals with the insulation of parts of a building either from internal or external noises. It is concerned with the elimination and reduction of unwanted sound.

37.9   Applications of rubber in packaging

The control of shock impulses on packaged goods has traditionally been a matter of incorporating into the pack shock absorbing or cushioning material, usually chosen by type and quantity and arranged according to the packer's knowledge of existing practice and of the handling and storage hazards expected.

37.10   Application of rubber in automobiles and allied industries

The application of rubber components in automobiles is on the rise. Used in practically all forms, including latex, cements, soft rubber, hard rubber or sponge rubber, almost every important characteristics of rubber is utilised. Given that rubber has a wider range of properties, the rubber industry has achieved greater advances that have effected in cost reduction and a vast improvement in quality. One of the most important applications, for example, is tyres. The modern breed of tyres (most of them are radial ply and tubeless) ensures greater safety and an ability to achieve higher speeds and exploit salient properties of the rubber to the fullest. This involves exploring properties of rubber like flexing endurance, with standing greater distortion without injury. Also, the tensile strength of elastomers like rubber is much larger than the proportionality limit. There's a greater degree of stretch adjustment – from soft rubber to hard rubber and higher capacity to absorb energy. Rubber flaunts very good resistance to chemicals and abrasion, has high dielectric strength and high coefficient of friction. Rubber is waterproof and has low permeability to most gases and liquids. Readily molded as a thermosetting plastic, rubber is low in material cost and available in a range of colours. Tyres, V-belts, toothed belts, hoses, cable sheathing, seals for engines, doors, windows and trunk lids are all applications that have become indispensable for today's mobile world and feature synthetic rubber. Yet tyres are still probably the best-known application.

Synthetic rubber therefore helps save fuel and cut CO_2 emissions from vehicles. Rubber as elastomers, over a period of time, is contributing to the requirement of weight reduction and the economy of vehicle at a time. A visit to an automobile manufacturing plant is enough to highlight the use of elastomer based components and parts.

The common ones include oil seals, O-rings, rubber pads on pedals, body mounts, etc. Elastomer is one of the types of polymers and are used depending on their properties like rebound, tensile strength, resistance to petroleum products, sustainability in cold and hot weather. A wide variety of elastomers like nitrile, EPDM, silicone, neoprene, HNBR, butyl, natural rubber, urethane rubber, fluorosilicone and fluorocarbon, etc., is available today. At the forefront are silicone elastomers, which find application in quick connectors for fuel system. Crank shaft seals and radiator seals are typically made of liquid silicone rubber. These offer elastic flexibility and with stand high temperatures. Rubber pads on accelerator, brake and clutch pedals are made up of natural rubber. Uses of nitrile base elastomers are common for sealing application. These O-rings and oil seals used on fuel system components, shock absorbers, where they have the good resistance to petroleum products and low temperature application. Exterior parts like front fascia, front and rear bumpers, fenders that are designed for the requirement of flexibility and energy absorption are made of polyurethane. They have high tear strength and high tensile strength. Also various types of grommets used on vehicle are made from polyurethane. Clutch cable outer cover, accelerator cable outer cover, seat mounted side air bag cover and seat cover are made with using neoprene. It has excellent resistance to ozone, abrasion resistance. Due to resistance to petroleum products it is also used for gaskets, oil seals purpose.

A wide area of application in auto is rubber coated metals, gaskets and seals to be precise. Fluoro elastomer can be used in diesel and gasoline engine head gaskets and other applications that require exposure to temperatures about 75°C higher than what HNBR can with stand without degradation. All set to play an important role in sealing, damping and protecting the world of tomorrow, rubber material developed for oil platforms can today with stand temperatures in excess of 1250°C for more than two hours. This durable material could perhaps be of use in other contexts tomorrow. As demands for performance increase, environmental requirements are also being tightened, such as in the automotive industry. What, for example, will replace gasoline. And how will the fuels of tomorrow affect the cars of the future. If the cars of tomorrow use alternative fuels, new, rubber-based distribution systems will be needed to bring these to consumers – new fuel filling stations, new hoses, new systems in the car itself and so forth. If, alternatively, the future brings fuel-cell powered vehicles with an electric motor at each wheel, requirements for rubber anti-vibration solutions will increase and change considerably. Not just that, research is being conducted towards the development of intelligent fenders intended for different types of vessels approaching the quay. With sensors located in rubber fenders, it is possible to have each vessel identify its weight, speed and angle of approach. This information can then be used to adjust the fenders' rigidity and energy-absorption capacity.

Similar intelligence can be built into the car's anti vibration systems, where electronically equipped rubber mounts alter their geometry according to the behaviour of the car and are able to predict movements and actively balance the vehicle's reactions.

Application of rubber in automobiles continues to be on the rise. The salient characteristics of rubber are taking it to places never dreamt of earlier.

37.10.1 Silicone rubber for automotive applications

Silicone rubber delivers the strength, resistance and durability needed in automotive applications. As consumers demand an increase in power and performance, the increase of heat that is generated as a result is a factor that engineers have to consider and manage.

Automotive manufacturers rely heavily on the quality and performance of rubber materials and with increasing heat present in new vehicles, silicone rubber comes into its own.

Due to the temperature rise, there has been an increase in silicone usage being used in new vehicles as EPDM gets replaced due to its inferior temperature performance (140°C max).

With high temperature resistance from 230 to –60°C and outstanding weathering properties, silicone rubber can extend the service life of automotive assemblies. Its resistance to rain, wind, salt, abrasion, ultraviolet radiation and chemicals make silicone the material of choice in the automotive sector.

Automotive applications of silicone rubber

Automotive applications of silicone rubber are summaries below in a wide ·range of car parts including:

1. Airbags.
2. Engine gaskets.
3. Headlamps.
4. Ignition cables.
5. HT cables.
6. Grommets.
7. Radiator seals.
8. Performance hoses.

9. Vibration dampening.
10. Shock absorbers.
11. Spark plug boots.
12. Ventilation flaps.

As well as being used on vehicle exteriors, silicone automotive products are also used in the interior providing cushioning, load bearing, vibration suppression and protective shock absorption qualities to car seats and dashboards - enhancing the comfort and driving experience.

Silicone engineering's automotive parts have gone on to be used by a number of the car manufacturing giants including BMW, Volkswagen and Ford.

37.11   Applications of rubber in construction and chemical industries

Prior to 1929, chemical process industries faced perplexing and growing problems. As more and more corrosive chemicals and compounds came into use, the need for a reliable and durable method of protecting mild steel and concrete storage tanks, process vessels, pickling lines, mixers, reactors, agitators, pipelines, tank trucks, railroad tank cars, ship tankers and exhaust gas scrubbers against corrosion became imperative and the use of rubber as a construction material began to be recognised universally. The rubber industry, which was hitherto engaged in the technology development of consumer rubber products starting from erasers, raincoats, footwear play balls, etc., to engineering products such as– rubber tyres and multifarious rubber components for all engineering applications, became alert to these problems of the chemical industry and introduced the first rubber lining in 1929.

Since the advent of the first rubber lining, research and development work continued and today's rubber technology took a different shape from eraser-to-tyre technology to a more sophisticated high technology discipline known as anticorrosive rubber or acid resistant rubber technology, which is eminently and reliably dependant upon by the process industries. Rubber as a material of construction was viewed with more seriousness than ever before by the chemical engineers and well recognised by the chlor-alkali, fertiliser and mining industries and the chemical processes in the oil well and nuclear industries. Today with higher temperature operations and with more complex and critical duty conditions such as nuclear radiation and high degree of thermal abuse as ablation in the rocket industry and mechanical abuse in ore mining industries and other severe environmental conditions prevailing in the oil well industries, the rubber technologist is faced with challenges in the art of developing newer rubber compositions, compounding and processing technologies which are quite different from practices adopted for conventional products. The chief characteristics which make rubber of great importance to the mechanical or structural and chemical engineers are its strength, the adhesion and strength of its bonding to metals and other substrates, its all-round deformability which enables it to be used in extension, compression, shear, torsion or combinations of these, its resilience, its resistance to fatigue, its resistance to attack by corrosive chemicals, its resistance to abrasion, its good electrical properties, its wide latitude of properties by compounding and its ease of molding or forming to any shape and size. Rubber however has some limitations such as swelling in oils, ageing, ozone attack and attack by flame, although most of them can be overcome to a great extent by compounding techniques and with the use of specialty synthetic rubbers. Acrylonitrile rubber can be used for oil resistance and the use of neoprene rubber and antioxidants in liberal doses greatly improves resistance to ageing and weathering, ozone attack and attack by flame.

Among the many reasons why rubbers are widely used in the process industries, three are considered as important. Firstly, rubber operates in a variety of environments and has usable ranges of deformity and durability and can be exploited through appropriate and more or less conventional equipment design principles. Secondly, rubber is an eminently suitable construction material for protection against corrosion in the chemical plant and equipment against various corrosive chemicals, acids and alkalies with minimum maintenance lower down time, negligible scale formation and a preferred choice for aggressive corroding and eroding environment. Thirdly, rubber can readily and rapidly and at a relatively lower cost, be converted into usable products, having complicated shapes and dimensions.

Rubber is also used for protection of other materials against fire, heat and wearing. Rubber gives excellent performance as a construction material, in vibration and shock dampening, in elimination of structural noise and is the ultimate material for sealing systems. All basic properties are more or less present in all types of products. For each application individual functional properties are optimised to meet specific requirements.

Rubber dampens: Rubber dampens by transforming kinetic energy into static energy. This basic rubber property is utilised in protection against explosion and impact and effectively reduces or eliminates noise, vibration and water hammer in pipelines and reaction tanks with agitators.

Rubber seals: Rubber has very good sealing properties due to its pliable and elastic behaviour and is the best choice to make oil, water and gas tight seals in the most demanding environments in the chemical and other engineering industries.

Rubber protects corrosion effects: Rubber is chemically resistant to most corrosive liquids, gases, salt water, ozone and UV light. These corrosive agents are commonly encountered in process industries, shipping and the offshore sector. Where steel is exposed to corrosion, it is protected with a rubber lining, or a total rubber or rubber inflatable structure itself is used. Rubber has very good wear resistant properties. It provides excellent protection for steel and other materials against abrasion and protects against solids and suspended particles.

Rubber gives thermal insulation: To prevent clogging of subsea oil and gas piping by wax and hydrate formation it is necessary to insulate subsea equipment with material with very good thermal properties, low k-value (the k-value, or heat transfer coefficient, is the measured value of the heat flow which is transferred through an area of 1 m^2 at a temperature difference of 1 K thermal conductivity, the time of rate of heat flow through unit area of a homogeneous material in a direction perpendicular to isothermal) also known as lambda the thermal conductance, it is the physical property of a material expressing its ability to conduct heat and high heat capacity. It is tough, impact resistant and has a very long service life.

Rubber gives passive fire protection: Certain types of synthetic rubbers such as neoprenes and Hypalons when suitably compounded with asbestos fillers are flame resistant and give passive fire protection. This safety aspect is a key priority in many chemical and engineering industries as well. These fire protection technologies are used to protect structures and equipment against all types of fires including the extreme conditions of a jet fire.

Rubber is ablative: Ablation means removal of material from the surface of an object by vapourisation, chipping, or other erosive processes. The term occurs in space physics. In space vehicle design, ablation is used to both cool and protect mechanical parts and/or payloads that would otherwise be damaged by extremely high temperatures. A low-density EPDM rubber is a fire stopping and fire proofing product that can be ablative in nature.

Rubber wears: Rubber wear products for the mineral processing, fertiliser and materials handling industries, such as scrubber linings, rubber screening panels, wear resistant sheets, etc., are well recognised and eminently suitable materials of construction.

Rubber bonds with metal: Rubber bonds well with metallic surfaces with suitable adhesives and this property is well utilised in many applications in the chemical industry, such as lining, metal rubber bonded anti corrosive molded components, diaphragms etc.

Rubber is impermeable: Rubbers like butyl, EPDM and neoprenes are unreactive to air and corrosive gases and are impermeable to them.

Table 37.1 summarised the main application of synthetic rubber.

Table 37.1: Main types and applications for synthetic rubbers.

Name	Type of rubber	Asphalt modifications	Footwear	Adhesives	Technical goods
E-SBR	Styrene butadiene in emulsion	–	X	X	X
S-SBR	Styrene butadiene in solution	X	X	X	X
BR	Polybutadiene	–	X	–	X
NBR	Nitrile	–	X	–	X
EPDM	Ethylene-propylene	X	–	–	X
IIR	Butyl	–	–	X	X
CR	Polychloroprene	X	X	X	X
TPR	Thermoplastic	X	X	X	–
Latex	Various types of latex	X	X	–	X

Name	Type of rubber	Tyres	Treads	Plastics modifications
E-SBR	Styrene butadiene in emulsion	X	X	–
S-SBR	Styrene butadiene in solution	X	X	–
BR	Polybutadiene	X	X	X
NBR	Nitrile	–	–	X
EPDM	Ethylene-propylene	X	–	X
IIR	Butyl	X	–	–
CR	Polychloroprene	–	–	–
TPR	Thermoplastic	–	–	X
Latex	Various types of latex	–	X	–

Health and safety aspects in rubber industry

38.1   Introduction

Rubber manufacturing generally comprises the following operations: raw materials handling, weighing and mixing, milling, extruding and calendering, component assembly and building, 'curing' or vulcanising, inspection and finishing, storage and dispatch.

Although the stages described below are applicable to the majority of rubber goods manufactured from solid polymer, a substantial proportion of rubber production involves the use of liquid latex. This applies to the manufacture of dipped rubber goods (such as rubber gloves and some footwear), foam-latex products (such as mattresses, cushions, etc.), and extruded thread products (such as elasticated fabrics and surgical hose).

All the materials required for the manufacture of the finished product are assembled. The raw polymer, either natural or synthetic is brought together at this stage with a variety of compounding chemical additives before being introduced into a mixer. The extensive range of chemicals required and the volume of raw material handled can give rise to substantial quantities of airborne dust.

The industry uses very powerful machinery with the potential to cause fatal and serious injuries. There are established industry safeguarding standards for two-roll mills, internal mixers and calenders. Many serious accidents take place during repairs or to clear blockages, etc., and there must be procedures in place to ensure that safe interventions take place.

38.2   Safety aspects in rubber compounding – powder handling, mixing and milling

38.2.1   Safety aspects in rubber compounding

Compounding involves the measuring and mixing together of raw rubber, process oils, carbon black, bulk fillers and rubber chemicals in pre-determined proportions, termed formulations.

A rubber compounder can typically use between 100 and 200 different ingredients to mix a range of formulations. The finished mixture is known as compound and is the material that is processed into rubber articles by molding, extrusion, calendering, etc.

Bale cutting: Before being added to the mixer the rubber may need to be cut into small pieces on a bale cutter or guillotine.

Bale handling: Most of the rubber industry use mechanical handling such as the vacuum bale lifter.

Fire and explosion hazards

Many of the rubber compounding additives are fire sensitive, particularly:
1. Sulphurs and organic peroxides (used as curing agents).
2. Azodicarbonamide (used as a blowing agent in some open celled rubber/ lattices).

Particular care is required in storage areas to make sure incompatible materials, such as– carbon black and sulphur, are adequately segregated.

There have been incidences of dust explosions in powder handling areas. Many rubber compounding additives such as azodicarbonamide, calcium and zinc stearates, are known to be highly explosive when in a finely divided state. Design dust extraction and collection systems for use with potentially explosive dusts. Good housekeeping will minimise the risk of secondary dust explosions.

38.2.2   Safety aspects in rubber mixing

Once the compound ingredients have been weighed out they are mixed together using specialised machines capable of dealing with the high stresses involved in shearing rubber. Mixing can be either by internal mixing or external/open mill mixing which have very different safety problems.

Main mechanical hazards

Feeding ingredients and collecting compound: The principal dangers and safeguards on a Banbury type internal mixer are:
1. The rotors, via, the feed opening (contact with rotors or falling in on larger machines).

   These risks can be guarded against by feed tables or conveyors placed in front of the feed opening, with additional fixed side guards, where necessary.
2. The floating weight trap with the fixed bridge casting from either the feed opening or the rear inspection door.

   There should be sufficient clearance between the bridge and the fixed bridge casting to prevent a finger-trapping hazard. Prevent access by using the same safeguards as for the rotors.
3. The floating weight and the lower edge of the front hopper door at the feed opening.

The trap at the bottom of the weight is more dangerous. Operators are at risk when sweeping down or when adding lubricant or other small ingredients. Where access to the trap is possible, interlock movement of the floating weight with the hopper door.

4. The front hopper door and frame as the door closes and the stops as it opens under power.

   Where powered movement of the door creates a trapping hazard, it's operation should be via a hold-to-run control, located out of reach of the door or a two-hand control on the hopper itself.

38.2.3   Cleaning and maintenance

The greatest hazard at internal mixers is during maintenance and cleaning, including, clearing of blockages. Operators may need access to the mixing chamber/rotors, normally via, the discharge opening. There is an incentive to do this quickly as the compound hardens as it cools and a possible temptation to take shortcuts.

Maintenance staff are more likely to use the rear inspection door.

A trapped key interlocking system is required at both the rear inspection and discharge openings: Because of the size of internal mixers, operators can be working at three levels: the rear inspection door on the feed platform, the discharge door at the motor platform and the dump chute access at floor level. Communication is therefore difficult with the potential to increase risks.

In addition, stored electrical, hydraulic and pneumatic energy as well as the potential energy from gravity fall of the floating weight and inertia from run down of the rotors, may be released if the correct run down procedures are not followed.

A trapped key interlocking system on all access doors, including a time delay and scotching of the floating weight is one means of preventing access to widely separated danger areas and is the accepted industry standard.

38.2.4   External or open mill mixing (two roll mills)

This refers to mixing operations using horizontal two roll mills. The operator (usually known as a mill man) places the various ingredients in the nip formed between the rolls and mixes the compound by cutting it off the rolls and re-feeding it into the nip until all the ingredients are added. Mills are used not only for blending of compound (open mill mixing) but also warming of pre-mixed compound (known as warming or cracker mills), or for cooling compound mixed in an internal mixer (known as dump mills).

The main mechanical hazard is the nip between the main mill rolls in forward (and reverse) motion. Depending on the design of the particular mill additional mechanical hazards may be created by ancillary equipment, including trapping between mill guides and rolls, cutting by mill knives, traps between the stock blender carriage and mill frame and nips between the stock blender rolls, ejection of moveable mill trays and trapping by movement of recovery belt conveyors (where fitted).

For effective safeguarding of the main nip between the rolls there should be a sensitive trip bar, fitted parallel to the mill rolls and positioned so that the operator works over it.

Operators drawn towards the nip will move the trip bar. This movement actuates the braking system and brings the rolls to rest before trapping the operator. Reliance on a single interlock switch for each trip bar is not acceptable. Each bar should have a positively operated switch at both ends.

38.3 Safety aspects in extrusion of rubber

Extrusion involves forcing uncured rubber through a die under pressure to form a shaped profile or sheet. Rotating knives (or die face cutters) can then convert extruded material into pellets or slugs for further processing.

Rubber can be dumped direct from an internal mixer into an extruder below it as an alternative to two roll mills for feeding calenders. An extruder feeding directly into a two-roll calender is known as a roller die extruder.

38.3.1 Mechanical hazards

Screw (or scroll) extruders

These machines convey uncured rubber forwards down the barrel and through the die by the action of a rotating screw. They produce continuous extruded sections such as cable covering. Single screw versions are most common but twin or triple screw extruders may be used for complex products such as cable sheathing where there may be several layers of compound (co-extrusion).

Screw extruders can be either hot fed, with warmed pre-mixed rubber, or cold fed. Temperature control is generally by water circulating in a jacket around the extruder barrel, screw and die head. The rubber is normally fed in strip form, taken off a two-roll mill.

The main mechanical hazard on a single screw rubber extruder is the trap between the rotating screw and the fixed parts of the machine at the feed opening. Feed systems can also create additional trapping hazards such as a crammer feed device.

Where rubber is fed in strip form, feed rollers are often used to guide the strip into the feed opening. These may be free running, or driven in contra-rotating directions so that the strip remains under constant tension. Crammer feeding with an additional secondary screw is also common.

Use adequate fixed or interlocking guards to prevent access to the screw and other dangerous parts, whatever the feeding arrangement. The simplest arrangement is a series of free running rollers to guide the rubber strip. These must be spaced to prevent finger access to the screw. Where hinged feed hoppers or feeding devices are used these should be positively interlocked with movement of the screw.

Haul-off units

A lot of reported injuries at extrusion lines occur at haul-offs downstream. These draw off the extruded profile continuously to be cured or cooled and cut to length or reeled up. They operate by pulling the profile by means of friction between the product and two counter-rotating tracks, belts, or rollers set one above the other. Trapping hazards typically occur between the tracks at the feed position, between the extruded profile and the tracks and between the drive pulleys and tension rollers and the tracks.

Ram extruders

Here, a roll of rubber compound or 'pig' is placed into the extruder barrel with the die swung out of position, the die is then closed, either manually or under power and the rubber is forced through the die by the forward action of a powered ram. The most common type is the Barwell extruder.

Because the extruder is loaded with individual batches of rubber, these machines cannot produce continuous profiles but produce a strip or pellet for further processing. Where pellets or slugs are produced, the die may have a rotating knife (die face cutter). The main mechanical hazard is the trap between the closing die and the barrel on machines where there is power assisted die closure. Accidents have occurred when a second member of the extrusion team has attempted to assist die closure by trimming back the rubber during closure.

Die closure should be under two hand control and safe systems of work adopted to exclude third parties. Interlock movement of the ram with the die, so that ram movement is not possible with the die swung out of position. Rotating knives of die face cutters should have an interlocking guard with appropriate time delay to account for run down.

38.4   Safety aspects in calendering of rubber products

A calender is a machine with a number of horizontal rolls (sometimes called bowls), heated or unheated, through which material and/or rubber is passed under pressure. They can create either a rubber sheet of a required thickness, or apply a thin layer of rubber onto a cloth liner. This is known as frictioning or combining. Calenders can have two, three or four, rolls of various sizes and configurations.

38.4.1 Mechanical hazards

The main mechanical hazards at calenders are caused by nips between the rolls, for both the rubber and fabric feed. This is a particular problem during threading up, cleaning and maintenance. There is also a running nip at the take-off or wind-up devices.

Depending on the design, there may be other mechanical hazards caused by:

1. Rubber feed nip.
2. Fabric feed nip.
3. Take-off and wind-up devices.
4. Threading up and maintenance.

Rubber feed nip

One of the simplest methods of preventing finger access to the nip is to use a fixed nip bar that extends the length of the rolls, positioned no more than 6 mm from the surface of the roll to prevent a trapping point. When rubber is fed manually into the nip access can also be prevented using the feed table as a distance guard, with integrated mesh guards at either end.

There is also a risk of the operator getting entangled with materials and dragged over the table towards the nip. To control this risk position a horizontal interlocking trip bar just below and slightly to the front edge of the feed table so that the operator cannot approach the nip between the rolls without hitting the trip bar. This should then operate two limit switches, one at each end of the bar and bring the rolls to rest in not more than an eighth of a revolution.

Fabric feed nip

When combining, fixed nip guards are often not practical because of the varying thicknesses of material being calendered and the need for lower roll adjustment. A 'limited movement' guard can therefore be used which consists of an interlocking guard of closely spaced bars or hoops/loops. The guard pivots around a horizontal axis and if a hand is drawn between the guard and a roll, the guard operates a position sensor, which stops the rolls. After the hand is removed the guard returns to its rest position but this does not cause restarting.

Take-off and wind-up devices

Running nips at various take-off and wind-up rollers, including the coated cloth wind-up mandrel, can create trapping and entanglement hazards. When assessing the risk it is important to take into account:

1. The machine speed.
2. Whether the roller or mandrel is powered or not.

3. The sheet tension.

4. The strength and tackiness of the material.

5. The need for approach.

Most rubber sheet and rubberised fabric will be very strong when under tension and capable of inflicting serious injury.

Use automatic tensioning equipment, as it should not be necessary for operators to feel or manually adjust the tension of the cloth whilst it is in motion. Use devices such as slipping clutches to regulate tension and 'banana' or bowed rollers to reduce creasing, as they will also reduce the need for approach.

Where there is a significant risk of injury, prevent access to the running nip by interlocking guards, pressure sensitive mats or photoelectric safety systems.

For start up, fix the cloth liners to the mandrel whilst it is still stationary.

Threading up and maintenance

Threading up and maintenance, including clearing of blockages, can create additional hazards. This is because a close approach to the nips may be necessary. Where possible threading up and maintenance should be with the calender stationary and the rolls separated.

Where it is necessary to reverse the calender drive, for example to clear blockages, it will create new nips. Reverse movement should be via a hold to run control and at a suitably slow speed, termed an inch device. All operators should also be excluded from the danger areas. If there is more than one operator there should also be flashing lights or an audible alarm. Also, provide sufficient emergency stop buttons on both sides of the calender and at the auxiliary feed and take-off devices.

38.4.2   Other mechanical hazards in calendering

Some of these hazards are given below:

1. Crushing at the trap between a moving calender roll surface and a fixed part of the machine.

2. Crushing between the stock guides and the calender roll.

3. Cutting hazard from the blades or trimming knifes.

4. Crushing between a swinging feed conveyor and the machine framework.

5. Trapping between contra-rotating cooling drums or between drums and fixed parts of the machine.

6. The roll nip adjustment mechanisms.

Appropriate procedures should be in place for cleaning and maintenance which are both potentially high risk activities.

38.5   Cloth coating

Cloth coating involves the impregnation of cloth, mostly synthetic rather than cotton, with a rubber solution to produce a waterproof or chemically resistant fabric for use in waterproof clothing, tarpaulins, sails, dinghy fabrics, etc. A range of rubbers and organic solvents are used to produce the rubber solution or 'dough'. Natural rubber, synthetic polymers such as chlorosulphonated polythene (Hypalon) and urethanes are the most widely used rubbers. Commonly used organic solvents include toluene, xylene, acetone, MEK and petroleum naphtha blends (n-hexane based).

Rubber compound, produced by internal or open mill mixing, is soaked in solvent before mixing to a solution with the consistency of light dough. Colours, other additives and more solvent may be added during mixing.

Some fabrics such as nylon require pre-treatment to achieve a good bond with the rubber. MDI is most widely used as a pre-dip, prior to the spreading process. Most fabrics are pre-dried, usually in a hot air oven, prior to coating.

Spreading machines evenly distribute the rubber solution onto the cloth as it is drawn under it, using an adjustable blade, known as a doctor blade. The coated fabric is then drawn into a drying oven to evaporate the solvent. The cloth is then passed over cooling rolls and wound up, either at the front or back of the machine. Embossing rollers positioned after the oven produce the desired surface finish. The process is usually repeated a number of times to achieve the required degree of proofing, the coating thickness at each pass is controlled by adjusting the height of the doctor blade.

The rubber solution is fed onto the cloth via a trough or tray. Solutions that are more viscous are added as a 'roll' of material directly onto the cloth and adjacent to the doctor blade. The speed of coating machines can vary considerably, from less than 10 metres per minute to over 100 metres per minute. The slower, usually older, spreaders often have steam-heated chests to drive off the solvent. High speed machines tend to use indirect gas-fired evaporating ovens.

The coated cloth can be vulcanised by either continuous curing, e.g., Rotocure, or more commonly by dusting the coated cloth with a starch based agent to prevent sticking and then re-rolling it and curing it in batches in a steam autoclave.

38.5.1   Mechanical and other hazards

The risk of entanglement and shear trapping hazards at dough mixers should be controlled by having interlocking lids, and/or interlocking enclosing guards.

As many of the solvents are highly flammable suitably protected electrical devices are required. Also, control the risk from static charges. Precautions at evaporating ovens should include explosion relief and flame failure devices.

Spreading machines can have nip and running nip hazards, particularly at the cloth wind-up, embossing and cooling rollers. Higher production speeds will increase the risk and fixed or interlocking guards or other protective devices will be required. At slower speeds and for inaccessible nips, trip wires fitted to emergency stops may be acceptable, if justified in the risk assessment.

Safeguarding may also be required for any running nips when the cloth is re-rolled after curing.

38.6 Fabrication (including tank lining, roller covering and hose winding)

Fabrication covers a range of operations where rubber and other materials are assembled to produce finished articles or components. Rubber lining operations and hose building present particular hazards.

38.6.1 Rubber lining operations

Rubber sheet may be bonded to metal components including valves, pipes, tanks and rollers and then vulcanised to provide a tough, corrosion resistant lining or covering. The main hazards common to all rubber lining operations are the health and fire hazards associated with the use of solvents for degreasing metal components and solvent-based bonding agents. Tank lining and roller covering have additional hazards.

Tank lining

Rubber lining of tanks or other enclosed structures can be either in a factory or on site, but in either case usually involves the following operations:

1. Clean out the vessel (usually by steam).
2. Prepare the vessel surface (usually by shot blasting).
3. Apply rubber solution, or bonding agent (usually toluene or *n*-hexane based).
4. Place sheets of rubber in position (sometimes pre-warmed).
5. Vulcanise (using hot water or steam).

These operations present a number of significant hazards, including those associated with entry into confined spaces, shot blasting, use of flammable and/or toxic solvents and use of pressure vessels.

For any work involving entry into confined spaces, strictly controlled safe operating procedures and high standards of operator training are essential.

Breathing apparatus will protect the operator against the health hazard associated with solvent use but a fire and explosion hazard may remain. Therefore, use a non-flammable solvent wherever possible. Forced ventilation should reduce the concentration of flammable vapours in air to below one quarter of its lower explosive limit. If this is not done then all potential ignition sources should be excluded and specially protected hand lamps used. Equipment should also be earthed. Small rubber-lined articles are usually cured in an autoclave. Larger tanks or vessels may be converted into temporary steam receivers by blanking-off the entry point and admitting steam up to 20 psi. The requirements of the pressure systems regulations will then apply. Establish the vessels safe operating limits and fit protective devices to protect against and warn of system failure.

Roller covering

Steel rollers are covered with rubber or polyurethane sheet as this will provide them with a controlled measure of resilience. They are then cured in autoclaves. In addition to health risks, there is also a danger of fire from the solvents and explosion from the fine cured rubber dust produced when grinding the covered roller to size. The majority of reported fires in rubber factories involve finely divided vulcanised rubber and dust. Extraction at rubber grinding machines, depending on the particle size, will generally need to be designed and installed for use with potentially explosive dusts. Consider where the dust collection unit is located and in particular where explosion relief vents to. Good housekeeping will minimise the risks of a secondary dust explosion. If the operator attempts to take too deep a cut, it will cause the rubber to overheat and increase the fire risk.

38.6.2   Hose winding

Rubber hose has a wide variety of applications such as car parts for power steering, air conditioning systems, etc. The most common types are nitrile and EPDM rubbers but silicones and flouroelastomers are more suited to high temperature applications.

The basic equipment used to produce hose consists of a metal mandrel supported in rotating chucks, one of which is driven. Standard metal working lathes are commonly used. To build up a hose, layers of rubber strip and fabric, with wire or rope used as reinforcing material, are wound onto the mandrel. The wire or rope is applied under tension. The final winding operation involves the application under slight tension of nylon webbing known as a spurle. The complete hose assembly on the mandrel is transferred to an autoclave for vulcanising. The cured hose is removed from the autoclave, put back in the machine and the spurle unwound by reversing the direction of the headstock – this is known de-poling. Light hose of small diameter is built up gradually by hand feeding materials along the length of the rotating mandrel. Heavy-duty hose is produced via a powered feed system, which automatically traverses along and in front of the rotating mandrel.

Mechanical hazards

The rotating mandrel and the fed material present a significant entanglement hazard. These have caused serious injuries to operators, particularly when wire and rope are used. There is also a running nip between the material and the mandrel. Operators should always wear close fitting clothing with cuffed or short sleeves and tie back long hair.

These machines are extremely difficult to safeguard and specialist help may be needed when devising solutions.

For very short lengths of hose, i.e., less than three metres, one option is to provide a hinged guard of horizontal bars, interlocked with a DC injection braking system, along the length of the mandrel and through which the various materials are fed. On machines where conventional guarding cannot be used the severity of an injury can be reduced by having a supported trip wire mounted above floor level, preferably so that the operator works over it , interlocked with a DC injection brake, combined with a monitored photoelectric safety device positioned above the mandrel. On very long mandrels there may be significant deflection so photoelectric devices may not be an option. A combination of other safety devices should be used such as trip wires, foot operated emergency stop controls and hold-to-run controls.

When producing wire reinforced hose, it has been common practice for operators (wearing a leather waistcoat and leather gloves) to pass the wire around their back and through their gloved hand before feeding it onto the mandrel. This allows them to control the tension in the wire by leaning away from the mandrel but significantly increases the risk of entanglement. An alternative method for feeding the wire should be used, such as a trolley-mounted reel or an automatic winding head, which traverses the length of the machine. If pressure testing is essential, it should be in a separate test facility capable of withstanding the effects of sudden energy release if a hose fails.

38.7 Vulcanisation (including presses, autoclaves and continuous vulcanisation)

38.7.1 Molding

All vulcanising methods can cause burns. Insulate hot machine surfaces to prevent accidental contact where possible. Otherwise, warning signs and protective clothing (for lower arms as well as hands) are likely to be required.

Compression molding: Compression molding is the most common molding technique used in the rubber industry. The presses are usually hydraulically or pneumatically powered and the molds can be heated electrically, by steam, or by oil.

Transfer molding: This is a variation on compression molding and involves loading a pre-formed blank of rubber in a cavity connected to the mold cavity by a runner. The blank is compressed when the mold closes and is forced under pressure into the mold cavity. The safety considerations are the same as for compression molding.

Injection molding: This produces precision moldings. The rubber is pre-heated and a rotating screw forces it into the mold cavity under pressure. Vertical injection molding machines are more common than the horizontal type generally used in plastics processing. Injection molding machines can have a manual mode where the operator removes the molded item at the end of each cycle. More common are semi-automatic or fully automatic machines where conveyor or pick and place robots remove the molded product. Injection molding machines tend to operate at faster speeds and on shorter cycle times than compression/transfer molding, which effectively increases the potential risk to the operator from mechanical hazards.

Tyre curing presses: These are usually down-stroking machines, most often electrically or pneumatically powered. They have a rubber bladder, which is inflated into the tyre once it is in the mold and then filled with steam or high-pressure hot water to aid curing inside the tyre. Bursting or leaking bladders can cause serious scalds, particularly during press opening. Pressure sensitive switches should prevent opening when internal pressure is present.

Mechanical hazards

These vary considerably with the press type and the operating procedure used. At most presses there are trapping hazards between:

1. The moving molds/platens, either under power and by gravity fall.
2. The moving platens and the press frame/press tables.
3. Mold loading and stripping devices.

When assessing the risks and precautions needed, consideration should be given to:

1. The closing speed.
2. The number of molds and amount of daylight.
3. Whether up-stroking or down-stroking.
4. Type of molds (loose or fixed).
5. Mode of operation, in particular, if worked at one or both sides and the number of operators.
6. The amount of body access between the molds.

Many incidents happen during maintenance and tool changing and adequate controls against preventing falls due to gravity are important for down-stroking presses. These could include scotching or having a pilot operated check valve and counterbalance valve assembly in the hydraulic circuit.

38.7.2 Autoclaves

Autoclaves, known as 'curing pans' or 'vulcanisers', are pressure vessels filled with steam that are used to cure rubber articles not contained in molds (hose, extruded section, coated cloth, retread tyres and small cable batches). Most are horizontal with vertical autoclaves now rare. They can have either quick opening doors or multi-bolted doors.

38.7.3 Continuous vulcanisation

Products produced on a continuous process such as rubber covered cable and strip are continuously cured, rather than on a batch basis. There are various methods but each one is sited just after the exit die of the extruder or calender.

Liquid salt baths: These are usually gas-fired troughs of molten sodium and potassium nitrate mixtures. This method is particularly suited for peroxide cured compounds, which must be cured in the absence of oxygen. The main hazard is fire and explosion associated with the use of molten salt at high temperature. Totally enclose salt baths as water in the molten salts could result in an explosive generation of steam. To prevent localised hot spots and overheating use effective temperature control. There is a potential danger of splashing with hot salt when leading the extrudate through the salt bath and during maintenance and cleaning so suitable protective clothing, including full-face protection, should be worn.

Pressurised liquid salt continuous vulcanising (PLCV): Cable making often uses this process. It consists of a long gently slopping pressure chamber or tube fixed to the extruder die. Vulcanising occurs by passing the cable through a molten mixture of salts in the pressurised chamber. The cured cable then passes through an enclosed water-cooled chamber to emerge cold at the end of the line. The main hazards associated with this process arise from the high temperatures and pressures involved. They include: (i) burns from accidental contact with the molten salt, (ii) fires from accidental mixing of salts with combustible materials and (iii) an explosive reaction, involving generation of steam and an associated pressure rise caused by accidental mixing of hot salt and cold water.

Control the risks by having regular maintenance, correct operator training and having the correct initial design. Interlocks are required between the salt and water pumps. Fit level controls to prevent water entering the molten salt in the curing chamber. There should also be pressure relief before access into the chamber is possible. Suitable protective clothing is required for maintenance operations.

38.8   Latex processing

Latex is a dispersion of rubber particles in an aqueous phase. Most types of rubber, both natural and synthetic, can be made into a latex form. Latex gets used in specialised applications such as those requiring oil, solvent or flame resistance. Different mixing techniques produce different grades of dispersions, i.e., ball milling or ultrasonics for fine particle dispersions and simple stirring or colloid milling for coarse dispersions. Latex dipping is used to make thin articles such as gloves, balloons, condoms, catheters, bladders, hot water bottles, etc., and very fine particle dispersions are needed. Coarser particle sized dispersion (slurry) is acceptable for latex foam, used mainly for carpet backing.

Various additives are mixed with the latex dispersion or emulsion:

1. Stabilisers, to maintain the rubber particles in a stable state of suspension. These include sulphonates, inorganic complex phosphates and soaps.
2. Curing agents, usually sulphur.
3. Accelerators, including dithiocarbamates, thiazoles, thiurams and xanthates.
4. Thickeners, including casein, glue, cellulose derivatives.
5. Anti degradants - including waxes, substituted phenols, amine based antioxidants.
6. PH adjusters, including ammonia, sodium and potassium hydroxide and formaldehyde.
7. Biocides, mostly halogen derivatives, for suppression of bacterial decay and fungal infection.
8. Coagulants and gelling agents, include calcium, magnesium and aluminium salts, acetic and formic acid, cyclohexylamine acetate and ammonium acetate.

38.8.1   Latex dipping and casting

This uses a porcelain or aluminium mold (or former) that is repeatedly dipped in the latex compound. Once set, the coated former is washed, air-dried and vulcanised in steam autoclaves before the product is removed. To increase the thickness of the deposit obtained per dip, the former is immersed in coagulant before and after each latex dip. Latex casting is a similar technique to dipping but is used to make hollow seamless articles. The product forms on the inside of the mold (in contrast to dipping) giving a better surface definition.

Mechanical hazards

Latex mixing plant and processing equipment such as carpet backing machines, may present a range of mechanical hazards that will require effective safeguarding by the techniques described in previous sections.

38.9 Polyurethanes (including PU foam production, re-constituted foam, foam conversion and rigid urethanes)

Polyurethanes are used in a wide variety of forms ranging from coatings and lacquers to flexible foam. Polyurethane elastomers can be divided into either foamed or rigid urethanes. However, both are based on a common chemical reaction between a long chain alcohol (polyol) and an isocyanate. Urethanes with differing physical and chemical properties are made by varying the type of polyol and/or isocyanate. Appropriate measures to control the health risk should be in place.

38.9.1 Polyurethane (PU) foam production

PU foam is produced by releasing gas into the reacting isocyanate and polyol so that when the reaction is complete the elastomeric material is formed around a network of fine gas bubbles. The process involves metering the liquid isocyanate and polyol into a mixing head with other chemicals (blowing agents, catalysts and surfactants). The main blowing agent used is carbon dioxide, generated by the reaction between water and isocyanate.

Fire hazard

Flexible PU foam, even combustion modified high resilience (CMHR) grades, is regarded as high fire hazard materials in manufacturing and conversion premises, unless there is evidence to demonstrate otherwise. Smoke from PU foam will contain highly toxic and irritant components including hydrogen cyanide and isocyanates and a high standard of fire precautions is required.

Mechanical hazards

Guard the trapping points at the mold closing and any between the fixed and moving parts of carousel molding machines. A further trap may exist between the injection nozzle and the mold in reaction injection molding. Apply the usual machinery guarding considerations.

38.9.2 Reconstituted foam

PU foam off-cuts can be reconstituted into 'recon' or 'chip foam'. The process involves grinding the off-cuts into small pieces (crumb) in either a granulator or hammer mill.

Fire and explosion hazards

The production and use of PU foam crumb can give rise to a very high fire hazard unless the operations are strictly controlled. Crumb can spread and act as a fuse for the spread of fire.

The dust and fine crumb produced in hammer mills and granulators can cause dust explosions.

These can be controlled by:

1. Fire resisting separation.
2. Enclosure and ventilation control at crumbling plant.
3. Good housekeeping.
4. Explosion relief on dust collection plant.

Mechanical and other hazards

Hammer mills and granulators are often fed by conveyor and the feed openings should be dimensioned so that operator access to dangerous parts is prevented. Access doors into the granulator/mill should be positively interlocked with time delay/braked motors, where necessary, to account for rundown times. High noise levels at granulators and hammer mills can be a problem so provide acoustic enclosures around the machines with interlocks where necessary to prevent access to dangerous parts. Otherwise, use localised acoustic cladding.

38.9.3   Foam conversion

Foam conversion involves cutting or shaping and fabrication (gluing and assembly) of block foam or reconstituted foam into product components, for example, shaped cushions for upholstery. It also includes finished products such as synthetic sponges and floor mats. Fire hazards area is a particular problem in cramped premises.

Mechanical hazards

Foam cutting band knives and other slitting machines are difficult to safeguard effectively. Because of the size of the foam blocks large sections of blade may have to remain exposed. Accepted safeguarding standards in foam conversion differ from those found in other industries. This is because foam is much easier to handle than wood, meat or textiles and there is there is little risk of kickback or slipping. There are low numbers of reported injuries on PU foam band knives. When assessing the risk and the safeguards required take account of the size, shape, range and throughput of the work, particularly the need for close approach. Consider also the type and age of the machine as well as the age and experience of the operator.

Fire hazards

Storage and handling of large volumes of PU foam, including raw materials, work in progress and finished goods, can present a significant fire hazard and a high standard of fire precautions is required in conversion premises.

Accumulations of fine dust inside band knife housings and from crumbling and grinding, can increase the fire hazard as well as causing a risk of dust explosions. Provision of separate fire resisting bulk stores and tight control over workflow, stacking and housekeeping are required.

38.9.4　Rigid urethanes

These are polyurethane elastomers in a high-density rigid or semi-rigid form. Although expensive they have a wide range of applications where their combination of properties, including abrasion resistance and dimensional stability, are superior to conventional rubber.

Hazards

These usually contain negligible amounts of isocyanate as the reaction has gone to completion. If processed at high temperatures there can however be some isocyanate release. Many solutions contain highly flammable solvents, which will require the usual occupational hygiene and fire safety precautions including LEV.

38.10　Tyre building

The processes and associated hazards are only briefly covered in this section as they are covered in greater detail in: compounding, extrusion, calendering, vulcanising and tyre retreading. Tyre building involves the assembly of various rubber-based components, by placing them in the correct sequence and position around a driven rotating drum, to form an uncured or 'green' tyre carcass, which is then cured. The basic process is similar for all pneumatic tyres, from passenger and truck tyres to earthmover and aircraft tyres. Aircraft tyres are often hand built by two or more operators, passenger and truck tyres are often mass produced on semi or fully automatic machines, usually with a single operator.

Tyre manufacture involves much more than tyre building. Upstream processes, i.e., production of the various tyre components, include:

1. Rubber compounding.
2. Calendering and cutting of rubberised cord fabric for breakers and plies.
3. Drawing, coating, cutting and forming of steel wire for beads, belts and plies.

Downstream processes include curing of the 'green' carcass in a tyre curing press and final inspection that involves inflation testing and X-ray examination.

All these stages can present significant hazards including:

1. Mechanical hazards, particularly at mixers, extruders, calenders, cutters and curing presses.
2. Health hazards, particularly exposure to rubber fume during compounding, calendering, extrusion and curing.

3. Noise, particularly at bias cutters, mills, mixers, wire twisting and braiding machines, bead making machines, tyre test rigs.
4. Hazards associated with use of steam at curing presses and compressed air at inflation testing machines.
5. Manual handling of tyre carcasses and cured tyres, particularly truck and earthmover tyres.

38.10.1 Mechanical hazards

Tyre building machines present a number of significant mechanical hazards including:

1. Entanglement on the drum and by drive shafts.
2. Trapping and crushing by various reciprocating parts and between the drum and stitching rollers.
3. Nips at conveyors feeding the servicers.
4. Cutting by knives.

Secure fencing in the conventional sense may not always be practicable at tyre building machines because of the close interaction between the operator and the machine, particularly the building drum. However, keeping the operator away from the hazardous movement by a safe system of work is unreliable and not acceptable as the only means of protecting the operator.

38.11　Tyre retreading

Tyre retreading is a specialist process carried out by small and medium sized companies as well as the major tyre manufacturers. Retreading is a means of recycling partially worn tyres, although the tread may have worn down, the rest of the tyre, if in good condition, can have a new tread molded or 'stuck' onto it. Recycled tyres are widely used for trucks, buses, aircraft and car rallies, but car retreads only have a small share of the market.

Most retreading operations will involve a number of basic stages such as: (i) inspection, (ii) buffing and peeling, (iii) skiving and repair, (iv) solutioning, (v) cold top capping, (vi) bead-to-bead remolding (or hot cure process) and (vii) inspection and inflation testing.

38.11.1 Mechanical hazards

Buffing machines

The main mechanical hazard during retreading is contact with the rotating rasp heads at buffing machines. The speed of rotation (up to 3000 rpm) combined with the fact that LEV is fitted to capture dust and fume, mean that loose clothing can easily be trapped and the potential for injury is significant.

Other potential mechanical hazards at buffing machines include:

1. Trapping by the tyre lift and by the flange plates or the expanding chuck when the tyre is being mounted.
2. Entanglement with the rotating tyre and with the tyre drive roller.
3. The nip between the tyre and the drive roller.

38.11.2 Other hazards

Noise

Tyre buffing machines typically generate noise levels of between 85–92 dB(A) and noise from skiving tools can exceed 100 dB(A). The main source of noise at the buffing machine is the action of the rasp on the tyre and control at source is not usually reasonably practicable at manually operated machines. However, fit silences to air exhausts and extraction systems.

Use of peeling to remove the bulk of the rubber reduces the amount of buffing required, which can significantly reduce operator exposures. Acoustic enclosures may be reasonably practicable at automatic buffing machines.

Reduce noise exposures at skiving by fitting silencers to the exhaust port of the skiving tool, or by use of electrical tools rather than pneumatic ones. Electric tools are however heavier and require proper support.

Tyre inflation testing machines can be noisy, with noise levels over 90 dB(A) particularly when air pressure is exhausted from the tyre. There is also a risk of a tyre bursting. Fit suitable silencers to the exhaust ports on these machines. Hearing protection will usually be necessary.

Fire and explosion

Fine rubber dust from skiving of the tyre casings is likely to be explosive when in a finely divided state. Rubber dust from buffing may also be explosive, depending on the particle size. The usual considerations for handling explosive dusts apply, including structure and sighting of dust collectors and provision of explosion relief.

38.12 Summary of safety and precautions in rubber industry

The general recommendations regarding safety precautions can be expressed briefly In the form of ten rules:

1. The inhalation of chemicals, whether as dusts, fumes or vapours, must be avoided. Impure air must be removed with air extracters and premises ventilated. Breathing apparatus should be worn if necessary. Those charged with sweeping or the removal of dust deposits should take care not to stir up clouds of dust.

2. In principle no chemical should be allowed to come into contact with the skin or get into the eyes. Persons who handle chemicals, especially for long periods or repeatedly, should wear such protective devices and clothing as gloves, goggles, overalls and aprons. Powder which gets onto the skin should be removed immediately, after which the skin should be washed with soap and water. If splashes get onto the skin, wash away with soap and plenty of water immediately. Should chemicals get into the eyes, rinse the eyes with running water, immediately, then if necessary, consult medical opinion.

3. As a matter of principle workers must refrain from everything which could enable chemicals to enter the body together with food or drink. Food should not be stored, prepared or consumed in rooms where chemicals are used. Should a chemical be swallowed, call the doctor immediately.

4. Occupational clothing should be changed and laundered or cleaned at frequent intervals. Badly soiled clothing should be changed immediately and discarded if necessary.

5. Normal and emergency washing facilities must be readily accessible from processing areas. Installations to enable the head or eyes to be rinsed with, plenty of water must be available in case of an emergency. Facilities to enable the hands to be washed before meals must also be readily accessible.

6. Chemicals should only be stored and weighed in well ventilated premises intended for the purpose. Packed chemicals must be stored tidily and within easy access, the packages must be clearly marked and must display any necessary danger warnings and the safety precautions to be taken during their handling. Regulations on the storage of flammable materials must be strictly observed.

7. Chemicals, except those in closed containers, should only be moved and processed by authorised persons who have been instructed how to handle them.

8. Spilt chemicals must be dealt with immediately and in accordance with the safety rules applicable to them.

9. Chemicals must be handled with suitable equipment shovels, buckets, etc. Such equipment must be available at all times and should not be used for any other purpose.

10. First aid supplies, fire extinguishers and other safeguards must be clearly visible and easy to reach. They must be inspected at frequent intervals by an employee responsible for safety. A telephone, clearly showing the telephone numbers of the ambulance unit and fire service, must also be within easy reach.

Testing of rubber products

39.1   Introduction

The testing of rubber, rubber products and elastomers is essentially a continuously changing process due to natural instability in testing procedures.

To achieve highest standards in rubber products, testing of rubber must contain not only a description of existing methods but also some indications of the directions in which changes are expected or needed. There has been a considerable progress, specimen equipment and procedures have been standardised for many tests to yield satisfactory results for specification purposes. These standard test methods are subject to revision and they are constantly being studied and refined since most of these testing methods are arbitrary. The greatest change in test laboratories in recent times and rubber is no exception, is the improvements made in apparatus by the introduction of more advanced instrumentation and automation, in particular the application of computers both to control tests and to handle the data produced.

The testing is necessary, the purposes of testing can be summarised as: (i) quality control, (ii) predicting service performance and (iii) design data

For quality control: The test should preferably be as simple, rapid and inexpensive as possible. Non-destructive methods and automation may be particularly attractive. The best tests will additionally relate to product performance.

For predicting product performance: The essence of test must be that it relates to product performance—the more relevant the test to service conditions, the more satisfactory, it is likely to be.

For producing design data: The need is for tests which give material property data in such a form that they can be applied with confidence to a variety of configurations. This implies very considerable understanding of the way material properties vary with geometry, time, etc. For complex and long running tests automation may be desirable.

Tests are usually classified by the parameters to be measured—mechanical, thermal, electrical, etc. These can be sub-divided to list the actual properties so that under mechanical, for example, there are strength, stiffness, creep. There are various test conducted on – rubber and rubber products. In this connection reader should refer – ASTM, ISO, DIN and BIS specifications. However, this chapter deals with testing of compounding and finished rubber products.

39.2 Testing of rubber compounds after mixing

Tests performed on mixed compound are usually carried out to determine how it will perform in the next step. That is, how will it extrude, calender, mold and cure? Such tests are usually referred to as tests of processability. Processability of elastomers and rubber compounds at each of the stages from raw polymer to finished product is and always has been, a problem for the processor. One difficulty is that rubber processability is a subjective concept in that it depends, to some extent, what you want to do with the material and in what equipment, whether it processes well or poorly. In addition, there is a lack of simple, easy tests, which can measure processability. Generally, if the test is simple, it doesn't correlate directly with processability and if it does correlate, it isn't simple.

A further complication is that flow of rubber compounds in processing equipment is always a combination of shear and elongational flows, with the added complication of slip of the material against metal surfaces. Yet a further complication is that rubber compounds are viscoelastic materials; that is, they exhibit both viscous and elastic responses to deformation. A viscous response is proportional to the rate of deformation, while an elastic response is proportional to the amount of deformation. In addition, viscoelastic properties of rubber vary with changes in applied strain, frequency and temperature. Testing at several different frequencies, strains and temperatures is usually necessary in order to characterise thoroughly the viscoelastic properties of a rubber sample.

Even if one has full information, from laboratory tests, on the elasticity and the flow behaviour of the material and on its curing characteristics, it is not usually possible to predict, from this, exactly how it will behave in an internal mixer, on a mill, in an extruder, injection molding machine, curing oven, or mold.

The meaning of processability depends on the stage of the process. In mixing, it means the ease with which the compounding ingredients can be incorporated and dispersed in the rubber. In extrusion, it means the ease with which the compound can be fluxed, moved by the screw and extruded, together with the dimensional stability, or die swell of the extrudate. In injection molding, it is the ease of flow and absence of excessive heat build-up in the runners and gates, together with the speed of vulcanisation.

Calendering, milling and other processes each have their own individual requirements. Therefore, clearly, there can be no such thing as a comprehensive processability test, which can examine the rubber at some stage in the process and define its processability in all other stages.

Many of the tests used in the rubber industry are empirical, even arbitrary, in that they require a specific, proprietary test apparatus and the test conditions and sample preparation have to adhere exactly to a written procedure or the results are invalid. Such tests give results in arbitrary units, which cannot readily be converted to fundamental properties, or compared to the results from another instrument. A prime example of this is the Mooney test.

Tests, to be of value in process control, must be rapid and amenable to operation and interpretation by plant personnel. The development of computer assisted test operation and interpretation has made it possible for more tests to meet these criteria. Thus, 'An ideal test would include fast operation, lack of operator dependence and simple sample preparation. In addition, the test should also give good differentiation between samples of the same grade of polymer with different processing characteristics, preferably by measuring some fundamental viscoelastic parameter and in basic units. The equipment should also be inexpensive and cheap to run and maintain. As well as giving a general guide to whether the processor will have problems with the material, the test would ideally give some indication of behaviour during the first stage of processing, the mixing. A further advantage would be that interpretation of test results requires no detailed understanding of the rheological basis of the test. To cap all this, it would be extremely convenient if such a test also gave a measure of how the material will behave in further downstream processing.'

Poor test quality, or lack of agreement between laboratories, in the rubber industry is a serious problem. ASTM committee D11 and ISO/TC-45 are actively addressing the problem of precision of test methods. Vieth defines the difference between accuracy and precision as follows:

The two measurement concepts of accuracy and precision are often used interchangeably—they should not be. Accuracy describes how well measured values agree with a reference or 'true' value: high accuracy implies good (close) agreement. Precision describes how well measured values agree with each other: high precision implies good (close) agreement.

The difference between the measured mean value and the reference value is the bias. A large bias implies an inaccurate measurement process, which may or may not be precise. A large systematic between-lab bias, with its value unique to each lab, is the root cause of poor inter-lab precision. Eliminating these large biases is a difficult job.

There are three sources of test differences:

1. Calibration of equipment.
2. Execution of the test.
3. Quality of the written standard test method.

Only if all three of these are corrected will test error and bias, be minimised. Even then, with empirical tests, such as Mooney viscosity, biases will still exist. Statistical methods have been developed for dealing with this situation.

There are four general types of tests performed on mixed compounds:

1. Tests to determine how well the carbon black has been distributed and dispersed. These are usually performed on cured sheets.
2. Tests on vulcanised test pieces, to determine whether the compound, after curing, gives the expected end product properties corresponding to a well-mixed compound.
3. Tests to measure certain basic properties of the compound such as Mooney viscosity, which are indicative of the degree of mixing.
4. Tests designed to simulate further processing.

In the first two types of test, although the starting material is the mixed compound, the actual tests are performed on cured material. This type of test will be discussed in Section 39.5.

As with most testing in the rubber industry, testing of mixed compound is performed in two basically different circumstances. The first is performed when specifying, characterising, or developing a new compound. In this case, test difficulty, test time and complexity of data interpretation are not important. However, most testing is performed in the plant for process control, or to predict behaviour in the next process. In this circumstance, as commented above, rapid results are needed and the data produced must be easily interpreted and unequivocal.

In most instances, a compound is not tested to provide specific information on how it will process. Rather, some property of the compound is tested to determine that it is as would be expected for the compound based on results from earlier batches of the same compound. If the property tested is known to be indicative of the degree of mixing, this provides information on how well the compound has been mixed this time.

This chapter will deal only with the most commonly used tests, most of which fit the criteria specified above for in-plant testing.

39.3   Processability test instruments

39.3.1   Mooney viscometer ASTM D1646, ISO 289

This instrument, initially developed 70 years ago, is the most common test instrument in the rubber industry. In the basic design, shown in Fig. 39.1, a flat serrated disk rotates in the rubber in a flat grooved or serrated cavity. There are national and international standards on its use and interpretation. ASTM D1646 specifies:

1. Fixed rotor speed of 2 rpm.
2. Large rotor (3.84 cm) for viscosity determinations.
3. Small rotor (1.28 cm) for scorch or high viscosity.

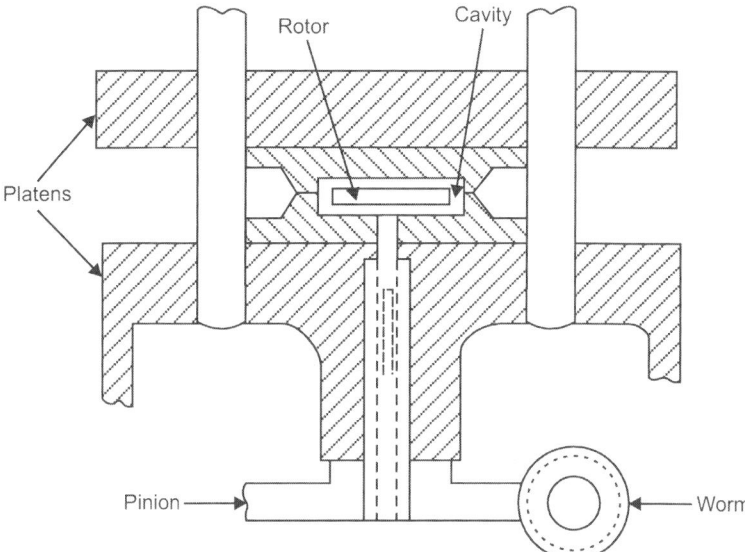

Figure 39.1: Basic design of the Mooney viscometer.

4. 100°C recommended test temperature.

5. 1 min pre-heat, 4 min test recommended.

Typical curves of Mooney viscosity versus time for low, medium and high viscosity rubbers are shown in Fig. 39.2. The maximum (a) observed in the viscosity curve is probably caused by the orientation of rubber molecules in the direction of flow. The energy required to untangle them causes the rise in torque. As temperature rises and disentanglement occurs, the required torque goes down and eventually, an equilibrium is reached.

While a fixed low speed is the standard of the Mooney machine, it has been run, in some research studies, over a range of speeds, generating up to 100 s^{-1} of shear.

Temperature control becomes difficult at these high shear rates. It is important to note that the Mooney machine is an empirical instrument and therefore, careful attention to technique is important. Also, the low shear rate (1.5 s^{-1} with the large rotor and 0.5 s^{-1} with the small rotor) are well below processing shear rates and therefore, it is of limited value in predicting processability. Its advantages are that it is easy to operate, the result is easy to interpret and the test time is short. Its major use is in specification testing of raw elastomers and compounds. Improvements that have been made to Mooney machines in recent years have been to improve temperature control, to improve torque measurement and recording, to make loading and discharging easier and to improve calibration.

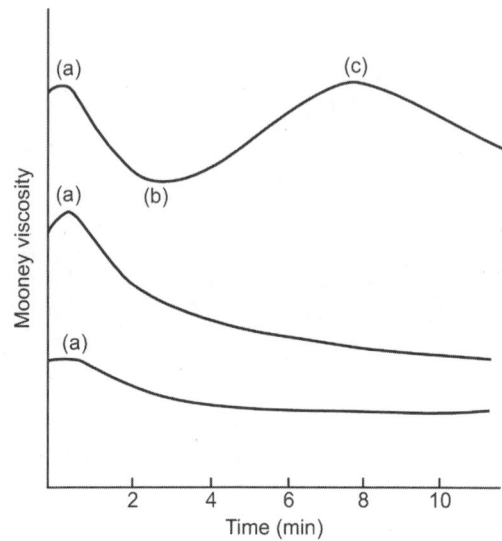

Figure 39.2: Typical Mooney curves.

39.3.2 Delta Mooney

ASTM D3346 describes the derivation of another parameter from the Mooney test known as Δ-Mooney. This is defined as the difference between two Mooney readings, either at two fixed times or between the early minimum (b) and subsequent maximum (c) of the high viscosity curves (see Fig. 39.2). The Δ-Mooney is of special value in predicting the processability of SBR and its compounds.

TMS Rheometer

Another interesting development is the TMS rheometer which is based on a Mooney machine, but in which the rubber is injected into the closed cavity by means of a transfer pot mounted directly above the cavity. By this means, rubber with freshly created surfaces can be made to fill the cavity and a control maintained on the pressure inside the cavity. This arrangement gives reproducible results at shear rates of up to 40 s^{-1}. Not only can shear stress for a given shear rate be measured, but stress relaxation and recovery can also be examined.

39.3.3 Capillary Rheometers ASTM D5099

The capillary rheometer is the oldest known type of instrument for measuring viscosity and has been the standard test instrument for measuring flow properties of plastics for many years. Developments in design and computer control, operation and calculation have led to an increased use of such instruments to measure flow properties of raw rubbers and compounds. The material is caused to flow through a capillary die, usually of circular cross-section, at a constant rate. The resulting laminar flow is similar to that in extruder dies and high shear rates typical of processing operations can be achieved.

Figure 39.3 shows a schematic diagram of a capillary rheometer used to measure polymer melt viscosity. Force (hydraulic or mechanical) is applied to the plunger to extrude melt. The extrusion rate is calculated from the drive speed for mechanical instruments and from velocity transducers, or from timed, weighed cuts for hydraulic instruments.

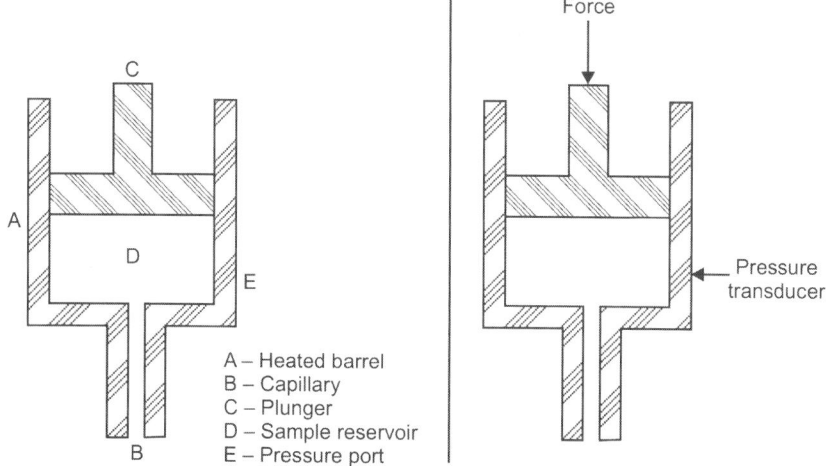

A – Heated barrel
B – Capillary
C – Plunger
D – Sample reservoir
E – Pressure port

Figure 39.3: Schematic diagram of a capillary rheometer.

The application to capillary rheometers requires the assumptions that there is no wall-slip and that the flow pattern does not alter along the length of the capillary and is not affected by the applied pressure. For raw elastomers, these are reasonable assumptions as long as the shear stress is not so high as to cause melt-fracture, i.e., non-laminar flow. Rubber mixes, which are micro-composites, display a much more complicated behaviour and the above assumptions do not always apply.

Pressure can be measured by using a pressure transducer mounted in the reservoir or calculated by dividing applied force by the cross-sectional area of the piston. Volumetric flow rate is calculated from piston speed and the cross-sectional area of the reservoir.

Modem capillary rheometers are automated, microprocessor controlled and the shear stress, shear rate and viscosity are calculated automatically. By using capillaries of differing diameters it is possible to operate at shear rates from 0.01 to 10000 s^{-1}.

This is a wider range of shear rate than is obtainable with any other type of rheometer and spans the range experienced by rubber in all types of processing equipment.

Figure 39.4 shows (top left) viscosity-shear rate data for neoprene compound at three different temperatures over much of the range. Clearly, shear rate has a much greater effect than does temperature.

1. High includes FKM and high hardness stocks.

2. Low includes silicones, EA and low hardness stocks.

3. Most rubber compounds are in between these.

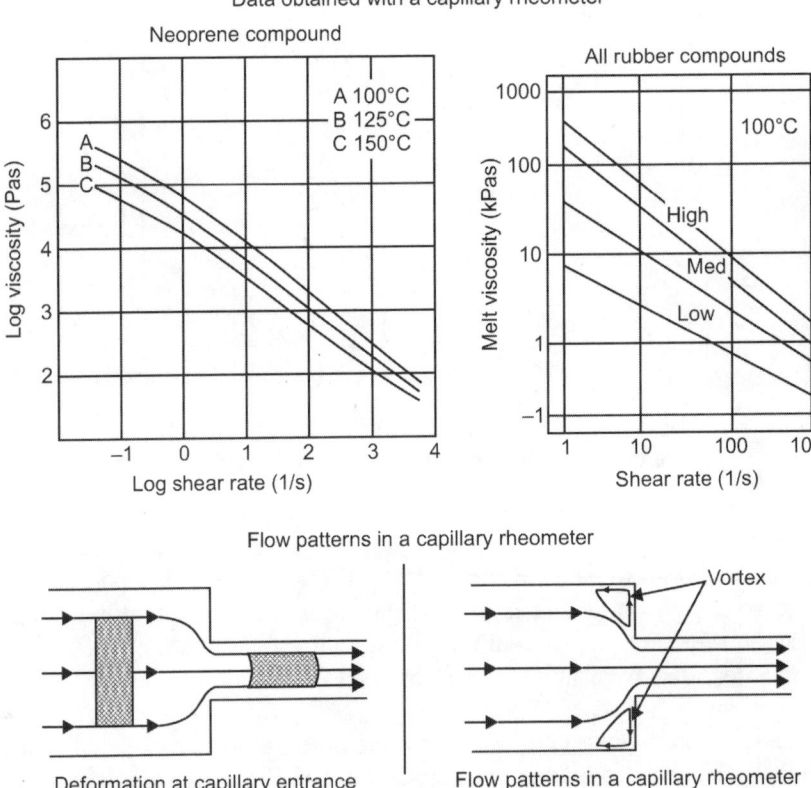

Data obtained with a capillary rheometer

Neoprene compound

All rubber compounds

Flow patterns in a capillary rheometer

Deformation at capillary entrance

Flow patterns in a capillary rheometer

Figure 39.4: Typical data obtained with a capillary rheometer.

Deformation in a capillary rheometer is mainly shear in the reservoir and a combination of shear and elongation flow in the capillary (see Fig. 39.4, bottom left). The flow pattern in reservoir and capillary can create a vortex or dead spots at the bottom of the reservoir (see Fig 39.4 bottom right).

This is a possible source of problems. Some of the, instruments use capillaries with a right angle included at the entrance to reduce this problem.

The ASTM Test Method for the measurement of processing properties of rubber using a capillary rheometer (ASTM-5099-93) should be referred to for fuller detail of technique and analysis of results. The total force or pressure to extrude a sample is more than that required to flow through the capillary as there are entrance and exit pressure drops too (see Fig. 39.5, top). The pressure drop through the capillary is due to viscosity. The exit pressure drop is due to elasticity; that at the entrance is due mainly to elasticity (Ca 95%). Thus, shear stress and viscosity values based on total force or pressure are inexact, but are sufficient for in-house control or correlations with processing. For a more precise use, as in engineering flow calculations, specification testing, or interchange of data between instruments and laboratories, corrections for entrance and exit effects are required. Figure 39.5 (bottom) shows uncorrected shear

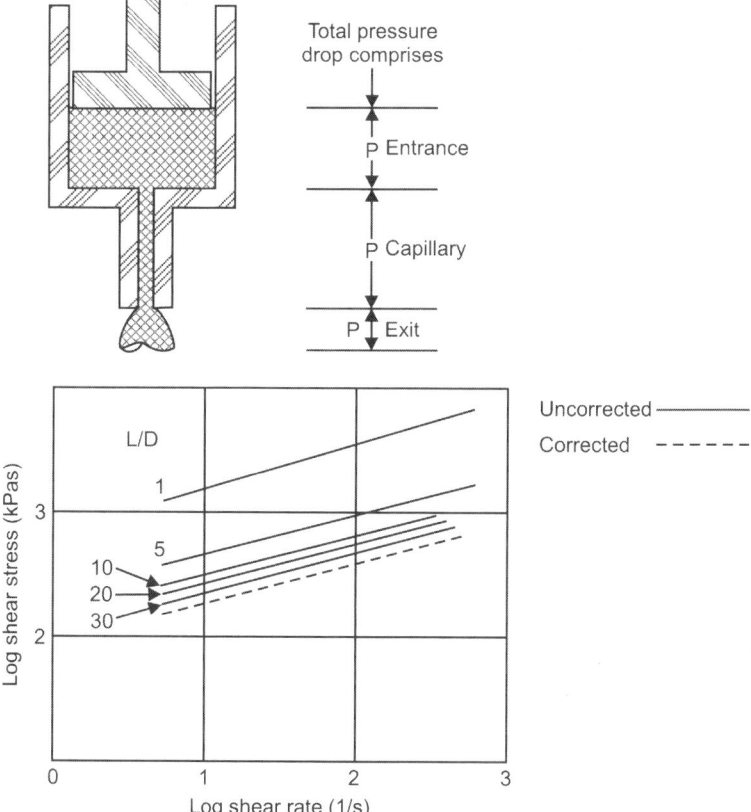

Figure 39.5: Entrance and exit pressure drops in capillary rheometers.

stress versus shear rate data for an EPDM compound using dies of five different length, to diameter (L/D) ratios (1, 5, 10, 20, 30), plus the same data corrected for systems effects. This shows the significant effect of short L/D capillaries.

39.3.4 Oscillating disk curemeters

Commonly called oscillating disk rheometers (ODRs) in the rubber industry, these are widely used to measure the curing behaviour of rubber compounds, by following the change in torque as the curing reaction proceeds. A rotor is embedded in the rubber sample, which is confined in a die-cavity under pressure and controlled at a predetermined temperature. The sample is subjected to an oscillatory shearing action of constant amplitude and the torque required to oscillate the rotor is measured. Since the rotor is straining the sample, the torque values are directly related to the shear modulus of the rubber.

Figure 39.6 shows typical curves from an oscillating rheometer. The use of this curemeter is described in ASTM D2084.

The values read from the curves are:

1. M_L minimum torque in Nm or lbf . in.
2. M_{HF} maximum torque where curve plateaus in Nm or lbf . in.
3. M_{HR} maximum torque of reverting curve in Nm or lbf . in.
4. M_H highest torque attained during specified period when no plateau or maximum torque is obtained in Nm or lbf . in.
5. t_{s1} min to 1 lbf . in rise above M_L – used with 1° arc.
6. t_{s2} min to 1 lbf . in rise above M_L – used with 3° arc.
7. t_x min to x% of maximum torque, t_x = min to x . $M_H/100$.
8. t'_x min to x% of torque increase, t'_x = min to $M_L + x(M_H - M_L)/100$.

Most rubbers provide one of the three general shapes illustrated in Fig. 39.6. The middle curve is typical of most natural rubber compounds, exhibiting reversion. Curemeters are used widely to test finished compound prior to the shaping and curing stage. They will detect changes in curing characteristics readily. They are also widely used in developing formulations. Changes, which affect viscosity and scorch are reflected in the first portion of the curve and the effect on rate of cure and the modulus of the resulting vulcanisate can be seen in the latter portion of the curve.

39.3.5 Rotorless curemeters

Monsanto introduced the moving die rheometer - MDR 2000, a rotorless curemeter, in which an eccentric linkage oscillates the lower die and a torque transducer measures the resistance to that oscillation as torque transmitted to the upper die. MDR stands for moving die rheometer. The MDR 2000 recovers

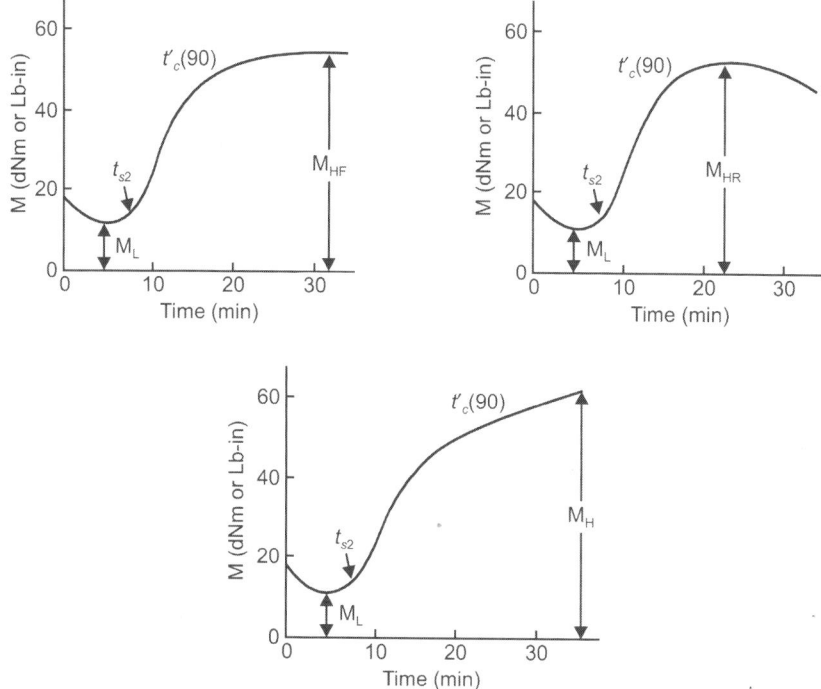

Figure 39.6: Typical curves from an ODR.

temperature faster, has tighter temperature control and a more uniform sample temperature than the ODR R-100 rheometer. It also records torque to a higher resolution ($\pm$ 0.01 dNm as opposed to $\pm$ 0.25 dNm). The applied oscillation is separated from the recorded torque in the MDR to allow recording of the in-phase, or elastic torque (s'), the viscous, or 90 degrees out-of-phase torque (s'') and the phase shift, or damping characteristic (tan δ) of the compound throughout the cure test. ASTM D5289-93 describes the test procedure and DiMauro and Sezna discuss the correlation of the results from the ODR (R-100) rheometer, the newer ODR 2000 instrument and the MDR 2000.

39.3.6   Dynamic mechanical rheological testers (DMRTs)

Monsanto (now Alpha technologies) introduced the rubber process analyser RPA 2000 in 1990. This instrument is an improvement on earlier DMRT machines, in that it has a sealed and pressurised sample chamber between two cone-shaped dies, which allows it to test with repeatability and reproducibility at very high strains. Another advantage is that no sample preparation is required. The RPA 2000 uses the same die cavity as the MDR 2000, with air cooling capability added. The test temperature can be varied from 40 to 200°C, the frequency from 2 to 2000 cpm and the arc of oscillation from ± 0.05 to 90 degrees arc. A PC, which is an integral part of the RPA operating system, controls the test procedure and handles data acquisition and processing. The instrument is extremely versatile; it can measure uncured dynamic properties over a range of frequencies and strains, at process temperature, then the temperature is increased to cure the specimen and finally the temperature is dropped for a frequency and/or strain sweep after cure. It can measure viscous and elastic responses up to a maximum shear rate of 30 s^{-1}. Its one limitation is, because it forms its own test piece, it cannot be adapted to test a finished part or sample, which has already been cured. The latest version has the ability to carry out stress relaxation tests.

39.3.7   Stress relaxation instruments

A pressure transducer mounted in the test chamber of a rheometer makes it possible to follow the decay in pressure, or stress, following the removal of the deforming force. A log-log plot of pressure versus time produces a straight-line relationship between these variables.

In all routine stress relaxation testing two conventions to improve accuracy and to simplify data handling are used:

1. The reference pressure or stress is arbitrarily taken 1.0 s after release of the applied stress. This is more accurately measurable than the maximum applied pressure or stress.

2. Time to a 50% reduction in pressure or stress is measured.

If the relaxation curve is rotated through 180°, the viscosity shear rate curve in Fig. 39.7 (right) results. From this latter curve it can be seen that most of the relaxation curve represents shear rates lower than those experienced in processing. Also, it can be inferred that variations in relaxation time between similar polymers and compounds are more sensitive to elasticity than to viscosity. That is similar to the situation with Mooney viscosity.

There are a number of instruments available to measure stress relaxation. The dynamic stress relaxometer (DSR) was developed in the late 1960s by BF Goodrich. It measures stress relaxation after a shear deformation.

The stress relaxation processability tester (SRPT), developed by RAPRA Technology Ltd, measures stress relaxation following compressive deformation. The monsanto processability tester (MPT), which is an automated capillary rheometer, can be programmed to apply one final pulse at the end of the flow tests. The high pressure developed in the barrel is allowed to decay by relaxation of the stresses in the stock and recorded, as shown in Fig. 39.7.

The Mooney machine can also be modified to measure stress relaxation. Improvements in equipment, including motor braking, reduced rotor friction and computer data analysis, have improved its sensitivity, repeatability and

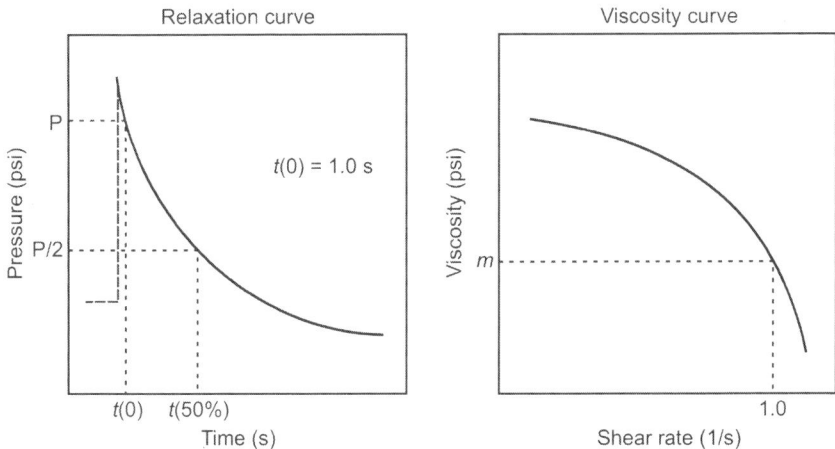

Figure 39.7: Stress relaxation curves.

reproducibility. The best summary of the technique and equipment, together with a clear exposition of the interpretation of stress relaxation data from the MV-2000 Mooney viscometer, is to be found in the paper by Burhin, Spreutels and Sezna. They point out that since elastomers and compounds are viscoelastic materials, they obey Hooke's Law for their elastic response and Newton's Law for their viscous response. Thus, a perfectly elastic material would not give a decrease in torque after the Mooney rotor is stopped, while a perfectly viscous material would give zero torque instantaneously. The defounation energy is stored in the material in the first case and dissipated in the second case. On a log/log plot of Mooney torque versus relaxation time, elastomers and rubber compounds have a response between these two extremes, usually a linear response. The shape of this linear function is a measure of the intrinsic elasticity of the material. A slope near zero indicates a more elastic material.

39.3.8   ODR cure times correlation with MDR

Monsanto instruments (now Alpha technologies) introduced the oscillating disc rheometer in the early 1960s. They have improved its ease of operation, reproducibility and reliability over the years. The latest model is the ODR 2000 which has microprocessor control, improved sample heating and reduced mechanical losses. It reduces test variation by half compared to the earlier R-100 instrument. It still contains a rotor, which acts as a heat sink and also makes sample removal more difficult.

The MDR 2000, a rotorless curemeter, was introduced in 1988. It has tighter temperature control and provides a more uniform sample temperature. The improved mechanical design, microprocessor control and the absence of a rotor with the MDR 2000 improve sensitivity and precision up to 5 times over the old R-100 rheometer. Both ODR and MDR follow the curing reaction by measuring the increasing resistance to deformation (stiffness) of the sample as curing proceeds.

The main difference between the ODR and the MDR is that the heat sink effect of the rotor means that ODR's have a significant thermal lag at temperatures above 150°C and this results in longer cure times than for an isothermal cure. Thus, the question for the processor in converting from the R-100 to the MDR 2000, is how to interpret the shorter MDR cure times.

The 'optimum cure time', interpreted as the time to reach 90% cure in an ODR has been used for many years as a guide to curing in extrusion and injection molding. It was shown early on that it correlated with tensile tests on oven-cured sheet specimens. However, as in many tests in the rubber industry, curemeter tests are performed under conditions rather different from those encountered in continuous vulcanisation lines, or injection molds. The main differences being in sample shape, size and thickness and in the temperature. Many processes today use curing temperatures of 200°C or higher. Thus the ODR 90% cure time gives a guide, but not a direct correlation.

After years of familiarity, processors are fairly confident in using ODR curves to predict curing performance in practical equipment. In essence, the processor has to develop the same degree of confidence in the MDR, perhaps by running it in parallel with the ODR. Sezna gives several examples of this and shows that cure times (at 50, 80 and 90% cure) and at 150, 175 and 200°C from MDR rheographs are significantly shorter than those from R-100 curves. He comments that correlations between curemeter tests (both ODR and MDR) and press cures have to be determined experimentally. Also, the 95% cure time from MDR tests is a better indication of optimum tensile properties than the ODR 90% cure time, because the correlation, for the MDR, is not temperature dependent in the range from 150 to 200°C.

39.4   Microprocessor control, automatic calibration

Some of the microprocessor control, automatic calibration series of instruments are given below:

1. MV2000 mooney viscometer with stress relaxation.
2. MPT monsanto processability tester – an automated capillary rheometer.
3. ODR2000 oscillating disc rheometer/curemeter.
4. MDR2000 moving die rheometer/curemeter.
5. RPA2000 rubber process analyser – a dynamic rheological tester/curemeter derived from the MDR.

In the introduction it was stated that, to be of value in process control, tests must be rapid, easy to perform and interpret and give good differentiation between samples of the same grade of polymer (compound) with different processing characteristics. Tests on a mixed compound obviously need to assess its flow behaviour and its curing behaviour, on one sample, in one test, in one instrument.

All the instruments in the above list have microprocessor control, automatic calibration, etc., that is we are considering the latest models. They can all give measurements of viscosity, elasticity and curing behaviour. The differences lies in the precision, repeatability, ease of operation and the range of conditions under which tests can be performed.

The following list summarises the capabilities of each instrument.

1. MV2000 frequency 2 cpm, shear rate 1.5 s⁻¹, temperature 100 or 121°C, (ASTM D 1646 and ISO 289). The torque resisting rotation is reported as viscosity in Mooney units. Can also be used to measure onset of vulcanisation or scorch at 121 or 135°C. Can perform a stress relaxation test at the end of the viscosity test. This indicates the relative levels of elastic versus viscous response exhibited by the material.

2. MPT Shear rates up to 25,000 s⁻¹, temperatures up to 200°C, die swell and have stress relaxation capabilities.

3. ODR 2000 gives indication of compound viscosity, cure rate and compound properties. ASTM D 2084 and low shear rate.

4. MDR 2000 in addition to above ODR capabilities, can measure both elastic and viscous responses and their ratio (tan δ).

5. RPA 2000 derived from the MDR but has much wider ranges of temperature (40 to 200°C), frequency (2 to 2000 cpm) and arc of oscillation (± 0.5 to 90 degrees arc). Can measure elastic and viscous torques, moduli (S′, S″, G′, G″, tan δ). Dynamic viscosity η′, η*, cure tests, cure simulation, post cure analysis (loss modulus, tan δ).

39.5 Testing of finished rubber product

When formulating a new compound the main requirement has to be that after mixing, shaping and curing, the compound must provide the properties required by the end application. Obviously, the set of properties needed varies considerably from one application to another. A tyre tread requires abrasion resistance, wet and dry traction, low heat build up due to hysteresis and so on, whilst an oil seal requires heat and oil resistance and low compression set and of course, most applications require the basic elasticity and strength of a cured rubber.

In developing a new compound for an application and in later production of the part in the plant, there are three main types of tests that are used.

1. Tests to determine how well the fillers, especially carbon black, have been distributed and dispersed.
2. Tests on cured test pieces, to determine whether the material gives the expected and required end product properties.

These tests are generally standard tests, ASTM, ISO, etc., are discussed. Such tests are often designed to subject the part to a lifetime of use in a relatively short length of time. One example is tyre testing on a vehicle, on a test track, subjecting the tyre to braking, acceleration, cornering and so on, on a variety of surfaces. Others are simulated tests on a laboratory test jig. For example, a door seal fitted to a prototype vehicle may be subjected to repetitive door slams and openings for a number of cycles representative of a ten-year vehicle life. To be realistic, such tests, whether on a test track or in a laboratory, have to be carried out over a wide range of environmental conditions. For instance, vehicles have to function at temperatures from –40 to +40°C and therefore tyres, seals, hoses, etc., have to function over the entire range. For each product, the range of tests has to be agreed between producer and customer and are reported in the Quality Assurance Reports.

39.6 Tests of filler distribution and dispersion

The importance of these in optimising cured product properties has been described earlier. Tests aiming to determine how well a filler has been distributed and dispersed microscopy and surface roughness.

39.6.1 Microscopy

The standard technique for determining the quality of a dispersion is by examining a razor-cut surface of the vulcanised compound under a low-power optical microscope and comparing the dispersion with a set of standards.

39.6.2 Surface roughness

This usually is carried out on a freshly cut surface as in the tests in Section 39.6.1 above. In one technique, the specimen surface is examined simultaneously with controls using split-field microscopy.

39.7 Tests on cured specimens

Most ASTM tests for rubber are on vulcanised test specimens in specified standard formulations, prepared and cured under specified standard conditions, described in ASTM 3183. Depending on the application these tests may include modulus, tensile strength, hardness, compression set, abrasion resistance, dynamic properties, oven ageing, ozone resistance, solvent swelling and oil resistance. In addition, there are tests which apply only to specific end applications. Sommer and Yeoh emphasise that it is important to differentiate between tests that yield data for engineering design and quality control tests, which confirm that one batch of mixed compound has the same properties as the preceding batch. They also make the point that tests useful for design purposes are usually more difficult and time consuming than quality control tests. As a result, in practice most of the standard test methods used in the rubber industry are quality control tests, which yield data not directly relatable to performance.

A further important point is that cross-linking can continue after the test piece has been molded, even when it has cooled to room temperature. Therefore, it is standard practice (ASTM D 3182) to leave molded specimens for a minimum of 16 hr before testing.

This section will outline the more important quality control tests and comment on their relevance and value in the field. Details of sample preparation, test procedures and the apparatus used can be found in the ASTM standards and reference.

39.7.1 Tensile tests ASTM D412

Rubber is rarely used in simple tension and never in practice is it stretched to its breaking point. However, one of the most common standard tests involves stretching dumbbell shaped specimens and measuring the forces at 100 and 300% elongation and at break. Tensile stresses are, by convention, calculated based on the original cross-section and are often referred to in the industry, incorrectly, as 100 and 300% modulus. The elongation at break is usually termed the ultimate elongation. The widespread use of this test for quality control can be attributed to the fact that it is relatively easy and quick to perform, the test results do not require interpretation and the results can be used as a general measure of quality. Tensile properties are sensitive to the amount of diluent fillers, added to reduce volume costs and more importantly, to poor filler dispersion or inadequate curing.

39.7.2 Hardness ASTM D2240, D531 and D1415

Hardness is defined as the resistance to indentation as measured under standard conditions using specified test instruments. These are, specifically, the Shore Durometer, a portable hand-held device and the International Rubber Hardness Tester, a bench top instrument. The latter is significantly more accurate than the Shore tester, which is routinely used because of its convenience. The depth of penetration of the indenter has been empirically related to the Young's modulus of the rubber.

39.7.3 Compression set ASTM D395

This is defined as the percentage of the deformation that is not recovered after a standard test piece has been subjected to a standard compressive load or deflection for a given period of time, at a specified temperature. The measurement is made 30 min after the compressive force has been removed.

The level of compression set required depends on the application. For instance most seals, in order to maintain their effectiveness over time, require a low compression set. Compression set decreases with increasing cross-linking and gives a good indication of the state of cure. However, overcuring in order to achieve low compression set is not a good practice, because other properties such as strength, flux resistance and ageing resistance can be adversely affected.

39.7.4 Solvent resistance ASTM D471

In some applications exposure to fluids, which could have adverse effects on rubber, occurs. Typical fluids are gasoline, oil, air-conditioning solvents, antifreeze solutions and aerosol propellants. The adverse effects that could occur are swelling of the rubber due to absorption of the fluid, extraction of some constituents such as plasticisers or antidegradants, or chemical reaction with the rubber or one of the compounding ingredients. The specific test applied depends on the nature of the solvent and expected effect on the rubber. The most generally applied test is to measure swelling after immersion in the liquid, for a specified time at a specified temperature. Swelling is usually measured by determining the volume change. It is quite common to measure also the effect of the immersion on tensile properties and hardness.

39.7.5 Ageing ASTM D573, D865, D572, D454

In many applications, rubber components are exposed to intermittent high temperatures in the presence of air. ASTM D573 gives the procedure for measuring the effect of combined oxidative and thermal ageing. Test specimens, usually dumbbells, are exposed in air at elevated temperatures for specified periods of time. Properties before and after ageing are measured and the rate or deterioration determined.

39.7.6 Ozone cracking ASTM D1149, D1171

Some unsaturated rubbers are susceptible to attack by ozone, even at miniscule levels, if they are also under tensile stress. The attack shows itself by the development of surface cracks at right angles to the direction of strain. In the test, specimens are stretched in an ozone-rich atmosphere at a specified temperature and the surface checked under magnification at several specified intervals.

39.8   Improving rubber testing with microcomputers

Rubber products can no longer be made as they once were. The demands of a global economy require increased product life and better product quality. The measurement of rubber properties has become an increasingly important part of rubber compound development and quality control to meet global standards. Tests that have been done for many years in the rubber industry are not adequate for today's needs. Many of these tests can still measure properties that are important to performance and quality, but the sensitivity of these tests is often no longer adequate. More information also needs to be collected from each test, because a single measurement is no longer considered adequate when determining material properties. There is also a need to quickly store and compare material property data to historical data.

These problems were difficult to solve until the invention of the microprocessor, which led to the development of the microcomputer. The microcomputer in turn has been applied successfully to instrumentation for rubber testing to improve the sensitivity of many test methods, to increase the amount of information from each test and to store results for quick comparison to historical data.

Today, microcomputers and their peripherals can be purchased off the shelf and assembled by an average laboratory manager at reasonable cost. This section describes how to use microcomputers and their peripherals to modernise the gathering, storage, retrieval and analysis of data. Once test information has been stored in a laboratory microcomputer, there is an almost immediate need to send it to other areas of a rubber factory. At one time, moving information stored in a microcomputer was difficult. Today microcomputers can be tied together into networks to share information almost instantaneously in a way competitive with a minicomputer using multiple terminals.

39.8.1   Microcomputers and hardware for data acquisition

A microprocessor, the heart of the microcomputer, is nothing more than a central processing unit (CPU) on a single chip or a few chips. A CPU is that portion of a microcomputer which is capable of decoding and following stored instructions. In addition, a CPU contains communication control between the microcomputer memory and its input/output channels, an arithmetic logic unit (ALU) for doing calculations, data storage registers for handling numbers and some programming stored on read-only memory (ROM). Input/output channels control the communications between the CPU and peripherals (disk drives, printers, displays, interfaces, etc.). Figure 39.8 is a simple diagram illustrating the flow of information between a CPU and microcomputer peripherals. Note that any information sent from one peripheral to another must go through

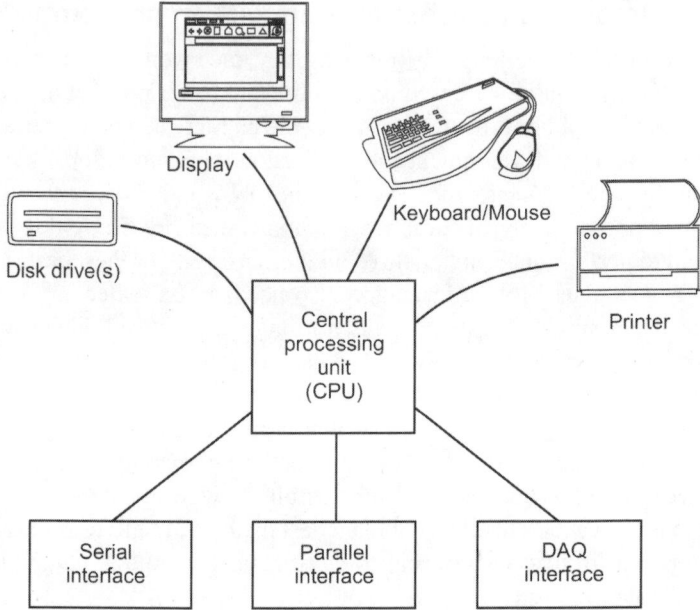

Display

Keyboard/Mouse

Disk drive(s)

Central
processing
unit
(CPU)

Printer

Serial
interface

Parallel
interface

DAQ
interface

Figure 39.8: Flow of data in a microcomputer. Note that all information that is passed from one peripheral to another must go through the CPU.

the CPU. A complete stand-alone microcomputer usually consists of the CPU and the following peripherals: fixed disk drive(s), printer(s), video display and keyboard. Initially there were many microcomputers produced by different manufacturers.

The application of microcomputers to rubber testing requires an interface. Figure 39.8 shows three types of interface used with microcomputers: serial interface, parallel interface and a data acquisition interface (DAQ). Rubber test data can be sent to a microcomputer from a sensor or instrument only via one of these three interfaces. Microcomputers are built with special areas called buses designed for optional printed circuit boards. Each interface consists of a printed circuit board, which is installed directly onto the microcomputer bus.

A rubber test apparatus cannot directly provide microcomputer interfaces with digital data. Rather, several pieces of additional hardware must be installed between the test apparatus and the microcomputer interface. Figure 39.9 illustrates the steps necessary to send information to microcomputer from a rubber test apparatus; these are conversion of a rubber measurement into a signal by a sensor or transducer, conditioning the resulting signal into a voltage for an analog-to-digital converter, analog-to-digital conversion and interfacing the digital output to a microcomputer. The steps illustrated in Fig. 39.9 are called a measurement system.

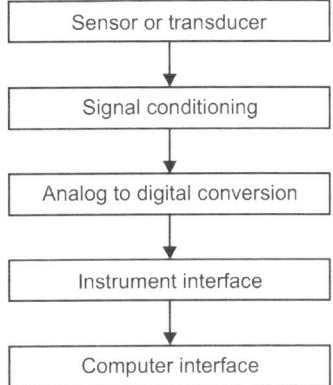

Figure 39.9: Hardware steps necessary to bring information from a test sensor to a microcomputer.

Sensors used in rubber testing

Many rubber tests use sensors to measure either force, position, or temperature to obtain data. The sensors used to make these measurements are discussed below.

Force measurement is used in many rubber tests to measure the resistance to a fixed amount of deformation of a rubber test piece. Some applications of force transducers to rubber testing include tensile tests, curemeters, stress relaxation and dynamic mechanical tests. There are a variety of methods to measure force. Not all these methods are suitable for electronic systems interfaced to microcomputers. One commonly used electronic method to measure force is the strain gage. A strain gage is a wire that has the special property whereby a change in its length produces a change in the resistance of the wire. Materials for strain gages are selected to produce a relatively large change in resistance with a given change in length. This change in resistance divided by the change in length is a strain gauge constant called the 'gage factor'. The higher the gage factor, the higher the sensitivity of the strain gage.

Strain gages measure the strain or deformation of an object. To measure the strain imposed on the object, the strain gage is first bonded to the object. Strain gages are directional and must be bonded in the proper direction to measure strain in the required direction. When a force deforms the object, the bonded strain gage will also be deformed. The deformation of the strain gage changes the resistance of the strain gage to a very small amount. To increase the magnitude of the resistance change, the strain gage wire length is increased and then wound back and forth numerous times.

A strain gage can be used to measure force, but not directly. If the strain gage is bonded to a spring, the force on the spring will be directly proportional to the strain measured on the spring. A strain gage attached to a spring can be converted to a force transducer by applying known forces and recording the measured strain signal. The resulting calibration factor can convert the measured strain into a force measurement. A force transducer also can be used to measure pressure by dividing the area over which the force acts into the measured force. Figure 39.10 illustrates a simple test apparatus using a force transducer. When the force transducer instantaneously compresses the sample to a fixed amount of strain, the instantaneous force measured will indicate the resistance of the sample to deformation. Since the actual measured force in Fig. 39.10 will decay as a function of time as a result of the nature of rubber, this system will measure the stress relaxation of the sample.

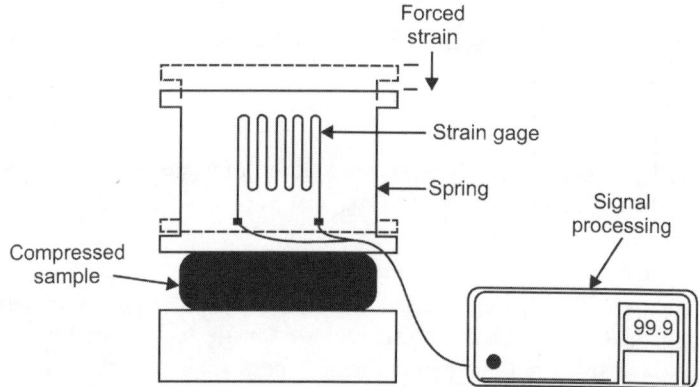

Figure 39.10: Example of a test apparatus designed to measure the force coming from a rubber sample at fixed strain. Note that the strain gage is bonded to a spring to create the force transducer.

Even with a long wound wire and a large gauge factor, the resulting resistance change is often very small and difficult to measure accurately. In a common method used to accurately measure the small strain gauge signal, the strain gage resistor is placed into one leg of a Wheatstone bridge (Fig. 39.11).

The imbalance of the bridge is then measured with better sensitivity than is afforded by the direct measurement of strain gage resistance. Strain gage also require compensation for changes in temperature. A proper arrangement of the strain gage in a Wheatstone bridge can compensate for temperature changes over the required range. Force transducers with strain gages can be purchased with attached signal processing electronics to simplify the assembly of a complete system. However, there are a wide variety of force transducers available and some knowledge of the application is necessary to make a selection.

Specifications include range of force measurement, operating temperature range and output voltage.

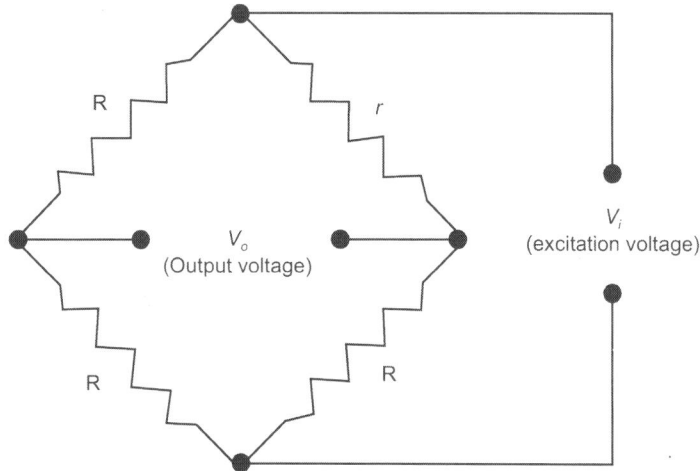

Figure 39.11: A Wheatstone bridge circuit used to measure the small changes in strain gage resistance.

Electronic position measurement can also be done in several ways. A common method uses a linear variable differential transformer (LVDT), consisting of a transformer with a movable magnetic core. The transformer contains a set of primary and secondary coils. The primary coils are energised with a fixed voltage. The voltage at the secondary coils is then measured while a central magnetic core is moved. The other end of the magnetic core is attached to the surface where the position is to be measured. As the surface moves, the magnetic core will also move. The movement of the magnetic core produces changes in the voltage at the secondary coils which are directly proportional to the movement of the magnetic core.

Calibration of the LVDT with fixed, known movements allows the calculation of position during a test. LVDTs are insensitive to temperature changes and do not have any hysteresis effect. LVDTs can also be purchased with electronics for a digital output or a computer interface. Applications of LVDTs to rubber testing include measurement of compression set and compliance.

There are two common methods used to measure temperature electronically: thermocouples and resistance temperature devices (RTDs). Thermocouples are used in harsh environments or where accuracy and stability are not a problem. RTDs are used in applications calling for good accuracy and stability.

Thermocouples are welded junctions of two wires made of dissimilar metals. This type of junction produces a small voltage. The resulting voltage varies with temperature and so can serve as a temperature sensor. Two junctions are actually needed to make one thermocouple for temperature measurement. One of the junctions is located in an area kept at a reference temperature, while the other is located in the area where the temperature is being measured. The difference in voltage between the two junctions indicates the temperature. A common thermocouple used in rubber testing is iron and constantan (a copper-nickel alloy) which is commonly called a type J thermocouple. A type J thermocouple covers the rubber test temperature range very well and has very good sensitivity to temperature changes (change in voltage with change in temperature). Building the electronics necessary to digitise thermocouple readings and interface them to a computer is difficult. Fortunately, standard thermocouple electronic modules can be purchased with a built-in reference junctions for a type J or other standard thermocouple.

The second method to measure temperature in rubber testing uses an RTD. Most RTDs contain a platinum resistance thermometer (PRT). The RTD is simply a rugged version of a PRT. The resistance of pure platinum wire increases with increasing temperature in a predictable way and closely follows a second-order polynomial called the Callendar-Van Dusen equation. The electronics used with an RTD must measure voltage accurately at low current levels to get good temperature measurements. This requires a great deal of electronic expertise. As with thermocouples, standard RTD electronic modules can be purchased. Errors in temperature measurement can be produced by the sensor or by the signal processing electronics. The temperature measurement specifications for either thermocouples or RTDs and their electronic modules must be checked to ensure that the measured temperature received at the microcomputer will be within required limits. In addition, the location of the temperature sensor may be critical to an application.

Some applications of temperature measurement in rubber testing include test sample environment and rubber processing equipment.

Signal conditioning

The signal coming from a strain gauge, LVDT, thermocouple, or RTD must first be conditioned to match the requirements of an analog-to-digital (A/D) conversion device. Commercially available standard signal conditioning modules can change the sensor signal to an output voltage that matches the input requirements of a standard A/D converter.

Some microcomputer products can even multiplex or accommodate up to 64 of these standard modules for 64 simultaneous inputs with any combination of force and temperature measurements. These modules can be very useful when instrumenting equipment while studying processing operations.

Analog-to-digital conversion

The next step is to change the analog signal into a digital value. This digital value can then be used by the microprocessor to indicate the transducer measurement. In the past, A/D converters had to be designed from scratch. The A/D converters are complete printed circuit boards, which can be installed directly into a microcomputer There are still some specifications that must be established before purchasing the proper A/D converter. The first is word length. The higher the word length, the better the digital resolution and the greater the cost. To determine word length, the required reading resolution must be determined. For example, if 10 V is the maximum input voltage expected and the required resolution is to the nearest 0.01 V, the A/D converter must be able to resolve at least 1000 (10/0.01) different readings.

Interfacing to microcomputers

After the sensor signal has been conditioned and digitised, it is ready to be sent to the microcomputer for further processing by the CPU. A computer interface, consisting of the hardware necessary to send and receive digital information between a computer and an instrument, is required to send the digitised signal to the microcomputer. The hardware in the interface must be set up properly in both the instrument and the microcomputer if information is to be transferred successfully.

39.8.2   Software for data acquisition and analysis

All microcomputers and their peripherals require software to do anything meaningful for the user. Software consists of stored instructions that direct the CPU to collect data, store data, analyse data and send output to printer(s), and fixed disk(s).

In some cases the microcomputer software can also control an instrument. Software has a wide range of complexity depending on its purpose.

Applications of data acquisition in a rubber factory

An earlier review of microprocessor use in the rubber industry emphasised the reduction of labour in testing. A decade ago, the microprocessors were incorporated into instruments produced by vendors for specific applications. The instrument reduced the data from a test and sent the results to a printer. Now, however, the potential uses of microcomputers in a rubber factory have increased.

One way a factory can utilise the data acquisition capabilities of a micro-computer is by monitoring factory processing equipment for force and temperature measurements. More robust microcomputer equipment is now available to handle environments of the type encountered in factories. The information gained can be used to adjust equipment parameters during operation or to release material to the next stage of production. Instrumented equipment can also help in development to determine the best conditions for equipment operation. Data gathered can be stored in microcomputers and then compared to earlier results. Computerised test instrumentation can provide historical databases for use with statistical process control (SPC) software to monitor materials or processes. SPC methods can help maintain good quality control in a factory. Computerised SPC can also reduce the amount of paper generated in a laboratory by printing only the final summary reports.

Stored data can also be used with experimental design software packages to help in development. Many experimental design software packages can utilise standard ASCII formats directly, which reduces the time required to enter data. Experimental design methods can reduce development time.

Device drivers

The most basic software programme necessary to collect data is called a device driver. The programme viewed by a microcomputer operator is normally referred to as the application programme. An application programme is usually written using a higher level computer language

To simplify programming in the higher level language, device drivers are used; these special microcomputer programmes are designed to operate peripherals such as an interface by allowing communication between the device and the CPU. Device drivers take care of the many details that must be correctly managed to successfully communicate a high level programme to a peripheral.

In this way, the programmer can produce an application programme in a relatively short period of time, since the software necessary to communicate with the peripherals is added as a device driver.

Menu-driven software

Menu-driven software requires no programming knowledge. The menu(s) in the software provide either a list of alternative selections or a list with blanks to be filled in. The operator makes choices in the menu and then runs the programme. One example of menu-driven software that can collect data is PROCOMM+. This software presents the operator several options in menu or list form. The operator makes selections from this menu until all the required choices have been made. The operator then tests the programme to see whether communication is taking place. If the wrong choices have been made the result input on the video screen will not be decipherable, whereupon the operator must go back to the menu and try some alternative selections.

'Virtual instrument' software

The current highest level of data acquisition software development is embodied in the 'virtual instrument' approach. The best example of this concept is LabVIEW, designed for MacIntosh computers. Instead of using menus,

LabVIEW uses icons to make a nearly self-explanatory picture of an instrument control panel on the video display. During the operation of this software, the operator manipulates the 'instrument controls' with a mouse to move 'levers,' turn 'dials', or press 'buttons'. The operator can observe the operation of the instrument by looking at 'panel meters', 'digital displays', or 'strip charts'. The system comes with device drivers for many instruments, or it can be programmed for other applications. Instead of writing software commands or code, the programmer creates a block diagram by arranging blocks into appropriate locations and joining them together. In this modulus system, parts of one instrument application can be applied to other applications.

Turn-key software

The software mentioned above requires a wide range of programming expertise or knowledge of microcomputer interfaces, some of which is quite easy by the standards of just a few years ago. However, some facilities lack the time or expertise to implement even the simplest application. These are the ideal customers for turn-key systems. A turn-key system is an application completely set up by a vendor. The user needs training in the operation of the software only.

Some positive points to look for in turn-key applications in rubber testing are as follows:

1. The fewer analog components the better. An A/D converter should digitise analog information as near to the transducer as possible.
2. The measurement electronics should be isolated, so that readings will be stable and not sensitive to the operation of neighbouring equipment.
3. The operation of the software should not slow down an operator.
4. A method should be available for checking calibration and recalibrating the apparatus.
5. Software algorithms for reducing data should not be easily fooled into producing meaningless information.
6. Error messages should be as literal as possible, so that anyone can understand them.

Data reduction software

In the switch from analog to digital instrumentation, the duties usually associated with an operator for instrument control, signal processing and data reduction were taken over by the software. In some cases, the software was also expected to send messages when there were problems in the instrument or test results. All these functions depend on software algorithms to make the instrument 'think'. These algorithms can either make an instrument worth the effort to upgrade or increase laboratory expenses (if poor decisions are made).

Errors in data acquisition and analysis

Problems with microcomputers in data acquisition are different from those encountered with analog systems and need to be understood. Errors in micro-computer data acquisition and analysis can be produced in four major areas: people, equipment (including any software), environment and methods. We discuss the influence of each of these on test results.

People errors are often due to poor training. Some of the more complex tests may require a higher skill level than is available in the test laboratory. Sloppy workmanship or poor motivation are other areas of concern. The best way to avoid people errors is to provide good training or to send people to be trained. Good training should provide a quick return on any investment, especially if complex hardware/software has been purchased. Software often uses special commands that are not obvious without training.

Equipment errors can be difficult to troubleshoot. There is a large pathway when going from the sensors to the microcomputer, with several distinct steps in between. If a problem is observed at the microcomputer, the exact source must first be located. Once the problem step has been found, the bad component can be sent out for repair or a replacement can be installed. Some components are so cheap that they are almost disposable. The labour cost of finding a small component failure on a circuit board can be quite high, however. There are many ways in which equipment can fail. Each piece of equipment can have its own set of problems.

Software problems can be just as tricky to solve as hardware problems. Many of today's software products come with an abundance of technical information and even the best efforts of software vendors cannot always prevent problems from occurring. Many software vendors provide technical assistance telephone numbers, which can be quite helpful. However, sometimes a problem cannot be resolved because the customer is using the software in a way not anticipated by the programmer.

Environment errors are produced by testing samples under test conditions significantly different from those expected. Rubber properties change with temperature, frequency and strain. All samples should be tested under identical conditions for proper comparison. Environment errors can also be produced by operating equipment under conditions of temperature, humidity and pressure for which it was not designed. Some microcomputer test equipment will not work next to equipment. Environment errors can also result from contamination of equipment by earlier samples.

Prevention of test method errors starts with keeping track of sample identification and ends with the final report. The test method should be the same every time from start to finish.

The method should of course measure the desired properties. Method errors can also be produced by poor equipment operation, such as lack of calibration.

39.8.3 Networking microcomputers

At first, the sharing of information in a factory should be done with a mainframe computer using multiple terminals. As the number of microcomputers in factories increases, the need to share information among them increases, as well. Initially, sharing can be done by copying files in a pan drive and then moving the pan drive to another computer. The next solution to the problem should commonly referred to as the local area network (LAN). In this approach, additional equipment must be added to physically link all microcomputers together. This equipment includes a file server, a hard disk and appropriate cabling. In addition, each computer on the LAN needs a special network communication board. Network cabling is connected to each computer via the network board. In some ways a LAN resembles a mainframe computer system with many terminals attached. However, there are large differences between a mainframe system and a LAN. The biggest difference is that in a mainframe system, all the computing takes place in a central computer. As the number of active terminals attached to the mainframe computer increase, the speed of computing at any one of the active terminals is reduced.

In a LAN, all the computing takes place in the microcomputers being used by individual operators. The file server and the hard disk provide only a central location for storage and retrieval of information that is accessible by all microcomputers on the LAN. The principal advantage of a mainframe is its higher speed processor; thus tasks with a large amount of processing will have significantly reduced time to completion compared with a microcomputer using a slower processor.

Various patterns are used in connecting computers into a LAN with the file server and hard disk. The best pattern for a particular application is usually determined with the help of LAN vendors who have already done many installations. A LAN is neither as standard nor as simple to put together as a single microcomputer. Very few factories have the technical expertise to produce a LAN without technical support from a LAN vendor.

The most commonly used LAN pattern in a factory is the linear bus topology or Ethernet shown in Fig. 39.12. In this layout, all the microcomputers have equal accessibility to the file server and the hard disk. This does not mean that all LAN users can actually look at all the data on the hard disk. Users are generally limited in their accessibility through the use of passwords. The principal advantages of Ethernet include standardisation of many parts, ease of installation and relatively low cost. Disadvantages of Ethernet include limits to the number of additional microcomputers and problems with troubleshooting.

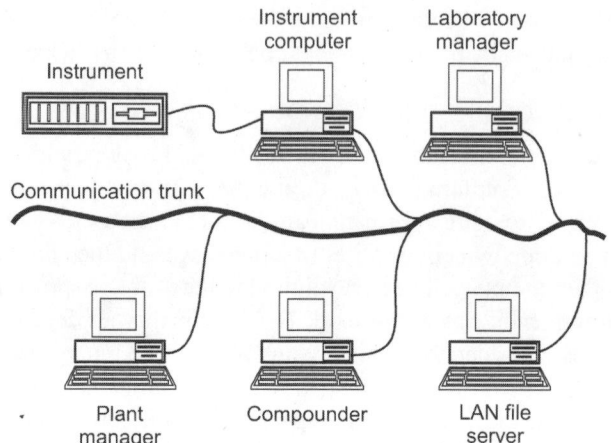

Figure 39.12: Example of a local area network in a rubber factory. All test information generated at the laboratory is stored on the hard drive controlled by the LAN file server, not at the laboratory microcomputer. Thus all other microcomputers on the LAN can immediately access the test information.

39.8.4   Application of LANs to rubber industry

When laboratory testing results were recorded with pencil and paper or with chart recorders, it was difficult for a factory manager located some distance from the laboratory to view the data. When the microcomputer is attached to a LAN, the data are available for viewing everywhere on the LAN. Anyone at a microcomputer located in a factory office can recall reduced tests results produced only seconds ago and compare them immediately to historical data without interfering with laboratory operations. Database-oriented programmes can be run at any microcomputer on the LAN to selectively look at data over time, batches, or compounds. A second major use for LANs is in sharing equipment. One commonly shared item is a special printer, such as a laser printer. Everyone on the LAN can have access to the laser printer with control at the individual's microcomputer.

Attaching a LAN to a laboratory information management system (LIMS)

A LAN can also be attached to a mainframe computer running LIMS software to tie a laboratory to other factory functions such as accounting, which normally are not on the LAN. The tasks a LIMS can do include scheduling tests, long-term storage of test results, integrating raw materials testing and compound testing and maintaining an audit trail on products from raw materials to final products. There are very few standards in LIMS. This means that nearly all installations require time-consuming and specialised programming.

Thus, no one test is adequate for all rubbers and with the exception of test results that can generate basic rheological parameters, there is no unique relationship between the readings of any two instruments, so that a given, reading on one does not correspond to a fixed and invariable reading on the other. This may often be due to the effects of slip of rubber on the walls of the instrument. Equally, lack of correlation between the results of the flow test and behaviour in processing equipment, such as extruders, may be due to this also. This may be complicated, in compounds, by the presence of processing aids. This is especially true of additives, which are less soluble/compatible in polymers at high strain.

Abrasion loss: Volume of rubber abraded from a specified test piece under specified conditions.

Abrasion resistance: Resistance to wear resulting from frictional action upon a surface. It is usually expressed by the abrasion resistance index.

Accelerated ageing test: Test in which an attempt is made to produce the effects of natural ageing in a shorter time by increasing the rate of degradation.

Accelerator: Compounding ingredient used in small amounts with a vulcanising agent to increase the speed of vulcanisation and/or enhance certain properties of the vulcanisate.

Acetone extract: Material extracted from rubber by acetone under specified conditions.

Activator: Compounding ingredient used in small amounts to increase the effectiveness of an accelerator.

Ageing: Irreversible change of material properties after exposure to an environment for a period of time.

Agglomeration: Reversible or irreversible joining together of latex particles.

Air oven ageing test: Test in which ageing proceeds in an oven in air at an elevated temperature and at atmospheric pressure in the absence of light.

Air trap: Void in a cellular polymer.

Alcoholic potash extract: Material removed from rubber by treatment with alcoholic potassium hydroxide solution after removal of the acetone extract and/or the chloroform extract.

Angle of braid: Acute angle between any strand of the braid and a line parallel to the axis of the hose.

Angle of helix/angle of lay/reinforcement angle: Acute angle between any strand of helical reinforcement and a line parallel to the axis of the hose.

Anklet fabric: Fabric interply to reinforce the lower leg.

Anklet rubber: Rubber interlayer to reinforce the lower leg.

Antioxidant: Compounding ingredient used to retard deterioration caused by oxidation.

Antistatic rubber: Rubber that is electrically conducting to the extent that, when earthed, it will prevent the build-up of electric charges and dissipate any charges applied, without thereby presenting a risk of fire or electric shock.

Applied skin: Thin solid layer, usually of rubber, applied to the surface of cellular material.

Armouring: Protective covering over a hose generally applied as a braid or helix to prevent mechanical damage or to support the reinforcement of a suction hose.

Artificial weathering: Exposure of material to laboratory conditions involving temperature, moisture and radiant energy, some or all of which may be varied cyclically, in an attempt to produce changes similar to those observed after long-term continuous outdoor exposure. The laboratory exposure conditions are usually intensified beyond those encountered in actual outdoor exposure in an attempt to achieve an accelerated effect. This term does not cover exposure to special conditions such as ozone, salt spray, industrial gases, etc.

Aspect ratio (H/S): Ratio of the section height to the section width of the tyre.

Autoclave press: Steam-pan incorporating means for applying mechanical pressure to a mold or molds.

Automobile belt/fan belt: V-belt to drive accessories on internal combustion engine.

Awl vent: Deliberate penetration of the sidewall rubber of a tubeless tyre usually in or near the bead area to prevent air pressure build-up in the carcass.

Backing cloth backer: Lining fabric of upper in a two-ply or three-ply material.

Backstrip: Internal or external strip of fabric or rubber covering the back centre line of the boot.

Bale cutter: Press in which a bale of raw rubber is forced against a knife, so cutting it into pieces for more convenient handling.

Ball mill: Rotating drum, generally mounted horizontally and containing loose balls of hard material to pulverise coarse particles.

Band pocket: Pre-assembly of some of the plies of the tyre carcass.

Bank: Accumulation of material at the opening between the rolls of a mill or calender or at a spreader bar or knife.

Bareness: Defect resulting from the failure of the rubber to fill out all the mold pattern detail.

Base rubber: Layer of rubber, usually of thickness 50–60% of the section height, situated below the tension element of a V-belt and acting as a compression member of the belt.

Batch: Product of one mixing operation in an intermittent process.

Bead: Part of the tyre that is shaped to fit the rim. It has a core made of one or several essentially inextensible strands with the plies wrapping around the core.

Bead core: Factory component for making the bead, consisting of rubber-covered wires in the form of a hoop.

Bead filler apex strip: Strip of relatively hard rubber mix placed on top of the bead core to prevent a void being formed by taut plies of fabric anchored round the bead core.

Bead heel: Part of the bead that fills the angle formed by the junction of the rim flange and the rim.

Bead toe: Part of the bead opposite the heel.

Bead wrapper: Tape of rubberised fabric to prevent distortion of the bead core during building, shaping or curing of the tyre.

Belt: Flexible band placed around two or more pulleys or sheaves for the purpose of transmitting motion, power or materials from one point to another.

Belt fastener: Device for securing the ends of a belt together.

Bench marks/Reference marks: Marks of known separation applied to a test piece and used to measure strain.

Bend radius: Radius of the innermost arc formed by bending a hose in a single plane.

Bias angle: Acute angle between the cutting line and the cords in the production of tyre cord fabric plies.

Bias belted tyre: Tyre structure of diagonal ply (cross ply) type in which the carcass is restricted by a belt comprising two or more layers of substantially inextensible cord material laid at alternate angles close to those of the carcass.

Bin curing: Unintentional vulcanisation of a mix during storage.

Binding: Strip covering an exposed fabric edge.

Blank slug: Piece of mix of suitable volume to fill the mold.

Blender: Vessel fitted with movable blades, to convert the contents, usually compounding ingredients, into a uniform, homogeneous mixture.

Blister bar: Bar parallel to a calender roll intended to minimise blistering during multiple bank calendering.

Bloom: Liquid or solid layer or deposit on the surface of a material that has been formed by the diffusion of a substance from within the material.

Blowing: Production of a cellular material by decomposition of an added ingredient.

Blowing agent: A compounding ingredient used in the manufacture of cellular articles to produce gas or vapour by chemical and/or thermal action.

Blowing down: Removal of excess ammonia from latex by stirring the latex while passing a stream of air across the surface.

Bound rubber: Portion of the rubber polymer in a mix that is so closely associated with compounding ingredients as to be insoluble in the usual rubber solvents.

Bowl: Term sometimes used for a roll of a calender.

Box/shell: Roller on which long lengths of rubber sheet or fabric are wound and temporarily stored during the processes of manufacture.

Braided hose: Hose in which the reinforcing material has been applied by braiding.

Break down: Preliminary softening, by mechanical work, of raw rubber or a mix to render it more suitable for masticating, mixing or further processing.

Breaker: Intermediate ply between carcass and tread.

Breaker pad mill bush: Member fitted into the frame of a mill or calender, designed to break preferentially in the event of overloading.

Breaker strip: Component used in making the breaker.

Breast roll/offset roll: Roll whose axis is arranged parallel to that of the other rolls but not in the same vertical plane.

Bridge: Two-spoked member supporting the centre in the head of the extruder.

Broad filling: Large piece of material placed under the narrow filling to secure the lasted edge of the upper end to provide an even base for the outsole.

Buckle shield/buckle patch: Rubber and fabric component to reinforce the joint of the buckle strap with the thigh extension.

Buckle strap: Strap securing the buckle to the thigh extension.

Bulk density: Mass per unit volume of material including both open and closed cavities normally associated with the material.

Bump breathe: Momentarily to open and close a mold at an early stage in the molding process to expel gas.

Burn: To undergo combustion.

Burst pressure test: Destructive pressure test of ultimate strength of a hose.

Buttress: Lower part of the shoulder area.

Cable cord construction V-belt: V-belt in which the tension element consists of a single layer of cord.

Calender: Machine with two or more rolls for converting rubber, or a combination of rubber with other materials, into sheet of a controlled thickness.

Calendered outsole: Outsole cut from a sheet of rubber having a pattern produced on a profile calender.

Camber crown: Convexity of the longitudinal section of a roll.

Camel back: Shaped strip of unvulcanised rubber of indefinite length used for retreading tyres.

Capped end: Rubber strip vulcanised to the end of a belt to protect the carcass.

Capped and sealed end: Hose end in which the wall section has been sealed with rubber to prevent the entry of extraneous material into the reinforcement.

Carcass: Impregnated fabric part of the belt which provides its main strength.

Carcass casing: Rubber-bonded cord structure of a tyre integral with the bead, which contains the inflation pressure.

Carrying face/top cover: Cover bonded to and protecting, the carcass on the side that carries the load.

Cavity/cell core: Blind hole deliberately made or formed in a rubber molded product, particularly a cellular rubber product.

Cell: Single small cavity surrounded partially or completely by walls.

Cellular: Consisting of a mass of cells.

Cellular material: Material having many cells (either open, closed, or both), dispersed through out its mass.

Cellular rubber: Mass of cells in which the matrix is rubber.

Centre cone/core: Part on the die which shapes the internal profile of an extrudate.

Centrifuged latex: Latex, the rubber concentration of which has been increased by the removal of serum by centrifugal force.

Chafer: Outer component of the bead to protect against rim chafing.

Chalk blower: Device for applying a powder to the inner surface of a tubular extrudate as it leaves the die.

Chalking: Formation of a powdery residue on the surface of a rubber resulting from surface degradation.

Chatter marks: Defect on calendered sheeting consisting of transverse narrow bands of alternately thicker and thinner material.

Chloroform extract: Material extracted from rubber by chloroform under specified conditions after removal of the acetone extract.

Cleat: Transverse bar or rib usually rubber molded or otherwise attached to the carrying face of the belt to prevent the slipping of the load on an incline.

Closed cell: Cell totally enclosed by its walls and hence non-interconnecting with other cells.

Closed cell cellular material: Cellular material in which practically all the cells are non-interconnecting.

Cloth marks: Impression left on rubber by a fabric.

Coagulant: Substance used for causing coagulation.

Coagulation: Irreversible agglomeration of particles originally dispersed in a rubber latex to form a continuous phase of the rubber and a dispersed phase of the serum.

Coagulum: Immediate product of coagulation of latex consisting of agglomerates of rubber particles.

Coiling diameter: Minimum diameter of coil to which a hose can be coiled without damage.

Cold checks/Cold shuts: Defect on calendered sheeting consisting of surface roughness.

Cold cure: Vulcanisation at room temperature, commonly by means of sulphur monochloride solution or vapour.

Collapse: Inadvertent densification of a cellular material during its manufacture, resulting from breakdown of its cell structure.

Combined sulphur: Sulphur remaining in a vulcanisate after extraction by a prescribed method. This term covers organically bound sulphur (bound e.g., in rubber) and inorganically bound sulphur (bound, e.g., in barium sulphate).

Combustion: Reaction of a substance with an oxidising agent, generally oxygen, with the release of heat generally accompanied by flaming and or emission of smoke.

Composition formula/recipe: Kinds and proportions of all ingredients contained in a mix.

Compounding ingredient: Substance added to a rubber polymer or latex to form a mix.

Compression hardness: Force required to produce a specified compression over all of the cellular material test piece under specified conditions.

Compression molding: Molding process in which the blank is placed directly in the mold cavity and compressed to shape by closure of the mold.

Compression set: Residual deformation of a test piece after removal of the compressive stress. Numerically, it is expressed as the percentage ratio of the residual deformation to the initial dimension of the test piece or to the compression strain.

Compression strain: Deformation produced in the test piece by a compression stress; it is the contraction in the direction of the stress expressed as a percentage of the original dimension in that direction.

Compression stress: Stress applied so as to cause shortening or contraction of the test piece in the direction of the stress; it is expressed as the average force per unit area of the original cross-section of the test piece.

Conducting rubber: Rubber with a sufficient degree of electrical conductivity to function as an electric current carrier. (*Note:* A product made from these materials has higher electrical conductance than one made from antistatic rubber).

Continuous vulcanisation/Continuous cure: Vulcanisation by the continuous passage of extruded, calendered or spread material over or through a unit which is capable of rapid and complete heat transfer (e.g., liquid or salt bath, fluid bed, microwave oven, rotary cure, hot air cure, steam cure).

Conveyor belt: Conveyor belt is the belt for transporting a load of articles or materials.

Convoluted hose: Hose manufactured, usually by molding, with regular annular bellows-like corrugations.

Cord: Textile or non-textile threads used in various components of the tyre carcass, plies, belts, breakers, etc.

Cord rubber: Soft rubber in which the tension element of a V-belt is fully embedded.

Core pin: Pin being part of a mold, to form a hole in a molded product.

Core/plug: Mold part that projects to form a cavity in the molded product, particularly for cellular products.

Cored cellular material: Cellular material containing a multiplicity of holes (usually, but not necessarily, cylindrical in shape), molded or cut into the material in some pattern, normally perpendicular to the largest surface and extending a part or all of the way through the piece.

Cover: External protective layer of polymeric material of a hose.

Cracker: Heavy-duty mill having two deeply corrugated or pyramid-cut rolls for breaking down a rubber or a mix, or for cutting rubber or a mix into pieces.

Creamed latex: Latex, the rubber concentration of which has been increased by creaming and removal of the separated serum.

Creaming: Process by which the particles in certain latices, especially natural rubber latex, due to their buoyancy, tend to concentrate at the surface of the latex.

Creaming agent: Substance which, when added to a latex, increases the rate of creaming probably by promoting flocculation of rubber particles and thus increasing their buoyancy.

Creel: Apparatus for holding a large number of bobbins or spools of textile or steel cord allowing the cords to be brought together to form a layer.

Crepe: Longitudinal movement of a belt relative to the periphery of the pulley, caused by its elastic stretching and contraction as it passes from the tight side to the slack side when running round pulleys, thus causing a loss of speed of the driven pulley.

Crescent tear test: Test for tear strength determination using a test piece of crescent shape with a nick of specified depth at the midpoint of the concave edge to initiate tearing.

Cross axis/Skew axis: Adjustment to the contour of the nip formed between two rolls of a calender by skewing the axis of one roll relative to the other.

Crown: Road-contacting area lying between the shoulders of a tyre.

Crown angle: Angle between tyre circumferential line and line of cords in the tyre at the crown of the tyre.

Crowsfeet: Small flow marks of V-shape on calendered sheeting.

Crystallisation: Orientation of the disordered long-chain molecules of a high polymer into repeating patterns.

Cure: Conditions necessary to produce a given state of vulcanisation.

Curemeter: Instrument designed to determine the increase in stiffness of a vulcanisable rubber mix during heating. The instrument is used for determining the appropriate time for vulcanisation or for assessing the tendency to premature vulcanisation.

Curing bag/air bag: Inflatable rubber container for exerting fluid pressure on a hollow article, e.g., pneumatic tyre during vulcanisation, to force it against a surrounding mold.

Curing diaphragm/curing bladder: Flexible component, which in modern process, fulfills a similar function to a curing bag.

Curing tube: Thin-walled curing bag.

Cushion rubber: Layer of rubber, usually softer than the base rubber and situated between the base rubber and the tension element of the belt.

Cut edge belt: Belt in which, subsequent to the molding operation, the belt is slit length-wise at the edges.

Cut growth test/crack growth: Test for the determination of the growth of a defined cut penetrating a test piece when subjected to cyclic flexing producing strains perpendicular to the length of the cut.

Daylight: Distance between adjacent platens of a daylight press in the open position.

Daylight press/single daylight press: Press having two superimposed platens, usually heated, between which molds are pressed.

Decorative rib: Circumferential raised pattern in the sidewall area for decorative purposes.

Delayed action accelerator: Accelerator whose effect markedly increases on reaching specific combinations of temperature and time.

Diagonal ply tyre/cross ply tyre: Tyre in which the ply cords extend to the beads and are laid at alternate angles substantially less than 90° to the centre-line of the tread.

Die: Detachable part of an extruder producing the profile of an extrudate.

Die holder nozzle/die box: Device for locating the die in relation to the extruder head.

Die line: Longitudinally raised identification line formed deliberately on an extrudate.

Dilatancy: Reversible ability of a fluid to become more viscous on agitation and to resume its more fluid state on standing.

Dipping: Process for manufacturing seamless articles by dipping a former into a rubber solution or latex.

Doubling machine: Machine with two rolls for building thicker sheeting from plies of thinner sheeting.

Dough: Paste-like mass of smooth texture consisting of a mix and solvent for spreading.

Dough mill: Mill for homogenising dough.

Draw: Taper on the member or members of a mold to facilitate the extraction of a molded product.

Drive ability: Ability of a properly tensioned belt to function without undue slip on the pulleys when operating at its recommended maximum effective belt tension.

Dry rubber content: Concentration of rubber in a latex usually expressed as a percentage by mass.

Dusting/chalking: Application of a powder to a rubber surface to prevent adhesion to another surface.

Dynamic properties: Properties exhibited under cyclic stressing.

Ebonite hard rubber: Hard, substantially inextensible vulcanisate whose glass transition lies above room temperature and which is produced by the vulcanisation of rubbers with a large proportion of sulphur, commonly in the order of 25–50% by mass on the rubber polymer.

Edge wheel: Hand tool comprising a wheel with a narrow smooth edge for consolidating a joint.

Effect of weathering: Combined detrimental influences of outdoor environment (e.g., sunlight, ozone, humidity, temperature) on a material.

Effective belt tension: Difference between the tension in the tight and slack parts of the belt as it enters and leaves the drives.

Elasticity: Tendency of a body to return to its original size and shape after having been deformed.

Elastomer: Macro-molecular material that returns rapidly to approximately its initial dimensions and shape after substantial deformation by a weak stress and release of the stress.

Elevator belt: Belt fitted with buckets or cups to raise a load from one level to another.

Elongation: Increase in length of a test piece produced by a tensile stress. This property is usually expressed as a percentage of the original test length.

Elongation at break: Percentage elongation of the test length of a test piece at the moment of rupture.

Embedding layer: Layer of rubber in which is embedded a reinforcing helix of wire or other material.

Embossing machine: Machine with two or more rolls for embossing sheeting.

Enlarged end: Soft end with the inside diameter larger than that of the main body of the hose.

Environmental conditioning: Storage of a test piece under specified conditions such as temperature, humidity, etc., for a specified time prior to testing.

Evaporated latex: Latex, the rubber concentration of which has been increased by evaporation of some of the water.

Expanded rubber/unicellular rubber: Cellular rubber made directly from solid rubber and usually composed of non-intercommunicating cells.

Extender: Organic material used as a replacement for a portion of the raw rubber required in a mix or for blending with the raw rubber (e.g., in oil-extended rubbers).

Externally corrugated hose: Hose containing a reinforcing helix in which the outer cover corrugated hose has been formed into the corrugations between the turns of the helix. Such a hose may be either rough bore, semi-embedded or smooth bore.

Extractable sulphur: Sulphur content of the rubber that can be extracted by a specified solvent under specified conditions.

Extruder head: Part of an extruder which houses the die or die holder.

Extrusion forcing: Continuous shaping of a material by passage through a die.

Extrusion mark/score line: Mark or line formed accidentally in an extruder on an extrudate.

Eyelet: Attachment through which laces may be fitted.

Eyelet stay: Reinforcement of the upper in the area where eyelets are secured.

Fabric back: Fabric interply to reinforce the quarters.

Fabric back stay: Fabric interply to reinforce the heel area of the upper.

Face cloth: Fabric treated to give a smooth finish and used for temporary support of sheet rubber.

Face cloth facer: Outside fabric of upper.

Fatigue life: Number of cycles of stress or strain of a specified character that a given test piece will sustain before failure.

Fatigue: Permanent deterioration in mechanical properties in a material subjected to fluctuating stresses and strains.

Field latex: Natural rubber latex as collected from the tree.

Filler: Solid particulate compounding ingredient normally added in relatively large proportions to the rubber.

Filler strip: Strip of rubber applied between wire turns.

Fire: Process of combustion characterised by the emission of heat accompanied by smoke and/or flame.

Fire resistance: Ability of an element of building construction (component or structure) to fulfill for a stated period of time the required stability integrity and/or thermal insulation specified in a standard fire resistance test.

First order transition: Change of state, usually synonymous with crystallisation or melting in a polymer.

Fissure: Split or crack in the cellular material.

Fitting line: Molded line or lines to serve as a guide to the tyre fitter that the fitted tyre is concentric with the wheel.

Flap: Profile ring of rubber to protect the base of an inner tube from chafing during service.

Flipper filler: Narrow strip of rubberised fabric enfolding bead components; its edges produce a gradual change in stiffness from bead to sidewall.

Floating platen: Platen suspended between the top and bottom platens of a multi-daylight press and capable of vertical movement.

Flocculant: Substance used for causing flocculation.

Flocculation: Formation (sometimes reversible) of loosely coherent, partially agglomerated rubber, distributed in the liquid phase of a latex.

Flow marks: Marks or lines on a molded product, caused by imperfect fusion of flowing fronts.

Foam rubber: Cellular rubber made directly from liquid starting materials and usually composed mainly of intercommunicating cells.

Folded edge belt: Belt where one or more plies are folded so that there are no fabric edges at the carcass edge.

Former: Shaped rigid object on or in which a rubber article is formed either by dipping, molding or manipulation and from which it is subsequently removed.

Formulation formula/recipe: Statement of all ingredients and their proportions to be contained in a mix.

Foxing: Reinforcing strip to protect the bottom edge of the upper from scuffing, incorporated along the bottom edge of the upper before, during or after the attachment of the sole.

Fractional coagulation: Deliberate coagulation of a portion of the rubber particles in latex.

Frame: Metal plate with a large shaped hole which, when used between flat plates, forms a mold.

Free sulphur: Sulphur present in the rubber in the elemental form irrespective of its origin.

Friction ratio: Ratio of surface speeds of two adjacent rolls (mill, calender or refiner).

Frictioning: Process of impregnating woven fabric with rubber using a calender whose rolls rotate at different surface speeds.

Front stay: Fabric or rubber interply to reinforce the front of the leg where the peak of the vamp joins the leg.

Gauge length: Initial distance between bench marks.

Gel rubber: Portion of raw rubber insoluble in a specified solvent.

Gelling: Formation of a uniform coagulum from which the aqueous phase has not separated.

Gelling agent: Substance used for causing gelling.

Glass transition: Second order transition between a brittle glassy condition and a rubber like condition. The midpoint of the temperature range over which this transition takes place is commonly called the 'glass transition temperature'.

Glowing: Combustion of a material without flame but with the emission of visible light from the surface.

Grain: Directional orientation of rubber and/or filler particles resulting in anisotropy of a material (e.g., calender grain, extruder grain).

Grown tyre: Tyre subjected to expansion or growth due to service conditions.

Green tyre/Raw tyre: Built tyre before vulcanisation.

Grown tyre overall diameter: Maximum overall diameter of an inflated tyre with an additional tolerance for service growth.

Grown tyre overall width: Maximum overall width of an inflated tyre with an additional tolerance for service growth.

Gum dipping: Process of impregnating textiles with rubber by immersion in rubber solution.

Gusset: Insert into the cut-out section of an upper designed to permit entry of the foot and to hold the footwear securely in position in wear.

Half moon reinforcement: Reinforcing material placed at the base of the stormguard in front of overshoes to prevent tearing of the upper.

Hardness: Resistance to indentation.

Hardwall hose: Hose containing a concentric supporting helix of wire or other material (primarily for suction applications).

Haul-off gear/take-up gear: Equipment for pulling an extrudate from an extruder.

Heat sensitiser: Gelling agent effective only at elevated temperature.

Heat stretching/heat stabilising: Process of adjusting the thermal and dimensional properties of a fabric under specific combinations of temperature, tension and time.

Heel: Bottom outside component providing the walking surface at the rear.

Heel piece: Reinforcement for stiffening the heel area of the upper.

Heel plug: Additional filler in the heel area.

Horseshoe breaker: Breaker to give additional edge protection by extending around the edges and into the non-carrying face.

Hot air cure: Vulcanisation in heated air.

Hot water cure: Vulcanisation in hot water or aqueous solution.

Hydrostatic stability test: None-destructive test in which the change in length and/or diameter and/or twist of a hose is measured at a specified pressure usually proof pressure.

Hysteresis: Ratio of energy lost to energy input in the rapid or instantaneous recovery of a deformed test piece.

Ignition: Initiation of combustion.

Impact resistance/impact strength: Resistance to fracture under mechanical shock force.

Impulse test: Pulsating pressure test, usually applied to high pressure hydraulic hose.

Indentation hardness: Force required to produce a specified indentation in the cellular material under specified conditions.

Injection molding: Molding process in which a predetermined quantity of mix is forced into a closed headed mold from a separate heating chamber.

Inner liner: Component of tyre lining the inner surface to retain air and/or to reduce chafing of the inner tube.

Inner vamp rubber: Rubber interlayer to stiffen the vamp.

Insert: Component made of metal or other material, which is to become or has become an integral part of a rubber molded product.

Insert pin/molding pin: Pin to place and hold in position an insert during a molding operation.

Insole: Bottom inside component adjacent to the bottom of the last.

Insole sock: Insole-shaped component placed in the foot-wear after vulcanisation.

Insulating ply: Layer of rubber between braided or spiralled plies of reinforcement.

Insulation squeegee: Additional thin layer of rubber to reduce the concentration of stress between plies.

Internal mixer: Mixer having specially shaped rotor or rotors operating in a closed chamber.

International rubber hardness degrees (IRHD): Measures of hardness, the magnitude of which is derived from the depth of penetration of a specified indentor into a test piece under specified conditions.

Jacket: Strip or strips of rubberised fabric usually applied on the base to the core of the belt so as to provide a wear resistance cover.

Kerbing rib: Molded rib in the upper sidewall region to protect the carcass from kerbstone abrasion.

Knee band: Strip of rubber or fabric around the joint between the thigh extension and the leg.

Knee cap: Rubber reinforcement to resist scuffing at the knee.

Knife bar: Bar carrying knives arranged for cutting sheeting longitudinally.

Knitted hose: Hose in which the reinforcing material has been applied by circular knitting.

Land: Contact area substantially normal to the direction of closing between two separate parts of a mold.

Latex: Colloidal aqueous dispersion of a polymeric substance.

Latex foam rubber: Foam rubber made from latex.

Lay: Distance advanced, parallel to the axis, by any strand in one complete turn.

Lease bars: Two or more metal bars interposed laterally between adjacent cords of a layer of cords to facilitate identification and tracing of individual cords.

Leg fabric: Fabric interply to reinforce the leg lining.

Leg lining: Fabric lining the whole leg and the back of the foot down to the insole.

Leg rubber: Rubber outer cover of the leg.

Let-off gear/take-up gear: Apparatus for releasing material from a reel or box under controlled tension.

Liner: Sheet material used for temporarily inter-leaving sheets that would otherwise stick together.

Lining: Inner material of the upper, i.e., adjacent to the foot or leg.

Lining tube: Innermost layer of a hose.

Locking strip: Strip of rubber around the sole to seal the sole-to-upper joint.

Lot: Definite quantity of some commodity manufactured or produced under conditions that are presumed uniform.

Low temperature cure: Vulcanisation at a lower temperature than usual for a particular product.

Machine-made hose: Mandrel-made hose made by machine instead of hand, particularly wrapped ply hose.

Mandrel-made hose: Hose built and vulcanised on a mandrel.

Masterbatch: Homogeneous mixture of rubber and one or more compounding ingredients in known proportions for use as a raw material in the preparation of the final mix. Masterbatches are used to facilitate processing or to enhance the properties of the final product, or both.

Mastication/break down: Process of plasticising raw rubber irreversibly by the combined action of mechanical work and oxygen, frequently at elevated temperature.

Maximum overall diameter: Overall diameter of a new inflated tyre plus an appropriate tolerance to provide for service growth and manufacturing tolerances.

Maximum overall width: Overall width of a new inflated tyre plus an appropriate tolerance to provide for service growth and manufacturing tolerances.

Mechanical conditioning: Prescribed programme of deformation of a test piece prior to testing.

Mechanical gasket: Deformable material clamped between essentially stationery faces to prevent the passage of matter through an opening or joint.

Mechanical stability: Ability of latex to resist coagulation under the influence of mechanical agitation.

Micelle: Submicroscopic aggregate of molecules of surface active materials usually in aqueous solution.

Micro-hardness: Measure of hardness using a scaled down version of the normal IRHD test, permitting the testing of thinner and smaller test pieces.

Middle backer interply: Material between the face cloth and backing cloth.

Mill: Machine with two driven rolls forming a nip for masticating, plasticising, mixing or sheeting.

Minimum bend radius: Smallest radius of the innermost arc to which a hose should be bent in service.

Mix compound: Intimate mixture of a polymer or polymers with all the ingredients necessary for the finished article.

Mixer: Machine used for mixing together polymers and compounding ingredients.

Mooney scorch: Time measure of incipient vulcanisation characteristics of a rubber mix using mooney shearing disc viscometer.

Mooney viscosity: Measure of the viscosity of a raw or unvulcanised rubber or a rubber mix, determined in a Mooney shearing disc viscometer.

Mother stock: Type of masterbatch in which the compounding ingredients are present in a higher proportion than those required in the final mix.

Mold finish: Surface finish of a mold.

Mold lid: Top part of a latex foam mold.

Mold lubricant: Material applied to a mold to prevent adhesion of the rubber to the mold.

Mold mark: Surface imperfection transferred to a molded product from corresponding marks on a mold.

Mold register: Means of correctly aligning the parts of a mold.

Molded finish: Surface finish of a molded product as it leaves the mold, no subsequent operations having been performed.

Molded hose: Hose usually made in long lengths, vulcanised in a rigid mold or inside a lead sheath which is subsequently removed.

Molded product: Object produced in a closed mold (e.g., by compression molding, injection molding or transfer molding).

Molded skin: Surface layer of a molded product which may differ slightly from the bulk of the material in the molded product.

Molding pressure: Force applied by the press divided by the projected area of the mold cavity or cavities.

Molding process: Process of shaping a material within a mold by applying pressure and usually heat.

Molding shrinkage: Difference in dimensions between a molded product and the mold cavity in which it was molded, both the mold and the molded product being at normal temperature when measured.

Mullins effect: Reduction in the elastic modulus of a vulcanised rubber as a result of deformation and recovery.

Multi-daylight press: Daylight press with more than one opening.

Narrow filling: Material under the insole to fill any cavity between insole and outsole.

Natural rubber: Rubber formed in a living plant, usually *Hevea Brasiliensis.*

Necking: Localised reduction in cross-section which may occur in a material under tensile stress.

New tyre: Tyre that has been neither used nor subjected to a retreading operation.

Nip/bite: Radial clearance between roll surfaces of a mill or calender on the centre lines, space between the rolls used for milling or calendering a material.

Nominal aspect ratio: One hundred times the ratio of the section height to the section width of the tyre on its theoretical rim.

Non-carrying face/pulley cover/bottom cover: Cover bonded to and protecting, the carcass on the side that does not carry the load.

Open cell: Cell not totally enclosed by its walls and hence interconnecting with other cells or with the exterior.

Open cell cellular material: Cellular material in which practically all the cells are interconnecting.

Open sided press/Swan-neck press: Cantilever press with columns on one side only and whose opening is, therefore, accessible for loading from three sides.

Open steam cure: Vulcanisation, without a mold in steam.

Optimum cure: Condition of vulcanisation to achieve an acceptable compromise between a number of desired properties or the optimum value of a selected property.

Outsole: Bottom outside component providing the walking surface in the forepart and the walking surface or the base for the heel at the rear.

Overall diameter: Diameter of an inflated tyre at the outermost surface of the tread.

Overall width: Linear distance between the outside of the sidewalls of an inflated tyre including elevations due to labelling (markings), decorations and protective bands or ribs.

Oxygen pressure/Oxygen bomb test: Accelerated ageing test in which ageing proceeds in a closed vessel containing oxygen at elevated temperature and pressure.

Peak cure: Vulcanisation during which the value of a major property proceeds to a maximum or minimum and then changes rapidly.

Pelletiser: Machine similar to a strainer for preparing pellets of rubber.

Peptiser: Compounding ingredient used in small amounts to accelerate by chemical action the permanent softening of raw rubber under the influence of mechanical action and/or heat.

Pigment: Insoluble compounding ingredient used to impart colour.

Pitch: Distance between adjacent corresponding points on a reinforcing helix, measured parallel to the axis.

Plain end: Soft end with the inside diameter the same as that of the main body of the hose.

Plasticise: To render less viscous.

Plasticity: Ability of a material to remain deformed after reduction of the deforming stress to or below its yield stress.

Plasticity number: Measure of plasticity that varies inversely with the ease of deformation.

Plasticity retention index (PRI): Measure of the resistance of raw natural rubber to oxidation. It is expressed as the percentage ratio of the plasticity measured after air oven ageing for 30 min. at 140°C to the plasticity measured before ageing using a parallel plate plastimeter.

Plastometer: Instrument for measuring plastic flow.

Plateau cure/flat cure: Vulcanisation during which the value of a desired property proceeds to a maximum and then remains nominally constant for a substantial period after the initial change.

Platen: Flat metal plate or chest to apply heat and pressure to a mold or molds in a press.

Plucking: Tearing out of isolated portions from a rubber surface during separation from a second surface.

Ply: Layer of rubber-coated parallel cords.

Ply rating: Arbitrary indication of carcass strength and load-carrying capacity originally related to the number of plies.

Ply-type belt: Belt whose carcass consists of more than one ply of fabric.

Polyether foam: Urethane foam based on a polyether-isocynate reaction product.

Polymer: Substance composed of very large molecules which consist essentially of recuring structural units.

Poromeric material: Man-made material that is generally similar in nature and appearance to shoe upper leather and in particular has a comparable permeability to water vapour.

Porosity: As applied to a defect, the accidental presence of numerous small cavities.

Post cure: Secondary cure at a controlled temperature, following a primary cure and carried out to enhance the physical properties of material.

Pot mold: Mold having a jacket through which a fluid may be circulated for controlling temperature.

Precoagulum: Coagulum resulting from partial inadvertent coagulation of a rubber latex.

Preform: Piece of mix of specified shape and size to fill the mold.

Pre-former: Member interposed between the extruder head and die to unify the rate of flow of rubber through the die.

Pre-molded outsole: Pre-molded sole-and-heel unit.

Preserved latex: Latex treated to inhibit putrefaction and accompanying coagulation during transport and storage.

Press cure: Vulcanisation in a mold in a press.

Pre-vulcanised latex/Vulcanised latex: Latex in which the rubber particles have been partially or completely vulcanised and from which vulcanised films can be produced by drying only.

Profiling machine: Machine with two or more rolls of which at least one carries one or more circumferential grooves for converting rubber into strip having a predetermined contour.

Proof pressure test: Pressure holding test to prove the structural integrity of a hose.

Protective breaker: Additional strip of ply material embodied circumferentially within the pneumatic tyre between the tread and the belt to minimise damage to the belt.

Quarters: Rear portion or portions of the upper that may be one piece or two pieces. The term originated from the use of four pieces to form the rear portions of a pair of shoes or boots.

Radial ply tyre: Tyre in which the ply cords extend to the beads and are laid substantially at an angle of 90° to the centreline of the tread, the carcass being stabilised by an essentially inextensible circumferential belt.

Raised edge: Upstanding border of a belt to prevent lateral spillage.

Raw edge belt: Belt where the textile reinforcing member is exposed at the edges of the belt.

Raw rubber: Natural or synthetic rubber of commerce forming the basic material for the manufacture of rubber articles. *(Note*: It is normally free from compounding ingredients but may also, for example, take the form of a masterbatch with oil and/or filler).

Reeling diameter: Minimum diameter of reel on which a hose can be coiled without damage.

Reinforcement: Stress-bearing member or members of a hose.

Reinforcing filler: Filler having a specific reinforcing effect, e.g., the ability to enhance abrasion resistance and mechanical strength properties.

Remolding: Process for replacing the tread, shoulder and sidewalls of a worn tyre.

Resilience: Ratio of energy output to energy input in a rapid or instantaneous recovery of a deformed test piece.

Retarder: Compounding ingredient used in small amounts to reduce the tendency of a mix to undergo vulcanisation prematurely.

Retracted spew/back-rinding: Defect in which the rubber adjacent to the spew line shrinks below the level of the molded product, the spew line often being ragged and torn.

Retreading: Generic term for reconditioning used tyres which covers replacement of the tread rubber only (shoulder to shoulder retreading, 'top capping' or 'recapping').

Reverse camber: Concavity of the longitudinal section of a roll.

Reverse stepped ply belt: Belt whose carcass contains fewer plies at the edges than in the middle to allow a thicker cover on the non-carrying face of the belt at the edges.

Rib: Wall between cavities.

Rip stop belt: Belt containing transverse high tensile cords, usually of textile materials or flexible steel cords, embedded in a rubber cushion in the middle of the carcass to limit longitudinal tearing.

Roll: Roller or hollow cylinder forming a major moving member of a rubber processing machine.

Roll bending: Adjustment to the contour of the nip formed between two rolls of a calender by blending one or both rolls.

Roll deflection: Blending of a moving rolls particularly when a nip is loaded.

Rough bore hose: Hose containing a concentric helix of wire or other material substantially exposed in the bore.

Rough top: Carrying face manufactured with irregularities, ridges or projections to produce an uneven surface to prevent slipping of the load, usually packages, on an incline.

Rubber back stay: Outer layer of rubber to reinforce the heel area of the upper.

Rubber hose: Flexible pipe made of rubber with a reinforcement, generally textile or metallic.

Rubber hydrocarbon: Rubber polymer based solely on carbon and hydrogen.

Rubber latex: Colloidal aqueous dispersion of natural or synthetic rubber.

Rubber polymer: Polymer or polymers constituting the basis of a rubber.

Rubber tubing: Flexible pipe made of rubber without reinforcement.

Rubber: A solid material with elastic properties made from latex of living plants, or synthetically and used in the manufacture of rubber products.

Saddle: Outside fabric to reinforce each side of the instep position from the base of the footwear to the binding at the top of the footwear.

Safety toe cap: Internal or external reinforcement, usually metal, in the toe area of the upper to protect the wearer's toes.

Sample: One or more items taken at random from a lot and intended to provide information on the lot and possibly to serve as a basis for a decision on the lot or on the process that had produced it.

Scorch: Premature vulcanisation of a rubber mix.

Screw/scroll worm: Rotating member with a helical groove to propel rubber along the barrel of an extruder.

Second order transition: Thermally induced physical change in the amorphous phase of a polymer which is associated with an abrupt change in the temperature coefficient of a physical property. (*Note*: The measured transition temperature is dependent on the rate of deformation and the rate of temperature variation).

Section height: Half the difference between the overall diameter and the nominal rim diameter.

Section width: Linear distance between the outsides of the sidewalls of an inflated tyre excluding elevations due to labelling (markings), decorations, or protective bands or ribs.

Self vulcanisation: Intentional vulcanisation at or near room temperature.

Semi-embedded hose: Hose containing a concentric helix of wire or other material embedded in the lining so that only the crests of the wire are exposed in the bore.

Serum: Aqueous dispersion medium of the rubber particles of a latex.

Set: Deformation remaining after complete release of the force producing the deformation.

Set after break: Tension set of a test piece stretched to rupture.

Shelf life/Storage life: Maximum storage time for which a material or product remains usable.

Shore D hardness: Measure of hardness of American origin, derived from the penetration of a specified indentor into the test piece under specified conditions. It is defined fully in ASTM D 2240, Standard Test method for rubber property - Durometer hardness and some details are given in BS 2719.

Shoulder: Transitional area between the sidewall and the tread.

Side stay: Additional fabric or rubber to reinforce the side of the leg.

Sidewall: Part of a tyre between tread and bead.

Sidewall rubber: Rubber layer on the sidewall of the tyre and over the carcass which may include ornamental or protective ribs and fitting lines.

Silking: Row of stitches on either side of the back seam.

Skin: Relatively dense layer at the surface of the cellular material.

Slab molding: Process of compression molding in which a multicavity mold is charged with a single slab or sheet of mix.

Smoke: Visible suspension in air of solid and/or liquid particles resulting from incomplete combustion.

Smooth bore hose/Fully-embedded hose: Hose containing a concentric helix of wire or other material fully embedded in the hose wall and which has a smooth bore.

Smoldering: Combustion of a material without light being visible and often evidenced by smoke. The term 'self-extinguishing' is often used in the context of fire terminology. It is a misleading term which may suggest that materials have a much better safety characteristic than they in fact have.

Sock lining: Fabric insole used in molding processes.

Soft end: End in which the rigid reinforcement, usually wire, of the body of the hose is omitted to permit the insertion and anchorage of fittings.

Softener: Compounding ingredient used in small proportions to soften the vulcanisate or to facilitate processing or incorporation of fillers.

Softwall hose delivery hose: Hose without a supporting helix of wire or other material as distinct from hardwall hose (primarily for delivery applications).

Spacing: Distance between adjacent turns of wire measured parallel to the axis of the helix, i.e., pitch minus the width of the wire.

Spew flesh: Surplus material forced from a mold on closure under pressure.

Spew line: Line on the surface of a molded product at the junction of the mold parts.

Spider: Member with three or more spokes supporting the centre in the head of the extruder.

Spider line: Radial line on a cross-section of an extrudate corresponding to the spokes of a spider or bridge.

Spiral wrapping: Method of applying external pressure to a hose during vulcanisation by using a narrow strip of cloth wound helically along the hose.

Spiralled hose: Hose reinforced with strands wound helically in layers in opposing directions.

Sponge rubber: Cellular rubber consisting predominantly of open cells and made from a dry rubber mix.

Spread sheet: Sheet made by spreading rubber dough on a support from which the sheet is finally removed.

Spreader: Machine for distributing rubber dough or latex on the surface of sheet material by means of a blade.

Spreader chest: Part of a spreader consisting of a heated table, the function of which is to assist evaporation of solvent as the coated fabric passes over it.

Stabilised latex: Latex treated to inhibit premature coagulation.

Steam-pan autoclave: Closed vessel for treating rubber in steam usually under pressure.

Steel cable conveyor belt/cable conveyor: Conveyor in which the tension member consists of two heavy high tensile steel cables. These support the belt which is reinforced with lateral right inserts for load support and is provided with molded channels to locate the steel cables.

Steel cord conveyor belt: Belt in which the tension member consists of a plurality of high tensile steel cords embedded in the rubber of the belt.

Stepped ply belt: Belt whose carcass contains more plies at the edges than in the middle to allow a thicker cover in the middle of the carrying face.

Stitch: Consolidation of a seam by rolling it with a stitch wheel.

Stitch wheel: Hand tool comprising a narrow wheel with a narrow serrated edge for stitching.

Stock blender: Device comprising one or more driven rollers mounted above and used in conjunction with a mill to improve blending of a mix.

Stormguard: Extension of the vamp of an overshoe to protect the instep.

Straight wrapping: Method of applying external pressure to a hose during vulcanisation by using a cloth laid along the length and then wrapping it several times round the hose.

Strain: Change due to force, in a dimension of a body referred to its initial size.

Strainer: Machine designed to force rubber through metal gauze screens to remove extraneous matter.

Strap: Strap running through the buckle to the wearer's belt to support the thigh extension.

Stress: Intensity, at a point in a body, of the forces or components of force that act on a given plane through the point. Stress is expressed as force per unit area. As used in tension, compression or shear tests, stress is normally calculated on the basis of the original dimensions of the appropriate cross-section of the test piece.

Sub-tread tread base: Layer of a two-component tread unit nearest the carcass usually composed of a softer rubber than the top cap.

Sulphide sulphur: Sulphur present in the rubber as inorganic sulphide.

Sunlight checking/sun checking: Crazing or cracking of the surface of rubber due to exposure of sunlight.

Sunlight crazing: Development of a random pattern of shallow surface cracks on a rubber surface due to exposure of sunlight.

Swelling: Change in volume or linear dimensions of a test piece immersed in a liquid or exposed to vapour.

Synthetic rubber: Man-made rubber prepared by polymerisation from lower relative molecular mass materials.

Tab sole: Bottom outside component of the outsole providing the walking surface in the forepart only.

Tack/tackiness: Ability of one rubber surface to adhere instantly to another on contact.

Take-off gear/wind-up gear: Apparatus for winding material on a reel or box from a calender or spreader under controlled tension.

Tear strength: Maximum force required to tear a specified test piece, the force usually acting substantially parallel to the major axis of the test piece.

Tensile strength: Maximum tensile stress applied during stretching a test piece to rupture.

Tensile stress at a given elongation: Stress required to stretch the test length of the test piece to the given elongation. In the rubber industry this definition is widely identified with the term 'modulus' and care should be taken to avoid confusion with the other use of modulus to denote the slope of the stress-strain curve at a given elongation.

Tensile stress: Stress applied to stretch a test piece; it is calculated by dividing the applied force by the original cross-sectional area of the test piece.

Tension element: Usually cores or fabric running parallel to the circumference, which resists the imposed tensile stresses in a V-belt.

Tension set/Tensile set: Extension remaining after a test piece has been stretched and allowed to retract in a specified manner, expressed as a percentage of the original test length or test strain.

Tension stand: Assembly of large diameter rollers for tensioning fabric.

Test piece test specimen: Piece of material of appropriate shape and size prepared so that it is ready for use in a test.

Test portion: Portion of a material set aside for the purposes of determining or estimating the identity of the material or some of its constituents or determining or estimating its ability to satisfy particular requirements.

T-head: Extruder head of T shape, to divert the flow of rubber to a direction at right angles to the axis of the screw, for direct extrusion round a core.

Thermoplastic rubber: Polymeric substance that undergoes reversible changes with temperature, to give rubber like properties at room temperature and thermoplastic properties at elevated temperature.

Thermo-plasticity: Ability to exhibit plastic flow with rise of temperature, to revert to comparative rigidity on cooling and to repeat the cycle indefinitely.

Thigh extension rubber: Rubber outer cover of the thigh extension.

Thixotropy: Characteristic of certain non-Newtonian fluids (e.g., latex mixes), in which the reformation of structural viscosity after agitation takes a finite time.

Through sole: Interlayer between the upper part of the footwear and either: (i) premolded sole and heel unit and (ii) a tab sole and molded heel.

Tip: Detachable conical end of a centre.

Toe filler: Fabric component fitted to the base of the lasted upper during fabrication to secure the lasted edges of the upper at the toe.

Toecap: Reinforcement of the upper to resist scuffing at the toe.

Toeguard: Rubber component fitted around the toe portion of the locking strip or foxing as an additional reinforcement.

Toepiece: Reinforcement on the inside of the footwear to resist abrasion of the upper by the toes.

Tongue: Shaped piece of material to protect the foot from chafing by the closure.

Tongue gusset/Bellows tongue: Rubber coated fabric joining the sides of the tongue to the upper to prevent entry of water.

Tongue protector: Rubber cover on the tongue.

Top binding: Narrow strip of rubber placed along the top edge of rubber boots or overshoes to give a finished appearance and to reinforce the edge.

Top cap/tread cap: Layer of a two-component tread unit into which the pattern is molded.

Top rubber: Rubber layer above the tension element of a V-belt.

Total solids: Proportion expressed as a percentage by mass, of matter not volatile at 100°C in a latex.

Total sulphur: All the sulphur present in a material, irrespective of its chemical form or origin.

Trader wheel: Hand tool comprising a roller with a knurled working face for consolidating a joint.

Transcord breaker: Breaker of high tensile cord fabric with the cord laid at right angles to the length of the belt.

Transfer mold: Mold for transfer molding.

Transfer molding/Injection molding: Molding process in which a predetermined quantity of mix is forced into a closed heated cavity or cavities from a heat chamber integral with the mold.

Transmission belt: Belt for transmitting power.

Tread: Part of a tyre that normally come into contact with the ground.

Tread groove: Drainage channel molded or cut in the tread.

Tread pattern: Pattern formed by the tread grooves.

Tread radius: Radius sub-tended by the tread.

Tread rubber: Factory component consisting of shaped strip of unvulcanised rubber eventually comprising the tread.

Tread sipe: Very narrow tread groove.

Tread width: Width of the tread measured between the points where the tread radius suddenly diminishes.

Treated liner/processed liner: Liner that has been specially treated to improve its smoothness or to ease its separation from the rubber.

Troughability/Transverse flexibility: Ability of a belt to bend transversely into an open-ended trough shape thus increasing the capacity of the belt and minimising spillage.

Turn up: Free end of a ply, chafer or filler which is folded around the bead core.

Twist: Rotation about its axis of a hose subject to internal pressure and usually measured in degrees per meter of hose length.

Undertread tread cushion: Thin layer of rubber, intermediate in hardness between the tread and the adjacent rubberised reinforcement ply, placed between them to provide a gradation in properties from tread to carcass.

Appendix I

Abbreviations of rubber

ABR	acrylate-butadiene rubber
ADC	azodicarbonamide
ADS	air-dried sheet
ACM	polyacrylic rubber
ASTM	American Society for Testing and Materials
AU	polyurethane (ester) rubber
AZDN	azoisobutyronitrile
BASRM	British Association of Synthetic Rubber Manufacturers
BR	polybutadiene rubber
BIS	Bureau of Indian Standards
BIIR	bromo-isobutene-isoprene rubber (brominated butyl rubber)
BS	British Standard
BSH	benzene sulphonyl hydrazide
BSI	British Standards Institution
CBS	cyclohexyl benzthiazyl sulphenamide
CCV	catenary continuous vulcanisation
CFM	polychlorotrifluoroethylene
CIIR	chlorinated butyl rubber
CM	chlorinated polyethylene
COD	cyclo-octadiene
CR	polychloroprene rubber
CSM	chlorosulphonated polyethylene
CV	continuous vulcanisation
CVNR	constant viscosity natural rubber
DBP	dibutyl phthalate
DCPD	dicyclopentadiene
DNPT	dinitroso-pentamethylene tetramine
EAM	ethylene-vinyl acetate co-polymer
EDTA	ethylene diamine tetra-acetic acid
ENB	ethylidene-norbornene
EP	ethylene propylene rubber

EPC	easy processing channel (black)
EPDM	ethylene—propylene terpolymer rubber
EPM	ethylene—propylene co-polymer rubber
ETU	ethylene thiourea mercaptoimidazoline
EU	polyurethane (ether) rubber
EV	efficient vulcanisation
EVA	ethylene vinyl acetate
FEF	fast extrusion furnace (black)
FF	fine furnace (black)
FPM	fluorocarbon rubber
FT	fine thermal (black)
GMF	quinone dioxime
GPF	general purpose furnace (black)
GPO	polypropylene oxide rubber
HAF	high abrasion furnace black
HCV	horizontal continuous vulcanisation
HMDA	hexamethylene diamine
HMF	high modulus furnace (black)
HMT	hexamethylene tetramine
HOFR	heat resisting, oil resisting and flame retardant
HPC	hard processing channel (black)
HS	high structure (black)
HSMB	hydrosolution masterbatch
IEC	International Electrochemical Committee
IIR	isobutylene – isoprene (butyl) rubber
IISRP	International Institute of Synthetic Rubber Producers
IM	polyisobutene
IR	polyisoprene rubber (synthetic)
IRHD	international rubber hardness degree
IRI	Institution of the Rubber Industry
ISAF	intermediate super abrasion furnace (black)
ISO	International Organisation for Standardisation
LM	low modulus (black)
LS	low structure (black)
LVN	limiting viscosity number
LVNR	low-viscosity natural rubbber
MB	masterbatch
MBTS	dibenzthiazyl disulphide
MDI	diphenylmethane-4,4′-diisocyanate
MOCA	4,4′-methyl-*bis*-2-chloroaniline
MPC	medium processing channel (black)

MPF	medium processing furnance (black)
MRPRA	Malaysian Rubber Producers Research Association
MRRDB	Malaysian Rubber Research and Development Board
MT	medium thermal (black)
NDI	naphthalene-1,5-diisocyanate
NIR	nitrile-isoprene rubber
NR	natural rubber
NS	non-staining (black)
OB	*pp'*-oxy-*bis*-benzene sulphonylhydrazide
OENR	oil-extended natural rubber
OEP	oil-extended polymer
OESBR	oil-extended styrene butadiene rubber
PAUS	pale amber unsmoked sheet
PBN	phenyl-β-naphthylamine
PBR	pyridine-butadiene rubber
PMMA	polymethylmethacrylate
PP	partially purified
pphr	parts per hundred of rubber
ppm	parts per million
PRI	plasticity retention index
PSBR	pyridine-styrene-butadiene rubber
PVMQ	phenyl vinyl methyl silicone rubber
PTFE	polytetrafluoroethylene
PVA	polyvinyl alcohol
PVC	polyvinyl chloride
RAPRA	Rubber and Plastics Research Association
RFL	resorcinol–formaldehyde latex
r.h.	relative humidity
RRIM	Rubber Research Institute of Malaysia
RSS	ribbed smoked sheets
SAF	super abrasion furnace (black)
SATRA	Shoe and Allied Trades Research Association
SBR	styrene butadiene rubber
SC	slow-curing (black)
SCF	super conductive furnace (black)
SCI	seal compatibility index
SMR	standard Malaysian rubber
SP	superior processing (NR)
SPF	super processing furnance (black)
SRF	semi-reinforcing furnace (black)
TCR	technically classified rubber

TDI	toluene diisocyanate
TDM	tertiary dodecyl mercaptan
TR	polysulphide rubbers
VCV	vertical continuous vulcanisation
VGC	viscosity gravity constant
VP	vinyl pyridine
VMQ	vinyl methyl silicone rubber
WLF	Williams–Landel–Ferry (equation)
XCF	extra conductive furnance black
XNBR	carboxylic-nitrile butadiene rubber (carboxynitrile rubber)
XSBR	carboxylic-styrene butadiene rubber
Y	prefix indicating thermoplastic rubber
YBPO	thermoplastic block polyether-polyester rubbers

Appendix II

Abbreviations for some common vulcanisation accelerators

BDTM	2-benzothiazole-dithio-*N*-morpholine
CBS	*N*-cyclohexylbenzothiazole-2-sulphenamide
DCBS	*N,N*-dicyclohexylbenzothiazole-2-sulphenamide
DOTG	Di-*o*-tolylguanidine
DPG	Diphenylguanidine
DPTT	dipentamethylene thiuram tetrasulphide
DTDM	4,4′-dithiodimorpholine
MBS	*N*-morpholinothiobenzothiazole-2-sulphenamide
MBT	2-mercaptobenzothiazole
MBTS	Dibenzothiazole disulphide
NOBS	*N*-oxydiethylbenzothiazole-2-sulphenamide
OTOS	*N*-oxydiethylenethiocarbamyl-*N*-oxydiethylene sulphenamide
PPD	piperidine pentamethylene dithiocarbamate
SDC	sodium diethyldithiocarbamate
SIX	sodium isopropylxanthate
SMBT	sodium mercaptobenzothiazole
TBBS	*N-t*-butylbenzothiazole-2-sulphenamide
TDEDC	tellurium diethyldithiocarbamate
TETD	tetraethylthiuram disulphide
TMTD	tetramethylthiuram disulphide
TMT	tetramethylthiuram disulphide
TMTM	tetramethylthiuram monosulphide
TPG	triphenylguanidine
ZBDP	zinc dibutyldithiophosphate
ZBX	zinc butylxanthate
ZDC	zinc diethyldithiocarbamate
ZEDC	zinc diethyldithiocarbamate
ZDBC	zinc dibutyldithiocarbamate
ZDMC	zinc dimethyldithiocarbamate
ZIX	zinc isopropylxanthate
ZMBT	zinc mercaptobenzothiazole

Appendix III

ASTM / International specifications of rubber and rubber products

Rubber—Testing

Test methods for rubber properties in tension.	ASTM D 412; 1983
Test methods for rubber property, adhesion to flexible substrates.	ASTM D 413; 1982
Test methods for rubber property, adhesion to rigid substrates.	ASTM D 429; 1981
Test methods for rubber deterioration, dynamic fatigue.	ASTM D 430; 1973
Test methods for rubber deterioration by heat and air pressure.	ASTM D 545; 1982
Test method for rubber property, effect of liquids.	ASTM D 471; 1979
Test method for rubber deterioration, surface cracking.	ASTM D 518; 1983
Methods of testing hard rubber products.	ASTM D 530; 1981
Test method for rubber property, Pusey & Jones indentation.	ASTM D 531; 1983
Test method for rubber deterioration by heat and oxygen pressure.	ASTM D 572; 1981
Test method for rubber, determination in an air oven.	ASTM D 573; 1981
Test methods for rubber properties in compression.	ASTM D 575; 1983
Test methods for rubber property, heat generation and flexing fatigue in compression.	ASTM D 623; 1978
Test method for rubber property, test resistance.	ASTM D 624; 1981
Test method for brittleness temperature of elastics and elastomers by impact.	ASTM D 746; 1979
Practice for rubber deterioration in carbon arc weathering apparatus.	ASTM D 750; 1974
Test methods for rubber property, Young's modulus at normal and sub-normal temperatures.	ASTM D 797; 1982
Test method for rubber deterioration, crack growth.	ASTM D 813; 1976
Test method for rubber property, vapour transmission of volatile liquids.	ASTM D 814; 1981

Test methods for rubber property, staining of surfaces (contact, migration and diffusion).	ASTM D 865; 1981
Test methods for rubbr property, staining of surfaces (contact, migration and diffusion).	ASTM D 925; 1983
Test methods for rubber properties in compression or shear (mechanical oscillograph).	ASTM D 945; 1979
Test method for measuring rubber deterioration, cut.	ASTM D 1052; 1976
Test method for rubber property, resilience using a rebound pendulum.	ASTM D 1054; 1979
Practice for rubber, standard temperatures and atmospheres for testing and conditioning.	ASTM D 1349; 1978
Test method for rubber property, stress relaxation in compression.	ASTM D 1390; 1976
Test method for rubber property, international hardness.	ASTM D 1415; 1983
Methods for testing vulcanised rubber part 1; determination of surface resistivity methods of testing plastics part 2; electrical properties method 231A; determination of surface resistivity.	BS 903; Part 1; 1982
Methods of test for raw isobutene isoprene rubbers (IIR or butyl).	BS 4470; 1980
Methods of test for evaluation of vulcanisation characteristics for raw natural rubbers.	BS 5738; 1980
Synthetic raw rubbers, determination of oil.	CSN 62 1133; 1982
Plastics moldings, tolerances and permissible deviation on dimensions (in english).	DIN 16 901; 1973
Testing of flexible cellular material, fatigue vibration test by constant load pounding in the indentation/pulsation range (in english).	DIN 53 174; 1977
Testing of cellular materials, dynamic modulus of elasticity and the loss factor using the vibrometer method (in english).	DIN 53 426; 1968
Testing of rigid cellular materials, tensile test (in english).	DIN 53 430; 1975
Testing of rubber and elastomers; conditioning of testpieces and test-conditions, times, temperatures and humidities.	DIN 53 500; 1979
Testing of rubber; determination of tear strength; trouser test piece.	DIN 53 507; 1983

Testing of rubber and elastomers accelerated test of ageing in elastomers by exposure to ozone, determination of ozone concentration (reference method) (in english).	DIN 53 509; Part 2; 1977
Testing of rubber; determination of rebound resilience.	DIN 53 512; 1981
Testing of rubber and elastomers, determination of the viscoelastic properties of elastomers, under forced vibration beyond resonance (in english).	DIN 53 513; 1978
Testing of rubber and elastomers, determination of the resistance to liquids, vapours and gases (in english).	DIN 53 521; 1979
Testing of rubber and elastomers, flexing endurance test, definitions, apparatus, preparation of test pieces (in english).	DIN 53 522; Part 1; 1979
Testing of rubber and elastomers, flexing endurance test, determination of resistance to flex cracking (in english).	DIN 53 522; Part 2; 1979
Testing of rubber and elastomers, flexing endurance test, determination of resistance to cut growth (in english).	DIN 53 522; Part 3; 1979
Testing of rubber; determination of volatile matter content in raw rubber.	DIN 53 526; 1982
Testing of rubber and elastomers measurement of vulcanisation characteristics (curometry).	DIN 53 529; Part 1; 1983
Testing of rubber and elastomers measurement of vulcanisation characteristics (curometry) operation.	DIN 53 529; Part 1; 1983
Testing of rubber; curemetry; determination of reaction during vulcanisation and evaluation of kinetic reaction of interiacing isotherms.	DIN 53 529; TEIL 2; 1983
Testing of rubber; determination of gas permeability.	DIN 53 536; 1980
Testing of rubber; staining of organic material by elastomers.	DIN 53 540; 1984
Rubber; determination of crystallisation effects by hardness measurements.	DIN 53 541; 1980
Testing of rubber; determination of the behavior at low temperatures (behaviour to cold), principles, test methods.	DIN 53 545; 1981

Testing of rubber; impact test for the determination of the low-temperature brittleness point.	DIN 53 546; 1980
Testing of rubbers; determination of crepe in compression.	DIN 53 547; 1981
Testing of flexible cellular materials; determination of compression set after constant deformation.	DIN 53 572; 1969
Statistical evaluation at off-hand samples with examples from testing of rubbers and plastics.	DIN 53 598; TEIL 1; 1983
Testing of rubber and elastomers, thin layer chromatographic analysis identification of accelerators.	DIN 53 622; TEIL 3; 1982
Testing of rubber and elastomers, testing of rubber in standard test mixes, natural rubber (NR) (in english).	DIN 53 670; Part 1; 1977
Testing of rubber and elastomers, testing of rubber in standard text mixes, natural rubber (NR) (in english).	DIN 53 670; Part 2; 1977
Testing of rubber and elastomers, testing of rubber in standard test mixes, styrene-butadiene rubber (SBR) (in english).	DIN 53 670; Part 3; 1977
Testing of rubbers and elastomers; testing of rubber in standard mixes; butadiene rubber (BR).	DIN 53 670; TEIL 4; 1981
Testing of rubber; testing of rubber in standard test mixes; chloroprene rubber (CR).	DIN 53 670; TEIL 5; 1983
Testing of rubber; testing of rubber in standard test mixes; acrylonitrile-butadiene-rubber (NBR).	DIN 53 670; TEIL 6; 1983
Testing of rubber and elastomers; testing of rubber in standard test mixes; isoprene rubber (IR)	DIN 53 670; TEIL 7; 1981
Testing of rubber; testing of rubber in standard test mixes; ethylene propylene-diene-rubber (EPDM).	DIN53 670; TEIL10; 1983
Testing of rubber and elastomers; testing of rubber in standard test mixes; butadiene rubber (BR).	DIN 53 670; TEIL 4; 1985
Testing of rubber; determination of total nitrogen content; Kjeldahi method.	DIN 53 625; 1984

Methods of test for natural rubber latex, part 1; determination of dry rubber content NRL; 1 (first).	IS 3708; PT 1; 1985
Methods of test for natural rubber latex, part 2; determination of sludge content NRL; 5 (first revision).	IS 3708; PT 2; 1985
Methods of test for natural rubber latex, part 6; determination of mechanical stability NRL; 9 (first).	IS 3708; PT 6; 1985
Methods of test for rubber latex; part V drawing of samples.	IS 9316; PT 5; 1979
Rubber, standard temperatures, humidities and times for the conditioning and testing of test pieces.	ISO 471; 1983
Rubber, raw, sample preparation.	ISO 1796; 1982
Polymeric materials, cellular flexible, determination of tensile strength and elongation at break.	ISO 1798; 1985
Polymeric materials, cellular flexible, accelerated ageing tests.	ISO 2440; 1983

Rubber standardisation

Rubber, raw natural, determination of plasticity retention index (PRI).	ISO 2930; 1981
Rubber, acrylonitrile-butadiene (NBR) test recipe and evaluation of vulcanisation characteristics.	ISO 4658; 1980
Rubber, determination of dynamic behaviour of vulcanisates at low frequencies, torsion pendulum method.	ISO 4663; 1984
Rubber compounding ingredients, carbon black, determination of tinting strength.	ISO 5435; 1981
Rubber, determination of styrene content, nitration method.	ISO 5478; 1980
Rubber and plastics, analysis of multi-peak traces obtained in determinations of tear strength and adhesion strength.	ISO 6133; 1981

Appendix IV

BIS / ISI specifications of rubber and rubber products

Title	*BIS/ISI*
Rubber, rubber chemicals and rubber products	
Activated calcium carbonate for rubber industry (first revision)	917-1976
Agricultural spray hose of rubber with braided textile reinforcement (second revision)	1677-1968
Aircraft fueling rubber hose, electrically bonded	5797-1982
Air hose of rubber with braided textile reinforcement (second revision)	911-1968
Ammonia preserved concentrated natural rubber latex (first revision)	5430-1981
Automotive hydraulic brake hose (first revision)	7079-1979
Petroleum, coal and related products	
Bale coating and marking of rubber, code of practice	5598-1969
Barytes for rubber industry (first revision)	1683-1973
Benzothiazyl-2-cyclohexyl sulphanamide	7069-1973
Calcium carbonate, activated, for rubber industry (first revision)	917-1976
Calcium silicate for rubber industry	9406-1980
Carbon blacks, methods of sampling and test	7498-1974
Cement grouting hose of rubber with braided textile reinforcement	5166-1969
Cement grouting rubber hose (first revision)	5137-1982
Cold polymerised oil-extended raw styrene-butadiene rubber	5188-1969
Cold polymerised raw styrene-butadiene rubber	5189-1969
Dibenzothiaxyl disulphide	8483-1976
Double centrifuged natural rubber latex	11001-1984
Ebonite	6693-1972
Electrically bonded road and rail tanker hose of rubber resistance to petroleum products	10733-1983
Fast extrusion furnace (FEF) carbon black	8135-1976
Fire fighting hose (rubber lined, or rubberised fabric lined, woven-jacketed) (second revision)	636-1979
Flexible rubber tubing for liquified petroleum gas	10908-1984
Formulae for evaluation of natural rubber (first revision)	7499-1981
Fumigation sheets and covers, rubberised	4810-1968

General purpose furnace (GPF) carbon black   10357-1982

Glossary of terms used in rubber industry:

    Part 1   7503 (Part 1)-1974

    Part 2   7503 (Part 2)-1976

    Part 3   7503 (Part 3)-1979

    Part 4   7503 (Part 4)-1979

    Part 5 Definitions relating to hoses   7503 (Part 5)-1981

High abrasion furnace (HAF) carbon black   7497-1974

Hose reel tubing for fire fighting   5132-1969

Hospital rubber sheeting without reinforcing fabric   8164-1976

Hospital rubber sheetings (first revision)   4135-1974

Hot-water hose of rubber (first revision)   5821-1979

Intermediate super abrasion furnace (ISAF) carbon black   8134-1976

Latex foam rubber products   1741-1960

Light magnesium oxide for rubber industry   9407-1980

Mercaptobenzothiazole   6918-1972

Natural red oxides of iron for rubber industry (first revision)   1684-1972

Natural rubber latex in drums, code packaging   5190-1969

Natural rubber latex, methods of test for :

    Part 1 dry rubber content, total solids, coagulum content, viscosity, sludge content, density, total alkalinity, KOH-number, mechanical stability, volatile fatty acid number, pH, total nitrogen, total copper, total iron, total manganese and total ash   3708 (Part 1)-1966

    Part 2 determination of boric acid and magnesium   3708 (Part 2)-1968

Natural rubber, methods of tests :

    Part 1 Determination of dirt, volatile matter, ash, total copper, manganese, iron, rubber hydrocarbon, viscosity (shearing disk viscometer), and mixing and vulcanising of rubber in a standard compound (first revision)   3660 (Part 1)-1972

    Part 2 Determination of solvent extract and nitrogen content   3660 (Part 2)-1968

    Part 3 Plasticity and plasticity retention index   3660 (Part 3)-1971

    Part 4 Determination of colour, accelerated storage hardening test and vulcanisation characteristics (mod test)   3660 (Part 4)-1979

Oil and solvent resistant hose of rubber with braided textile reinforcement (first revision)   3418-1968

Part 5 Drawing of samples	9316 (Part 5)-1979
Part 6 Determination of pH	9316 (Part 6)-1982
Rubber mats for electrical purposes	5424-1969
Rubber protective sheaths (condoms)	3701-1966
Rubber, raw, natural (second revision)	4588-1977
Rubber sand blast hose with woven textile reinforcement	6417-1971
Rubber sealing rings for domestic fruit and vegetable preserving jars	5193-1969
Rubber sealing rings for gas mains, water mains and sewers	5382-1969
Rubber stream hose	10655-1983
Rubber tubing for medical use	5680-1969
Rubber tubings for general purposes (revised)	637-1965
Rubber value-tubing for cycle tube valves	5079-1969
Rubber ward-dressing and porter's gloves	5783-1970
Rubber water hose (third revision)	444-1980
Rubbers for the dairy industry	6450-1971
Semi reinforcing furnace (SRF) carbon black	10387-1982
Sheet rubber jointing and rubber insertion jointing (second revision)	638-1979
Storage of vulcanised rubber, code of practice	6713-1972
Styrenated phenol	7351-1974
Styrene-butadiene rubber (SBR) latices, methods of test:	
Part 1 Determination of dry polymer, total solids, coagulum, pH, surface tension, density, viscosity, residual styrene, bound styrene and soap content	4511 (Part 1)-1967
Styrene-butadiene rubbers (SBR), methods of tests :	
Part 1 Determination of volatile matter, total ash, organic acid, soap, antioxidants, bound styrene and Mooney viscosity	4518 (Part 1)-1967
Part 2 Determination of solvent extract and oil content	4518 (Part 2)-1971
Suction hose of rubber for fire services	2410-1963
Sulphur for rubber industry	8851-1978
Super abrasion furnace (SAF) carbon black	10358-1982
Surgical rubber gloves	4148-1967
Symbols for rubbers and latices	6611-1972
Synthetic red oxide of iron for rubber industry (excluding venetian red)	8219-1976

Tetramethyl thuram disulphide	8978-1978
Titanium dioxide (anatase type) for rubber industry	8862-1978
Vulcanised natural rubber based compounds (first revision)	5192-1975
Vulcanised rubbers, methods of test:	
Part 1 Tensile stress-strain properties (first revision)	3400 (Part 1)-1977
Part 2 Hardness (first revision)	3400 (Part 2)-1980
Part 3 Abrasion resistance—Du Pont constant load method	3400 (Part 3)-1965
Part 4 Accelerated aging (first revision)	3400 (Part 4)-1978
Part 5 Adhesion of rubbers of textile fabrics	3400 (Part 5)-1965
Part 6 Resistance to liquids (first revision)	3400 (Part 6)-1983
Part 7 Resistance to flex-cracking	3400 (Part 7)-1967
Part 8 Resistance to crack-growth (first revision)	3400 (Part 8)-1983
Part 9 Density (first revision)	3400 (Part 9)-1978
Part 10 Compression set at constant strain (first revision)	3400 (Part 10)-1977
Part 11 Determination of rebound resilience	3400 (Part 11)-1969
Part 12 Tear strength—crescent test piece	3400 (Part 12)-1971
Part 13 Tension set (first revision)	3400 (Part 13)-1984
Part 14 Adhesion of rubber to metal	3400 (Part 14)-1971
Part 15 Volume resistivity of electrically conducting antistatic rubbers	3400 (Part 15)-1971
Part 16 Measurement of cut growth of rubber by the use of the ross flexing machine	3400 (Part 16)-1974
Part 17 Tear strength—angular test piece	3400 (Part 17)-1974
Part 18 Stiffness at low temperature (Gehman test)	3400 (Part 18)-1976
Part 19 Permeability to gases (constant volume method)	3400 (Part 19)-1976
Part 20 Resistance to ozone	3400 (Part 20)-1977
Part 21 Determination of permeability to gases-constant pressure method	3400 (Part 21)-1980
Part 22 Chemical analysis	3400 (Part 22)-1984
Vulcanised styrenebutadiene rubber (SBR) based compounds	7450-1974
Vulcanised vegetable oils (factice) for rubber industry	10130-1982
Water hose of rubber with braided textile reinforcement (second revision)	913-1968
Water suction and discharge hose of rubber, heavy duty (second revision)	3549-1983

References

Alfrey, T., *Testing of Polymer,* Interscience, New York, USA.

Allen, N.S., *Degradation and Stabilisation of Polyolefins,* Elsevier, USA.

Alliger, G., *Vulcanisation of Elastomers,* Reinhold, New York.

Bark, L.S., and Allen, N.S., *Analysis of Polymer Systems*, Elsevier, USA.

Bateman, L., *Chemistry and Physics of Rubber like Substances,* Applied Science, London.

Blow, C.M., *Rubber Technology and Manufacture,* Butterworth Scientific, London.

Brennan, J.J., *Toxicity and Safe Handling of Rubber Chemicals,* Academic Press, New York.

Bruins, P.F., *Plasticiser Technology,* Reinhold, New York.

Ceresa, R.J., *Block and Graft Copolymerisation,* Wiley Interscience, New York.

Collyer, H.J., *Environmental Aspects of Polymer and Rubber,* Interscience, New York.

Craig, A.S., *Dictionary of Rubber Technology,* Philosophical Library, New York.

Donnet, J.B., *Carbon Black,* Marcel Dekker, New York.

Dorian, A.F., *Dictionary of Plastics and Rubber Technology,* Iliffe Books, London.

Ferdinard Rodriguez, *Principles of Polymers Systems,* McGraw-Hill, International Edition.

Ferry, J.D., *Viscoelastic Properties of Polymers,* Wiley Interscience, New York.

Gardon, J.L., *Encyclopaedia of Polymer Technology,* Interscience, New York.

Georgeodian, K.M., *Principles of Polymerisation,* John Wiley and Sons, New York.

Harper, C.A., *Handbook of Plastic and Elastomers,* McGraw-Hill, New York.

Holden, G., *Chemistry and Technology of Rubber,* Reinhold, New York.

Iwen, E.D., *Degradation and Stabilisation of PVC,* Elsevier, USA.

Khromov, M.K., *Mechanical Testing of Unvulcanised and Vulcanised Rubber,* Maclaren, London.

Killeffer, D.H., *Banbury—The Master Mixer,* Palmerton, New York.

Kraus, G., *Reinforcement of Elastomers,* Interscience, New York.

Kuryla, W.C., *Flame Retardancy of Polymeric Materials,* Maclaren, London.

Lee, S.T., *Foam Extrusion,* Technomic Publishing Co., Inc., USA.

Lewis, F.W., *High Polymer Series,* John Wiley and Sons, New York.

MacDermott, C., *Selection Rubber for Engineering Applications,* Marcle Dekker, Reinhold, USA.

Matthews, F.G., *Twin Screw Extruders: A Basic Understanding,* Van Nostrand Reinhold, USA.

McKelvey, J.M., *Polymer Processing,* Wiley Interscience, New York.

Morton, M., *Rubber Technology,* Van Nostrand Reinhold, New York.

Nauntron, W.J., *The Applied Science of Rubber,* Arnold, London.

Norman, R.H., *Conductive Rubber,* Maclaren, London.

Nourry, A., *Reclaimed Rubber,* Maclaren, London.

Paul F. Bruins, *Polyurethane Technology,* Interscience Publishers, New York.

Penn, W.S., *Injection Moulding of Rubbers,* Maclaren, London.

Saunders, J.H., *Polyurethanes–Chemistry and Technology,* Interscience, New York.

Scott, J.R., *Physical Testing of Rubber,* Maclaren, London.

Smith, C.K., *Polymer Mixing Technology,* Elsevier, New York.

Sprague, G.R., *The Vanderbilt Rubber Hand Book,* Vanderbilt, New York.

Stamberger, P., *The Colloid Chemistry of Rubber,* Oxford University Press, Oxford.

Stannett, V., *Natural and Synthetic Rubber,* Academic Press, London.

Stern, H.J., *Rubber: Natural and Synthetic,* Maclaren, London.

Thompson, D.C., *Encyclopaedia of Polymer Science,* John Wiley and Sons, New York.

Treloar, L.R., *Polymer Technology,* Clarendon Press, London.

Wall, L.A., *Fluoropolymers,* Interscience, New York.

Werner Hofman, *Rubber Technology,* Hanser Publishers, Munish, New York.

Wheelans, M.A., *Injection Moulding of Elastomers,* Maclaren, London.

Index